Cyanobacterial Blooms: Ecology, Evolution and Biogeochemical Impacts

Shah Fahad · Shah Saud · Jinxiao Song ·
Taufiq Nawaz · Ruanbao Zhou
Editors

Cyanobacterial Blooms: Ecology, Evolution and Biogeochemical Impacts

Microbial Dynamics and Global Implications

Springer

Editors
Shah Fahad
Department of Agronomy
Abdul Wali Khan University Mardan
Mardan, Pakistan

Shah Saud
College of Life Science
Linyi University
Linyi, Shandong, China

Jinxiao Song
College of Life Science
Linyi University
Linyi, Shandong, China

Taufiq Nawaz
Department of Biology and Microbiology
South Dakota State University
Brookings, SD, USA

Ruanbao Zhou
Department of Biology and Microbiology
South Dakota State University
Brookings, SD, USA

ISBN 978-3-032-06041-9 ISBN 978-3-032-06042-6 (eBook)
https://doi.org/10.1007/978-3-032-06042-6

This Springer imprint is published by the registered company Springer Nature Switzerland AG
The registered company address is: Gewerbestrasse 11, 6330 Cham, Switzerland

Preface

Cyanobacterial blooms, a recurring phenomenon in freshwater, brackish and marine ecosystems worldwide, represent a complex ecological challenge with far-reaching consequences. These blooms, often dominated by toxin-producing species, pose significant risks to human health, aquatic biodiversity and ecosystem services. The increasing frequency and intensity of these events in recent decades, exacerbated by anthropogenic activities and climate change, have underscored the urgent need for a comprehensive understanding of their underlying mechanisms. This book, *Cyanobacterial Blooms: Ecology, Evolution and Biogeochemical Impacts*, aims to address this critical need by providing a timely synthesis of current knowledge, spanning the disciplines of ecology, evolution, biogeochemistry and management.

Our motivation for undertaking this project stemmed from a recognized gap in the existing literature. While numerous publications address specific aspects of cyanobacterial blooms, a holistic perspective integrating the diverse facets of their biology, environmental interactions and societal impacts has been lacking. We envisioned a volume that not only summarizes established findings but also highlights emerging research directions and identifies key knowledge gaps. This book seeks to fill this void by bringing together leading experts in their respective fields to contribute chapters that delve into the intricate processes driving bloom formation, their evolutionary context and their profound biogeochemical consequences.

This book is distinguished from existing literature by its comprehensive approach, encompassing the evolutionary history of cyanobacteria, the intricate interplay of environmental drivers, the biogeochemical transformations mediated by blooms and the cascading effects on food webs and ecosystem services. We also place considerable emphasis on the influence of human activities, particularly agriculture and climate change, on bloom dynamics. Furthermore, the book explores cutting-edge research methodologies, such as metagenomics, and delves into the physiological and evolutionary adaptations that enable cyanobacteria to thrive in bloom conditions. Finally, we address the pressing need for effective management strategies and explore the potential of cyanobacteria in bioremediation and environmental restoration. We believe this book will serve as a valuable resource for researchers, students, policymakers and environmental managers seeking a deeper understanding of these complex and increasingly prevalent ecological events.

Mardan, Pakistan Shah Fahad

Contents

Abbreviations

2-OG	2-Oxoglutarate
ABC	ATP-Binding Cassette
ACh	Acetylcholine
AgNPs	Silver Nanoparticles
AHLs	N-acyl Homoserine Lactones
AI	Artificial Intelligence
ALS	Amyotrophic Lateral Sclerosis
AMGs	Auxiliary Metabolic Genes
AML	Machine Learning
Amt	Ammonium Transporter
APase	Alkaline Phosphatase
APC	Allophycocyanin
ARISA	Automated Ribosomal Intergenic Spacer Analysis
ASP	Amnesic Shellfish Poisoning
ATP	Adenosine Triphosphate
B12	Cobalamin (Vitamin B12)
BMPs	Best Management Practices
BOD	Biological Oxygen Demand
C	Carbon
C/N	Carbon-to-Nitrogen ratio
CA4, CA6	Chromatic Acclimation Systems (Types 4 and 6)
CAFOs	Confined Animal Feeding Operations
CBB	Calvin-Benson-Bassham
CCA	Complementary Chromatic Adaptation
CcaS/R	Chromatic Acclimation Sensor/Regulator System
CCM	Carbon-Concentrating Mechanism
CCMs	Carbon-Concentrating Mechanisms
CE-QUAL-W2	A Two-Dimensional Hydrodynamic and Water Quality Model
CFP	Ciguatera Shellfish Poisoning
CFU/mL	Colony-Forming Units per Milliliter
CH_4	Methane

CHL	Chlorophyll
CNNs	Convolutional Neural Networks
Co	Cobalt
CO_2	Carbon Dioxide
COR	Collinear Orthologous Regions
CPDs	Cyclobutane Pyrimidine Dimers
CRISPR	Clustered Regularly Interspaced Short Palindromic Repeats
CRISPR-Cas9	Clustered Regularly Interspaced Short Palindromic Repeats Associated protein 9
CSIA	Compound-Specific Isotope Analysis
Cu	Copper
CWA	Clean Water Act
CyAN	Cyanobacterial Assessment Network
CyanoHABs	Cyanobacterial Harmful Algal Blooms
DAGs	Diacylglycerols
DDN	Diazotroph-Derived Nitrogen
ddPCR	Digital Droplet PCR
DIP	Dissolved Inorganic Phosphorus
DNA	Deoxyribonucleic Acid
DOC	Dissolved Organic Carbon
DOM	Dissolved Organic Matter
DOP	Dissolved Organic Phosphorus
DPPH	2,2-Diphenyl-1-Picrylhydrazyl
DRP	Dissolved Reactive Phosphorus
DSP	Diarrheal Shellfish Poisoning
eDNA	Environmental DNA
EDX	Energy-Dispersive X-ray
EII	Ecological Integrity Index
ELISA	Enzyme-Linked Immunosorbent Assay
EPA	Environmental Protection Agency
EPS	Extracellular Polymeric Substances
EQIP	Environmental Quality Incentives Program
ETC	Electron Transport Chain
EU WFD	European Union Water Framework Directive
FaRLiP	Far-Red Light Photoacclimation
Fe	Iron
Fe(II)	Ferrous Iron
Fe(III)	Ferric Iron
FeoB	Ferrous Iron Transporter B
FLIM	Fluorescence Lifetime Imaging Microscopy
FRET	Förster Resonance Energy Transfer
FTWs	Floating Treatment Wetlands
Ga	Giga-Annum (billion years ago)
GCF	Green Climate Fund
GEF	Global Environment Facility

GHG	Greenhouse Gas
GLEON	Global Lake Ecological Observatory Network
GlnA	Glutamine Synthetase
GLWQA	Great Lakes Water Quality Agreement
GOE	Great Oxidation Event
GPX	Glutathione Peroxidase
GR	Glutathione Reductase
GS-GOGAT	Glutamine Synthetase–Glutamate Synthase cycle
GWP	Global Water Partnership
HAB	Harmful Algal Bloom
HABs	Harmful Algal Blooms
HCBs	Harmful Cyanobacterial Blooms
HELCOM	Helsinki Commission
HGT	Horizontal Gene Transfer
HLIPs	High-Light-Inducible Proteins
HSPs	Heat Shock Proteins
InVEST	Integrated Valuation of Ecosystem Services and Tradeoffs
IoT	Internet of Things
ISO	International Organization for Standardization
ITS	Internal Transcribed Spacer
IWRM	Integrated Water Resources Management
km	Michaelis Constant (Denoted as Km in Transport Kinetics)
LA-ICP-MS	Laser Ablation Inductively Coupled Plasma Mass Spectrometry
LC-MS	Liquid Chromatography Mass Spectroscopy
LHCs	Light-Harvesting Complexes
LTEM	Long-Term Ecological Monitoring
MAAs	Mycosporine-like Amino Acids
MCDA	Multi-Criteria Decision Analysis
McF	McFarland Standard
MFS	Major Facilitator Superfamily
MH	Mueller-Hinton
MIC	Minimum Inhibitory Concentration
Mn	Manganese
Mo	Molybdenum
MoCo	Molybdenum Cofactor
MODIS	Moderate Resolution Imaging Spectroradiometer
mRNA	Messenger Ribonucleic Acid
MTF-1	Metal-Response Transcription Factor 1
N	Nitrogen
N_2	Dinitrogen (Molecular Nitrogen)
N_2O	Nitrous Oxide
$NADP^+$	Nicotinamide Adenine Dinucleotide Phosphate (Oxidized Form)
NADPH	Nicotinamide Adenine Dinucleotide Phosphate (Reduced Form)
NbS	Nature-Based Solutions
NDCs	Nationally Determined Contributions

NGS	Next-Generation Sequencing
NH_3	Ammonia
NH_4^+	Ammonium
Ni	Nickel
Nif	Nitrogen Fixation Gene Cluster
nifH	Molybdenum Nitrogenase Gene
Nir	Nitrite Reductase
NMPs	Nutrients Management Plans
NO_2^-	Nitrite Ion
NO_3	Nitrate
NO_3^-	Nitrate Ion
NOAA	National Oceanic and Atmospheric Administration
NOE	Neoproterozoic Oxidation Event
NO_x	Nitrous Oxides
NPQ	Non-Photochemical Quenching
NRCS	Natural Resource Conservation Services
NRPS	Non-Ribosomal Peptide Synthetase
NSP	Neurotoxic Shellfish Poisoning
NtcA	Nitrogen Control Transcription Factor A
O_3	Ozone
OCP	Orange Carotenoid Protein
OCPs	Orange Carotenoid Proteins (Plural Form, Same Definition as OCP)
OD	Optical Density
OM	Organic Matter
OMVs	Outer Membrane Vesicles
P	Phosphorus
P(II)	Signal Transduction Protein P(II)
PA@AuNPs	Plant Extract-Assisted Gold Nanoparticles
PacR	Photosynthesis-Associated Carbon Regulation
PAHs	Polycyclic Aromatic Hydrocarbons
PAM	Pulse-Amplitude Modulated (used in PAM fluorometry)
PAOs	Polyphosphate-Accumulating Organisms
Pb^{2+}	Lead Ion
PBS	Phycobilisomes
PC	Phycocyanin
PCR	Polymerase Chain Reaction
PE	Phycoerythrin
PES	Payment for Ecosystem Services
PET/CT	Positron Emission Tomography / Computed Tomography
PG	Peptidoglycan
PGPR	Plant Growth-Promoting Rhizobacteria
PHAs	Polyhydroxyalkanoates
Pi	Inorganic Phosphate
PO_4^{3-}	Orthophosphate

POC	Particulate Organic Carbon
PolyP	Polyphosphate
PPKs	Polyphosphate Kinases
PSI	Photosystem I
PSII	Photosystem II
PSMA	Prostate-Specific Membrane Antigen
PSP	Paralytic Shellfish Poisoning
qPCR	Quantitative Polymerase Chain Reaction
QS	Quorum Sensing
RA	Regenerative Agriculture
RcaE	Regulator of Chromatic Acclimation E (Sensor Kinase)
rDNA	Ribosomal DNA
RISA	Ribosomal Intergenic Spacer Analysis
RNA	Ribonucleic Acid
RNNs	Recurrent Neural Networks
ROS	Reactive Oxygen Species
rpm	Revolutions Per Minute
rRNA	Ribosomal Ribonucleic Acid
RuBisCO	Ribulose-1,5-Bisphosphate Carboxylase/Oxygenase
SAV	Submerged Aquatic Vegetation
SEEA	System of Environmental-Economic Accounting
SIP	Stable Isotope Probing
SOD	Superoxide Dismutase
SRB	Sulfate-Reducing Bacteria
SRP	Soluble Reactive Phosphorus
TAGs	Triacylglycerols
TCA Cycle	Tricarboxylic Acid Cycle
TDMLs	Total Maximum Daily Loads
TF	Transcription Factor
TMDL	Total Maximum Daily Load
TMDLs	Total Maximum Daily Loads
T-RFLP	Terminal Restriction Fragment Length Polymorphism
UAVs	Unmanned Aerial Vehicles
UNEP	United Nations Environment Programme
US	United States
USDA	United States Development Authority
UV	Ultraviolet
WHO	World Health Organization
WQI	Water Quality Index
ZIP	Zrt/Irt-Like Proteins
Zn	Zinc
ZS	Zetasizer
ZSE	*Zaleya pentandra* Stem Extract
α-KG	Alpha-Ketoglutarate (another name for 2-OG)
σ (sigma) factor	Transcription Initiation Factor

Chapter 1
Introduction to Cyanobacterial Blooms and Global Perspective

Muhammad Anas, Waseem Ahmed Khattak, Erdogan E. Hakki, and Shah Fahad

Abstract This chapter provides a comprehensive overview of the ecological, health-related, and socio-economic dimensions of cyanobacterial blooms in freshwater systems. Cyanobacteria are ancient photosynthetic microorganisms essential to aquatic ecosystems; however, under certain environmental conditions—such as nutrient enrichment, warming temperatures, and hydrological changes—they can proliferate excessively, forming harmful blooms. These blooms not only disrupt ecological balance by reducing light penetration, oxygen levels, and biodiversity, but also pose serious risks to human and animal health through the release of potent cyanotoxins. This chapter explores key drivers of bloom formation, including climate change, anthropogenic nutrient loading, land-use transformations, and altered mixing regimes in water bodies. It also delves into the morphology and taxonomy of bloom-forming genera and the environmental factors that enhance their growth and dominance. Field-based and remote sensing techniques for bloom detection are examined alongside emerging tools such as environmental DNA and machine learning-based predictive models. The discussion further highlights the wide-ranging impacts of cyanobacterial blooms—from ecosystem degradation to human illness, economic losses, and diminished recreational opportunities. Management strategies are reviewed, focusing on nutrient reduction, biological and chemical interventions, and governance frameworks that promote adaptive, evidence-based decision-making. Future directions emphasize the integration of advanced technologies,

M. Anas (✉) · E. E. Hakki
Department of Soil Science and Plant Nutrition, Selçuk University. Campus, Konya, Türkiye
e-mail: muhammed.anas@selcuk.edu.tr

E. E. Hakki
e-mail: eehakki@selcuk.edu.tr

W. A. Khattak
Department of Plant Sciences, Faculty of Biological Sciences, Quaid-I-Azam University, Islamabad, Pakistan
e-mail: waseemkhattack@bs.qau.edu.pk

S. Fahad
Department of Agronomy, Abdul Wali Khan University Mardan, Mardan, Khyber Pakhtunkhwa, Pakistan

S. Fahad et al. (eds.), *Cyanobacterial Blooms: Ecology, Evolution and Biogeochemical Impacts*, https://doi.org/10.1007/978-3-032-06042-6_1

climate-resilient approaches, and global collaboration. Recognizing that cyanobacterial blooms reflect deeper systemic issues, this chapter calls for interdisciplinary action and long-term commitment to safeguard freshwater resources in an era of accelerating environmental change.

Keywords Cyanobacterial blooms · Eutrophication · Cyanotoxins · Water quality management · Climate change adaptation

1.1 Introduction

1.1.1 Importance of Cyanobacteria in Aquatic Ecosystems

Cyanobacteria, often referred to as blue-green algae, are among the most ancient organisms on Earth, with a fossil record dating back over three billion years. These microscopic, photosynthetic prokaryotes have played a pivotal role in shaping the Earth's atmosphere and aquatic ecology (Lobus & Kulikovskiy, 2023). They were among the first life forms to perform oxygenic photosynthesis, contributing significantly to the rise in atmospheric oxygen and the development of aerobic life. In modern aquatic ecosystems, cyanobacteria continue to play a central role as primary producers. They convert sunlight into energy, fix carbon dioxide, and in some species, even fix atmospheric nitrogen, thus enriching water bodies with essential nutrients (Raimi et al., 2021). This nitrogen-fixation capability is especially vital in nutrient-poor waters, supporting productivity where other organisms may struggle to thrive. In both freshwater and marine environments, cyanobacteria form the base of the food web, supporting zooplankton, invertebrates, and fish species. They also participate in biogeochemical cycles, influencing the balance of elements like nitrogen, phosphorus, and carbon. Despite their ecological importance, cyanobacteria are also associated with harmful phenomena, especially when environmental conditions allow their populations to grow uncontrollably (Paerl & Otten, 2013).

1.1.2 Definition and Characteristics of Cyanobacterial Blooms

A cyanobacterial bloom refers to a sudden and extensive increase in cyanobacterial biomass within an aquatic environment, often resulting in visible surface scums or discoloration of water (Lakshmikandan et al., 2024). These blooms typically occur in nutrient-enriched waters under warm, stable, and stagnant conditions. They are most common in eutrophic lakes, reservoirs, and slow-moving rivers, particularly during the summer months. Bloom formation is driven by a combination of environmental factors, including high nutrient availability (especially phosphorus and nitrogen), elevated water temperatures, ample sunlight, and limited water movement. Cyanobacterial blooms can appear green, blue-green, or even reddish-brown depending on the dominant species and their pigments (Wehr & van Vuuren, 2024). Genera such as *Microcystis*, *Dolichospermum* (formerly Anabaena), *Aphanizomenon*, and *Planktothrix* are among the most common bloom-forming cyanobacteria. One of the most concerning aspects of these blooms is the production of toxins, collectively known as cyanotoxins. These include microcystins (liver toxins), anatoxins (neurotoxins), cylindrospermopsin (a general cell toxin), and saxitoxins (another group of neurotoxins). These compounds pose risks to human and animal health when ingested, inhaled, or contacted through recreational or drinking water. In addition to their toxic potential, blooms can lead to severe ecological disruptions. The massive biomass can block light penetration, reducing photosynthesis in submerged plants and altering aquatic habitats. When blooms collapse, the decomposition of large amounts of biomass consumes dissolved oxygen, often resulting in hypoxic or anoxic conditions that cause fish kills and disrupt aquatic life. These blooms are now recognized as a significant threat to water quality, biodiversity, and public health, prompting increased monitoring and management efforts worldwide (Brooks et al., 2016).

1.1.3 Historical Context and Early Observations

The occurrence of cyanobacterial blooms is not exclusive to the modern era. Historical accounts from various cultures describe events that likely involved bloom conditions, even if the scientific understanding at the time was limited. Ancient Chinese texts from the Tang dynasty (seventh to tenth centuries) describe water that "smelled like rotten greens" and appeared "as green as spinach," likely referencing bloom events. In the nineteenth century, naturalists and early public health officials in Europe began documenting green water in lakes and reservoirs, particularly during warm months. These early observations noted the foul odors, altered water taste, and potential connections to fish deaths and illness in livestock or humans. However, it was not until the late nineteenth century that cyanobacteria were properly identified as the culprits behind these phenomena. Ferdinand Cohn, a German botanist, was among the first to describe and name the genus *Microcystis* in 1878, isolating it from bloom-contaminated water. The scientific tools available during that time allowed for basic microscopy but lacked the analytical power to detect toxins or understand bloom dynamics fully. Throughout the early twentieth century, bloom reports increased alongside urban expansion and agricultural development, though the link between nutrient enrichment and blooms was not yet clearly understood. The real turning point came in the 1970s, when livestock deaths and human health incidents were directly linked to cyanotoxin exposure. A notable case in Australia in 1979 involved sheep deaths near Lake Alexandrina, which was later found to contain toxic *Microcystis* strains. These and similar events led to more structured studies, linking nutrient loading (especially phosphorus) and climate conditions to bloom frequency (Glibert & Burford, 2017). Since then, a growing body of research has confirmed that anthropogenic activities are largely responsible for the intensification and geographical spread of cyanobacterial blooms. Expanding agricultural runoff, urban wastewater discharge, and warming global temperatures create ideal conditions for bloom development. Monitoring programs have been established globally, and international health bodies, such as the World Health Organization, have issued guidance on cyanotoxin limits in drinking and recreational waters. Although bloom events were observed centuries ago, they have now become more frequent, widespread, and complex, prompting new challenges in management, public health, and ecosystem conservation (Berdalet et al., 2016).

1.2 Ecology and Biology of Cyanobacteria

1.2.1 Morphology and Taxonomy

Cyanobacteria exhibit a diverse range of morphologies, which have contributed to their successful colonization of nearly all aquatic and terrestrial environments (Pathak et al., 2022). They can exist as unicellular organisms, colonies, or filamentous chains, with or without specialized cells. Some species form complex structures like mats or surface scums, while others remain suspended in the water column. Common shapes include spherical (*Synechococcus*), rod-shaped (*Oscillatoria*), or spirally coiled filaments (*Spirulina*) (Jana, 2024). Filamentous forms may develop heterocysts—specialized thick-walled cells that provide an anaerobic environment for nitrogen fixation—distinguishing them from other photosynthetic microorganisms. Another cell type, the akinetes, serves as a dormant, resistant stage that allows cyanobacteria to survive adverse conditions. Taxonomically, cyanobacteria are classified as prokaryotes within the domain Bacteria (Hugenholtz et al., 2021). Although their photosynthetic abilities resemble those of algae, cyanobacteria differ fundamentally in structure and genetics. Unlike eukaryotic algae, they lack a true nucleus and membrane-bound organelles (Gruber & Kroth, 2024). Modern classification is based on molecular tools, particularly 16S rRNA gene sequencing, which has greatly refined our understanding of cyanobacterial diversity. Taxonomic groups commonly associated with freshwater blooms include *Microcystis*, *Dolichospermum* (formerly *Anabaena*), *Planktothrix*, and *Aphanizomenon*. These genera are often dominant in eutrophic systems and have been widely studied for their role in toxin production and ecological impacts (Li et al., 2016).

1.2.2 Environmental Conditions Favoring Bloom Formation

Cyanobacterial blooms do not occur randomly; they are strongly influenced by environmental conditions that support rapid growth and dominance over other phytoplankton. Several physical, chemical, and biological factors interact to create favorable conditions for bloom development. One of the most critical factors is temperature. Most bloom-forming cyanobacteria thrive in warm water, typically above 25 °C, and their growth rates often surpass those of other algae at elevated temperatures. This gives them a competitive edge in warm, stratified water bodies during the summer months. Light availability also plays an important role. Cyanobacteria can regulate their buoyancy using gas vesicles—tiny intracellular structures that allow them to float or sink in the water column. This ability enables them to access optimal light conditions near the surface or move downward to obtain nutrients when needed (Ramanna et al., 2017). In stratified lakes or reservoirs, where the water column separates into layers, cyanobacteria often dominate the surface layer, benefiting from stable temperatures and light exposure while minimizing competition. Hydrological conditions such as low turbulence, slow water movement, and thermal stability further promote bloom formation. In stagnant or slow-flowing systems, cyanobacteria are less likely to be dispersed, giving them time and space to accumulate biomass. Moreover, low flushing rates in reservoirs and lakes reduce the physical removal of blooms. Biologically, reduced grazing pressure from zooplankton and low competition from other phytoplankton species can also contribute to bloom dominance. Some cyanobacteria release compounds that inhibit the growth of other algae (allelopathy), giving them further ecological advantage (Mohamed, 2017).

1.2.3 Nutrient Dynamics and Eutrophication

The availability of nutrients, especially nitrogen (N) and phosphorus (P), is one of the most influential drivers of cyanobacterial bloom formation (Anas et al., 2025a, 2025b). In many freshwater ecosystems, phosphorus is the limiting nutrient, meaning its concentration controls the growth of primary producers. When phosphorus inputs increase due to agricultural runoff, sewage discharge, or industrial effluents, cyanobacteria can respond rapidly, leading to blooms. Nitrogen also plays a vital role, particularly in systems where cyanobacteria lacking nitrogen-fixation abilities dominate (Mishra et al., 2021). Some genera, however, like *Dolichospermum* and *Aphanizomenon*, can fix atmospheric nitrogen, allowing them to flourish even when inorganic nitrogen is scarce (Kramer, 2022). Eutrophication refers to the process by which water bodies become enriched with nutrients, often resulting in excessive plant and algal growth (Mishra, 2023). This process is largely driven by human activities such as intensive farming, urbanization, and deforestation. Fertilizers, livestock waste, and detergents are common sources of nutrient pollution (Anas et al., 2025a, 2025b). Once these nutrients enter water bodies, they stimulate the growth of

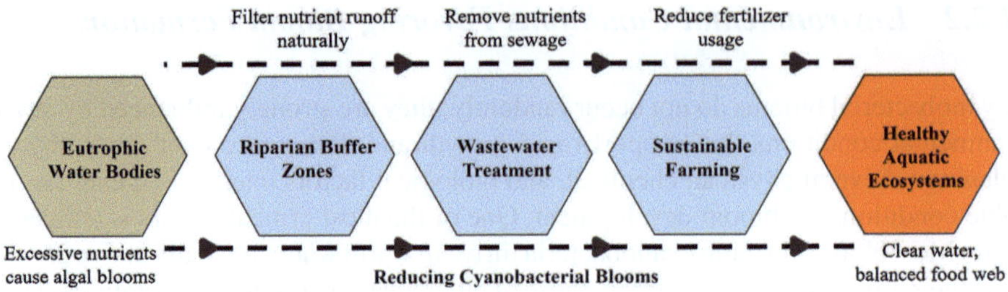

Fig. 1.1 Reducing cyanobacteria bloom through the process

cyanobacteria and other phytoplankton. The resulting increase in biomass can reduce water clarity, alter oxygen dynamics, and disrupt aquatic food webs (Fig 1.1). A positive feedback loop often exacerbates eutrophication. As cyanobacteria grow and die, their decomposition consumes dissolved oxygen, leading to hypoxic conditions that can kill fish and other aquatic organisms. Dead zones may form, especially in deep or stratified waters (Orita et al., 2015). Additionally, internal nutrient loading, where nutrients stored in bottom sediments are released under low-oxygen conditions, can sustain or re-trigger blooms even after external nutrient inputs are reduced. Effective management of eutrophication and bloom formation requires a reduction in both point-source and non-point-source nutrient loads. Strategies such as riparian buffer zones, improved wastewater treatment, and sustainable farming practices are critical to restoring ecological balance and preventing bloom recurrence (Cole et al., 2020).

1.3 Drivers of Cyanobacterial Blooms

1.3.1 Climate Change and Rising Temperatures

One of the most influential global drivers of cyanobacterial blooms is climate change, particularly the steady rise in average surface water temperatures. Cyanobacteria are generally well adapted to warm environments and often outcompete other phytoplankton species under elevated temperature conditions (Paerl & Paul, 2012). Many bloom-forming genera—such as *Microcystis* and *Dolichospermum*—demonstrate higher growth rates and metabolic activity at temperatures above 25 °C. As global temperatures continue to rise, these thermally tolerant species are likely to gain competitive dominance in a broader range of freshwater ecosystems. Additionally, climate change contributes to longer growing seasons and earlier onset of thermal stratification in lakes and reservoirs. Warmer temperatures reduce vertical mixing in water columns, leading to stable, stratified conditions that favor buoyant cyanobacteria capable of regulating their position to remain near the sunlit surface (Visser et al., 2016). Warmer winters and fewer ice-covered days on lakes in temperate regions are also allowing blooms to emerge earlier and persist longer. These extended bloom seasons increase the duration and intensity of exposure to cyanotoxins for both humans and aquatic life. Moreover, climate-driven extreme weather events—such as heatwaves, droughts, and intense rainfall—further amplify bloom risks. Drought conditions reduce water volume and flow, promoting stagnation and concentration of nutrients. On the other hand, heavy rain can rapidly flush nutrients from surrounding landscapes into water bodies, seeding sudden bloom events. Thus, climate change not only raises the baseline risk but also intensifies the variability and unpredictability of bloom dynamics (Mészáros et al., 2021).

1.3.2 Anthropogenic Nutrient Loading

Nutrient enrichment from human activities, especially excessive inputs of nitrogen and phosphorus, is one of the most direct and well-documented drivers of cyanobacterial blooms (Mohanty et al., 2024). Agricultural runoff, urban wastewater discharge, industrial effluents, and atmospheric deposition introduce large quantities of nutrients into lakes, rivers, and reservoirs. Fertilizers applied to crops and lawns are a major source of phosphorus and nitrogen, and without effective management, these nutrients are easily carried into aquatic ecosystems by rainfall and irrigation (Anas et al., 2024). Increased nutrient availability, particularly phosphorus, stimulates the rapid growth of phytoplankton. While many types of algae benefit from this enrichment, cyanobacteria are especially capable of exploiting these conditions, often dominating the planktonic community. Some cyanobacteria also possess nitrogen-fixing capabilities, allowing them to grow even when nitrogen becomes limiting, further ensuring their persistence under imbalanced nutrient conditions. Urban development compounds the problem by replacing permeable land with impervious surfaces like roads and buildings, reducing the natural filtration of runoff and increasing the volume of nutrient-rich water reaching aquatic environments (Deksissa, 2023). Combined sewer overflows, leaking septic systems, and inadequate wastewater treatment infrastructure contribute significantly to the nutrient load, particularly during storm events. Efforts to reduce nutrient loading—through improved land management, wastewater treatment upgrades, and nutrient trading programs—are critical but often slow to take effect. Until nutrient inputs are sufficiently controlled, cyanobacterial blooms will remain a recurring feature in many human-impacted water bodies (Basu et al., 2023).

1.3.3 Hydrological Modifications and Land-Use Changes

Alterations to the natural hydrology of a watershed have far-reaching implications for the frequency and intensity of cyanobacterial blooms. The construction of dams, reservoirs, canals, and irrigation networks often disrupts the natural flow of water, creating artificial environments with characteristics highly favorable to bloom development. Reduced flow velocity, increased water residence time, and thermal stratification in impounded systems provide ideal conditions for buoyant, slow-growing cyanobacteria (Olsson, 2021). Reservoirs are vulnerable, as they often collect nutrient-rich runoff from large catchment areas and hold water long enough for blooms to develop and persist (Oliver et al., 2019). Moreover, drawdown and refill cycles in reservoirs can resuspend nutrients from sediments, further fueling bloom formation. Land-use changes in the surrounding watershed, such as deforestation, agriculture, and urbanization, also play a critical role. Clearing of vegetation increases erosion and sediment transport into water bodies, which can carry bound nutrients and organic matter that enrich aquatic systems. Livestock grazing near waterways adds organic waste and nutrients directly into streams and ponds, while poor stormwater management in urban areas leads to unfiltered runoff carrying fertilizers, oils, and debris into lakes and rivers. These changes often act synergistically. For example, a newly built dam in an agricultural region may trap nutrient-rich runoff and create warm, stratified conditions, rapidly accelerating bloom formation (Wang et al., 2022). Understanding and managing land–water interactions is therefore crucial in the broader context of cyanobacterial bloom prevention.

1.3.4 Light Availability and Mixing Regimes

Light is a critical resource for photosynthetic organisms, and cyanobacteria have evolved various adaptations to optimize their access to it. Light availability, both in quantity and quality, significantly influences bloom dynamics. In stratified water bodies, where light penetrates the upper layers but is blocked below, cyanobacteria equipped with gas vesicles can migrate vertically to position themselves at the optimal light level near the surface (Wang et al., 2022). This ability gives them a distinct advantage over heavier, non-buoyant algae that sink and are exposed to reduced light intensities. Furthermore, cyanobacteria can thrive in conditions of fluctuating light, using their photosynthetic apparatus efficiently even at low irradiance levels. This means that in turbid waters, where light penetration is limited, cyanobacteria may still perform better than competing phytoplankton. Species with specialized pigments, such as phycobiliproteins, are particularly efficient at capturing light across a broad spectrum, further enhancing their adaptability (Glibert, 2024). Water column mixing influences both nutrient availability and light exposure. In shallow or polymictic lakes that mix frequently, light and nutrients are distributed more evenly, sometimes limiting the development of dense surface blooms (Fig. 1.2). However, in thermally stratified systems, mixing is reduced, and cyanobacteria can dominate the upper layer, monopolize light, and shade out other organisms. Wind patterns, rainfall, and reservoir operations all influence mixing regimes. Calm weather promotes stratification and bloom accumulation at the surface, while strong winds and storms can break stratification, disperse cells, and delay bloom formation. Therefore, understanding local light and mixing dynamics is essential in predicting bloom behavior and developing targeted mitigation strategies (Yan et al., 2024).

Fig. 1.2 Cyanobacterial bloom dynamics influenced by light and mixing

1.4 Global Distribution and Case Studies

1.4.1 North America

Cyanobacterial blooms (cyanoHABs) in North America have now become ubiquitous, particularly in massive freshwater systems such as the Great Lakes, Lake Okeechobee (Florida), and numerous other reservoirs in the western United States and Canada (McKindles et al., 2020). Lake Erie has had repeated intense cyanoHABs, often with *Microcystis aeruginosa* being the dominant organism, largely due to agricultural runoff and phosphorus loading from surrounding agricultural forces. As a result, this has led to the shutdown of drinking water facilities (with the most prominent cases being Toledo, Ohio) and phytoplankton monitoring programs in US states such as Florida and Ohio. Additionally, federal agencies, including the EPA and NOAA, have also developed satellite-based systems for early warning and drinking water health advisories (Bayler et al., 2024).

1.4.2 Europe

Cyanobacterial blooms in freshwater ecosystems have been a long-standing concern in Europe, particularly among nutrient-rich lakes and reservoirs in Central and Eastern Europe, such as Lake Balaton in Hungary, Baltic Sea coasts, and Lake Vomb in Sweden. Agricultural runoff and wastewater are still key contributors to bloom risks despite multiple nutrient-control initiatives, including national catchment management efforts from the EU Water Framework Directive (WFD). Germany has utilized high-resolution modeling and some long-term data to support adaptive management (Belleflamme et al., 2023), while the Netherlands and the United Kingdom have embraced more integrated catchment strategies to manage risks related to blooms (Bloemen et al., 2018). Studies in Europe are often viewed as global examples of linking land use, nutrient dynamics, and bloom response (Costa et al., 2022).

1.4.3 Asia

Asia, particularly China and Southeast Asia, has experienced some of the most extreme and persistent cyanobacterial blooms on Earth. Lake Taihu and Lake Dianchi, both located in China, have often been cited as examples of waterbodies where rapid urbanization, aquaculture, and excessive fertilizer use have led to pronounced *Microcystis* blooms that have caused significant ecological degradation and issues in public health. China's government has invested resources into large-scale remediation interventions, including diversion of water, dredging of sediments, and constructed wetlands (Liu et al., 2020). In India and Bangladesh, smaller-scale lakes and reservoirs are also increasingly afflicted, but monitoring and intervention frameworks are still limited because of poor infrastructure and funding (Cons & Eilenberg, 2019).

1.4.4 Africa

The number of documented cases of blooms and toxin production by cyanobacteria in African freshwater bodies is increasing, especially in areas with water scarcity and where such bodies of water are overused and polluted. Lake Victoria and Lake Naivasha in East Africa and South Africa's reservoirs, including Bloemhof, Lesotho Highlands, and Rustenburg, have reported bloom events dominated by *Microcystis* and *Anabaena* (Ndlela et al., 2016; Omara et al., 2023). Blooms threaten drinking water supplies, as well as the fishing and agricultural sectors. Knowledge gaps, limited resources, poor water governance, and threats of blooms rather than effective response plans negatively influence the timely detection of cyanobacterial blooms and toxin production. In some regions of Africa, building partnerships with international institutions and the emerging uptake of satellite technology should improve data availability and understanding of these events (Froehlich et al., 2022).

1.4.5 *Australia and Oceania*

Cyanobacterial bloom events have a lengthy history in Australia, with significant blooms found in the Murray-Darling Basin and lakes such as Lake Alexandrina (Tibby et al., 2020). A notable bloom occurred in the late 1990s and extended over 1000 km of riverine habitat. A national response strategy was delivered through an inquiry. To respond to the risk of blooms, Australian authorities now have a system in place that utilizes a combination of field surveillance, remote sensing, and public health advisories. Various factors contribute to blooms, including nutrient runoff from agriculture, prolonged periods of drought, and warm temperatures. In New Zealand, several small blooms have occurred, particularly in inland lakes vulnerable to impacts from dairy farm catchments and land clearance in general (Scarsbrook & Melland, 2015).

1.4.6 *Comparative Global Trends*

The frequency, duration, and geographic extent of cyanobacterial blooms are increasing globally. While the drivers of blooms, such as nutrient enrichment, climate change, and land-use change, are consistent across regions, the severity of blooms and the capacity to manage them are highly variable (Weber et al., 2020). Countries with higher incomes usually have better monitoring systems, policies and regulations, and public engagement; whereas, lower-income areas frequently struggle to monitor, access, and utilize data effectively and provide resources for mitigation. What is evident from the case studies of cyanobacterial blooms globally is that harmful algal blooms are becoming a regular systemic response to environmental pressures on a global scale, rather than isolated events or seasonal occurrences. A comparative approach shows the need for local adaptation and collaborative global responses, such as shared databases, technology transfer (including information networks), and guided and integrated watershed governance across the world (Wang et al., 2016). Experiences and lessons learned from one region can and should be translated to inform prevention and control in other areas, emphasizing the commonality of the problem and the advantages of global responses.

1.5 Impacts of Cyanobacterial Blooms

1.5.1 Ecological Consequences

Cyanobacterial blooms affect aquatic ecosystems in several substantial ways. The excess cyanobacteria grow quickly and float on the water surface, leading to less light penetrating the water column and indirect effects on submerged aquatic macrophytes and other algae by reducing the amount of light available for photosynthesis (Lakshmikandan et al., 2024). This reduced light changes the structure of aquatic communities in several ways, largely by reducing primary production at lower depths and then limiting habitat quality for many aquatic species. One of the most important negative effects occurs when cyanobacterial blooms collapse. The breakdown of a mass of cyanobacterial biomass through decomposition consumes vast amounts of dissolved oxygen, often producing hypoxic (low oxygen) or anoxic (no oxygen) environments (Zhang et al., 2020). Many fish and invertebrates do not survive in these low- or zero-oxygen environments, leading to fish kills and a loss of biodiversity. Some cyanobacteria also produce allelopathic compounds, which are chemical substances that inhibit the growth of competing organisms, and these compounds again destabilize the aquatic food web. In combination with nutrient loading, blooms can also change zooplankton community structure. Many bloom-forming cyanobacteria, including those that are toxic, may be unpalatable to grazers such as *Daphnia*, leading to reduced grazing pressure on the bloom and allowing it to persist. This is particularly relevant for higher trophic levels in the food web, especially fish that depend on zooplankton as a food source. Ultimately, over time, excessive cyanobacterial blooms can lead to permanent shifts in ecosystem structure that change the relationship of clear-water systems to turbid eutrophicated systems where cyanobacteria dominate (Burch et al., 2021).

1.5.2 Human and Animal Health Risks

Cyanobacterial blooms can present direct and serious risks to animal and human health through the delivery of cyanotoxins. There are a variety of toxins with different characteristics and effects, but all cyanotoxins can pose a risk, and even those with lower toxicity levels cannot be considered benign. *Microcystins* are one of the most detected cyanotoxins because they are powerful liver toxins, probable carcinogens, and can exist in such extreme concentrations that they can easily act at toxic levels. Anatoxins and saxitoxins can directly affect the nervous system, and cylindrospermopsin can affect multi-organ systems, including the liver and kidneys (Metcalf et al., 2021). Humans will be exposed to cyanotoxins through drinking contaminated water, participating in recreational activities such as swimming or launching a boat, or eating fish or shellfish from the affected water. Acute exposure to cyanotoxins can result in symptoms of skin irritation, as well as gastrointestinal issues, liver damage,

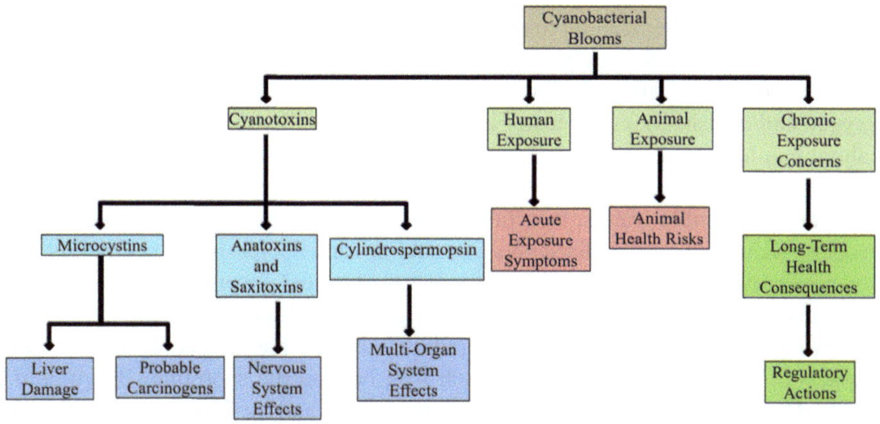

Fig. 1.3 Flow sheet diagram shows the cyanobacterial blooms and health risks

or neurological symptoms, depending on the cyanotoxin type, concentration, and level of exposure. Livestock and pets are at higher risk, and the number of reported cases of animals succumbing after drinking from water sources with toxic blooms is substantial. In general, dogs are at risk of being affected while swimming in or drinking from lakes and, in some cases, ponds affected by blooms. Wildlife, particularly birds, and aquatic mammals are also at risk and may experience sublethal and lethal effects from drinking cyanotoxins. Chronic exposure to low levels of cyanotoxins is a growing concern, especially for regions where blooms regularly affect drinking water sources. While long-term health and safety consequences are still being researched, there are growing bodies of evidence that repeated, low-dose exposure is associated with liver damage and offers a neurological risk (eg, motor neuron disease, ALS, or Parkinson-like syndromes) (Fig. 1.3). Some upstream impacts have led regulators to develop new regulations and prompted other countries or regions to enhance a health surveillance program (Adini et al., 2019).

1.5.3 Economic and Recreational Impacts

Aside from environmental and health-related impacts, cyanobacterial blooms can create serious economic burdens and limit the recreational use of waterways. Drinking water treatment plants may have to implement advanced and expensive filtration systems to reduce cyanotoxins, taste-and-odor compounds, and suspended biomass (Shang et al., 2018). These costs are eventually transferred to municipalities and consumers, especially when blooms persist or recur. Similarly, tourism and recreation suffer. Many lakes and reservoirs that allow boating, swimming, or fishing can have short- or long-term closures during bloom episodes based on health advisories. Even when toxins may not be present, the discolored waters with scums or unpleasant odors deter public use, leading to revenue losses for local tourism-related businesses such as marinas, hotels, restaurants, or recreational services. The commercial and recreational fisheries may also be impacted when hypoxic events deplete dissolved oxygen and/or fish stocks are lost. Fish kills, water quality degradation, and loss of habitat can lead to decreased catch and reduced economic returns. In some areas, consumers may also develop a loss of confidence in the safety of the fish caught in bloom-susceptible waters, resulting in loss of the market (Nicklay et al., 2025). Property values along the lakeshores can decline with decreasing water quality and fewer recreational opportunities. In many cases, the local government is responsible for managing these problems, including designing and funding cleanup operations, observing blooms, and educating the public—this all has a cost. Ultimately, the impacts of cyanobacterial blooms stretch well beyond the shoreline (Bargu et al., 2023). They change ecosystems, they threaten health, and they confer real economic costs, making their prevention and control a matter of public concern.

1.6 Detection, Monitoring, and Modeling

1.6.1 Field-Based Monitoring Techniques

Field-based monitoring is still the foundational means of identifying and tracking cyanobacterial blooms in aquatic systems. Under this approach, sampling in lakes, reservoirs, or rivers typically consists of regular water sample collection for one to several chlorophyll-a concentration, phycocyanin (the pigment molecule specific to cyanobacteria), chemical analysis (nutrients—nitrogen and phosphorus), physical parameters (temperature, dissolved oxygen, pH, turbidity), and microscopic identification of the dominant species of cyanobacteria (Khungwayo, 2021). Identifying the dominant cyanobacterial species is important for understanding toxicity risk, bloom progression, or longevity. Additional "advances" in field methods include handheld fluorometers and multi-parameter measuring probes, which can detect cyanobacterial pigments and other properties for more real-time and rapid environmental reading. This type of equipment is operated from boats, shore, or even buoyed floating platforms and allows for vertical profiling or spatial mapping of blooms (Beckler et al., 2019). More "comprehensive" monitoring approaches are automated sampling stations that utilize telemetry to continuously monitor a range of environmental variables and send alerts when special bloom thresholds are exceeded. These systems, which generate high-frequency data streams, enable tracking bloom onset, peak intensity, and bloom decline—providing perspective into short-term changes in condition that may not have been affordable, using manual sampling techniques. Although field-based monitoring provides detailed and accurate local information, it is resource-intensive and limited in spatial coverage. To complement these efforts, broader-scale surveillance and predictive modeling are increasingly being integrated into monitoring frameworks (Soubeyrand et al., 2024).

1.6.2 Remote Sensing and Satellite Applications

Remote sensing, particularly via satellite plant images and aerial drones, has revolutionized the capacity for the satellite-based detection and monitoring of large-scale cyanobacterial blooms (Wang & Qin, 2025). There are a multitude of satellite platforms currently transmitting images to Earth, including NASA's MODIS (Moderate Resolution Imaging Spectroradiometer), ESA Sentinel, and Landsat missions, which continue to provide relatively ubiquitous near-real-time imagery surveillance of surface waters. Satellite platforms indicate reflectance (or radiance) on the water surface, and, more importantly, satellite platforms can delineate pigments in cyanobacteria, including phycocyanin and chlorophyll-a, from other algal biomass and other suspended material (Hamed et al., 2025). Remote sensing technology, at the end of the day, can be of significant additional value in terms of bloom extent identification, tracking the movement of blooms, and locating areas that are nearly impossible to assess from the surface of the water for boats. By processing images with image processing algorithms, maps can be created to measure the extent of bloom density, measure spatial change and time, and measure spatial risk of toxicity based on previous empirical measures of toxicity based on measures of pigment concentrations. Some nations/agencies operationally have used satellite data to develop an operational platform, like the US EAP's Cyanobacterial Assessment Network (CyAN) program, to allow means to provide week-to-week bloom forecasts and warnings to the public and water managers (Grasso et al., 2019). In addition to satellites, drones fitted with multispectral or hyperspectral cameras are also being used for high-resolution mapping of smaller water bodies or smaller localized bloom events. All these technologies allow for near-real-time data to improve the ability to effectively respond to a bloom event and enable field investigations to determine algal species identification (Zahir et al., 2024a).

1.6.3 Predictive Models and Early Warning Systems

Predictive modeling provides a useful mechanism to assist in managing and mitigating cyanobacterial blooms. These models provide assessments of bloom dynamics based on environmental conditions, including nutrient inputs, temperature, rainfall, winds, light, and history of previous blooms, to predict future risk of blooms. A predictive modeling approach identifies patterns regarding relationships among variables that provide guidance to decisions based upon complex factors and allows for early actions. Models can be simple (e.g., empirical "formula" models) or complex (e.g., process-based models) (Beck et al., 2015). Empirical approaches as described above often use a statistical relationship (correlation) between variable values associated with the presence and absence of blooms (e.g., phosphorus levels, water temperature), which are often easier to use but generate limited usefulness because potentially important biological and physical interactions that influence bloom behavior are not considered in the modeling process. More complex models called mechanistic models can include aspects of hydrodynamics, nutrient cycling and supply, phytoplankton competition, and the impacts of weather conditions, to allow some limited level of dynamic prediction modeling. Examples of predictive models include CE-QUAL-W2 for stratified reservoirs, AQUATOX for ecosystem response models, and agent-based modeling (to model several organisms interacting with their environment). Researchers, environmental agencies, and utilities use these tools to anticipate bloom events days to weeks ahead of time, evaluate alternative management scenarios, or assess and quantify long-term impacts of land-use and climate change. Early warning systems are also increasingly being designed around real-time sensor networks, various forms of remote sensing data, and predictive algorithms. These systems can utilize the results of predictive algorithms and alert the user and the public after a particular threshold has been reached, allowing authorities to adaptively make treatment adjustments, recreational closures, or communication campaigns as required in a timely manner. With the rise in availability of data from multiple sources, the modern modeling and warning systems provide a hedge against the major risk the cyanobacterial threat represents (Zahir et al., 2024b).

1.7 Management and Mitigation Strategies

1.7.1 Nutrient Load Reduction

Reducing nutrient inputs into water bodies represents one of the most effective long-term approaches to cyanobacterial bloom management since excess nitrogen and phosphorus are the primary contributors to eutrophication, and affecting their sources is necessary for restoring and delivering ecological balance. This approach considers point-source pollution (discharge from a wastewater treatment plant), as well as non-point sources (e.g., runoff from agriculture, urban stormwater, and septic discharge) (Xia et al., 2020). Improving agricultural practices is a necessary step in nutrient management. For example, precision application of fertilizers, cover cropping, conservation tillage, and buffer strips along waterways can help limit farm runoff. Retention basins, constructed wetlands, and vegetated swales deliver filtration of nutrients prior to entering lakes or rivers. In urban areas, upgrading the wastewater infrastructure to include nutrient removal stages is equally important. Green infrastructure practices, like permeable pavements and rain gardens, can help manage the flow of nutrient-rich stormwater from entering waterways. On the scale of a larger watershed, nutrient management plans that promote good land use zoning, include the community, and promote monitoring plans can have longer-lasting impacts. While results from these actions may take time to arrive at a change in bloom intensity or frequency monitoring, nutrient load reductions remain a more ecologically sound approach for directly addressing the occurrence of blooms at their source instead of merely managing the symptoms after the fact (Glibert & Burford, 2017).

1.7.2 Biological and Chemical Control Methods

In situations where immediate action is warranted, especially during bloom events that are ongoing, we occasionally use biological or chemical control strategies to suppress bloom intensity, water column toxins, or bloom development. However, we must use these tools but be cautious to prevent damaging other portions of the ecosystem. Biological controls usually focus on the enhancement or introduction of natural grazers (e.g., specific zooplankton taxa) that graze on cyanobacteria. A few experimental approaches have evaluated using bacteria or fungi that are specific to inhibiting growth/ or degrading cyanotoxins in cyanobacteria. Another approach that has been studied is to biomanipulate fish populations to generate a food web demonstrating smaller zooplankton that will graze more effectively on cyanobacteria. Chemical controls, or algaecides, such as copper sulfate or hydrogen peroxide, can provide rapid removal of blooms. Alternatively, Phoslock (lanthanum-modified clay) can bind phosphorus in the sediments and make it unavailable for uptake by cyanobacteria. These interventions may provide short-term protection, but their effectiveness tends to rely on their presence, may need repeated use, and can potentially kill fish, kill beneficial plankton, or lead to toxins being released by lysed cyanobacteria. For these features, chemical interventions are often treated as emergency or last resort measures, and they can be maximally effective when combined with long-term controls of nutrients (Shortle et al., 2020).

1.7.3 Policy Frameworks and Governance

Successful governance and policy development are critical for achieving sustainability in managing cyanobacterial blooms. Effective regulatory frameworks will need to be put in place to set nutrient thresholds, govern land-use activities, and govern monitoring and response activities. These frameworks will need to detail regulatory standards for water quality, permitting systems for pollutant discharges, and best management practices across various sectors. Internationally, organizations such as the World Health Organization (WHO) provide guidance on appropriate and safe cyanotoxin concentrations in drinking and recreational waters. National governments will typically provide standards based on WHO recommendations. In some countries, this may include the use of “total maximum daily load” (TMDL) regulations for watersheds, which can regulate how much of a nutrient contributes to a water body (Harding, 2015). Collaboration between sectors will be essential because managing cyanobacterial blooms requires shared responsibility. Water utility managers will need to connect with agriculture, urban planning, industry, etc. Public engagement is also important. Educating communities about how blooms occur and the risks they pose will inform actions such as limiting fertilizer use or reporting polluted or suspicious conditions in treated waters. Finally, it is critical that governance strategies include an adaptive management framework. Because blooms may respond to climate change and land-use assessment, it is important that policies are flexible, science-based, and responsive to new data. Funding for research, innovation, and real-time monitoring technologies can further strengthen resilience against cyanobacterial threats (Saleem et al., 2023).

1.8 Future Outlook and Research Directions

1.8.1 Emerging Technologies for Bloom Detection

As cyanobacterial blooms become increasingly complex and widespread due to climate and land-use changes, so does the need for more rapid, accurate, and scalable methods of detection. Future investigations will inherently gravitate toward technological advancements that will further resolve nebulous spatio-temporal measures of blooms. Recent technological advances now include autonomously operated in situ sensors capable of continuous monitoring of important bloom parameters (e.g., phycocyanin, chlorophyll-a, turbidity, and gaseous saturation) in real time. These sensors can be fixed to buoys or robotic platforms for both autonomous data collection and sampling at different depths in situ. Likewise, recent developments in environmental DNA (eDNA) are yielding early instructions on identifying bloom-forming taxa and toxin-producing genes, thus providing us with pre-symptomatic responses prior to their visual manifestation. At the same time, we are fine-tuning machine learning algorithms to manage large datasets from satellites, drones, and field sensors to improve bloom predictions. Portable biosensors are under development, which will facilitate accurate screening of cyanotoxins in the field in about 30 min. These user-friendly devices can provide local stakeholders (e.g., drinking water authorities, fishery managers, recreational site operators) with the ability to successfully manage their respective activities without the need for laboratory reliance. As these technologies mature, the emphasis will be on cost-efficiency, field adaptability, and integration into broader monitoring networks.

1.8.2 Climate Resilience and Adaptive Management

Increased temperatures altered precipitation regimes, and the occurrence of extreme weather events is reshaping lakes and rivers physically and chemically, providing opportunities for cyanobacterial growth. It will be imperative to revisit adaptive management as we look to manage this evolving threat. Specifically, we will need to explore potential threshold conditions experienced under scenarios of future climate, mixing regimes, prolonged stratification periods, and rapid cycling of nutrient forms of N and P. Watershed modeling layered across climate scenarios will assist managers with anticipating potential new hotspots and prepare specifically targeted interventions. Nature-based solutions that center on effective nutrient filtration along with climate change adaptation—such as wetland restoration and riparian buffers—have gained traction due to these dual advantages. In addition to nutrient filtration, nature-based approaches function to absorb runoff, stabilize water flows, and are more ecologically resilient in times of blooms and other climate-induced stresses. A major challenge for researchers and decision-makers alike will be finding the correct policy framework that values flexible, resilient, and thoughtfully designed long-term benefits for ecosystems.

1.8.3 Interdisciplinary Approaches and Global Collaboration

A strategy approach to effectively addressing cyanobacterial blooms necessitates coordination across disciplines, geographies, and communities of practice. Each area of future research in this domain would be enhanced by more cross-collaborative partnerships between ecology, hydrology, microbiology, climatology, public health, economics, and governance. Collaborative research networks that exchange data, methods, and lessons learned across geographic regions will play an important role in understanding the variability of blooms and responses to management. On the policy front, clear linkages fostered by international initiatives can be used to facilitate the harmonization of monitoring protocols, toxin thresholds, and actions to be taken in the event of blooms. It will be important to transfer knowledge and best-practice methods to developing countries, as the need to respond to bloom threats with science will be essential, but especially to assist them in building capacity, knowledge, and tools. Another compelling area for future research is public engagement, as low-cost, place-based citizen science programs that engage and equip local communities with sampling, surveillance, and data interpretation can broaden the reach of monitoring programs while strengthening public awareness and stewardship of watersheds. Innovative ways to co-develop solutions may emerge from platforms that combine scientific knowledge and information with local community knowledge. Ultimately, the future of the research surrounding cyanobacterial blooms will require innovation, flexibility, and collaboration on a global scale. As we respond to their evolution, re-examining our tools, policies, and collaborative social structures will be important in planning for future responsive actions.

1.9 Conclusion

Cyanobacterial blooms have become a complicated, worldwide problem with important implications for ecosystem health, human health, and economic stability. Blooms have gone from a seasonal nuisance to a precursor demonstrating broader environmental impacts that are the result of nutrient enrichment, climate change, hydrological changes, and land-use impacts. As this chapter has shown, cyanobacterial biology and ecology, toxin production, and exploitation of anthropogenic environmental conditions give cyanobacteria a huge advantage as agents of water quality decline. The ecological impacts of blooms—ranging from oxygen depletion and habitat destruction to altering food webs—are extensive, as are the risks to human health from contact with cyanotoxins in drinking (potable) water as well as recreational sites. The economic burden for the cost of treatment and monitoring, as well as the loss of recreational or commercial use of water, can financially burden government and resident communities. Addressing cyanobacterial blooms requires a holistic, multi-level approach. Although field-based monitoring and remote sensing have improved our ability to detect blooms, future efforts must be preventative—particularly through nutrient load reduction and catchment management. New technologies, including biosensors, eDNA screening, and AI-enabled forecasting, will provide us with new tools, but their success will rely on integrating them into well-structured governance regimes. Perhaps most importantly, the fight against cyanobacterial blooms cannot be done in siloed efforts. It requires transdisciplinary research, stakeholder engagement, and dusting off international collaboration. Access for climate adaptation, enhancing public understanding, and accommodating change will be necessary for the viability and safety of our water systems for generations to come. Addressing cyanobacterial blooms is not only an ecological issue, but also a societal one. It represents how we interact with natural systems and what matters to us in our development and environmental stewardship. With the complexity of modern water-feeding systems, continuous, science-based approaches to sustainability will assist in combining collaborative management approaches with science so we may better address the growing issue of harmful cyanobacterial blooms in a rapidly changing world.

References

Adini, B., Singer, S. R., Ringel, R., & Dickmann, P. (2019). Earlier detection of public health risks—Health policy lessons for better compliance with the International Health Regulations (IHR 2005): Insights from low-, mid-and high-income countries. *Health Policy, 123*(10), 941–946.

Anas, M., Falak, A., Hassan, S., Khattak, W. A., Saleem, M. H., Khan, K. A., Khalid, A., & Fahad, S. (2025a). Microbial interactions and bacterial responses to metal stress in plants: Mechanisms, adaptations, and applications for sustainable agriculture. *Journal of Crop Health, 77*(1), 36.

Anas, M., Khalid, A., Saleem, M. H., Ali Khan, K., Ahmed Khattak, W., & Fahad, S. (2025b). Symbiotic synergy: Unveiling plant-microbe interactions in stress adaptation. *Journal of Crop Health, 77*(1), 1–21.

Anas, M., Khattak, W. A., Fahad, S., Alrawiq, N., Alrawiq, H. S., Abdelsalam, N. R., Jaremko, M., & Quraishi, U. M. (2024). Mitigating heavy metal pollution in agriculture: A multi-omics and nanotechnology approach to safeguard global wheat production. *Journal of Hazardous Materials Advances*, 100584.

Bargu, S., Skaggs, B., Boudreaux, M., Hammond, C. N., Snow, C., Aw, T. G., & Stumpf, R. (2023). Water quality and toxic cyanobacteria in oligohaline estuary beaches during the longest Mississippi River Basin flood event in 2019. *Estuaries and Coasts, 46*(7), 1865–1879.

Basu, N. B., Dony, J., Van Meter, K., Johnston, S. J., & Layton, A. T. (2023). A random forest in the great lakes: Stream nutrient concentrations across the transboundary great lakes basin. *Earth's Future, 11*(4), e2021EF002571.

Bayler, E., Chang, P. S., De La Cour, J. L., Helfrich, S. R., Ignatov, A., Key, J., Lance, V., Leuliette, E. W., Byrne, D. A., & Liu, Y. (2024). Satellite oceanography in NOAA: Research, development, applications, and services enabling societal benefits from operational and experimental missions. *Remote Sensing, 16*(14).

Beckler, J. S., Arutunian, E., Moore, T., Currier, B., Milbrandt, E., & Duncan, S. (2019). Coastal harmful algae bloom monitoring via a sustainable, sail-powered mobile platform. *Frontiers in Marine Science, 6*, 587.

Belleflamme, A., Goergen, K., Wagner, N., Kollet, S., Bathiany, S., El Zohbi, J., Rechid, D., Vanderborght, J., & Vereecken, H. (2023). Hydrological forecasting at impact scale: The integrated ParFlow hydrological model at 0.6 km for climate resilient water resource management over Germany. *Frontiers in Water, 5*, 1183642.

Berdalet, E., Fleming, L. E., Gowen, R., Davidson, K., Hess, P., Backer, L. C., Moore, S. K., Hoagland, P., & Enevoldsen, H. (2016). Marine harmful algal blooms, human health and well-being: Challenges and opportunities in the 21st century. *Journal of the Marine Biological Association of the United Kingdom, 96*(1), 61–91.

Bloemen, P., Reeder, T., Zevenbergen, C., Rijke, J., & Kingsborough, A. (2018). Lessons learned from applying adaptation pathways in flood risk management and challenges for the further development of this approach. *Mitigation and Adaptation Strategies for Global Change, 23*, 1083–1108.

Brooks, B. W., Lazorchak, J. M., Howard, M. D., Johnson, M. V., Morton, S. L., Perkins, D. A., Reavie, E. D., Scott, G. I., Smith, S. A., & Steevens, J. A. (2016). Are harmful algal blooms becoming the greatest inland water quality threat to public health and aquatic ecosystems? *Environmental Toxicology and Chemistry, 35*(1), 6–13.

Burch, M., Brookes, J., & Chorus, I. (2021). Assessing and controlling the risk of cyanobacterial blooms: Waterbody conditions. In *Toxic Cyanobacteria in Water* (pp. 505–562). CRC Press.

Cole, L. J., Stockan, J., & Helliwell, R. (2020). Managing riparian buffer strips to optimise ecosystem services: A review. *Agriculture, Ecosystems & Environment, 296*, Article 106891.

Costa, D., Sutter, C., Shepherd, A., Jarvie, H., Wilson, H., Elliott, J., Liu, J., & Macrae, M. (2022). Impact of climate change on catchment nutrient dynamics: Insights from around the world. *Environmental Reviews, 31*(1), 4–25.

Deksissa, T. (2023). Urban stormwater runoff for potable use: potential and challenges. In *Alternative water sources for producing potable water: Advances in research & technology* (pp. 71–96). Springer.

Froehlich, A., Ringas, N., & Wilson, J. (2022). How space can support African civil societies: Security, peace, and development through Efficient Governance Supported by space applications. *Acta Astronautica, 195*, 532–539.

Glibert, P. M. (2024). Capturing light and the diversity of pigments. *Phytoplankton Whispering: An Introduction to the Physiology and Ecology of Microalgae*, 85–103.

Glibert, P. M., & Burford, M. A. (2017). Globally changing nutrient loads and harmful algal blooms: Recent advances, new paradigms, and continuing challenges. *Oceanography, 30*(1), 58–69.

Grasso, I., Archer, S. D., Burnell, C., Tupper, B., Rauschenberg, C., Kanwit, K., & Record, N. R. (2019). The hunt for red tides: Deep learning algorithm forecasts shellfish toxicity at site scales in coastal Maine. *Ecosphere, 10*(12), Article e02960.

Gruber, A., & Kroth, P. G. (2024). Translocation of proteins into four membrane-bound complex plastids of red algal origin. In *Endosymbiotic organelle acquisition: Solutions to the problem of protein localization and membrane passage* (pp. 433–463). Springer.

Harding, W. R. (2015). *A feasibility evaluation of the total maximum daily (pollutant) load (TMDL) approach for managing eutrophication in South African dams: Report to the Water research commission*. Water Research Commission Pretoria, South Africa.

Hugenholtz, P., Chuvochina, M., Oren, A., Parks, D. H., & Soo, R. M. (2021). Prokaryotic taxonomy and nomenclature in the age of big sequence data. *The ISME Journal, 15*(7), 1879–1892.

Jana, B. (2024). *Aquatic sciences in the tropics: Plankton*. CRC Press.

Khungwayo, A. N. (2021). *Investigating the effects of nutrients and chlorophyll concentrations on the water quality of the Nuwejaars River, Cape Agulhas, Western Cape*

Lakshmikandan, M., Li, M., & Pan, B. (2024). Cyanobacterial blooms in environmental water: Causes and solutions. *Current Pollution Reports, 10*(4), 606–627.

Li, X., Dreher, T. W., & Li, R. (2016). An overview of diversity, occurrence, genetics and toxin production of bloom-forming Dolichospermum (Anabaena) species. *Harmful Algae, 54*, 54–68.

Liu, L., Wang, H., & Yue, Q. (2020). China's coastal wetlands: Ecological challenges, restoration, and management suggestions. *Regional Studies in Marine Science, 37*, Article 101337.

Lobus, N. V., & Kulikovskiy, M. S. (2023). The co-evolution aspects of the biogeochemical role of phytoplankton in aquatic ecosystems: A review. *Biology, 12*(1), 92.

McKindles, K., Frenken, T., McKay, R. M. L., & Bullerjahn, G. S. (2020). Binational efforts addressing cyanobacterial harmful algal blooms in the Great Lakes. *Contaminants of the Great Lakes*, 109–133.

Mészáros, L., Van der Meulen, F., Jongbloed, G., & El Serafy, G. (2021). Climate change induced trends and uncertainties in phytoplankton spring bloom dynamics. *Frontiers in Marine Science, 8*, Article 669951.

Metcalf, J. S., Tischbein, M., Cox, P. A., & Stommel, E. W. (2021). Cyanotoxins and the nervous system. *Toxins, 13*(9), 660.

Mishra, A., Rajput, S., Gupta, P. S., Goyal, V., Singh, S., Sharma, S., Shukla, S., Singh, A., Shukla, K., & Varma, A. (2021). Role of cyanobacteria in Rhizospheric nitrogen fixation. In *Soil Nitrogen ecology* (pp. 497–519). Springer.

Mishra, R. (2023). The effect of eutrophication on drinking water. *British Journal of Multidisciplinary and Advanced Studies, 4*(1), 7–20.

Mohamed, Z. A. (2017). Macrophytes-cyanobacteria allelopathic interactions and their implications for water resources management—A review. *Limnologica, 63*, 122–132.

Mohanty, S., Paul, A., & Sahoo, P. K. (2024). A strong driving force for cyanobacterial bloom. *Handbook of aquatic microbiology*, 28.

Ndlela, L. L., Oberholster, P. J., Van Wyk, J., & Cheng, P.-H. (2016). An overview of cyanobacterial bloom occurrences and research in Africa over the last decade. *Harmful Algae, 60*, 11–26.

Nicklay, H., Filstrup, C., Reavie, E., Birschbach, P., Reinl, K., Widiker, J., Schuldt, N., & Erickson, D. (2025). *Building a collaborative monitoring strategy for a changing St. Louis River Estuary: A recommendation report*.

Oliver, S., Corburn, J., & Ribeiro, H. (2019). Challenges regarding water quality of eutrophic reservoirs in urban landscapes: A mapping literature review. *International Journal of Environmental Research and Public Health, 16*(1), 40.

Olsson, F. (2021). *Impacts of water residence time on lake thermal structure: Implications for management and climate change*. Lancaster University (United Kingdom).

Omara, T., Nagawa, C. B., Kyarimpa, C., Böhmdorfer, S., Rosenau, T., Lugasi, S. O., Matovu, H., Odongo, S., & Ssebugere, P. (2023). Lacustrine cyanobacteria, algal blooms and cyanotoxins in East Africa: Implications for human and ecological health protection. *Phycology, 3*(1), 147–167.

Orita, R., Umehara, A., Komorita, T., Choi, J.-W., Montani, S., Komatsu, T., & Tsutsumi, H. (2015). Contribution of the development of the stratification of water to the expansion of dead zone: A sedimentological approach. *Estuarine, Coastal and Shelf Science, 164*, 204–213.

Paerl, H. W., & Otten, T. G. (2013). Harmful cyanobacterial blooms: Causes, consequences, and controls. *Microbial Ecology, 65*, 995–1010.

Paerl, H. W., & Paul, V. J. (2012). Climate change: Links to global expansion of harmful cyanobacteria. *Water Research, 46*(5), 1349–1363.

Pathak, J., Singh, P. R., Sinha, R. P., & Rastogi, R. P. (2022). Of cyanobacteria. *Ecophysiology and Biochemistry of Cyanobacteria.*

Raimi, M. O., Abiola, I., & Omini, D. E. (2021). Exploring how human activities disturb the balance of biogeochemical cycles: Evidence from the carbon, nitrogen and hydrologic cycles. *Nitrogen and Hydrologic Cycles (July 30, 2021).*

Ramanna, L., Rawat, I., & Bux, F. (2017). Light enhancement strategies improve microalgal biomass productivity. *Renewable and Sustainable Energy Reviews, 80*, 765–773.

Saleem, F., Jiang, J. L., Atrache, R., Paschos, A., Edge, T. A., & Schellhorn, H. E. (2023). Cyanobacterial algal bloom monitoring: Molecular methods and technologies for freshwater ecosystems. *Microorganisms, 11*(4), 851.

Scarsbrook, M., & Melland, A. (2015). Dairying and water-quality issues in Australia and New Zealand. *Animal Production Science, 55*(7), 856–868.

Shang, L., Feng, M., Xu, X., Liu, F., Ke, F., & Li, W. (2018). Co-occurrence of microcystins and taste-and-odor compounds in drinking water source and their removal in a full-scale drinking water treatment plant. *Toxins, 10*(1), 26.

Shortle, J. S., Mihelcic, J. R., Zhang, Q., & Arabi, M. (2020). Nutrient control in water bodies: A systems approach. *Journal of Environmental Quality, 49*(3), 517–533.

Soubeyrand, S., Estoup, A., Cruaud, A., Malembic-Maher, S., Meynard, C., Ravigné, V., Barbier, M., Barrès, B., Berthier, K., & Boitard, S. (2024). Building integrated plant health surveillance: A proactive research agenda for anticipating and mitigating disease and pest emergence. *CABI Agriculture and Bioscience, 5*(1), 72.

Tibby, J., Haynes, D., & Muller, K. (2020). The predominantly fresh history of Lake Alexandrina, South Australia, and its implications for the Murray-Darling Basin Plan: A comment on gell (2020). *Pacific Conservation Biology, 26*(2), 142–149.

Visser, P. M., Ibelings, B. W., Bormans, M., & Huisman, J. (2016). Artificial mixing to control cyanobacterial blooms: A review. *Aquatic Ecology, 50*, 423–441.

Wang, S., & Qin, B. (2025). Application of optical remote sensing in harmful algal blooms in lakes: A review. *Remote Sensing, 17*(8).

Wang, X., Chen, Y., Yuan, Q., Xing, X., Hu, B., Gan, J., Zheng, Y., & Liu, Y. (2022). Effect of river damming on nutrient transport and transformation and its countermeasures. *Frontiers in Marine Science, 9*, 1078216.

Weber, S. J., Mishra, D. R., Wilde, S. B., & Kramer, E. (2020). Risks for cyanobacterial harmful algal blooms due to land management and climate interactions. *Science of the Total Environment, 703*, Article 134608.

Wehr, J., & van Vuuren, S. J. (2024). Algae and cyanobacteria communities. In *Wetzel's limnolopgy* (pp. 463–510). Elsevier.

Xia, Y., Zhang, M., Tsang, D. C., Geng, N., Lu, D., Zhu, L., Igalavithana, A. D., Dissanayake, P. D., Rinklebe, J., & Yang, X. (2020). Recent advances in control technologies for non-point source pollution with nitrogen and phosphorous from agricultural runoff: Current practices and future prospects. *Applied Biological Chemistry, 63*, 1–13.

Yan, Z., Kamanmalek, S., Alamdari, N., & Nikoo, M. R. (2024). Comprehensive insights into harmful algal blooms: A review of chemical, physical, biological, and climatological influencers with predictive modeling approaches. *Journal of Environmental Engineering, 150*(4), 03124002.

Zahir, M., Su, Y., Shahzad, M. I., Ayub, G., Rehman, S. U., & Ijaz, J. (2024a). A review on monitoring, forecasting, and early warning of harmful algal bloom. *Aquaculture*, 741351.

Zahir, M., Su, Y., Shahzad, M. I., Ayub, G., Rehman, S. U., & Ijaz, J. (2024b). A review on monitoring, forecasting, and early warning of harmful algal bloom. *Aquaculture*, 741351.

Zhang, W., Gu, P., Zhu, W., Jing, C., He, J., Yang, X., Zhou, L., & Zheng, Z. (2020). Effects of cyanobacterial accumulation and decomposition on the microenvironment in water and sediment. *Journal of Soils and Sediments, 20*, 2510–2525.

Chapter 2
The Evolutionary History of Cyanobacteria and Their Role in Earth's Biogeochemistry

Muhammad Nauman Khan, Barkat Ullah, Nasir Assad, Hansa Gul, Taleeha Roheen, Muhammad Mustaqeem, Maryam Bibi, Arafat Abdel Hamed Abdel Latef, Alevcan Kaplan, Abdul Razzaq, Tabarak Malik, and Shah Fahad

Abstract Cyanobacteria are among the oldest photosynthetic organisms on Earth, playing a crucial role in shaping the planet's biogeochemical cycles. Their evolutionary history dates back over 3.5 billion years, making them key contributors to the Great Oxygenation Event (GOE) around 2.4 billion years ago. Through oxygenic photosynthesis, cyanobacteria transformed Earth's early anoxic atmosphere, leading to significant ecological and evolutionary changes, including the rise of aerobic

M. N. Khan (✉) · B. Ullah
Department of Botany, Islamia College Peshawar, Peshawar, Pakistan
e-mail: nomiflora@uop.edu.pk

M. N. Khan
Biology Laboratory, University Public School, University of Peshawar, Peshawar, Pakistan

N. Assad · T. Roheen · M. Mustaqeem
Institute of Chemistry, University of Sargodha, Sargodha, Pakistan

H. Gul
Department of Zoology, University of Mianwali, Punjab, Pakistan

M. Bibi
Department of Botany, University of Malakand, Chakdara, Khyber Pakhtunkhwa, Pakistan

A. A. H. A. Latef
Department of Botany and Microbiology, Faculty of Science, South Valley University, Qena, Egypt

A. Kaplan
Department of Crop and Animal Production, Sason Vocational School, Batman University, Batman, Turkey

A. Razzaq
Qarshi Herb Research Center, Qarshi Industries (Pvt.) Ltd. Hattar, Haripur, Pakistan

Qarshi University, Lahore, Punjab, Pakistan

T. Malik
Department of Biomedical Sciences, Institute of Health, Jimma University, Jimma, Ethiopia

S. Fahad et al. (eds.), *Cyanobacterial Blooms: Ecology, Evolution and Biogeochemical Impacts*, https://doi.org/10.1007/978-3-032-06042-6_2

life. Their ability to fix nitrogen further enhanced their impact on nutrient cycles, influencing primary productivity in both marine and terrestrial ecosystems. Over time, cyanobacteria diversified into various ecological niches, evolving specialized metabolic pathways to adapt to changing environmental conditions. They established symbiotic relationships with eukaryotes, ultimately giving rise to plastids in algae and plants through endosymbiosis. This evolutionary milestone had profound implications for the biosphere, leading to the expansion of oxygen-dependent life forms and the development of complex ecosystems. In modern biogeochemistry, cyanobacteria continue to play a vital role in carbon and nitrogen cycling, influencing global climate regulation. However, anthropogenic factors such as nutrient pollution and climate change have led to the proliferation of harmful cyanobacterial blooms, posing ecological and economic challenges. Understanding the evolutionary history and biogeochemical significance of cyanobacteria provides insight into their resilience and adaptability, offering potential applications in biotechnology, biofuel production, and environmental management. This chapter explores the evolutionary trajectory of cyanobacteria and their lasting influence on Earth's biosphere, highlighting their fundamental role in shaping the planet's atmospheric and ecological dynamics.

2.1 Introduction

Cyanobacteria, long called blue green algae, are one of the most ancient and consequential lineages of photosynthetic prokaryotes, with their fossil history taking it back at least 3.5 billion years. Early fossil records of stromatolites and microfossils (3.48 Ga Dresser Formation, Western Australia) display morphological and laminar patterns that form one of the characteristic features of cyanobacterial mats, setting their early dominance in shallow marine ecosystems. By changing their oxygenic photosynthesis, which was the only metabolic process capable of splitting water to produce molecular oxygen, cyanobacteria initiated the Great Oxidation Event (GOE) at 2.45 Ga, turning the Earth's atmosphere and surface oceans from anoxic to oxic conditions and allowing the emergence of aerobic organisms (Demoulin et al., 2019; Schirrmeister et al., 2015).

Other than the production of oxygen, cyanobacteria have supported global biogeochemical cycles for billions of years. Their capacity to fix atmospheric nitrogen into bioavailable forms undergirds productivity in nitrogen-limited ecosystems, while their endosymbiotic progeny the chloroplasts of algae and plants drive most of current primary production (Archibald, 2015, Sánchez-Baracaldo et al., 2017). To this day, cyanobacteria control carbon and nutrient fluxes in both aquatic and terrestrial environments, and anthropogenic nutrient loading and climate warming continue

S. Fahad
Department of Agronomy, Abdul Wali Khan University Mardan, Mardan, Khyber Pakhtunkhwa, Pakistan

to drive blooms that threaten ecosystem and human health. Known formerly as "blue-green algae," cyanobacteria are among the most ancient and ecologically important prokaryotic organisms. Those organisms are responsible for creating oxygenic photosynthesis one of the most crucial evolutionary innovations in the history of the Earth. This metabolic activity not only made it possible for the cyanobacteria to utilize the sun's energy to convert water and carbon dioxide to organic compounds and oxygen but also established the biochemical basis of Earth's atmospheric oxygenation during the Great Oxidation Event (GOE) some 2.45 billion years ago. This incident was the turning point in the changing of the planet from a world devoid of oxygen to a habitable one for sophisticated and aerobic life (Aiyer, 2022).

Cyanobacteria are exceptional in terms of their adaptability and ecological range. They are found in an astonishing variety of habitats, such as freshwater in lakes and rivers, marine environments, deserts, rocks, the polar ice caps, hot springs, and hyper-saline lakes. Their robustness is reflected in their ability to prosper in extreme environments like geothermal vents, sub-zero polar places, and highly salty or nutrient-deficient waters. Their ecological versatility is one of the factors that make them successful as a primary producer in numerous ecosystems, especially nutrient-deficient or extreme ecosystems where other species find it hard to survive.

As a concept in taxonomy, cyanobacteria are very diverse and complex. It is estimated that the phylum may have up to 8000 cyanobacterial species, though only approximately 5185 are formally identified and classified to date. These species are systematized into a number of principal taxonomic orders with respect to morphological as well as molecular phylogenetic analyses. The most significant ones include *Chroococcales*, *Gloeobacterales*, *Nostocales*, *Oscillatoriales*, *Pleurocapsales*, *Spirulinales*, and *Synechococcales*. Each group plays a distinct role in ecological roles such as nitrogen fixation, toxin production, and symbiotic relations with plants, along with fungi (Dvořák et al., 2015).

Cyanobacteria have been equally important in the unraveling of events leading to the evolution of photosynthesis in eukaryotes. Endosymbiosis resulted in the incorporation of a cyanobacterial forebear into early eukaryotic cells and the consequent rise of chloroplasts in plants and algae. This evolutionary breakthrough further accentuated the global significance of photosynthetic phenomena, creating the biological foundation for the bulk of the primary production on the contemporary Earth. Our knowledge of the evolutionary inheritance of cyanobacteria is corroborated by several geochemical and fossil lines. The geochemical proxies indicate that the oxygen may have been locally produced in so-called "oxygen oases" as early as 2.7 billion years ago; therefore, cyanobacteria have shaped Earth's atmosphere much earlier than the Earth became globally oxygenated. These results provoke questions concerning the timing and environmental relations of the early oxygenic photosynthesizers, of which many are objects of current science (McFadden, 2014).

2.2 Origin and Evolutionary History

The cyanobacteria are one of the oldest and most influential groups in the history of the Earth, and their history spans at least 3.5 billion years and comes from the Archean eon. They are well known to be the first organisms that evolved oxygenic photosynthesis, which was the metabolic revolution that changed the cosmos' atmosphere, chemical makeup, and evolutionary course. Based on the molecular clock analyses, it is suggested that the core protein subunits of Photosystem II, which were essential for the oxygen production, emerged between 3.6 and 3.2 Ga in the Paleoarchean, with the crown group of cyanobacteria coming about in the Mesoarchean, 3.2–2.8 Ga (Sánchez-Baracaldo et al., 2022). These estimates are predating the Great Oxidation Event (GOE) by hundreds of millions of years, pointing to a drawn-out period during which oxygenic photosynthesis was present before the filling of the atmosphere with oxygen. Remarkably, the large majority of current cyanobacterial lineages appear to have radiated only after the GOE, meaning that oxygenation stayed the hand of one of the most dramatic evolutionary radiations in the history of life (Aiyer, 2022).

Fossil evidence augments genomic results and offers real evidence of early cyanobacterial activity. Those laminated microbialites, such as 3.48 billion-year-old stromatolites collected from the Dresser Formation near Yilgarn granites in Western Australia, prove the existence of strongly growing mat-forming microbial communities, which might have been mostly represented by cyanobacteria. Although the cyanobacterial morphology related to microbiota impedes the ability to recognize the ancient microfossils within the deep fossil record, the unambiguous cyanobacterial fossil is identified in the 1.89–1.84 billion-year-old stromatolites from the Belcher Supergroup. These organisms grew successfully in shallow marine environments, building massive stromatolitic reefs which not only served as sinks for carbon but also played a significant role in locally regulating the oxygen levels, thus providing temporary "oxygen oases" well before the global oxygenation of the atmosphere (Demoulin et al., 2019).

During the Archean and early Proterozoic, cyanobacteria played a dominating ecological role and influenced the early microbial ecosystems and surface environment oxidation. Their biochemical novelties, especially oxygenic photosynthesis, not only facilitated the appearance of aerobic life but also controlled a number of geochemical cycles. Notably, one of the first endosymbiotic events in which a eukaryotic host engulfed a cyanobacterium led to the origin of plastids, whose initial developments marked the evolution of chloroplasts in algae and plants. This event has fixed cyanobacterial photosynthetic apparatus into the eukaryotic lineage, so that their legacy became a part of almost all modern photosynthetic life. Since principles of paleontology, geochemistry, and molecular biology all have important roles to play in defining timescale and events, the most efficient way to understand the evolutionary history and environmental impact of the cyanobacteria the actual builders of Earth's rich, oxygen-imbued biosphere is surely by integration (Garcia-Pichel et al., 2019).

2.3 The Great Oxygenation Event (GOE)

We are so adjusted to the existence of oxygen on Earth that we tend to take it for granted. However, in the process of about 4.5 billion years that Earth has existed, for almost half of this period, its atmosphere had practically no free oxygen. The appearance of the Earth at the early stage consisted of a reducing atmosphere and had gases such as carbon dioxide, methane, and water vapor. Any produced molecular oxygen, for example, that breaks off water vapor under photodissociations, reacts instantaneously with methane and other reductants, getting locked in the crust and oceans of the planet. Free oxygen was almost nonexistent (Blaustein, 2016; Hannah et al., 2004; Oakeshott, 2023).

About 2.7 billion years ago, a new division of microorganisms appeared: cyanobacteria. These prokaryotes possessed an amazing ability to perform oxygenic photosynthesis, which involved the use of light to convert water and carbon into organic matter and evolve oxygen as a by-product. Their origin is supported by the studies of the 16S and 23S rRNA sequences, genomic reconstruction, and records with fossils (stromatolites) a very ancient criterion. In the opinion of the biochemist Leslie Orgel and others, life originated approximately 3.8 billion years ago, but stayed anaerobic till the presence of the cyanobacteria (Shih et al., 2017).

When it oxidized water, cyanobacteria made the world's biogeochemistry fundamentally different. Although early oxygen was quickly taken by reduced compounds in the oceans and crust, cyanobacteria were releasing oxygen for hundreds of millions of years. This gradual accumulation later overwhelmed buffering systems to lay the ground for atmospheric change. About 2.4 billion years ago, a tipping point was attained The Great Oxygenation Event. This dramatic shift in the atmosphere is recorded by geochemical markers, in particular, sulfur isotope fractionation and oxidized iron formations (banded iron formations). The accumulation of free oxygen in the atmosphere appeared for the first time in the history of the Earth. Such a transformation not only alters the redox status of the oceans but also has great effects on the climate of the Earth, where it was linked to the Huronian glaciation via methane oxidation (Bekker et al., 2004; Lyons et al., 2014).

The consequences of the GOE were profound. Oxygen allowed for the development of aerobic respiration, a far more efficient form of energy production than anaerobic metabolism. It also led to the formation of an ozone (O_3) layer, which protected early life from harmful ultraviolet radiation, enabling colonization of shallower waters and eventually land. At the same time, rising oxygen levels were toxic to many anaerobes, leading to widespread microbial extinctions and ecological restructuring (Fournier et al., 2021).

2.4 Metabolic Capabilities and Ecological Roles

Cyanobacteria are one of the most metabolically diverse organisms on the planet, with a wonderful variety of physiological processes that established ecosystems for billions of years. The capacity to carry out oxygenic photosynthesis, which entails the conversion of water and carbon dioxide into organic substances at the expense of releasing oxygen, is one of their characteristics. This process was crucial during the Great Oxygenation Event, which drastically altered Earth's atmosphere and allowed the domination of aerobic life. Today, the cyanobacteria continue to play a major role in global carbon fixation, providing up to 25–30% of net primary productivity. Prominent oceanic genera like *Prochlorococcus* and *Synechococcus* are particularly important, fixing billions of tons of CO_2 each year in the surface areas of the ocean known as the euphotic zone, and forming the food base along which many aquatic food chains rely.

At the same time, cyanobacteria are also major contributors to nitrogen fixation, a task paramount to maintaining the productivity of nitrogen-limited ecosystems. With the aid of the nitrogenase, cyanobacteria convert atmospheric nitrogen (N_2) into ammonia (NH_3), making it bioavailable. This activity helps in primary production in nutrient-deficient settings, including the open oceans, lakes, and soils. Under anaerobic conditions, nitrogen fixation is achieved by the formation of heterocyst cells by filamentous cyanobacteria like *Anabaena* and *Nostoc* to maintain the requirement of low oxygen by nitrogenase. Such organisms are often in symbiotic relationships with plants, lichens, and marine diatoms, enhancing nutrient cycling and ecosystem productivity. Cyanobacterial N-fixation, then, has thus a prime position in the nitrogen economy of the biosphere, especially in oligotrophic/marginal environments (Pandey et al., 2025).

Apart from carbon and nitrogen metabolism, cyanobacteria have the ability to synthesize a variety of secondary metabolites that have critical ecological and biotechnological functions. Some of these are the mycosporine-like amino acids (MAAs), which offer protection from UV. Carotenoids and phycobiliproteins, which aid in photosynthesis efficiency, and cyanotoxins, including microcystins and anatoxins, could be used as an allelopathic agent or deterrent against grazers. Not only are these compounds important in ecological interactions, but they also have potential applications in pharmaceuticals, agriculture, and biotechnology. The formation of such metabolites is stringently controlled and is frequently even influenced by environmental conditions like light intensity, availability of nutrients, and oxidative stress (Kim et al., 2024).

Cyanobacteria are highly regulatory in a complex organization, where the genomes contain elaborate sensing and response systems. These include two-component signal transduction systems, redox-sensitive regulators, and even circadian clocks to temper metabolic activity to light cycles of a day. Cyanobacteria are largely metabolically flexible and can switch between autotrophic and heterotrophic modes of living or between aerobic and anaerobic respiration based on the environment. Such metabolic flexibility allows them to exist in an enormous variety of environments, from hot springs through polar ice, to hypersaline lagoons or arid deserts (Latifi et al., 2009).

Their ecological dominance is also underlined by the fact that in both freshwater and marine settings, cyanobacterial blooms develop. Although the blooms represent toxin production and oxygen depletion, they are also an indication of how cyanobacteria are adaptable and competitive successes. Cyanobacteria are important components of biological soil crusts in terrestrial ecosystems, which are important for soil stabilization, accumulation of organic matter, and nutrient cycling, especially in the desert and semi-arid regions (Pandey et al., 2025).

The biotechnological potential of the cyanobacteria is continuously being realized with the help of the current development in synthetic biology and metabolic engineering. Their strong carbon-fixing equipment and secondary metabolite paths make them perfect candidates for being used as sustainable bioreactors. To illustrate, a study by Kanno et al. (2017) showed how genetic re-wiring of cyanobacterial metabolism could be used to maximize CO_2 uptake and recast carbon flux toward bio-based chemicals such as 2,3-butanediol. Such innovations are an indication of future use of cyanobacteria in carbon–neutral bioprocesses, production of renewable energy, and synthesis of valuable compounds (Kanno et al., 2017).

In conclusion, the ecological and metabolic properties of cyanobacteria are the basis of life on Earth, and they shape the global biogeochemical cycles. Their metabolic flux, regulatory sophistication, and the nature of their biosyntheses not only constitute them as important constituents of natural ecosystems but also unquestionable candidates for sustainable technology in energy, agriculture, and the environment.

2.5 Diversification and Adaptation

Cyanobacteria are some of the oldest and environmentally adaptable progenies that have the magnificent ability to differentiate and endure in a wide spectrum of environmental settings. Their success as a species has allowed them to colonize almost every imaginable environment on Earth—from freshwater and marine ecosystems to deserts, polar ice, volcanic rocks, hot springs, and hypersaline lakes. This is due to a set of structural innovations, physiological flexibility, genetic adaptation, and biochemical resilience.

An important factor in their survival in harsh circumstances is their ability to make complex and synchronized responses to simultaneous stressors. Cyanobacteria can tolerate high levels of ultraviolet (UV) radiation, desiccation, intense saltiness, nutrient restriction, and enormous changes in temperature. For example, desert-resident species like *Leptolyngbya ohadii* have developed genetic means to switch on protective gene groups under high light and heat conditions to make them more resistant to dehydration and to oxidative stress. Analogically, filamentous strains such as *Nostoc flagelliforme* synthesize secondary metabolites that help to protect them against the UV radiation, scytonemin, and mycosporine-like amino acids (MAAs) protect their photosynthetic machinery. The loss of scytonemin compromises the structural integrity of the cells and decreases the Photosystem II desiccation recovery capacity, which speaks to its functional role in adaptation to the environment.

Morphological plasticity is another characteristic feature attributed to the cyanobacterial adaptability. These facilitate the animals concerned to switch between unicellular and filamentous forms, develop specialized nitrogen-fixing heterocysts, produce dormant akinetes, and develop motile hormogonia for dispersal. These remodelings arise from cytoskeletal proteins like MreB and RodZ and elongasome-associated enzymes that modify the peptidoglycan (PG), altering the peptidoglycan cell wall, allowing alterations in the shape and structure in response to environmental signals (Bataeva & Grigoryan, 2024; Kanno et al., 2017).

Cyanobacteria also have pronounced metabolic versatility, with which they can switch from photoautotrophic, mixotrophic, and heterotrophic life based on the availability of nutrients and surrounding light conditions. They have vast pigmentation capabilities; they have chlorophylls a, d, and f, which help them utilize light in different spectral ranges. The FaRLiP (far-red light photoacclimation) explains their tendency to alter their photosynthetic set-up in an efficient energy-capturing manner, either in a shaded or a far-red light situation, to enhance their ecological range (Ho et al., 2017).

Genetically, cyanobacteria also assimilate very well to changing environmental conditions in ways that include genome streamlining, horizontal gene transfer, and stress-responsive operon activation. They have highly evolved regulatory systems that detect and respond to changes in the environment, such as variation in temperature, light, salinity, and oxidative stress. In heterocyst-forming cyanobacteria, nitrogen fixation genes are tightly regulated to regulate the activity of oxygen-sensitive nitrogenase enzymes. By the biological form of nitrogen fixation, the cyanobacteria can supply 25–150 kg of nitrogen per hectare for a year, increasing the soil fertility and helping to develop the ecosystem in impoverished habitats.

Moreover, cyanobacteria show wonderful plasticity in the symbiotic relations with various organisms, namely fungi (e.g., lichens), plants (e.g., hornworts and cycads), and marine invertebrates, such as sponges and corals. These forms of relationship contribute to the survival of each of the two partners in arduous environments and represent the adaptive flexibility of the cyanobacteria in co-evolutionary systems (Chittora et al., 2020).

Some studies indicate the importance of the cyanobacteria sourced from the extreme environments in bioprospecting, since they quickly produce peculiar metabolites and enzymes that are useful in pharmaceuticals, agriculture, and biotechnology. Strains of cyanobacteria obtained from hot springs, salty lakes, and polar ice covers have been found to produce new bioactive compounds with curative and industrial properties, thus validating their notion as a resource for innovative biotechnological search. A summary of the key adaptation strategies that enable cyanobacteria to thrive in extreme environments is presented in Table 2.1.

Table 2.1 Adaptation strategies of cyanobacteria across diverse environments

Adaptation category	Mechanism/Feature	Adaptive function	Representative taxa
Stress tolerance	Production of UV-protective compounds like scytonemin and MAAs	Shields cellular components from UV-induced damage	*Nostoc flagelliforme*
	Activation of protective genes under high light and heat conditions	Enhances resistance to dehydration and oxidative stress	*Leptolyngbya ohadii*
Morphological plasticity	Differentiation into heterocysts, akinetes, and hormogonia	Facilitates nitrogen fixation, survival under adverse conditions, and dispersal	*Nostoc* spp.
	Transition between unicellular and filamentous forms	Allows adaptation to varying environmental conditions	Various genera
Metabolic versatility	Ability to switch between photoautotrophic, mixotrophic, and heterotrophic modes	Enables survival under fluctuating nutrient and light conditions	*Synechocystis* spp., *Cyanothece* spp.
	Utilization of diverse pigments, including chlorophylls a, d, and f	Expands the range of usable light spectra for photosynthesis	Various genera
Genetic adaptation	Genome streamlining and horizontal gene transfer	Enhances adaptability and efficiency in resource utilization	Various genera
	Activation of stress-responsive operons	Facilitates rapid response to environmental changes	Various genera
Symbiotic flexibility	Formation of symbiotic relationships with fungi, plants, and marine invertebrates	Provides mutual benefits such as nutrient exchange and enhanced survival in harsh environments	*Nostoc* spp., *Anabaena* spp.
Biotechnological potential	Production of unique bioactive compounds in extreme environments	Offers potential applications in pharmaceuticals, agriculture, and biotechnology	Strains from hot springs, saline lakes, and polar regions

2.6 Endosymbiosis and Origin of Plastids

The best explanation for a source of plastids—important organelles for performing photosynthesis in algae and plants—is given by the endosymbiotic theory. Based on this theory, a plastid has evolved from a free-living cyanobacterium, which formed a symbiotic relationship with the ancestral eukaryotic heterotroph more than 1.5 billion years ago. The process is shown in Fig. 2.1. The cyanobacterium was turned into a photosynthetic organelle by the host, rather than being digested, via a slow process of genetic and cellular assimilation. A number of structural and molecular evidence support this cyanobacterial origin, such as double membrane envelope structure, circular DNA, 70S ribosomes, and thylakoid envelopes in plastids. Such features are typical of the prokaryotes, being reinforced by the fact that plastids, similar to bacteria, reproduce by binary fission (Lhee et al., 2024). Parts of the cyanobacterial genome were transmitted to the host nucleus due to endosymbiotic gene transfer during the course of time, which limited the autonomy of the endosymbiont while maintaining its photosynthetic function. From this first endosymbiotic process, the plastids of the Archaeplastida lineage are derived, i.e., green algae, red algae, glaucophytes, and land plants. The procuring of plastids allowed these organisms to undergo oxygenic photosynthesis; this metabolic innovation had such effects on riverine biospheres as causing atmospheric oxygenation and the development of complex multicellular organisms. After this, other events which are referred to as secondary endosymbioses took place, whereby other non-photosynthetic eukaryotes digested eukaryotic algae containing primary plastids (Huisman et al., 2018; Stadnichuk & Kusnetsov, 2023).

These secondary endosymbioses led to plastids that we have encapsulated with three- or four-layered membranes as a consequence of this layered evolutionary history. Excellent examples include chlorarachniophytes, with a vestigial nucleus (nucleomorph) in their plastids, and other groups like euglenids and some dinoflagellates that acquired their plastids from other green/red algae. Such events allowed photosynthesis to diversify in the different eukaryotic groups and broaden the ecological and evolutionary potency of algae and protists in the aquatic habitat. Since the diversification, plastids have developed specialized forms including chloroplasts for photosynthesis, chromoplasts for the synthesis and storage of pigments, and

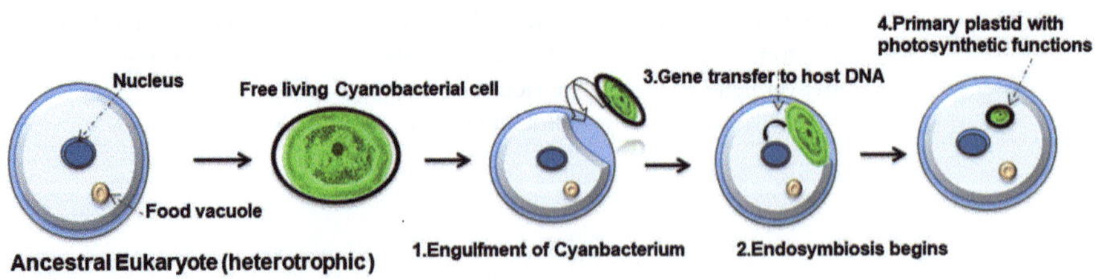

Fig. 2.1 Evolution of plastids via primary endosymbiosis between cyanobacteria and ancestral eukaryotes

leucoplasts for storage of starches and other non-photosynthetic molecules in non-photosynthetic tissues. With this functional diversity, all plastids have a common origin from cyanobacteria, demonstrating the seminal role of endosymbiosis in driving photosynthetic eukaryotes' metabolic abilities and ecological supremacy. The emergence of autotrophs from heterotrophic cells as a result of taking plastids is one of the major transition events in the history of the Earth, which has long affected the global biogeochemical cycles and productivity in various ecospheres (Sato, 2021).

2.7 Modern Roles in Biogeochemical Cycles

Modern biogeochemical cycles would not be where they are without the cyanobacteria, which have strongly regulated the flow of carbon, nitrogen, phosphorus, and methane on both land and water bodies. For being ancient oxygenic phototrophs, cyanobacteria are a foundation of global primary production and still drive elemental fluxes that govern the ecological and atmospheric processes. During the Great Oxidation Event, their emergence some 2.5 billion years ago played a key role in fundamentally altering Earth's atmosphere forever, by introducing molecular oxygen into the atmosphere and helping the emergence of life based on molecular oxygen. Nowadays, they are ecologically invaluable, accounting for up to half of all microbial biomass within a system while supporting the biodiversity of microbes through symbiotic and competitive relations. The ubiquity of cyanobacteria (from polar deserts to tropical oceans) reflects their physiological flexibility and the role in maintaining life-supporting processes at several trophic levels (Sánchez-Baracaldo et al., 2022).

In the carbon cycle, cyanobacteria act as dominant primary producers in the marine photic zone, freshwater bodies, and terrestrial biomes. Marine taxa such as *Trichodesmium*, *Prochlorococcus*, and *Synechococcus* contribute almost 25% of total global carbon fixation, while *Prochlorococcus* contributes about 4 gigatons of carbon annually (Sato, 2021). Their high photosynthetic efficiency and low cellular needs make them quite successful in the oligotrophic oceans poor in nutrients. By photosynthesis, they incorporate atmospheric CO_2 into biomass, which supports pelagic food webs and long-term carbon sequestration through the biological pump (sinking and burial of organic particles in the ocean sediments). In terrestrial systems, the cyanobacteria components of the biological soil crusts, more commonly known as biocrusts, may contribute to as much as 70% of the carbon input in arid and semi-arid environments. These crusts not only stabilize the soil surfaces and prohibit erosion but also increase water retention and micursorization that stabilizes resiliency and success of these delimitations. Their contribution to the global carbon cycle is now gaining recognition in climate models that aim toward predicting carbon flux in response to perturbations of the environment (Rajaneesh et al., 2020).

Cyanobacteria are also highly essential to the nitrogen cycle. Some genera, such as *Nostoc*, *Anabaena*, *Aphanizomenon*, and *Cylindrospermopsis*, are diazotrophs with the capability to assimilate atmospheric nitrogen (N_2)-fixing, through a complex of nitrogenase enzymes, into ammonium (NH_4^+). This process (which happens in specialized heterocysts in the case of nitrogen fixation in phototrophs, or via temporal separation of these two processes in the case of nitrogen fixation and photosynthesis in non-heterocystous strains) is a crucial contributor to the global nitrogen inputs especially in nitrogen-deficient conditions. In marine systems, the cyanobacterial fixation of nitrogen influx is estimated to be around 100–200 teragrams of nitrogen a year, a significant part of the total oceanic nitrogen budget. On land, the cyanobacteria are associated symbiotically with hornworts, ferns (e.g., Azolla), cycads, and lichens to augment nitrogen in the soil microhabitats. These contributions are beneficial to plant growth and soil fertility, especially in the early successional environments and degraded areas, therefore making cyanobacteria prominent agents in both natural and agricultural nitrogen cycling (Issa et al., 2014; Mishra et al., 2021).

Their contribution to the cycle of phosphorus also deserves mentioning. In the eutrophic and mesotrophic waters, cyanobacterial blooms frequently cause internal phosphorus loading releasing phosphorus from sediments and circulating it in the water column. Many bloom-forming species, such as *Microcystis* and *Planktothrix*, have extracellular alkaline phosphatases, which hydrolyze dissolved organic phosphorus (DOP) into bioavailable orthophosphate (PO_4^{3-}), thereby enabling them to out-compete other phytoplankton under depleted conditions. Studies have demonstrated how alkaline phosphatase activity increases exponentially at low availability of phosphorus as a biochemical feedback to drive the persistence of the bloom. Furthermore, certain benthic cyanobacteria are able to solubilize mineral-bound phosphorus through excretion of organic acids, thereby contributing to nutrient dynamics. These processes are critical in nutrient recycling and influence the phytoplankton succession, eutrophication process, and phosphorus management in terms of aqua-ecosystems (Wei et al., 2020).

Recent studies have also implicated cyanobacteria in methane cycling, albeit indirectly. Although cyanobacteria are not traditional methane producers or consumers, they contribute to methane fluxes by modulating local oxygen concentrations and influencing methanotrophic bacterial activity (Rumeau et al., 2007). In stratified and hypoxic water columns, the oxygen generated during cyanobacterial photosynthesis facilitates aerobic methanotrophy in adjacent microbial communities, which can reduce methane emissions by oxidizing CH_4 before it escapes into the atmosphere. For instance, in another study demonstrated that oxygen microzones produced by benthic cyanobacteria in anoxic sediments can sustain aerobic methanotrophs even under predominantly anaerobic conditions. Additionally, emerging evidence suggests that under specific conditions, certain cyanobacteria may possess the enzymatic capacity to produce trace amounts of methane via methylphosphonate degradation or demethylation pathways, although the ecological relevance of this pathway remains under investigation (Sánchez-Baracaldo et al., 2022). These insights reveal the nuanced, often overlooked, roles of cyanobacteria in greenhouse gas regulation, with implications for global methane budgets and climate feedback mechanisms.

In summary, cyanobacteria are not only evolutionary pioneers of oxygenic photosynthesis but also enduring regulators of modern biogeochemical processes. Their contributions to carbon and nitrogen fixation, phosphorus turnover, and methane mitigation emphasize their ecological centrality and relevance in the context of environmental change. As anthropogenic pressures continue to alter climate and nutrient cycles, understanding cyanobacterial metabolic pathways, ecological interactions, and environmental feedbacks becomes crucial for accurate ecosystem modeling and sustainable management. Investing in advanced omics tools, long-term monitoring, and synthetic biology approaches will help unlock the full ecological and biotechnological potential of cyanobacteria in safeguarding planetary health.

2.8 Harmful Cyanobacterial Blooms (HCBs)

In the freshwater systems, the interest in researching cyanobacterial communities has increased, following the increasing rate of harmful cyanobacterial blooms (HCBs), which are, in turn, one of the most pressing emerging threats to aquatic ecosystems. The HCBs are defined by the rapid, wide-scale development of the harmful cyanoproducts cyanobacteria, which too often lead to ecological degradation, public health hazards, and social and economic distortions. These blooms have been found to be increasingly deadly in the current climate change scenarios, persistent, and geographically widespread (Schostag et al., 2022).

A variety of cyanobacterial taxa suspected in causing HCBs are able to synthesize potent secondary metabolites, cyanotoxins, which have hepatotoxin, neurotoxin, cytotoxic/daphneotoxin, or dermatotoxic activity. Among them, the microcystins are the most studied because of their strong prevalence and serious impacts on health. Other, however, equally dangerous toxins such as saxitoxins, anatoxins, cylindrospermopsin, lyngbyatoxins, and aplysiatoxins are concerns as well. These compounds are not only associated with acute toxicity but also have chronic health implications with long-term exposure at low levels of concentration. For example, cylindrospermopsin is known to be cytotoxic and neurotoxic in nature and has a long-term guideline as low as 0.7 μg/L in drinking water, compared with that for microcystins. The presence of these toxins in the aquatic system, the capacity of the toxins to bioaccumulate in the aquatic life and crops irrigated with the contaminated water, and the likelihood of exposure to humans through ingestion, dermal contact, or inhalation of aerosols highlight the importance of the toxins in the context of Sukenik and Kaplan (2021).

There is a wide range of identified cyanobacterial genera with toxigenic strains, i.e., *Microcystis*, *Dolichospermum* (formerly *Anabaena*), *Aphanizomenon*, *Cuspidothrix*, *Sphaerospermopsis*, *Raphidiopsis* (formerly *Cylindrospermopsis*), and *Oscillatoria*. Over the past two decades, there has been a significant increase in the number of recorded toxic blooms, while in frequency, intensity, and spatial spread, the effect of climate change has increased its pace. High temperatures, increased CO_2 contents in the atmosphere, and hydrological disruptions (e.g., floods and droughts) provide good conditions for the formation of blooms. Moreover, tropical bloom-forming taxa, *Raphidiopsis raciborskii*, *Cuspidothrix issatschenkoi*, and *Sphaerospermopsis aphanizomenoides* are being reported more often in the temperate zones, indicating the change in the biogeography of HCBs (Huisman et al., 2018). A summary of key mitigation strategies addressing these blooms is presented in Table 2.2.

Importantly, not all strains in the toxin-producing species are toxigenic. Bloom measurements are confounded because different strains of cyanobacteria produce different amounts of toxins that vary geographically, and even morphologically similar cyanobacteria can play very different ecological roles and present very different health risks. This variation highlights the difficulty in tracking HCBs using the traditional morphological techniques alone. While several approaches related to satellite remote sensing, microscopy, pigment analysis, and molecular diagnostics have been developed for the detection of blooms, these techniques are rarely applied in conjunction and fail to relate to high-resolution taxonomic or functional data.

The previous classification of cyanobacteria was based on morphological features, but obtaining good taxonomy results of bloom-forming genera (especially Nostocaceae) has been proven impossible through this method. Genera like *Anabaena*, *Aphanizomenon*, *Cuspidothrix*, and *Sphaerospermopsis* show significant phenotypic plasticity and co-dominance coupled with overlapping traits during blooms, making it difficult for species-level identification. Molecular approaches such as 16S rRNA gene sequencing have indicated that many of the traditional genera are polyphyletic and stimulated a shift in taxonomy based on polyphasic integration of phenotypic, archaeobiosynthetic, and molecular data. This integrative framework is very important for accurate diagnosis of toxic strains and knowledge of bloom dynamics, especially in the absence of diagnostic characters such as heterocysts or akinetes.

Table 2.2 Mitigation strategies targeting cyanobacterial blooms: hydrological and nutrient approaches

Strategy	Approach	Outcomes	Limitations	References
Hydrological manipulation	Decrease water retention time to reduce cyanobacterial doubling	Decrease biomass and internal P loads	Difficult in large, seasonally variable systems	(Watson et al., 2016)
Watershed nutrient management	Reduce nutrient (N, P) input from upstream and land sources	Long-term bloom reduction	Time/resource intensive, impact varies with climate change	(Huisman et al., 2018; Wood, 2016)
Hypolimnion flushing	Replaces nutrient-rich bottom waters with nutrient-poor water	Lowers internal phosphorus, limits bloom formation	Ineffective for large lakes or where retention is non-uniform	(Visser et al., 2016)
Dilution Rate adjustment	Adjust residence time based on nutrient load and lake morphology	Controls spatial heterogeneity in phytoplankton	Requires accurate modeling and hydrological data	(Watson et al., 2016)

Cyanobacterial HCBs often include consortia of organisms of similar morphology displaying different ecological niches and physiological responses. Some cyanobacteria have a preference for a high nutrient environment, while others dominate in a low nutrient or disturbed thermal environment. Climate change, particularly the increase in temperatures and CO_2 levels, is a known proponent of nitrogen-fixing cyanobacteria such as *Raphidiopsis raciborskii* and *Sphaerospermopsis aphanizomenoides*, leading to changes in blooms and their resilience. Besides, microevolutionary mechanisms and genetic plasticity in species confer quick tolerance to environmental stressors, which, in most cases, enhances blood tenacity and increases toxin production.

The distribution of toxin biosynthetic gene clusters in a species may vary in different strains; therefore, not all strains within the same species will have the cyanotoxin biosynthesis capacity. Environmental triggers (e.g., light intensity, temperature, ratios of nutrients, interactions between microbes) can regulate these genes. An increase in the dominance of toxigenic strains and upregulation of the toxin production have been associated with warmer temperatures specifically. As such, toxin gene screening at the molecular level, along with identification at the strain level, is essential for proper risk assessment, management of blooms, and amelioration of harm to human and environmental health (Sukenik & Kaplan, 2021).

Conclusively, HCBs pose a complex and multifactorial phenomenon that requires integrative monitoring strategies, superior molecular diagnosis, and a profound study of cyanobacterial ecology and evolution. As global circulations of climate continue to alter, these blooms will increase in both frequency and intensity and will necessitate urgent science-based policy and management measures to ensure ecosystems and public health are protected (see Fig. 2.2 for a summary of causes, consequences, and strategies to manage harmful cyanobacterial blooms).

2.9 Applications in Biotechnology and Environmental Management

Cyanobacteria are now an important tool of modern biotechnology, particularly in providing sustainable solutions in energy and materials. While numerous biological platforms are available for the bioplastic and biofuel production, cyanobacteria have been under growing attention as the microorganisms that are capable of accumulating energy-rich compounds such as triacylglycerols (TAGs) and diacylglycerols (DAGs), which are biochemical precursors of biodiesel. Their potential in the production of biofuel and bioplastic lies in their innate nature of carrying out oxygenic photosynthesis and the ability to fix atmospheric carbon dioxide. When nutrient-limited, the cyanobacteria store lipid bodies and carbohydrates that can be processed into bioethanol, biodiesel, or biogas. Some of the species, like *Synechococcus* and *Arthrospira*, have undergone genetic modification to increase biofuel production by overexpressing major enzymes in the synthesis of lipids and the carbon flux pathways.

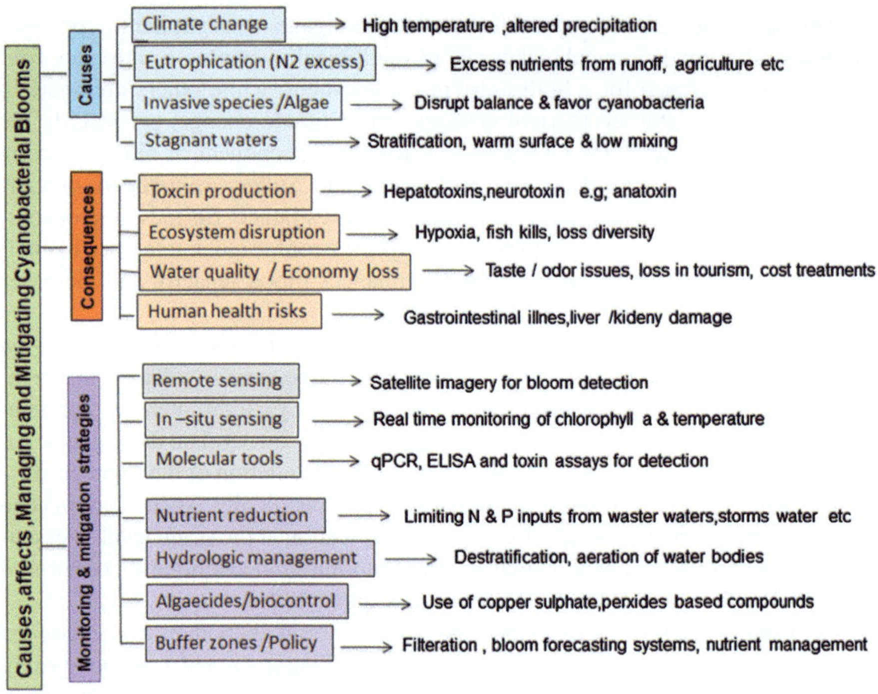

Fig. 2.2 Cyanobacteria blooms (HCBs): causes, consequences, and mitigation strategies (Sukenik & Kaplan, 2021)

In addition, methods toward the production of a kind of biodegradable plastic, polyhydroxyalkanoatcs (PHAs), have been proven in strains such as the *Synechocystis* sp. PCC 6803. These PHAs are naturally biodegradable in the environment, providing a sustainable replacement to petroleum plastics and enabling the circular bioeconomy (Schostag et al., 2022). Besides plastics, cyanobacteria are also the renewable platforms for the next-generation biofuels. The traditional production of fuels based on petroleum, like gasoline, diesel, and jet fuel, heavily burdens the environment through greenhouse gas emissions and toxic derivatives. On the other hand, engineered cyanobacteria are a cleaner substitute. Genetically engineered strains like *Synechocystis* sp. have been genetically modified to produce biofuels such as 2-methyl-1-butanol, isobutanol, 2,3-butanediol, ethanol, and ethylene. Notably, private companies engaged in the biotechnology industry have advanced technology in the commercialization of biofuels from cyanobacteria (Joule Unlimited and Algenol), and it contributes to seeing a shift in the direction toward carbon–neutral energy solutions (Tiwari & Tiwari, 2020).

Cyanobacteria are known for being resilient and metabolically diverse in the bioremediation of polluted soils and water. They are able to absorb and detoxify a wide assortment of pollutants, including heavy metals (such as cadmium, lead, and arsenic), pesticides, hydrocarbons, and industrial dyes. Cyanobacterial mats and biofilms, and in particular, those established by *Nostoc*, *Oscillatoria*, and *Anabaena*, act as a biological filter in constructed wetlands and wastewater treatment schemes. The occurrence of metal-chelating compounds such as phycobiliproteins and extracellular polymeric substances (EPSs) makes the sequestration and immobilization of pollutants easy. Also, cyanobacteria can be used in the reclamation of saline or barren soils through increasing the nitrogen and organic carbon content, which leads to improving the structure and fertility of the soils.

Cyanobacteria are a widespread source of bioactive metabolites used in pharmaceuticals and nutraceuticals. These include diverse classes of secondary metabolites that include cyanotoxins, polyketides, alkaloids, and nonribosomal peptides, among many other secondary metabolites, many of which display powerful biological functions. *Microcystis aeruginosa*, for example, synthesizes cytotoxic and anticancer microcystins, *Lyngbya majuscula* produces compounds like curacin A and majusculamide C, which are active against antimicrobial and anticancer. Cryptophycins from *Nostoc* spp. are studied for possible use in chemotherapy because they are capable of disrupting the assembly of microtubules in tumor cells. Some of the cyanobacteria, such as *Spirulina platensis*, are commercially cultivated in the nutraceutical world because of their high nutritional value of proteins, essential amino acids, omega-3 fatty acids, phycocyanin, and vitamins B12 and E. Such compounds are useful in immune support, antioxidant defense, and overall health, which explains why they find their way in dietary supplements and functional foods across the world (Tan, 2007; Žegura et al., 2011).

Cyanobacteria equally exhibit immense promise in carbon capture and sustainability in agriculture, which is part of the major biological entities of climate remediation measures. By carbon fixation through the Calvin–Benson cycle, they take large quantities of CO_2, which can be sequestrated in the form of biomass or even converted to value-added products (Kanno et al., 2017). The application in photobioreactors for capturing carbon from industrial emissions has been suggested as an eco-friendly way of relaxing the atmospheric carbon load (Gonçalves, 2021).

In agriculture, cyanobacteria (*Anabaena*, *Nostoc*, and *Tolypothrix*) act as biofertilizers that convert atmospheric nitrogen into bioavailable forms, hence there is increased crop yield without the use of synthetic fertilizers. These cyanobacterial inoculants also enhance the texture of the soil, add organic matter, and enhance microbial diversity. Furthermore, their potential in triggering plant resistance against pathogens and abiotic stress has been studied, and that makes them useful in climate-resilient agriculture (Bhuyan et al., 2023).

Cyanobacteria form a multifunctional spectrum of microbes that have manifold utilities in biotechnology and in environmental science. From becoming green factories for renewable energy and materials to being biological agents for pollutants control, health improvement, and climate adaptation, their potential is enormous. As new breakthroughs continue in the field of synthetic biology, metabolic engineering, and systems biology, the future foresees an even more site-specific and efficient use of cyanobacteria in addressing global challenges of energy, environment, and the latest concern for health.

2.10 Future Perspectives

Besides their biotechnological prospects, cyanobacteria have enormous prospects as bioindicators and biosensors in monitoring the environment. Their fast physiological reaction to pollutants and environmental challenges, while having a tractable genome, enables the creation of living biosensors capable of measuring heavy metals, toxins, and deficiencies in the amounts of nutrients on the go. Such functionality can be mobilized to enhance the surveillance of the ecosystem and control of pollution, especially in freshwater and marine ecosystems that are susceptible to anthropogenic stress. In addition, cyanobacterial systems integration into sustainable agriculture is gaining momentum, with engineered strains being investigated for biofertilizer development, nitrogen fixation improvement, and biopesticide production—environmentally friendly alternatives to the heavy chemical-intensive practices.

The current endeavors also investigate the utility of extremophilic and thermophilic cyanobacterial strains for industrial biotechnology, especially bioprocessing under adverse conditions where traditional microbial systems are incompetent. These strains have tolerance capacities for hot environments and saline or acidic conditions, and they therefore increase the range of industrial usage. Synthetic ecology methods, where cyanobacteria are engineered to perform in microbial communities with heterotrophs, are also improving productivity and stability in sophisticated bioprocesses such as the production of bioplastics and the recycling of carbon.

The combination of both omics technologies, machine learning, and systems biology is transforming our view of cyanobacterial regulatory networks and metabolic fluxes. This integrated data-driven approach allows for rational strain engineering, precise metabolic re-wiring, and optimization of culture conditions for maximum yield and ecological safety. Looking ahead, the development of standardized genetic parts, modular toolkits, and open-source platforms will democratize cyanobacterial biotechnology, enabling broader global participation in addressing challenges related to food security, energy demands, and environmental resilience. As climate change continues to intensify the frequency of harmful cyanobacterial blooms, these innovations are both timely and necessary to shift from reactive management to predictive and preventive strategies.

2.11 Conclusion

Cyanobacteria have been central to the evolutionary, ecological, and biogeochemical history of Earth. As the first organisms to perform oxygenic photosynthesis, they were responsible for initiating the Great Oxygenation Event (GOE), fundamentally altering the composition of Earth's atmosphere and enabling the rise of aerobic and multicellular life. Their metabolic innovations—particularly carbon fixation and nitrogen assimilation—continue to sustain ecosystems across both aquatic and terrestrial domains. The ability of cyanobacteria to thrive in extreme environments and adapt morphologically and physiologically to shifting conditions underlines their evolutionary resilience. Equally significant is their role in the origin of plastids through endosymbiosis, an event that transferred photosynthetic capacity to eukaryotes and catalyzed the diversification of algae and land plants. Through this event, the photosynthetic machinery of cyanobacteria became a driving force behind the planet's primary productivity and the development of complex food webs. In the modern biosphere, cyanobacteria remain pivotal in carbon and nitrogen cycles and increasingly in phosphorus and methane dynamics. However, anthropogenic pressures, such as nutrient pollution and climate change, have led to the emergence of harmful cyanobacterial blooms (HCBs), posing substantial threats to water quality, biodiversity, and public health. These blooms highlight the dual nature of cyanobacteria as both ecosystem engineers and ecological disruptors. Simultaneously, cyanobacteria hold immense promise in biotechnology. Their capacity for biofuel and bioplastic production, environmental remediation, and use in sustainable agriculture positions them as key players in the pursuit of green technologies. As advances in synthetic biology, genomics, and metabolic engineering progress, cyanobacteria are poised to contribute meaningfully to environmental management and climate resilience. Understanding their evolutionary legacy and ecological significance is essential—not only to appreciate their past contributions, but also to harness their full potential in shaping a sustainable future.

References

Aiyer, K. (2022). The great oxidation event: how cyanobacteria changed life. *American Society for Microbiology.*

Archibald, J. (2015). Endosymbiosis and eukaryotic cell evolution. *Current Biology, 25*, R911-R921.

Bataeva, Y., & Grigoryan, L. (2024). Ecological features and adaptive capabilities of cyanobacteria in desert ecosystems: A review. *Eurasian Soil Science, 57*, 430–445.

Bekker, A., Holland, H. D., Wang, P. L., Rumble, D., Stein, H. J., Hannah, J. L., Coetzee, L. L., & Beukes, N. J. (2004). Dating the rise of atmospheric oxygen. *Nature, 427*, 117–120.

Bhuyan, P. P., Nayak, R., Jena, M., & Pradhan, B. (2023). Convoluted role of cyanobacteria as biofertilizer: an insight of sustainable agriculture. *Vegetos, 36*, 309–321.

Blaustein, R. (2016). The great oxidation event: evolving understandings of how oxygenic life on earth began. *Bioscience, 66*, 189-195.

Chittora, D., Meena, M., Barupal, T., & Swapnil, P. (2020). Cyanobacteria as a source of biofertilizers for sustainable agriculture. *Biochemistry and Biophysics Reports, 22*, Article 100737.

Demoulin, C. F., Lara, Y. J., Cornet, L., François, C., Baurain, D., Wilmotte, A., & Javaux, E. J. (2019). Cyanobacteria evolution: Insight from the fossil record. *Free Radical Biology & Medicine, 140*, 206–223.

Dvořák, P., Poulíčková, A., Hašler, P., Belli, M., Casamatta, D. A., & Papini, A. (2015). Species concepts and speciation factors in cyanobacteria, with connection to the problems of diversity and classification. *Biodiversity and Conservation, 24*, 739–757.

Fournier, G., Moore, K., Rangel, L., Payette, J., Momper, L., & Bosak, T. (2021). The Archean origin of oxygenic photosynthesis and extant cyanobacterial lineages. *Proceedings of the Royal Society B, 288*, 20210675.

Garcia-Pichel, F., Lombard, J., Soule, T., Dunaj, S., Wu, S.H., & Wojciechowski, M. F. (2019). Timing the evolutionary advent of cyanobacteria and the later great oxidation event using gene phylogenies of a sunscreen. *mBio, 10.*

Gonçalves, A. (2021). The use of microalgae and cyanobacteria in the improvement of agricultural practices: A review on their biofertilising, biostimulating and biopesticide roles. *Applied Sciences, 11*, 871.

Hannah, J. L., Bekker, A., Stein, H. J., Markey, R. J., Holland, H., & Letters, P. S. (2004). Primitive Os and 2316 Ma age for marine shale: implications for Paleoproterozoic glacial events and the rise of atmospheric oxygen. *Earth and Planetary Science Letters, 225*, 43–52.

Ho, M. Y., Gan, F., Shen, G., Zhao, C., & Bryant, D. A. (2017). Far-red light photoacclimation (FaRLiP) in Synechococcus sp. PCC 7335: I. Regulation of FaRLiP gene expression. *Photosynthesis Research, 131*, 173–186.

Huisman, J., Codd, G. A., Paerl, H. W., Ibelings, B. W., Verspagen, J. M., & Visser, P. (2018). *Cyanobacterial Blooms., 16*, 471–483.

Issa, A. A., Abd-Alla, M. H., & Ohyama, T. (2014). Nitrogen fixing cyanobacteria: future prospect. *Advances in Biology and Ecology of Nitrogen Fixation.* IntechOpen.

Kanno, M., Carroll, A. L., & Atsumi, S. (2017). Global metabolic rewiring for improved CO_2 fixation and chemical production in cyanobacteria. *Nature Communications, 8*, 14724.

Kim, D. S., Moreno-Cabezuelo, J. Á., Schulz, E. N., Lea-Smith, D. J., & Sagaram, U. (2024). Recent advances in engineering fast-growing cyanobacterial species for enhanced CO_2 fixation. *Frontiers in Climate, 6*, 1412232.

Latifi, A., Ruiz, M., & Zhang, C.-C. (2009). Oxidative stress in cyanobacteria. *FEMS Microbiology Reviews, 33*, 258–278.

Lhee, D., Bhattacharya, D., & Yoon, H. S. (2024). *The evolutionary origin of primary plastids, endosymbiotic organelle acquisition: solutions to the problem of protein localization and membrane passage* (pp. 3–38). Springer.

Lyons, T. W., Reinhard, C. T., & Planavsky, N. J. (2014). The rise of oxygen in Earth's early ocean and atmosphere. *Nature, 506*, 307–315.

McFadden, G. I. (2014). Origin and evolution of plastids and photosynthesis in eukaryotes. *Cold Spring Harbor Perspectives in Biology, 6*, Article a016105.

Mishra, A., Rajput, S., Gupta, P. S., Goyal, V., Singh, S., Sharma, S., Shukla, S., Singh, A., Shukla, K., & Varma, A. (2021). *Role of cyanobacteria in rhizospheric nitrogen fixation, soil nitrogen ecology* (pp. 497–519). Springer.

Oakeshott, A. C. (2023). *The influence of changing temperatures on cyanobacterial oxic methanogenesis.*

Pandey, P., Pandey, D., Gupta, A., Gupta, R., Tiwari, S., & Singh, S. (2025). Cyanobacterial green chemistry: A blue-green approach for a sustainable environment, energy, and chemical production. RSC Sustainabilty.

Rajaneesh, K. M., Naik, R. K., Roy, R., & D'Costa, P. M. (2020). Cyanobacteria in tropical and subtropical marine environments: Bloom formation and ecological role. *Advances in cyanobacterial biology* (pp. 35–46). Elsevier.

Rumeau, D., Peltier, G., & Cournac, L. J. P. (2007). Chlororespiration and cyclic electron flow around psi during photosynthesis and plant stress response. *Plat and CEll, Environment, 30*, 1041–1051.

Sánchez-Baracaldo, P., Raven, J. A., Pisani, D., & Knoll, A. (2017) Early photosynthetic eukaryotes inhabited low-salinity habitats. *Proceedings of the National Academy of Sciences, 114*, E7737–E7745.

Sánchez-Baracaldo, P., Bianchini, G., Wilson, J.D., & Knoll, A.H. (2022). Cyanobacteria and biogeochemical cycles through Earth history. *Trends in Microbiology, 30*, 143–157.

Sato, N. (2021). Are cyanobacteria an ancestor of chloroplasts or just one of the gene donors for plants and algae? *Genes, 12*, 823.

Schirrmeister, B. E., Gugger, M., & Donoghue, P. (2015). Cyanobacteria and the Great Oxidation Event: Evidence from genes and fossils. *Palaeotology, 58*, 769–785.

Schostag, M. D., Gobbi, A., Fini, M. N., Ellegaard-Jensen, L., Aamand, J., Hansen, L. H., Muff, J., & Albers, C. N. (2022). Combining reverse osmosis and microbial degradation for remediation of drinking water contaminated with recalcitrant pesticide residue. *Water Research, 216*, Article 118352.

Shih, P. M., Hemp, J., Ward, L. M., Matzke, N. J., & Fischer, W. (2017). Crown group Oxyphotobacteria postdate the rise of oxygen. *Geobiolgy, 15*, 19–29.

Stadnichuk, I. N., & Kusnetsov, V. (2023). Phycobilisomes and phycobiliproteins in the pigment apparatus of oxygenic photosynthetics: from cyanobacteria to tertiary endosymbiosis. *International Journal of Molecular Sciences, 24*, 2290.

Sukenik, A., & Kaplan. A. (2021). Cyanobacterial harmful algal blooms in aquatic ecosystems: A comprehensive outlook on current and emerging mitigation and control approaches. *Migroorganisms, 9*, 1472.

Tan, L. T. (2007). Bioactive natural products from marine cyanobacteria for drug discovery. *Phytochemistry, 68*, 954–979.

Tiwari, A. K., & Tiwari. B. (2020). Cyanotherapeutics: An emerging field for future drug discovery. *Appiled Phycology, 1*, 44–57.

Visser, P. M., Verspagen, J. M., Sandrini, G., Stal, L. J., Matthijs, H. C., Davis, T. W., Paerl, H. W., & Huisman, J. (2016). How rising CO_2 and global warming may stimulate harmful cyanobacterial blooms. *Harmful Algae, 54*, 145–159.

Watson, S. B., Miller, C., Arhonditsis, G., Boyer, G. L., Carmichael, W., Charlton, M. N., Confesor, R., Depew, D. C., Höök, T. O., & Ludsin, S. (2016). The re-eutrophication of Lake Erie: Harmful algal blooms and hypoxia. *Harmful Algae, 56*, 44–66.

Wei, F., Shahid, M.J., Alnusairi, G. S., Afzal, M., Khan, A., El-Esawi, M. A., Abbas, Z., Wei, K., Zaheer, I. E., & Rizwan, M. (2020). Implementation of floating treatment wetlands for textile wastewater management: A review. *Sustaniability, 12*, 5801.

Wood, R. (2016). Acute animal and human poisonings from cyanotoxin exposure—A review of the literature. *Environmental International, 91*, 276–282.

Žegura, B., Štraser, A., & Filipič, M. (2011). Genotoxicity and potential carcinogenicity of cyanobacterial toxins—A review. *Mutation Research/reviews in Mutation Research, 727*, 16–41.

Chapter 3
Ecological Drivers of Cyanobacterial Bloom Formation

Muhammad Nauman Khan, Barkat Ullah, Abdul Razzaq, Alevcan Kaplan, Sana Wahab, Nasir Assad, Hansa Gul, Ayesha Bibi, and Shah Fahad

Abstract Cyanobacterial blooms are increasingly frequent in aquatic habitats globally, driven by a complex interaction of ecological variables. This chapter addresses the key ecological forces that lead to bloom production, persistence, and toxicity. Nutrient enrichment, especially excess nitrogen and phosphorus from anthropogenic sources, is recognized as a crucial component driving cyanobacterial development. Additionally, physical variables such as water temperature, stratification, and hydrodynamics play key roles in shaping bloom dynamics. Rising global temperatures and changed precipitation patterns related to climate change further increase bloom

M. N. Khan (✉) · B. Ullah
Department of Botany, Islamia College Peshawar, Peshawar, Pakistan
e-mail: nomiflora@uop.edu.pk

M. N. Khan
Biology Laboratory, University Public School, University of Peshawar, Peshawar, Pakistan

A. Razzaq
Qarshi Herb Research Center, Qarshi Industries (Pvt.) Ltd. Hattar, Haripur, Pakistan

Qarshi University, Lahore, Punjab, Pakistan

A. Kaplan
Department of Crop and Animal Production, Sason Vocational School, Batman University, Batman, Turkey

S. Wahab
Department of Plant Sciences, Quaid-I-Azam University, Islamabad, Pakistan

N. Assad
Institute of Chemistry, University of Sargodha, Sargodha, Pakistan

H. Gul
Department of Zoology, University of Mianwali, Mianwali, Pakistan

A. Bibi
Department of Botany, Qurtuba University of Science & Information Technology, Hayatabad, Peshawar, Pakistan

S. Fahad
Department of Agronomy, Abdul Wali Khan University Mardan, Mardan, Khyber Pakhtunkhwa, Pakistan

S. Fahad et al. (eds.), *Cyanobacterial Blooms: Ecology, Evolution and Biogeochemical Impacts*, https://doi.org/10.1007/978-3-032-06042-6_3

occurrences. Light availability and competition with other phytoplankton species also impact cyanobacterial dominance, favoring species with effective light-harvesting systems. Biological interactions, including grazing pressure from zooplankton and microbial community dynamics, impact bloom growth and collapse. Furthermore, cyanobacteria adopt different adaptation techniques, such as buoyancy control, toxin synthesis, and nitrogen fixation, which boost their survival and competitive advantage. Understanding these ecological factors is crucial for anticipating and controlling blooms, as well as influencing management measures for lowering fertilizer inputs, enhancing water quality, and minimizing ecosystem disturbances.

Keywords Aquatic ecosystem · Bloom formation · Cyanobacteria · Ecological adaptation · Nitrogen fixation

3.1 Introduction

Blue-green algae, which are another name for cyanobacteria, are ancient cells that are essential for the cycling of nutrients and the primary production in water bodies (Jassim et al., 2023). They have continued to exist for over 2.5 billion years mainly because they can produce their own food, fix atmospheric nitrogen, and adapt to many environmental conditions. Lately, a rise in cyanobacteria known as harmful algal blooms (HABs) has led to serious concerns both for nature and for people's health. When these blooms occur, cyanobacteria accumulate quickly and become a major problem by causing scums, turning the water green or blue, making it smell unpleasant, and producing toxins that affect fish, animals on land, and humans.

All over the world, incidents of cyanobacteria blooming in both freshwater and marine environments are on the rise. As nutrients from people have increased in the water and as the climate has changed, their frequency, the length of time, and the range over which they occur have become much larger. It is common for the genera *Microcystis*, *Dolichospermum* (previously *Anabaena*), *Planktothrix*, and *Aphanizomenon* to cause toxic blooms. These cyanotoxins, microcystins, saxitoxins, anatoxins, and cylindrospermopsin can seriously harm both nature and human beings. Cyanotoxins have been found to cause problems with the liver, brain, skin, and cancer in both humans and animals. The EPA points out that microcystins are frequently found in US water systems and may reach unsafe concentrations when there is a major bloom (Padilla, 2021).

The impacts of cyanobacterial blooms on ecosystems are very serious. These events may hinder the function of aquatic food webs, covering the vegetation underwater and taking over other phytoplankton due to the substances they release. Falling bloom levels also use a lot of dissolved oxygen, which may result in fish dying from lack of oxygen and a loss of living things on the riverbed. In addition, blooms can disrupt the movement of nutrients, give cyanobacteria a competitive advantage, and continue to enhance eutrophic conditions. These problems in the environment are made worse by the social and economic consequences they cause. For example, when the 2014 microcystin contamination of Toledo, Ohio's water occurred because of a *Microcystis* bloom in Lake Erie, it impacted over 400,000 citizens and brought attention to the risk HABs pose to municipal water systems (Wells et al., 2020). The same thing happened in Lake Taihu (China), the Vaal Dam (South Africa), and the Murray-Darling Basin (Australia), showing that this is a global problem.

The reason cyanobacterial blooms are more common now is due to a mix of ecological and environmental elements. Primary reasons include nutrient enrichment, especially from the phosphorus and nitrogen waste produced by farming, cities, and industries. An increase in nutrients, called eutrophication, causes the environment where phosphorus exceeds certain levels to become favorable for cyanobacteria. Moreover, many cyanobacteria have the ability to stay buoyant and to capture nitrogen, making them stronger competitors when the supply of nutrients is uneven. As temperatures rise due to climate change, the chances of blooms occurring are increased since cyanobacteria can easily cope with the temperatures and are favored over most phytoplankton. Temperature increases cause metabolic rates to rise and water layers to become more distinct, which prevents vertical mixing and maintains the ideal top layer for cyanobacteria. The latest research by scientists Paerl and Otten (2016) and Huisman et al. (2018) has demonstrated that changes in precipitation patterns triggered by climate can introduce more nutrients to water and modify the hydrologic regimen, both of which encourage bloom events.

Besides abiotic aspects, the growth of a population is also affected by other organisms. While grazing from zooplankton may curb phytoplankton growth, many cyanobacteria's large or unsafe forms prevent top-down controls from being effective. In addition, contact with heterotrophic bacteria, viruses, and fungi can change the growth and death of cyanobacterial populations, shaping the structure and longevity of blooms. Cyanobacteria are complex because their microbiome acts in nutrient cycling and in the production of toxins linked to blooms.

The goal of this chapter is to understand the different ecological reasons behind the buildup of cyanobacteria. The emphasis will be on how nutrient changes, weather conditions, water and land changes, and plant and microbe interactions each affect the onset and maintenance of algal blooms. The authors will support their explanations with recent case studies and evidence from experiments of such drivers working together. The purpose of this chapter is to use current knowledge to inform techniques for tracking, forecasting, and addressing blooms of cyanobacteria in water systems. This knowledge should help policymakers, water resource managers, and environmental scientists work toward solutions for coping with cyanobacterial blooms as the world continues to change environmentally.

3.2 Nutrient Enrichment and Anthropogenic Inputs

It is widely accepted that the enhanced supply of nutrients is a major reason for cyanobacterial blooms in water bodies. Blooms of cyanobacteria depend heavily on the presence of nitrogen (N) and phosphorus (P), among the macronutrients, and they are capable of responding to their amounts. Phosphorus generally limits growth in freshwater, and when more of it is available, cyanobacteria begin to dominate the phytoplankton community. Nitrogen, in addition, can increase the seriousness of a bloom and enhance the production of toxins. How N and P interact often decides both the strength of a bloom and the types of cyanobacteria present. *Microcystis* expands well in areas with plenty of nitrogen, while genera such as *Dolichospermum* grow in places with little nitrogen but lots of phosphorus.

Most of these nutrients come from things humans create. Runoff of fertilizers and animal wastes during storms and from poorly handled fields is a major way agriculture leads to nutrient pollution. The most common method for transporting nitrogen is as nitrate, although phosphorus usually sticks to the soil and can become free during anoxic states. Wastewater treatment plants release nutrients into rivers and lakes, and in much of the world, treated wastewater still has a small amount of nitrogen and phosphorus that can encourage algae to grow. The problem is made worse by urban runoff, since the water rushes from solid areas into nearby lakes and rivers without getting filtered. Together, these diffuse and point sources of pollution cause freshwater to contain more nutrients, moving the systems toward eutrophic and sometimes hypereutrophic conditions. As a result of these conditions, there are more cases of harmful algae blooms, both regular and cyanobacterial-induced ones (difficult changes in weather are driving cyanobacteria to spread worldwide).

Thanks to the high levels of nutrients flowing into water, eutrophication is the main cause of cyanobacterial blooms. It boosts the main role of phytoplankton in the ecosystem, destabilizes ecosystems and weather patterns, and encourages more cyanobacteria growth than that of other phytoplankton. In these conditions, cyanobacteria work well, so they can often be seen as surface scums, limiting how much light gets to others and giving themselves an advantage. Additionally, lots of cyanobacteria are able to stay at a certain depth, keep extra resources, and endure tough environmental situations, which gives them an important place in nature. They may also produce toxic substances, microcystins, that endanger both drinking water, fish, and other aquatic life and human health. When oxygen levels drop, producing anoxic conditions, excess phosphorus from bottom sediments usually helps to keep P blooms occurring even after external phosphorus has decreased.

Controlled loading of nutrients inside the lake can be achieved by using nutrient sequestering metals (e.g., iron (Fe), aluminum (Al), lanthanum bentonite), as well as with dredging, mixing, aeration, biomanipulation, and planting of macrophytes. Cyanobacteria are now being controlled using chemicals, physical methods, and biological strategies like straws. Eutrophication and nutrient threshold levels are very important in controlling how blooms develop and change. It has been demonstrated in recent studies that there are nutrient levels above which bloom-forming conditions are much more likely. In many cases, when total phosphorus is greater than 0.02–0.05 mg/L and total nitrogen is greater than 0.3–1.0 mg/L, such water conditions can promote the spread of cyanobacteria (Kosten et al., 2012; Paerl & Otten, 2016).

Still, the thresholds can change depending on the shape, flow patterns, and climate of the lake. Modest nutrient increases in lakes or areas with warm climates can frequently cause persisting blooms since recycling and stratification persist for longer periods. Therefore, management strategies should consider these thresholds as guidelines rather than absolute limits and adapt them to local ecosystem characteristics. Implementation of these thresholds into water quality regulations and watershed policies is crucial for preventing ecosystem degradation and protecting freshwater resources (Kosten et al., 2012).

Science has focused on figuring out what nutrient caps can stop blooms from forming. Experience shows that when the total phosphorus in water reaches 0.02–0.05 mg/L or the total nitrogen amounts to 0.3–1.0 mg/L, blooms are likely to form. However, these thresholds are not suitable for everyone, since the factors affecting blooms can vary sharply from one place to another. This highlights the importance of using watershed management across land-use activity, farm work, riparian buffers, and strengthened wastewater processing.

Cyanobacteria are influenced by how nutrients are available and by changes in many environmental features. Yet, most previous studies have overlooked other areas, so we still know little about bloom events in warm climates. While the bulk of lakes lie in the north temperate zone, getting access to freshwater and threats to that water are mainly local issues, since more than 71% of the planet's population lives in the tropics and subtropics. For this reason, examining how warm-climate lakes function is vital for avoiding threats to the water sources of most humans and for gaining information on likely future effects elsewhere as Earth warms up. Since understanding these drivers includes many different environmental conditions, it is necessary to conduct studies across a wide range of regions (Paerl et al., 2011a, 2011b).

For a better insight into global patterns driving cyanobacterial blooms, we need to study them in more regions. The Americas are about 14,000 km long north to south, support a population of almost a billion, take up 28% of Earth's land, and have a major part of the total freshwater resources. From equatorial areas at the center to polar places along the coast, the region covers the major climate zones. Thus, it makes it possible to learn about the relationships between higher nutrients, different climate conditions, and the appearance of blooms in many parts of the ocean.

In short, although enhancements of nutrients from society continue to largely promote cyanobacterial blooms, a detailed approach is becoming more necessary. Part of this process is recognizing that different areas deal with distinct challenges, considering more than one factor influencing blooms, and expecting that climate change can cause variations in bloom behavior. Protection of water resources relies on combining scientifically planned actions to limit nutrients and ensure the general environment around blooms remains healthy.

3.3 Physical and Climatic Factors

Physical and climatic elements greatly determine the way cyanobacterial blooms happen, how intense they are, and how long they last in freshwater ecosystems. Cyanobacterial physiology and the behavior of their surrounding communities are often most affected by water temperature. Many cyanobacteria are most active and grow well when the temperature is at least 25 °C, whereas other groups of phytoplankton prefer cooler waters. Warm weather improves how much cyanobacteria grow and metabolize, while encouraging strains like *Microcystis aeruginosa* and *Dolichospermum* spp. to proliferate. Seasonal conditions with hotter summers and gentler winters let cyanobacteria grow and delay senescence. Because most of their life cycles span additional months, these organisms cause more environmental and health risks (Matthews, 2014).

It is likely that the number of CyanoHABs will grow because of causes such as global warming, altered rainfall patterns, and how land is used. As a result, the nutrient runoff, modifications to watershed water, and changes to lake temperatures and water levels raise the risk of algal blooms. Different layers of water help cyanobacteria to thrive. In thermally stratified lakes, fewer vertical currents cause nutrients to settle in a lower layer, forming an epilimnion rich in oxygen and nutrients and a hypolimnion that lacks oxygen but contains plenty of nutrients. Because of the presence of gas vesicles, cyanobacteria rise or fall in the water column, aiming to get the optimum light and nutrients. Due to this buoyancy regulation, cyanobacteria outcompete many other phytoplankton when the water is stable (Visser et al., 2016). Research has recently discovered that the rise in surface water temperature because of climate change makes thermal stratification last longer, so that overturn of surface and deep water happens later in the season, which means cyanobacteria are able to dominate even longer (Dasari et al., 2020). Conversely, many mixing events, regardless of how they occur, can reduce blooms by disturbing their arrangement in the water and exposing them to more unfavorable light levels. However, in places with plenty of nutrients, mixing the water may actually bring up phosphorus from sediments, boosting the potential for blooms. It was found using high-quality lake surveillance that higher thermal stability goes hand in hand with more and stronger cyanobacterial blooms, generally in lake systems that provide nutrients (Metzgen et al., 2021).

Changes in water movement and hydrodynamics are important but not usually given enough attention in cyanobacterial bloom studies. As cyanobacterial biomass grows, strong mixing conditions may help the growth of aggregates and result in blooms. At moderate wind speeds (e.g., 3 m per second), experiments reveal the first upward movement of water, which increases surface cyanobacteria and promotes the start of blooms (Cousino et al., 2015; Kim et al., 2021; Li et al., 2024). Slow currents (between 0.02 and 0.04 m per second) have been associated with *Microcystis* sticking together, which helps blooms develop (Gao et al., 2018; Li et al., 2018). Should waves or water turbulence increase, cyanobacterial blooms may actually be reduced because the cells could be damaged (Gu et al., 2023; Pandey et al., 2023). *Microcystis* splits into small fragments when there is strong turbulence, which lowers its main productivity by a factor of ten (Xu et al., 2013). It was also observed that cellular harm and lipid peroxidation were induced in cyanobacteria by water flow at 0.2 m s^{-1}, which led to their failure to increase. The results support the notion that manipulating hydrodynamic parameters may help stop bloom outbreaks. As for previous research, it has largely relied on average wind and flow data rather than real-time monitoring data. Also, our knowledge of the underlying mechanisms behind hydrodynamic thresholds and cyanobacteria growth is not fully developed, especially from the point of view of fluid dynamics (Gondo et al., 2023).

The way water flows and how long it spends in an area can change the nature of a bloom. Areas with water that does not flow easily often collect living matter and give food for blooms to thrive. Higher flow rates may help remove cyanobacteria and prevent blooms from lasting for long. Because of the changing flows in reservoirs and impoundments, blooms may suddenly become unpredictable. Different ways people use water resources, such as building dams or changing river channels, can seriously affect cyanobacteria (Achaga et al., 2022; Huisman et al., 2018) (Table 3.1).

In water bodies all around the world, things such as more nutrients in the water or warmer temperatures support blooms of cyanobacteria. Unfortunately, relatively little is understood about particular drivers in river systems. Cyanobacterial blooms in the stands of large trees tend to be studied mostly in lakes and the sea, although rivers face them more frequently. Rivers in Australia, North America, and Europe, along with tropical and subtropical rivers, often experience cyanobacterial blooms when flow and turbulence are low. When there is ample flow and discharge, cyanobacteria growth may not develop even under rich nutrient conditions (Kleinteich et al., 2024). Changes in climate and the actions of people often cause major changes in the ecosystems of large, inland rivers.

Studies of Western and Central Europe suggest that drier summers, higher temperatures, and more frequent.

It has been widely found that nutrient availability is related to the proportions, locations, and amounts of cyanobacteria in aquatic systems (Granéli & Turner, 2006). But climate change also affects these connections. Global warming, modified rainfall, and differences in how water bodies send or flush their water impact the growth, makeup, and cycles of cyanobacteria (Joehnk et al., 2008; Paerl & Huisman, 2008, 2009; Paerl et al., 2011a, 2011b; Paul, 2008). More heat on the surface area is ideal for these species, which are common cyanobacteria preferred by warm temperatures, showing their most growth at temperatures above 25 °C (Foy et al., 1976; Robarts & Zohary, 1987). In high temperatures, cyanobacteria usually do better competing with eukaryotic algae. When eukaryotic taxa reach their highest or lowest growth rates because of higher temperatures, cyanobacteria show fastest growth rates (Weyhenmeyer, 2001).

3.4 Biological Interactions

Cyanobacteria are bacteria in the cyanobacteria phylum that help fix nitrogen in sunlight and can form relationships with many types of marine organisms. Particularly, it is their ability to make nitrogenase that makes a significant difference in their symbiotic relationships. Diazotrophs are a group of organisms that make diazotroph-derived nitrogen (DDN) available to their host organisms in places lacking nitrogen (Bonnet et al., 2023).

Cyanobacteria found in symbiosis are also known to make a wide range of secondary metabolites and toxins (Mutalipassi et al., 2021). Such plants make a

Table 3.1 Physical and climatic drivers of cyanobacterial blooms in freshwater ecosystems

Factor	Mechanism/ influence	Impact on cyanobacteria	Examples/species affected	References
Water temperature	Warmer temperatures (≥25 °C) increase metabolic activity and growth rates	Promotes bloom formation, delays senescence, and extends bloom duration	*Microcystis aeruginosa*, *Dolichospermum* spp.	(Matthews, 2014)
Vertical mixing	Disrupts stable water columns, brings up deep nutrients (esp. phosphorus)	Can suppress or intensify blooms depending on nutrient levels and frequency of mixing	*Microcystis*, other bloom-formers	(Metzgen et al., 2021)
Thermal stratification	Stable layers (epilimnion and hypolimnion) separate nutrients and oxygen	Buoyancy regulation via gas vesicles lets cyanobacteria access optimal light and nutrients	Buoyant genera (e.g., *Microcystis*)	(Visser et al., 2016)
Water currents	Slow flows (0.02–0.04 m/s) enhance cell aggregation	Promote bloom persistence; higher flows reduce bloom stability	*Microcystis* spp.	(Gondo et al., 2023)
Climate change	Increases in surface water temperature; delayed mixing	Enhances bloom frequency and dominance; longer periods of thermal stratification	Many species	(Metzgen et al., 2021)
Residence time/ stagnation	Long residence times promote nutrient buildup and cell accumulation	Increases bloom formation likelihood; stagnant water supports persistence	Cyanobacteria in reservoirs and lakes	(Huisman et al., 2018)

variety of health-promoting substances, including agents that guard against harm from light and from being eaten by animals (Bonnet et al., 2023). To give something back, hosts protect cyanobionts from tough external conditions and from being eaten by predators, and they also allow cyanobionts places to multiply where there is not much competition. In addition, some cyanobiont secondary metabolites are controlled by their host, as proved in the case of *Nostoc* and terrestrial plants such as *Gunnera* and *Blasia*. In *Nostoc punctiforme*, the presence of a host changes how the nostopeptolide synthetase gene is regulated and the amounts of different nostopeptolide variants secreted. Those metabolic changes probably help form motile filaments (hormogonia), which are responsible for beginning symbiosis and may also directly contribute to the infection and communication between the host and the cyanobacteria (Adams et al., 2013).

They are able to form many forms of symbiosis, depending on how much they are connected to their hosts. It is highly likely that each mutualistic arrangement came about on its own and has its own unique features. Cyanobionts are transmitted from one generation to another through both inheritance and environmental methods (Fig. 3.1). Cyanobacteria are less reliant on living with their hosts because of heterocysts and systems that cut down the amount of oxygen in their cellular fluid (Bright & Bulgheresi, 2010).

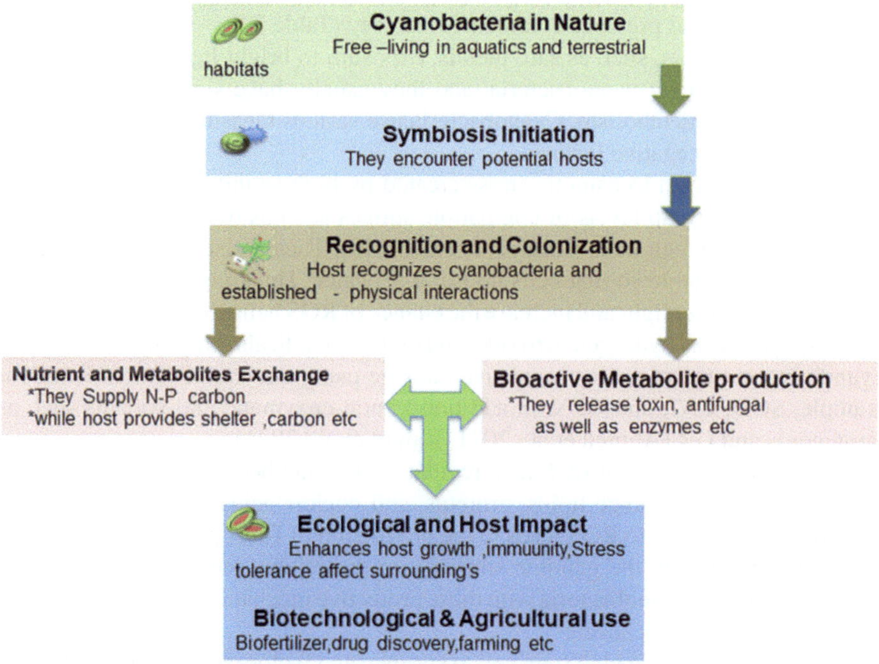

Fig. 3.1 Biological interaction of cyanobacteria

One of the most common groups of symbionts is those species of *Nostoc* which form heterocysts, fix nitrogen, and make hormogonia. Their nitrogen-fixing ability is supported by several oxygen-sensitive enzymes, most notably molybdenum nitrogenase (nifH), vanadium nitrogenase (vnfH), and iron-only nitrogenase (anfH) (Mutalipassi et al., 2021).

A lot of multicellular organisms and symbiotic microbes have changed together, influencing each other's body functions. In some cases, the host and its microbiota act as a single, combined biological system. Since bacteria and host organisms have coevolved, they now communicate through using infochemicals, receptors, and signals. They influence things such as which microbes are found in the body, host defenses, growth, shape, and the way the host acts, and they are frequently dose-dependent and have remained the same over evolutionary time (Weiland-Bräuer, 2021).

Finally, it is important to note that cyanobacterial symbioses are complex agreements between species, where important exchanges of metabolic and molecular substances take place. Not only do these interactions help organisms in their environment, but they also generate a variety of bioactive compounds suitable for pharmaceutical and biotechnological use.

Heterotrophs in cyanobacterial consortia use lysed cell contents, osmolytes, and fermentation byproducts, among other ecological interactions. Additionally, they use limited phototrophy, direct carbon transfer, and outer membrane vesicles to occupy particular niches and preserve community stability (Table 3.2).

Different stressors, such as salt, metals, exposure to bright light, and dry conditions, are common for cyanobacteria, and they usually have significant effects on their biology and metabolism. Cyanobacteria use various measures and pathways in response to these negative factors.

Plants often adjust to osmotic stress created by high salinity, freezing, or dehydration by building up levels of compatible solutes such as trehalose, sucrose, and glycine betaine in their cells (Rezayian et al., 2019). They assist in keeping proteins and membranes stable so that cells do not lose water. Thick layers of snow reflect more radiation in sunlight and increase the chance of ROS being formed, for example, superoxide radicals, hydrogen peroxide, and hydroxyl radicals (Hegedűs et al., 2021). Cyanobacteria protect against oxidative damage using enzymatic antioxidants, for example, SOD, CAT, and GPX, in addition to non-enzymatic antioxidants such as carotenoids and GR (Ahmed et al., 2021; Rajput et al., 2021).

While ordinary levels of ROS help regulate genes and become a defense system, extreme levels can damage lipids, proteins, and nucleic acids (Latifi et al., 2009). Remarkably, exposure to ROS leads to the production of DNA photoproducts called CPDs, which damage the genome similarly to the effects seen during UV-B and desiccation stress. Cyanobacteria with these protective mechanisms can recover from light damage that is not severe.

The cell membrane is vital for a cell to survive, and factors such as major changes in temperature and salt can affect its fluidity and ability to let nutrients in or out. Cyanobacteria respond to changing temperatures by either softening their membranes by introducing more unsaturated fatty acids in cold conditions or stiffening their

Table 3.2 Ecological interactions and niches of heterotrophs in cyanobacterial consortium (Ataeian et al., 2022)

Ecological niche	Metabolite source	Associated heterotrophic groups	Interaction description
Fermentation products	Formate, acetate, lactate, ethanol, H_2 (night excretion)	*Gammaproteobacteria*, *Bacteroidota*	Utilization of byproducts from the cyanobacterial fermentation process
Outer membrane vesicles	DNA, RNA, proteins, lipids	*Planctomycota, Verrucomicrobiota*	Utilization of complex macromolecules from OMVs for growth and gene transfer
Osmolyte release	Amino acids, sugars	*Alphaproteobacteria, Bacteroidota*	Heterotrophs uptake osmolytes released by cyanobacteria as a stress response
Cyanobacterial cell lysis	Intracellular contents released post-lysis	*Bacteroidota, Planctomycetota*	Breakdown of complex carbon sources using glycoside hydrolyases and sulfate
Direct carbon transfer	13C-labeled bicarbonate fixed by cyanobacteria	*Wenzhouxiangella*	Strong coupling of carbon flow from cyanobacteria to heterotrophs confirmed via stable isotope probing (SIP)
Phototrophy by non-cyanobacterial members	Light energy and limited phototrophic capability	*Alphaproteobacteria, Archaea*	Limited phototrophic metabolism—niche stability and light-based energy harvesting
Exopolysaccharide and high molecular weight DOM	Released polysaccharides and organic matter	*Verrucomicrobiota, Planctomycetota*	Breakdown of complex carbon sources using glycosides

membranes in high temperatures and salt. Because of these various defense and function changes, cyanobacteria are well adapted to cope with extreme environments (Hegedűs et al., 2021).

3.5 Adaptive Strategies of Cyanobacteria

Cyanobacteria move up or down in the water column by controlling the presence of intracellular gas vesicles. With this regulation, they can easily change according to how much light, heat, and nutrients are around them, which allows them to outperform other types of phytoplankton.

Tiny gas bubbles in cyanobacteria reduce the cell density, so they float to the top at night and sink when there is too much light or minerals farther down. Thanks to this movement, algae can more easily have enough light for photosynthesis and obtain minerals, mainly in water bodies that are divided geographically (Ranjbar et al., 2021).

How a cell looks and how large it is play an important role in its buoyancy. The rapidity of vegetative cells to either float or sink depends on their size and density. When *Microcystis aeruginosa* or other organisms bloom, the surface may become dominated by large cells, further increasing both the bloom and its effects. Also, akinetes may respond to changes in buoyancy that assist them in their move between the bottom and upper layers, but this was mostly studied in studies of non-dormant vegetative cells.

Light, nutrient availability, and cellular metabolism are important factors that control the environmental regulation of gas vesicle formation. Should nutrients be scarce or the light too bright, gas vesicles help a cyanobacterium choose a better place to live and thrive (Aparicio Medrano et al., 2013).

Cyanobacteria produce a group of poisons known as cyanotoxins, for example, microcystins, nodularins, saxitoxins, anatoxins, and cylindrospermopsin. These toxins are mostly formed through NRPS and PKS routes, and their production is under careful control due to light, temperature, nutrients, and salinity. Microcystins are the most important group and are produced by *Microcystis*, *Anabaena*, and *Planktothrix* species. In addition to toxicity, these substances provide ecological functions. These scaffolds help cyanobacteria by stopping phytoplankton growth near them, repelling grazing from *Daphnia magna*, and encouraging more colonies to be formed (Park et al., 2024). Certain compounds, such as microcystins, are valuable, as they help the cells escape harm from danger in the environment. Anatoxins and saxitoxins are serious neurotoxins that cause aquatic organisms' nerve systems to fail, which leads to a swift and devastating collapse of ecosystems during toxic blooms. Cyanotoxins affect entire ecosystems, influencing microbes, lowering biodiversity, and changing how organisms relate in the water system. Besides, some cyanotoxins prevent the germination and limit the growth of both aquatic plants and competing algal species. The growth of these organisms is boosted by similar mechanisms to quorum sensing, where there is more bloom production when there are more cells. Importantly, these toxins are slowly released by sediment when conditions allow, enabling bloom cycles to last. This shows that toxin synthesis is necessary for cyanobacteria to be dominant, survive under challenging conditions, and live for a long time in a variable environment (Dittmann et al., 2013).

Akinetes are dormant, thick-walled cells that form when *Anabaena*, *Nostoc*, and *Aphanizomenon* cyanobacteria encounter unfavorable environmental conditions. Thanks to these structures, cyanobacteria can tolerate severe stresses like drying out, low temperature, nutrient deficiency, and a lot of bright light.

New research points to how well akinetes cope with difficult conditions. Heat tolerance in akinetes of *Nostoc* sp. HK-01 is credited to the accumulation of betaine, glucosylglycerol, and glycine. In addition, akinetes have shown the ability to survive extreme drying for many years and can usually germinate after being stored in darkness for up to 5 years at 27 °C (Klemke et al., 2014; Sukenik et al., 2012).

Thanks to their structure, akinetes are highly hardy organisms. Akinetes in *Anabaena cylindrica* can stand harsh drying, proving capable of germination following heat-drying at 60 °C for 50 h as well as drying in sunlight for 10 h. In *Anabaena variabilis*, mature akinetes with glycolipid-rich envelopes are necessary to defend them against environmental challenges such as freezing, drying out, stress from free radicals, and enzyme attack. Intriguingly, before akinetes mature, some cyanobacteria can handle osmotic stress, freezing, and desiccation better than they do once they mature. As a result, over time, akinetes can manage and survive in various changing environmental conditions (Sukenik et al., 2012).

Akinetes are significant at the ecological level for more purposes than just surviving. They help explain why cyanobacteria tend to grow in blooms every year at specific times. After settling into the sediments, akinetes become a long-term seed bank and help kick off blooms when the weather conditions improve again. All in all, the production of akinetes gives cyanobacteria a key means to adapt and live for a long time in various and frequently extreme environments.

3.6 Booms Dynamics: Formation, Persistence, and Collapse

A variety of environmental, physiological, and ecological factors work together to shape when and how cyanobacteria grow and die off. Initiation of blooms often occurs when cyanobacterial growth is favored by increased temperature, high brightness levels, vertical layering of water, and an abundance of phosphorus or nitrogen in the water column. Calm winds maintain low turbulence, which also encourages *Microcystis* and *Aphanizomenon* to concentrate at the surface. Due to their ability to convert atmospheric nitrogen (found in Anabaena species), some cyanobacteria have a clear advantage when nitrogen is scarce, helping them to outcompete phytoplankton.

After being established, a number of elements help keep blooms lasting. Among these are nutrient recycling in the body by microorganisms, picking up dilute nutrients in the water, searching for light and nutrients by rising in the water column, and making cyanotoxins and allelopathic chemicals that harm other species and deter herbivores. Building mucilaginous colonies and making gas vesicles allow cyanobacteria to reach a position close to the water's surface, where they can use light efficiently. The group behavior of colonies, similar to quorum sensing, may help decide when to bloom and which genes to activate for continued development (Gerphagnon et al., 2015). The warming temperatures from climate change and more nutrients reaching water from farms have made harmful blooms longer and more likely everywhere (O'Neil et al., 2012).

A range of biotic and abiotic stressors typically causes the collapse of a bloom. Among them are nutrient loss, sudden temperature drops, increased mixing from wind or storms, stress from ultraviolet light, and damage from related chemicals. Lysis by viruses, infections with fungi, and attacks from protozoans and zooplankton are important in reducing bloom populations. At times, cyanobacteria experience programmed cell death and an increase in damaging oxygen species, which help break up algal blooms. After blooms break down, the microbes there cause the big biomass to rot, which can make the water too oxygen-poor and push oxygen levels down to nothing, causing fish kills and other major disturbances to the ecosystem. For this reason, learning how blooms develop is vital for designing useful techniques and strategies aimed at reducing cyanobacterial blooms (Huisman et al., 2018; Paerl et al., 2018; Qu et al., 2015).

3.7 Implications for Ecosystem and Health Risks

Cyanobacterial blooms affect both the health of ecosystems and water quality, changing food webs in water, biodiversity, and biogeochemical cycles. They can have a number of different effects that spread to all areas within aquatic ecosystems. A major impact of cyanobacterial blooms is an upset in aquatic food webs. Too many dense blooms can lower the amount of sunlight in the water, stopping many photosynthetic creatures from growing. These changes make habitats simpler and most often help invasive or less valuable species. Furthermore, cyanobacteria are little help as a food source to zooplankton thanks to their poor nutritional value and the presence of toxins and mucilage. For this reason, the energy is not passed through the food chain efficiently, so zooplankton decline, and the fish they provide food for also decrease. Also, a surplus of cyanobacteria can make it so that therapeutic phytoplankton decrease in number, leading to lasting alterations in the types of species present and less biodiversity (Müller-Navarra et al., 2000; Zilliges et al., 2011).

Toxication caused by toxins from cyanobacteria endangers both aquatic creatures and humans. Among cyanotoxins, microcystins, cylindrospermopsin, saxitoxins, and anatoxins can collect in the tissues of fish, shellfish, and other aquatic animals. Bioaccumulation results in biomagnification, so contaminated seafood can harm apex predators and humans who eat it. Being exposed to cyanotoxins may lead to gastrointestinal disturbances, liver damage, and brain and nerve issues and often can cause cancer in humans (Ghernaout & Elboughdiri, 2020). Wild animals, domesticated animals, and pets can be exposed to infection through contaminated water. In addition, blooms can block out recreational water activities, requiring beaches to close and raising questions about the safety of drinking water. While the World Health Organization accepts a limit of 1 μg/L microcystins in drinking water, many affected areas still have problems with reliable monitoring and treatment (Ghernaout & Elboughdiri, 2020; Hiskia et al., 2020).

An important result of blooms is that they reduce the amount of oxygen dissolved in the water, which causes ocean "dead zones." At the end of their bloom, cyanobacteria gather, and their huge mass is consumed in a process that needs a lot of oxygen. A lack of oxygen may cause fish and similar aquatic organisms to die, disturb the proper balance in nature, and result in significant losses for fisheries. In particularly serious situations, dead zones stay around for a long time, leading to issues with nutrient recycling and preventing the environment from recovering. Phosphorus trapped in sediments is more likely to be released in anoxia, which adds to eutrophication and continues the bloom process (Diaz & Rosenberg, 2008).

Overall, cyanobacterial blooms can seriously harm both the health of aquatic ecosystems and the quality of their water. There are many results of pollution, such as reduced variety of living creatures, problems with the food chain, harmful chemicals, and air lacking oxygen. These results highlight the requirement for approaches that address nutrients and climate change and repair damaged habitats to ensure freshwater resources and their living things are secure.

Cyanobacteria die-offs both upset food chains and can produce poisons that harm the health of living beings, while also making water become polluted and less safe. Because of these changes, ecosystems could become unstable, fisheries may decline, water supplies could be harmed, and climate effects might change.

Cyanobacteria frequently win out in waters rich in nutrients by using their skills of controlling buoyancy, fixing nitrogen, and resisting grazing. Because of this, the number of phytoplankton decreases, and valuable services from diverse aquatic communities are lessened. Loss of differentiating genera of primary producers can destabilize ecosystems, making them more receptive to threats brought by invasive species and changes in climate. Furthermore, shifts in which species live in a habitat can cause predator and prey to mismatch in numbers, which can change the timing of aquatic animals' reproduction and increase or decrease their numbers. There are other ways for toxins in the environment to influence people, besides eating them. Cyanotoxins in mist or aerosols may reach the lungs of people and animals and cause breathing problems. Cyanotoxins are dangerous for agriculture since contaminated water may harm crops and possibly contaminate land. In places where people depend on unsafe surface water, exposure to cyanotoxins is a persistent public health problem that mostly affects children, as well as elderly and immunocompromised members of the community (Backer et al., 2010).

The growth of cyanobacteria can cloud the water, cause unpleasant smells, and release substances harmful to water quality. Having high organic loading increases the use of oxygen and may affect microbes in ways that can help pathogenic bacteria. Therefore, water treatment is both harder and more expensive to perform. The current methods cannot always remove cyanotoxins, meaning many communities cannot afford the advanced technology options that are required.

Also, bloom events can be connected to changes in greenhouse gas emissions. Through their decay, cyanobacteria release methane and nitrous oxide, which add to climate change. Consequently, more blooms caused by hotter weather add to the climate change problem.

Sediment behavior is also influenced by these changes. As blooms die, the phosphorus and other nutrients tied up in sludge can again be introduced into the water column when the water does not contain enough oxygen, a process known as internal loading. This allows eutrophic conditions to persist and helps repeated blooms even if other nutrients are no longer supplied. On top of that, the collection of dead biological material can alter the physical environment, creating difficulties for bottom-feeding fish and many other invertebrates (Li et al., 2007).

All in all, cyanobacterial blooms reduce the beauty, fun, and health of water bodies, as well as their ability to support ecosystems and last into the future. They demonstrate how nutrient management, responding to climate change, and ecological restoration are all important for sustainable agriculture. To solve these blooms, efforts should be made by many experts who work on monitoring, managing watersheds, setting public health rules, and acting globally on climate change to shield freshwater resources and sustain the ecosystem for years to come.

3.8 Management and Mitigation Strategies

Effective control of cyanobacterial blooms is mostly achieved by reducing the nitrogen and phosphorus levels in the water, because their presence causes eutrophication and promotes these blooms. Untreated wastewater, stormwater from towns, and excess nutrients from farms are important factors that cause the increase in blooms seen in freshwater. Researchers find that dealing with both nitrogen and phosphorus together is necessary because managing nitrogen alone may not stop blooms, as phosphorus and cyanobacteria are also important factors. Managing nutrients by building vegetative zones, finding the right application rates for fertilizer, and upgrading wastewater plants works well in cutting off nutrients to aquatic systems.

Other than making nutrients less available, mixing the water and increasing aeration stop dense layers that separate the warmer surface from the colder deeper water. This deprives cyanobacteria of their preferred sunlight and temperature, which makes them occur less often. After hydraulic flushing to eliminate cyanobacterial cells, adding modified clay to the sediments further limits phosphorus recycling in the internal system. For example, many lakes use copper sulfate and hydrogen peroxide to control algae, but you have to manage these chemicals as cyanobacteria break down and release harmful substances that may pollute the water and harm organisms. More physical approaches, including mechanical mixing, lowering reservoirs, surface collection, and ultrasound therapy, are used to break up blooms, but their effectiveness depends on each waterbody's features. In addition, approaches that include barley straw deployment and coagulation or flocculation work to reduce blooms by either controlling their growth or causing them and their nutrients to settle. The mix of these two approaches supplies various ways to manage cyanobacterial blooms, underlining that only targeted site-based solutions are truly effective (Carpenter et al., 2025).

The two methods biocontrol agents use to control cyanobacteria are called direct attack and indirect attack. In a direct attack, the biocontrol agent comes in contact with a cyanobacterial cell and penetrates it, letting the agent either eat nutrients or prevent the cell from growing. Some biocontrol agents come together to wrap around and catch cyanobacterial cells (Papenfus et al., 2020). Without touching the victim, indirect attack depends on extracellular lytic products like enzymes, antibiotics, biosurfactants, and proteases that damage cyanobacterial membranes and interfere with photosynthesis.

New studies of transcriptomes have helped explain why certain bacteria become more mobile, responsive, and lytic as they interact with cyanobacteria, who respond by lowering their photosynthetic activities, showing an active form of antagonism. While we know more about morphology and physiology, additional studies are needed to better understand the genes and stress responses involved during suppression by both cyanobacteria and biocontrol agents. Using filtering zooplankton or fish is a biological method to control cyanobacteria, as is using viruses or antibacterial microbes that only affect the poor types (Wijesooriya et al., 2021).

Setting and upholding limits on nutrient discharge is necessary to prevent nutrients from reaching water bodies. Water quality guidelines should encourage the use of best management strategies and help develop facilities for removing nutrients. Getting local government, citizens, and farmers to join forces is essential to thoroughly managing where nutrients come from. Using remote sensing and sensors on site, it is possible to quickly catch a bloom as it begins, so that action can be taken to stop it. Spreading awareness and involving people in decisions are necessary to prevent blooms and help conserve land and water. When used together, these approaches can greatly lower the number and effects of cyanobacterial blooms and support the well-being of freshwater ecosystems (Michalak et al., 2013).

3.9 Conclusion

Cyanobacterial blooms are driven by a synergistic interplay of ecological, climatic, and anthropogenic factors. The intensification of eutrophication, fueled by excessive nitrogen and phosphorus inputs from agricultural runoff, wastewater discharge, and urbanization, has significantly altered freshwater ecosystems, making them more susceptible to harmful algal outbreaks. Rising global temperatures, altered precipitation regimes, and extended thermal stratification due to climate change further amplify bloom frequency, duration, and toxicity. Cyanobacteria have evolved a suite of adaptive strategies—such as buoyancy regulation, nitrogen fixation, toxin production, and akinete formation—that enhance their ecological dominance under fluctuating environmental conditions. Moreover, their interactions with heterotrophic microbes, zooplankton, and host organisms form intricate networks that influence bloom persistence and collapse. These biological relationships not only sustain bloom dynamics but also contribute to the production of ecologically significant metabolites. The consequences of cyanobacterial blooms extend beyond ecological disruptions to include severe socio-economic and public health risks, including oxygen depletion, biodiversity loss, bioaccumulation of toxins, and compromised water security. Effective management must therefore adopt an integrated approach—targeting nutrient reduction, ecosystem restoration, and climate adaptation strategies—while leveraging real-time monitoring and community engagement. Understanding the ecological drivers of cyanobacterial blooms is essential for predicting future trends, designing targeted mitigation strategies, and safeguarding freshwater ecosystems in an era of rapid environmental change.

References

Achaga, R., Gogorza, C., Irurzun, M. A., Goguitchaichvili, A., Mestelán, S., Fernández, A. R., Cabeza, J. S., Bettucci, L. S., Sinito, A., Morales, J., & Martínez, D. (2022). Monitoring the environmental evolution and its relationship with anthropogenic activities using magnetic and geochemical proxies on Lake Melincué sediments. *Journal of South American Earth Sciences, 116*, 103827.

Adams, D. G., Bergman, B., Nierzwicki-Bauer, S. A., Duggan, P. S., Rai, A. N., & Schüßler, A. (2013). Cyanobacterial-Plant symbioses. In E. Rosenberg, E. F. DeLong, S. Lory, E. Stackebrandt, & F. Thompson (Eds.), *The Prokaryotes: Prokaryotic biology and symbiotic associations* (pp. 359–400). Springer Berlin Heidelberg.

Ahmed, H., Pathak, J., Rajneesh, Sonkar, P. K., Ganesan, V., Häder, D. P., & Sinha, R. P. (2021). Responses of a hot spring cyanobacterium under ultraviolet and photosynthetically active radiation: Photosynthetic performance, antioxidative enzymes, mycosporine-like amino acid profiling and its antioxidative potentials. *3 Biotech, 11*(1), 10. https://doi.org/10.1007/s13205-020-02562-1

Aparicio Medrano, E., Uittenbogaard, R., Dionisio Pires, M., Wiel, B., & Clercx, H. (2013). Coupling hydrodynamics and buoyancy regulation in Microcystis aeruginosa for its vertical distribution in lakes. *Ecological Modelling, 248*, 41–56. https://doi.org/10.1016/j.ecolmodel.2012.08.029

Ataeian, M., Liu, Y., Kouris, A., Hawley, A. K., & Strous, M. (2022). Ecological interactions of cyanobacteria and heterotrophs enhances the robustness of cyanobacterial consortium for carbon sequestration. *Frontiers in Microbiology, 13*, 780346.

Backer, L. C., McNeel, S. V., Barber, T., Kirkpatrick, B., Williams, C., Irvin, M., Zhou, Y., Johnson, T. B., Nierenberg, K., Aubel, M., & LePrell, R. (2010). Recreational exposure to microcystins during algal blooms in two California lakes. *Toxicon, 55*(5), 909–921.

Bonnet, S., Benavides, M., Le Moigne, F. A. C., Camps, M., Torremocha, A., Grosso, O., Dimier, C., Spungin, D., Berman-Frank, I., Garczarek, L., & Cornejo-Castillo, F. M. (2023). Diazotrophs are overlooked contributors to carbon and nitrogen export to the deep ocean. *The ISME Journal, 17*(1), 47–58. https://doi.org/10.1038/s41396-022-01319-3

Bright, M., & Bulgheresi, S. (2010). A complex journey: Transmission of microbial symbionts. *Nature Reviews Microbiology, 8*(3), 218–230. https://doi.org/10.1038/nrmicro2262

Carpenter, K. D., Rosen, B. H., Donahue, D., Duncan, K., Hilbrandt, B., Lewis, C., Swan, K., Triplett, T., & Welch, E. (2025). Reservoir and riverine sources of cyanotoxins in oregon's cascade range rivers tapped for drinking water supply. *Phycology, 5*(2), 16.

Cousino, L. K., Becker, R. H., & Zmijewski, K. A. (2015). Modeling the effects of climate change on water, sediment, and nutrient yields from the Maumee River watershed. *Journal of Hydrology: Regional Studies, 4*, 762–775.

Dasari, S., Andersson, A., Stohl, A., Evangeliou, N., Bikkina, S., Holmstrand, H., Budhavant, K., Salam, A., & Gustafsson, O. (2020). Source quantification of South Asian black carbon aerosols with isotopes and modeling. *Environmental Science & Technology, 54*(19), 11771–11779.

Diaz, R. J., & Rosenberg, R. (2008). Spreading dead zones and consequences for marine ecosystems. *Sciences, 321*(5891), 926–929.

Dittmann, E., Fewer, D. P., & Neilan, B. A. (2013). Cyanobacterial toxins: Biosynthetic routes and evolutionary roots. *FEMS Microbiology Reviews, 37*(1), 23–43.

Foy, R., Gibson, C., & Smith, R. (1976). The influence of daylength, light intensity and temperature on the growth rates of planktonic blue-green algae. *British Phycological Journal, 11*(2), 151–163.

Gao, X., Kong, B., & Vigil, R. D. (2018). Multiphysics simulation of algal growth in an airlift photobioreactor: Effects of fluid mixing and shear stress. *Bioresource Technology, 251*, 75–83.

Gerphagnon, M., Macarthur, D. J., Latour, D., Gachon, C. M., Van Ogtrop, F., Gleason, F. H., & Sime-Ngando, T. (2015). Microbial players involved in the decline of filamentous and colonial cyanobacterial blooms with a focus on fungal parasitism. *Environmental Microbiology, 17*(8), 2573–2587.

Ghernaout, D., & Elboughdiri, N. (2020). Dealing with cyanobacteria and cyanotoxins: Engineering viewpoints. *Open Access Library Journal, 7*(5), 1–20.

Gondo, T. F., Jönsson, M., Karlsson, E. N., Sandahl, M., & Turner, C. (2023). Extractability, selectivity, and comprehensiveness in supercritical fluid extraction of seaweed using ternary mixtures of carbon dioxide, ethanol, and water. *Journal of Chromatography A, 1706*, Article 464267.

Granéli, E., & Turner, J. T. (2006). *Ecology of harmful algae* (Vol. 189): Springer.

Gu, P., Wang, Y., Wu, H., Chen, L., Zhang, Z., Yang, K., Zhang, Z., Ren, X., Miao, H., & Zheng, Z. (2023). Efficient control of cyanobacterial blooms with calcium peroxide: Threshold and mechanism. *Science of The Total Environment, 882*, 163591.

Hegedűs, C., Juhász, T., Fidrus, E., Janka, E. A., Juhász, G., Boros, G., Paragh, G., Uray, K., Emri, G., Remenyik, É., & Bai, P. (2021). Cyclobutane pyrimidine dimers from UVB exposure induce a hypermetabolic state in keratinocytes via mitochondrial oxidative stress. *Redox Biol, 38*, 101808. https://doi.org/10.1016/j.redox.2020.101808

Hiskia, A. E., Triantis, T. M., Antoniou, M. G., Kaloudis, T., & Dionysiou, D. D. (2020). Water treatment for purification from cyanobacteria and cyanotoxins.

Huisman, J., Codd, G. A., Paerl, H. W., Ibelings, B. W., Verspagen, J. M., & Visser, P. M. (2018). Cyanobacterial blooms. *Nature Reviews Microbiology, 16*(8), 471–483. https://doi.org/10.1038/s41579-018-0040-1

Jassim, Y. A., Awadh, E. F. A., & Al-Amery, S. M. H. (2023). A review of general properties of blue-green algae (Cyanobacteria). *Biomedicine and Chemical Sciences, 2*(2), 143–148.

Joehnk, K. D., Huisman, J., Sharples, J., Sommeijer, B., Visser, P. M., & Stroom, J. M. (2008). Summer heatwaves promote blooms of harmful cyanobacteria. *Global Change Biology, 14*(3), 495–512.

Kim, S., Mehrotra, R., Kim, S., & Sharma, A. (2021). Probabilistic forecasting of cyanobacterial concentration in riverine systems using environmental drivers. *Journal of Hydrology, 593*, Article 125626.

Kleinteich, J., Frassl, M. A., Schulz, M., & Fischer, H. (2024). Climate change triggered planktonic cyanobacterial blooms in a regulated temperate river. *Science and Reports, 14*(1), 16298. https://doi.org/10.1038/s41598-024-66586-w

Klemke, F., Beyer, G., Sawade, L., Saitov, A., Korte, T., Maldener, I., Lockau, W., Nürnberg, D. J., & Volkmer, T. (2014). All1371 is a polyphosphate-dependent glucokinase in Anabaena sp. PCC 7120. *Microbiology, 160*(12), 2807–2819.

Kosten, S., Huszar, V. L., Bécares, E., Costa, L. S., van Donk, E., Hansson, L. A., Jeppesen, E., Kruk, C., Lacerot, G., Mazzeo, N., & De Meester, L. (2012). Warmer climates boost cyanobacterial dominance in shallow lakes. *Global change biology, 18*(1), 118–126.

Latifi, A., Ruiz, M., & Zhang, C.-C. (2009). Oxidative stress in cyanobacteria. *FEMS Microbiology Reviews, 33*(2), 258–278. https://doi.org/10.1111/j.1574-6976.2008.00134.x

Li, M., Wei, K.-S., Xue, Y., Wu, S.-J., Liu, Y.-J., Chen, D.-M., Yan, X. F., & Kang, C. (2024). Microscopic spatiotemporal changes in cell wall cellulose and pectin during Nicotiana tabacum L. leaf growth and senescence based on label-free Raman microspectroscopic imaging combined with multivariate curve resolution. *Industrial Crops and Products, 222*, 119865.

Li, M., Xiao, M., Zhang, P., & Hamilton, D. P. (2018). Morphospecies-dependent disaggregation of colonies of the cyanobacterium Microcystis under high turbulent mixing. *Water Research, 141*, 340–348.

Li, Z., Lee, K., Cobanli, S. E., King, T., Wrenn, B. A., Doe, K. G., Jackman, P. M., & Venosa, A. D. (2007). Assessment of sediment toxicity during anaerobic biodegradation of vegetable oil using Microtox® and Hyalella azteca bioassays. *Environmental Toxicology: An International Journal, 22*(1), 1–8.

Matthews, M. W. (2014). Eutrophication and cyanobacterial blooms in South African inland waters: 10 years of MERIS observations. *Remote Sensing of Environment, 155*, 161–177.

Metzgen, A. D., Dahmke, A., & Ebert, M. (2021). Anaerobic corrosion of zero-valent iron at elevated temperatures. *Environmental Science & Technology*, *55*(12), 8010–8019.

Michalak, A. M., Anderson, E. J., Beletsky, D., Boland, S., Bosch, N. S., Bridgeman, T. B., Chaffin, J. D., Cho, K., Confesor, R., Daloğlu, I., & DePinto, J. V. (2013). Record-setting algal bloom in Lake Erie caused by agricultural and meteorological trends consistent with expected future conditions. *Proceedings of the National Academy of Sciences, 110*(16), 6448–6452.

Müller-Navarra, D. C., Brett, M. T., Liston, A. M., & Goldman, C. R. (2000). A highly unsaturated fatty acid predicts carbon transfer between primary producers and consumers. *Nature, 403*(6765), 74–77.

Mutalipassi, M., Riccio, G., Mazzella, V., Galasso, C., Somma, E., Chiarore, A., de Pascale, D., & Zupo, V. (2021). Symbioses of Cyanobacteria in marine environments: Ecological insights and biotechnological perspectives. *Mar Drugs, 19*(4). https://doi.org/10.3390/md19040227

O'Neil, J. M., Davis, T. W., Burford, M. A., & Gobler, C. J. (2012). The rise of harmful cyanobacteria blooms: The potential roles of eutrophication and climate change. *Harmful Algae, 14*, 313–334. https://doi.org/10.1016/j.hal.2011.10.027

Padilla, S. (2021). *Cyanobacteria harmful algal blooms: Addressing critical policy gaps.* College of Environmental Science.

Paerl, H. W., Hall, N. S., & Calandrino, E. S. (2011a). Controlling harmful cyanobacterial blooms in a world experiencing anthropogenic and climatic-induced change. *Science of the Total Environment, 409*(10), 1739–1745.

Paerl, H. W., & Huisman, J. (2008). Blooms like it hot. *Science, 320*(5872), 57–58.

Paerl, H. W., & Huisman, J. (2009). Climate change: A catalyst for global expansion of harmful cyanobacterial blooms. *Environmental Microbiology Reports, 1*(1), 27–37.

Paerl, H. W., & Otten, T. G. (2016). Duelling 'CyanoHABs': Unravelling the environmental drivers controlling dominance and succession among diazotrophic and non-N2-fixing harmful cyanobacteria. *Environmental Microbiology, 18*(2), 316–324.

Paerl, H. W., Otten, T. G., & Kudela, R. (2018). Mitigating the expansion of harmful algal blooms across the freshwater-to-marine continuum. *Environmental Science & Technology, 52*(10), 5519–5529. https://doi.org/10.1021/acs.est.7b05950

Paerl, H. W., Xu, H., McCarthy, M. J., Zhu, G., Qin, B., Li, Y., & Gardner, W. S. (2011b). Controlling harmful cyanobacterial blooms in a hyper-eutrophic lake (Lake Taihu, China): The need for a dual nutrient (N & P) management strategy. *Water Research, 45*(5), 1973–1983.

Pandey, S., Narayanan, I., Vinayagam, R., Selvaraj, R., Varadavenkatesan, T., & Pugazhendhi, A. (2023). A review on the effect of blue green 11 medium and its constituents on microalgal growth and lipid production. *Journal of Environmental Chemical Engineering, 11*(3), Article 109984.

Papenfus, M., Schaeffer, B., Pollard, A. I., & Loftin, K. (2020). Exploring the potential value of satellite remote sensing to monitor chlorophyll-a for US lakes and reservoirs. *Environmental Monitoring and Assessment*, *192*(12), 808.

Park, C.-H., Kwon, D.-R., Kim, K., Sim, Y.-B., & Hwang, S.-J. (2024). Temperature-dependent akinete formation strategies of the harmful cyanobacterium Dolichospermum circinale. *Harmful Algae, 139*, Article 102722. https://doi.org/10.1016/j.hal.2024.102722

Paul, V. J. (2008). Global warming and cyanobacterial harmful algal blooms. *Cyanobacterial harmful algal blooms: state of the science and research needs*, 239–257.

Qu, B., Su, J., Wang, Z., Wang, Y., Han, X., Wang, H., Liu, Y., Jia, Y., Pan, J., & Ren, G. (2015). Effect of H. pylori infection on cytokine profiles and oxidative balance in subjects with chronic alcohol ingestion. *PLoS One, 10*(6), e0129352.

Rajput, V. D., Harish, Singh, R. K., Verma, K. K., Sharma, L., Quiroz-Figueroa, F. R., Meena, M., Gour, V. S., Minkina, T., Sushkova, S., & Mandzhieva, S. (2021). Recent developments in enzymatic antioxidant defence mechanism in plants with special reference to abiotic stress. *Biology (Basel), 10*(4). https://doi.org/10.3390/biology10040267

Ranjbar, M. H., Hamilton, D. P., Etemad-Shahidi, A., & Helfer, F. (2021). Individual-based modelling of cyanobacteria blooms: Physical and physiological processes. *Science of the Total Environment, 792*, Article 148418. https://doi.org/10.1016/j.scitotenv.2021.148418

Rezayian, M., Niknam, V., & Ebrahimzadeh, H. (2019). Stress response in cyanobacteria. *Iranian Journal of Plant Physiology, 9*(3), 2773–2787.

Robarts, R. D., & Zohary, T. (1987). Temperature effects on photosynthetic capacity, respiration, and growth rates of bloom-forming cyanobacteria. *New Zealand Journal of Marine and Freshwater Research, 21*(3), 391–399.

Sukenik, A., Hadas, O., Kaplan, A., & Quesada, A. (2012). Invasion of Nostocales (cyanobacteria) to subtropical and temperate freshwater lakes–physiological, regional, and global driving forces. *Frontiers in Microbiology, 3*, 86.

Visser, P. M., Verspagen, J. M., Sandrini, G., Stal, L. J., Matthijs, H. C., Davis, T. W., Paerl, H. W., & Huisman, J. (2016). How rising CO_2 and global warming may stimulate harmful cyanobacterial blooms. *Harmful Algae, 54*, 145–159.

Weiland-Bräuer, N. (2021). Friends or Foes-Microbial Interactions in Nature. *Biology (Basel), 10*(6). https://doi.org/10.3390/biology10060496

Wells, M. L., Karlson, B., Wulff, A., Kudela, R., Trick, C., Asnaghi, V., Berdalet, E., Cochlan, W., Davidson, K., De Rijcke, M., & Dutkiewicz, S. (2020). Future HAB science: Directions and challenges in a changing climate. *Harmful Algae, 91*, 101632.

Weyhenmeyer, G. A. (2001). Warmer winters: Are planktonic algal populations in Sweden's largest lakes affected? *AMBIO: A Journal of the Human Environment, 30*(8), 565–571.

Wijesooriya, M., Widanagamage, S., & Masakorala, K. (2021). Biological control of freshwater cyanobacterial blooms: A review. *Journal of the University of Ruhuna,9*(1).

Xu, H., Yu, G., & Jiang, H. (2013). Investigation on extracellular polymeric substances from mucilaginous cyanobacterial blooms in eutrophic freshwater lakes. *Chemosphere, 93*(1), 75–81.

Zilliges, Y., Kehr, J.-C., Meissner, S., Ishida, K., Mikkat, S., Hagemann, M., Kaplan, A., Boerner, T., & Dittmann, E. (2011). The cyanobacterial hepatotoxin microcystin binds to proteins and increases the fitness of Microcystis under oxidative stress conditions. *PloS One, 6*(3), e17615.

[illegible]

Chapter 4
Nutrient Dynamics and Stoichiometry in Cyanobacterial Blooms

Nasir Assad, Muhammad Nauman Khan, Muhammad Naeem-ul-Hassan, Sharia Fatima, Barkat Ullah, Fethi Ahmet Ozdemir, Tabarak Malik, Sarah Abdul Razak, and Shah Fahad

Abstract Cyanobacterial blooms are an escalating global concern, driven by nutrient imbalances, climate change, and human activities. This chapter examines the underlying nutrient dynamics and stoichiometric principles that govern bloom formation, focusing on the critical roles of nitrogen (N) and phosphorus (P). Deviations from the classical Redfield ratio (C:N:P = 106:16:1) are explored to understand how altered nutrient stoichiometry favors bloom-forming cyanobacteria and affects toxin production. The role of micronutrients such as iron (Fe), sulfur (S), and trace metals is also discussed, particularly in relation to nitrogen fixation and bloom physiology. External nutrient inputs—via agricultural runoff, wastewater discharge, and atmospheric deposition—are identified as key contributors to eutrophication. Internal nutrient recycling, including sediment release and biomass decomposition, further sustains bloom persistence. Adaptive traits such as buoyancy regulation, nitrogen fixation, phosphorus scavenging, and resting stages enhance cyanobacterial competitiveness and promote feedback loops that reinforce bloom dominance. These blooms

N. Assad · M. Naeem-ul-Hassan · S. Fatima
Institute of Chemistry, University of Sargodha, Sargodha, Punjab, Pakistan

M. N. Khan (✉) · B. Ullah
Department of Botany, Islamia College Peshawar, Peshawar 25120, Pakistan
e-mail: nomiflora@uop.edu.pk

M. N. Khan
Biology Laboratory, University Public School, University of Peshawar, Peshawar 25120, Pakistan

F. A. Ozdemir
Department of Molecular Biology and Genetics, Faculty of Science and Art, Bingol University, Bingol, Turkey

T. Malik
Department of Biomedical Sciences, Institute of Health, Jimma University, Jimma, Ethiopia

S. A. Razak
Institute of Biological Sciences, Faculty of Science, Universiti Malaya, Kuala Lumpur, Malaysia

S. Fahad
Department of Agronomy, Abdul Wali Khan University Mardan, Mardan, Khyber Pakhtunkhwa, Pakistan

S. Fahad et al. (eds.), *Cyanobacterial Blooms: Ecology, Evolution and Biogeochemical Impacts*, https://doi.org/10.1007/978-3-032-06042-6_4

disrupt food webs, deplete oxygen, and produce toxins harmful to aquatic life and human health. Long-term ecosystem changes, including sediment–water nutrient fluxes and biodiversity loss, complicate recovery. Effective management requires a multifaceted approach involving nutrient reduction strategies, biological control (biomanipulation), and innovative remediation methods such as Phoslock application, floating wetlands, and emerging biotechnological tools. The chapter emphasizes the importance of integrating stoichiometric understanding with remote sensing, nutrient modeling, and ecosystem-based management. A transdisciplinary framework is essential for predicting bloom dynamics and developing sustainable solutions to mitigate their environmental and public health impacts.

4.1 Introduction

Referred to commonly as blue-green algae, cyanobacteria are distinctive prokaryotes that have played a major role in the environment for 3.5 million years or more (Magana-Arachchi & Gunathilaka, 2025). Oxygenic photosynthesis and their historical part in shaping the early Earth's atmosphere are features for which they are well recognized. Among primary producers in water bodies today, cyanobacteria are highly abundant, especially where nutrients are abundant. As a result, adverse ecological conditions can trigger massive growths of cyanobacteria, causing dangerous blooms, also known as CyanoHABs. The broad scale ecological, economical, and human health problems resulting from these blooms have made them a pressing global matter (Madany et al., 2025). The number, extent, and location of cyanobacterial blooms have grown significantly over the last few decades (Qiang et al., 2024). This growing trend is generally attributed to the interacting actions of factors, including increased nutrient pollution from human activities, shifts in climate, temperature rise, modification of water regimes, and changes in how land is used. Specifically, water body eutrophication resulting from agricultural, municipal, and industrial N and P pollution establishes an environment favorable for cyanobacterial dominance over various phytoplankton species (Lakshmikandan et al., 2024). While many eukaryotic algae face challenges in such conditions, cyanobacteria remain highly competitive in low-light, warm, and vertically stratified waters, a benefit especially notable in summer and subtropical parts of the world (Welbara et al., 2024).

A major focus in explaining the emergence and long-term appearance of cyanobacterial blooms involves investigating their nutrient requirements, centers on studying nutrient cycling, together with the balances among nutrient elements, in aquatic habitats. These relationships are summarized in Fig. 4.1, which illustrates the key nutrient dynamics and stoichiometric factors influencing cyanobacterial bloom development. This classical Redfield ratio (C:N:P = 106:16:1), obtained from the average elemental composition of marine phytoplankton, forms the basis for analyzing nutrient patterns in aquatic ecosystems. Values observed in the elemental content of marine phytoplankton have long been considered a main standard for evaluating nutrient ratios in aquatic systems, investigating the equilibrium between nutrients present in aquatic systems. However, deviations from this ratio—with regard to eutrophic lakes and reservoirs—a frequent finding is the occurrence of smaller than normal N:P ratios that have frequently been found in connection with cyanobacterial dominance (Xu et al., 2024). For example, nitrogen-fixing species within Aphanizomenon, Dolichospermum, and Nodularia are frequently abundant where phosphorus is abundant. Because they are able to transform atmospheric nitrogen into bioavailable forms, they succeed in environments deficient in N. Such production of bioavailable N allows these organisms to dominate conditions where nitrogen is scarce (Mioni et al., 2011). Such nutrient imbalances are made more complex by the contribution of micronutrients. Cases include iron (Fe), sulfur (S), and minor metals such as zinc, copper, and manganese. Iron, for example, is necessary for both the activity of nitrogenase enzymes in converting atmospheric nitrogen and for maintaining functional electron transport, maintaining photosynthetic electron transport. An insufficiency in iron or supporting cofactors may diminish bloom intensity or alter algal community composition and the abundance of the bloom or the species making it up. Alternatively, when trace metal concentrations rise above safe levels, they may suppress non-cyanobacterial algal groups. Toxicity caused by a few metals may limit the success of competing phytoplankton, thereby favoring cyanobacterial growth indirectly. cyanobacterial proliferation (Manic et al., 2024). Along with nutrient supply, cyanobacteria exhibit a range of adaptations at physiological and ecological levels, attributes that help them succeed in variable and occasionally tough conditions. Their use of gas vesicles to control buoyancy helps cyanobacteria achieve superior placement in the water column. In the water, their efficient placement enhances both light absorption and acquisition of surface nutrients. A number of cyanobacterial species can produce resting akinetes that enable them to endure hardships and regrow when environmental improvement allows. When conditions improve, cyanobacteria also release extracellular enzymes, such as, for example, alkaline phosphatase to take up bioavailable phosphorus; cyanobacteria rely on alkaline phosphatase under circumstances where phosphorus supply is restricted. Enhanced capability to access nutrients under scarcity helps cyanobacteria endure in environments where nutrients are limited. These adaptations are additionally supported by sophisticated feedback responses that maintain their persistence and bloom persistence. For example, when a bloom happens, plentiful cyanobacterial biomass tends to reduce sunlight available to other organisms. Multiplying biomass can limit how much light other microalgae receive and thereby hinder their development. Upon senescence, completion of the cyanobacterial decay process frequently returns nutrients to the water column. Soluble nutrients reinforce ongoing bloom activity. Bloom failure can, in addition, produce environmental conditions. Specifically in stratified aquatic systems, hypoxia or anoxia tends to trigger internal phosphorus loading from sediments is an additional mechanism that perpetuates bloom behavior (Erratt & Freeman, 2024). The impacts of cyanobacterial blooms are much greater than simply increasing productivity. Some bloom-forming cyanobacteria release secondary metabolites that can cause serious damage to the liver. Microcystins are hepatotoxic (example given); anatoxins cause

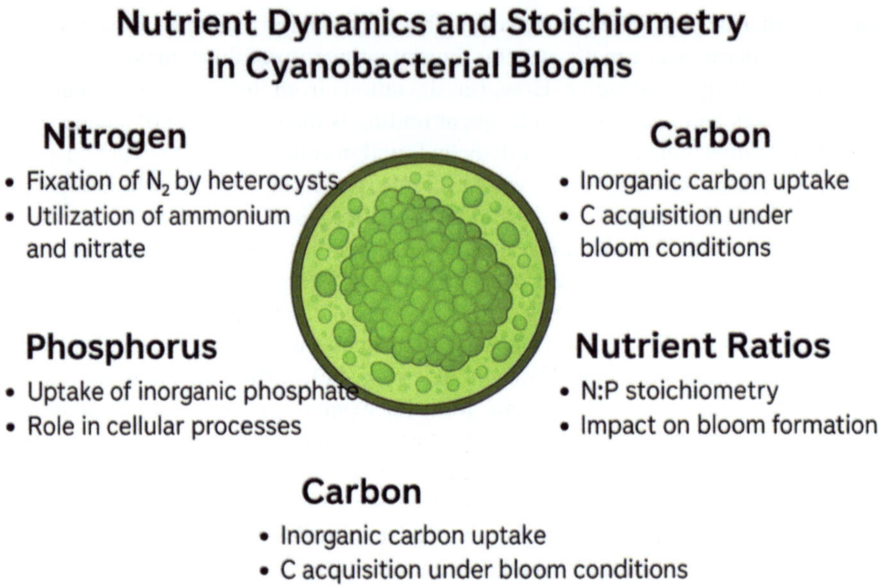

Fig. 4.1 Diagram illustrating the nutrient dynamics and stoichiometric relationships in cyanobacterial blooms, highlighting the roles of nitrogen, phosphorus, carbon, and nutrient ratios in bloom development and sustainability

neurotoxicity, and cylindrospermopsin is a cytotoxin. Toxins produced by bloom-forming cyanobacteria can cause major threats to animals living in water and to people. In particular, the problems are exacerbated if blooms form in water used for drinking or in areas designated for recreational purposes.

Moreover, blooms can disrupt aquatic food webs by reducing biodiversity, altering grazer populations, and modifying predator–prey relationships. They also impact the biogeochemical cycling of carbon, nitrogen, and phosphorus, influencing sediment–water interactions and long-term ecosystem stability (Villalobos et al., 2025). To effectively mitigate cyanobacterial blooms, it is essential to address their root causes—primarily nutrient loading and stoichiometric imbalances. Traditional management approaches have focused on reducing external nutrient inputs through improved agricultural practices, wastewater treatment, and land use regulations. However, emerging evidence suggests that internal nutrient cycling, legacy phosphorus in sediments, and climate-induced hydrological changes must also be considered in long-term bloom control strategies (Lakshmikandan et al., 2024). Recent advances in remote sensing, nutrient modeling, and ecological forecasting have enabled the development of more precise and predictive bloom management tools.

By including stoichiometric principles in these models, it becomes possible to improve the prediction of bloom events caused by variations in nutrient proportions and environmental settings. Furthermore, developments in bloom management include the use of phosphorus-binding agents, one form of innovative remedy. For example, Phoslock may be used, or floating wetlands installed, as well as genetically engineered microbial consortia. Developing sustainable bloom control strategies may be achieved through the use of genetically modified microbial consortia, as indicated in (Wang et al., 2024). As climate change goes on influencing how much and when it rains as well as affecting both surface water temperature and the way bodies of water mix. Changes in stratification trends together with nutrient dynamics are forecasted to make bloom events more likely scenarios. The rate of cyanobacterial growth and activity is increased with rising water temperatures, favoring thermophilic species. In parallel, changes in how and when rainfall occurs and water drains influence the environment. Changes in the ways and times that nutrients enter and persist in water bodies may develop new locations for harmful blooms and bloom formation (Amin et al., 2024). Consequently, integrating research across several disciplines, including hydrology, ecology, microbiology, and environmental management, is needed and combines knowledge from limnology, biogeochemistry, microbial ecology, hydrology, and environmental management. Exploring the stoichiometric interactions and dynamics that control cyanobacterial bloom stages is important. The cyclic behavior of blooms from initiation to collapse can provide insight for more efficient interventions. Successful ecosystem management balances knowledge of nutrient budgets with on-site governance. Engaging stakeholders and using adaptive governance are needed to accomplish stable water quality in the long-run quality goals. In brief, this chapter investigates the fundamental influence of nutrient dynamics and stoichiometry in cyanobacterial bloom development. We examine the importance of stoichiometry during cyanobacterial bloom formation by referring to the recent advances in ecology theory and applied water management. The chapter investigates the roles played by major and minor nutrients in bloom dynamics. The influence of micronutrients, the significance of nutrient ratios, and the cellular and physiological responses of cyanobacteria to changing environmental conditions play a crucial role in determining their growth, metabolism, and ecological adaptability. By linking biogeochemical processes with the integration of bloom ecology, the chapter introduces a strategy for both understanding and predicting cyanobacterial blooms and mitigating harmful cyanobacterial events.

4.2 Key Nutrients Driving Bloom Formation

Essential macronutrients, such as nitrogen and phosphorus, which are most important, create a leading reason for the formation of cyanobacterial blooms. The number of nitrogen and phosphorus supplies that go into an ecosystem plays a great part in timing, strength, and ultimate ecological outcome of cyanobacterial blooms. The manner in which blooms alter adaptive dynamism of aquatic systems. Together with their metabolic functions, these nutrients also determine the pattern of phytoplankton assemblages and their interactions. Cyanobacteria need nitrogen for their basic functions, and moreover, nitrogen is also important for structuring aquatic communities. Direct groupings of phytoplankton in the ecosystem, in terms of how the groups assemble and how they engage the system, assemblages. Animals need nitrogen for the development of amino acids, nucleic acids, and chlorophyll and cellular enzymes. Forms of nitrogen in water bodies range from ammonium among others. Such environments promote the presence of ammonium (NH_4^+), nitrate (NO_3^-), and various organic nitrogen compounds. The availability of nitrogen forms is closely evaluated by both physical–chemical characteristics of the system and by the rate at which organisms acquire these forms. Primarily, bioavailability is variable due to the function of nitrification and denitrification by biological process. Several cyanobacterial species possess the outstanding ability to access nitrogen from both inorganic and organic nitrogen sources showing capacity for assimilation of inorganic as well as organic forms of nitrogen. Diazotrophic cyanobacteria, such as apart from these genera (Anabaena, Dolichospermum, Nodularia), the activity of nitrogenase enzymes is observed. In particular, this feature provides the capacity to transform atmospheric nitrogen (N_2) to ammonia with the help of activity of nitrogenases, especially under low nitrate conditions. Such a capacity therefore endows cyanobacterial ecosystems with a competitive edge over other microorganisms. This capability provides this company with a competitive advantage. Nitrogen-deficient and phosphorus-enriched settings regularly account for why N_2 overwhelms and slip in species during bloom episodes in this type of setting (Yan et al., 2024).

Phosphorus, on the other hand, is widely recognized as the most common limiting nutrient in freshwater systems. It is essential for energy transfer via ATP, the synthesis of nucleic acids, and membrane phospholipids. Unlike nitrogen, phosphorus does not have a gaseous phase, making its availability more dependent on geological and anthropogenic sources. In natural systems, phosphorus inputs occur through weathering of rocks, decomposition of organic matter, and sediment–water interactions. However, anthropogenic sources—particularly agricultural runoff containing fertilizers, untreated sewage, detergents, and animal waste—now represent the predominant vectors of phosphorus loading in aquatic environments. The bioavailability of phosphorus is further influenced by pH, redox conditions, and interactions with metals such as iron, aluminum, and calcium, which can precipitate phosphorus as insoluble complexes. Under anoxic or hypoxic conditions, such as those caused by microbial decomposition during bloom senescence, iron-bound phosphorus in sediments can be released back into the water column—a process known as internal loading—which can fuel subsequent bloom events and prolong eutrophic conditions (Feng et al., 2024). The relationship between nutrient enrichment and bloom formation is both direct and synergistic. Numerous field studies and mesocosm experiments have demonstrated that increased inputs of nitrogen and phosphorus lead to greater phytoplankton biomass, with cyanobacteria often outcompeting other groups under such conditions due to their physiological traits. Notably, the relative availability of nitrogen to phosphorus, typically expressed as the N:P ratio, serves as a critical determinant of community composition (Luong et al., 2025). Key nutrient roles, thresholds, and their impacts on bloom formation are summarized in Table 4.1. The classical Redfield ratio suggests an optimal C:N:P molar ratio of 106:16:1 for balanced phytoplankton growth. Deviations from this ratio can lead to selective advantages for different groups. For example, when N:P ratios are low, diazotrophic cyanobacteria that can fix atmospheric nitrogen tend to dominate, as seen in many eutrophic lakes and reservoirs. Conversely, in high N:P scenarios, non-diazotrophic cyanobacteria such as Microcystis may proliferate, particularly in systems where phosphorus is abundant due to anthropogenic loading. Additionally, nitrogen availability can influence the production of secondary metabolites such as microcystins, which have been shown to increase under high nitrogen regimes. This link between nutrient availability and toxicity further underscores the public health relevance of nutrient management (Klip, 2023). Furthermore, nutrient enrichment not only promotes cyanobacterial growth but also alters ecosystem processes. Elevated nutrient levels can suppress biodiversity, alter trophic dynamics, and increase the risk of hypoxia as dense blooms decay and microbial decomposition depletes oxygen levels in bottom waters. The impacts are not merely localized; nutrients can be transported across watersheds and into estuarine and coastal systems, triggering large-scale hypoxic zones such as those observed in the Gulf of Mexico and the Baltic Sea (Devlin & Brodie, 2023). These "dead zones" are direct consequences of nutrient-driven primary production followed by oxygen depletion, often exacerbated by stratification and reduced mixing. In addition to external nutrient inputs, internal nutrient cycling within the water column and sediments plays a pivotal role in sustaining cyanobacterial blooms. Sediments can act both as sinks and sources of nutrients, particularly phosphorus. During periods of stratification and low oxygen, internal phosphorus release can be substantial, supporting bloom growth even in the absence of new external inputs. This internal loading complicates management efforts, as reducing surface runoff may not immediately translate into improved water quality if legacy phosphorus stores in sediments remain bioavailable (Molot et al., 2021). In summary, the intricate relationship between nitrogen and phosphorus availability and cyanobacterial bloom dynamics is well established. Nutrient enrichment—whether through input direct or internal recycling—represents the primary inducing driver of bloom onset. proliferation, and persistence. The potential for taking advantage of a variety of nutrient available to cyanobacteria sources in addition to their physiology modification along with ecological feedbacks allows them to dominate under nutrient-rich conditions. Consequently, managing nutrient inputs—especially from agricultural and urban sources—is crucial to the control of frequency and intensity of as well as ecological effects of poisonous cyanobacterial blooms in water environments.

4.3 The Redfield Ratio and Stoichiometric Imbalances

An established concept in aquatic biogeochemistry, the Redfield ratio describes the average elemental composition of marine phytoplankton and the corresponding nutrient proportions required for their balanced growth. It is given as a molar ratio of Carbon, Nitrogen:Phosphorus (C:N:P = 106:16:1), and it means that 106 mol of carbon are taken up; in photosynthesis, phytoplankton will ordinarily need 16 mol of nitrogen and 1 mol of phosphorus. This ratio has been used as a parameter to use as a benchmark to evaluate the nutrient limitation. Nutrient balance in aquatic

Table 4.1 Key nutrient parameters influencing cyanobacterial blooms

Parameter	Role in cyanobacterial blooms	Typical thresholds/ observations	Impacts on bloom dynamics
Nitrogen (N)	Essential for amino acids, nucleic acids, and chlorophyll synthesis	Elevated concentrations (>0.3 mg/L) promote bloom formation	High N levels can enhance cyanobacterial growth; N-fixing species dominate under low N conditions
Phosphorus (P)	Critical for ATP, nucleic acids, and membrane phospholipids	Concentrations >0.1 mg/L often lead to eutrophication	Excess P supports bloom initiation; internal loading from sediments can sustain blooms
N:P ratio	Determines nutrient limitation status	Ratios <16:1 (Redfield ratio) indicate potential N limitation; >16:1 suggest P limitation	Low N:P ratios favor N-fixing cyanobacteria; high ratios can lead to dominance of non-N-fixing species like *Microcystis*
Iron (Fe)	Cofactor for nitrogenase and photosynthetic enzymes	Availability varies; often limited in freshwater systems	Fe limitation can restrict N fixation; adequate Fe supports bloom development
Internal nutrient loading	Release of nutrients from sediments under anoxic conditions	Significant during stratification and low oxygen events	Sustains blooms even when external inputs are reduced; complicates management efforts
Anthropogenic inputs	Nutrient additions from agriculture, wastewater, and urban runoff	Varies by region; often exceeds natural background levels	Major driver of eutrophication and bloom events; management focuses on reducing these inputs
Climate change factors	Influences temperature, precipitation patterns, and stratification	Rising temperatures and altered rainfall patterns observed globally	Warmer temperatures and increased stratification enhance bloom frequency and intensity

ecosystems is popularly used to interpret phytoplankton, dynamics, nutrient cycling, and trends of productivity (Inomura et al., 2022). However, in freshwater and coastal ecosystems specifically those impacted by anthropogenic nutrient loading observed. Elemental ratios are often far away from the Redfield standard. These deviations, known as stoichiometric imbalances, are primarily caused by disproportionate inputs of nitrogen or phosphorus from agricultural runoff, industrial effluents, and urban wastewater. For example, watersheds dominated by extensive practice of an intensive livestock production; fertilizer application frequently enters high phosphorus-to-nitrogen ratio leading to low N:P ratio despite the high N:P ratios in the wastewater. On the other hand, urban or urbanized areas with nitrogen-rich effluents or influenced by an exemplary discharge and high nitrogen-to-phosphorus ratios are likely to be observed in atmospheric nitrogen deposition. These shifts in nutrient implications for the composition of phytoplankton communities are gravely significant through stoichiometry bloom dynamics. Low N:P ratios are in most cases connected with proliferation of nitrogen in eutrophic systems and restoration of cyanobacteria, i.e., Aphanizomenon, Dolichospermum (formerly Anabaena). Nodularia. These species have special heterocysts through which the nitrogenase enzyme is contained and complicated, which allows them to convert and make atmospheric nitrogen (N_2) bioavailable ammonia. This capacity is a valuable ecological advantage for conditions where nitrogen is depleted, but numerous alkaline buffers are present. Phosphorus-sufficient environment enables dominance of the phytoplankton assemblages by these taxa. Cyanobacteria, 30:1, tend to dominate over non-diazotrophic species of cyanobacteria, e.g., Microcystis, which are subject to inputs from outside. Nitrogen sources do very well in a phosphorus-rich environment. Notably, Microcystis is one of the most cosmopolitan bloom-forming genera and is renowned for being a source of the powerful microcystins that represent significant threats to water safety and ecosystem health (Kramer et al., 2022). The variance from the Redfield ratio also affects production of toxin apart from bloom formation, the nutrient uptake efficiency, and interspecies competition. Research has shown that under nitrogen-rich conditions, and some cyanobacteria are diverting excess nitrogen toward the secondary metabolites synthesis, synthesis of hepatotoxins, and neurotoxins. This not only increases the toxicity of blooms, but also increases their level of grazing resilience to zooplankton. In addition, unbalanced stoichiometry can change the way frictional power in competition is played. Algal groups usually suppress diatoms and green algae which need well-balanced nutrients conditions, thereby reinforcing cyanobacterial dominance. This opening can create long-term effects, changes in community structure, decreases biodiversity and impacts the higher trophic level (Manic et al., 2024). Moreover, the continued existence of modified nutrient ratios throughout time can induce a feedback loop, in which cyanobacterial blooms actually increase the imbalances of nutrients. For example, the decay of biomass during bloom senescence liberates organic types of nitrogen and phosphorus to the water column. Nonetheless, the rate and stoichiometry of this release may promote additional cyanobacterial uptake, especially in systems where internal loading from sediments replenishes phosphorus under hypoxic conditions. These feedback mechanisms can reinforce eutrophic states, making ecosystem recovery difficult even after external nutrient reduction measures are implemented. Essentially, Redfield ratio is an important lens for viewing the nutrient ecology of the industries. Aquatic systems can be assessed. Its departures from the norms in eutrophic spaces are not merely logical algebraic relationships, statistical anomalies, and flag dramatic ecological shifts and the controlling factor of cyanobacterial, proliferation of species, composition of species, toxin production, and the resilience. Understanding these stoichiometric imbalances is essential for identifying bloom behavior and formulating nutrient industries. Management approaches restore balance and safeguard the health of aquatic ecosystem.

4.4 Role of Micronutrients in Bloom Dynamics

Although occurring in trace amount, micronutrients are of prime importance in regulating the growth, metabolism, and bloom dynamics of cyanobacteria. Among these, iron (Fe) stands as a very important component, especially given its active role in two important physiological processes: photosynthesis and nitrogen fixation. Iron is the main ingredient of many proteins. Electron transport proteins include cytochromes and ferredoxins, which are mandatory for trapping and converting solar energy into a chemical form, when undertaking photosynthesis. In nitrogen fixing cyanobacteria, iron cannot be dispensed with in the activity of nitrogenase enzyme complex, which facilitates the transformation of atmospheric nitrogen, (N_2) to ammonia. The nitrogenase enzyme on its own contains iron-sulfur clusters in some species and in some molybdenum vanadium cofactors, all of which are to be bioavailable for best enzyme functions. A shortage of iron is, therefore, consequently able to critically hamper both nitrogen fixation and photosynthetic efficiency sufficiency, thus checking cyanobacteria growth and capping лето success of iron-depleted ecosystems by bloom-forming species (Wei et al., 2023). In addition to iron, other micronutrients such as sulfur (S) and trace metals—zinc (Zn), copper (Cu), and others—are also of concern. Copper (Cu), manganese (Mn), cobalt (Co), and molybdenum (Mo) are also essential for cyanobacterial physiology (Yadav et al., 2024). Sulfur is used in the synthesis of some amino acids, (e.g., Cysteine and Methionine (coenzymes, and iron-sulfur proteins that play a part in metabolism pathways and electron transport. Manganese for its part, however, is an essential cofactor for the oxygen-evolving complex of photosystem II, enabling the photolysis of water and oxygen. Zinc plays a vital role in the structure and activity of many enzymes, including those involved in DNA replication and antioxidant defense. Copper serves as a cofactor plastocyanin another major electron carrier of the photosynthetic process apparatus, on the other hand, while molybdenum is required for nitrate reductase and nitrogenase enzymes, especially in diazotrophic species (Manic et al., 2024). Despite that, these micronutrients are in low concentrations in natural waters. Their bioavailability varies very much and subjected to various physicochemical factors, including pH, redox and the organic complexation, and competition with other ions. For example, in stratified and anoxic bottom waters, the solubility and mobility of metals such as iron are significantly increased. There is a possibility of manganese increasing resulting in localized enrichment. Conversely, in oxygenated surface waters, iron can precipitate out of solution as insoluble ferric hydroxides, and this restricts the access of the iron to the atmosphere availability to phytoplankton. In addition, organic ligands released by cyanobacteria or decaying biomass can bind metals, affecting their uptake dynamics (Khoshru et al., 2023). The concept of micronutrient limitation, and more importantly, co-limitation, has gained recognition in recent years as a critical factor influencing bloom development. Co-limitation refers to situations where the simultaneous deficiency of more than one nutrient typically a macronutrient and one or more micronutrients constrains biological productivity. For instance, even in systems with adequate phosphorus and nitrogen levels, insufficient iron or molybdenum can limit nitrogen fixation and overall bloom intensity. Similarly, in some oligotrophic systems, trace metal availability may dictate whether cyanobacteria can outcompete other phytoplankton, especially in competition for light and nutrients. Field experiments and laboratory cultures have shown that the addition of iron, either alone or in combination with nitrogen or phosphorus, can stimulate bloom formation and increase toxin production in certain cyanobacterial strains. Moreover, different cyanobacterial species are observed to have different micronutrient needs affecting the species composition and succession during blooms. For instance, Aphanizomenon and Dolichospermum could succeed under iron-, nitrogen-poor, enriched conditions because of their nitrogen fixing abilities, but Microcystis can be dominant in nutrient-rich of low content in iron, where diazotroph is weak competitive. Such differences in micronutrient demand and strategies of uptake can form the mús. ecological trajectory of blooms toxicity, duration, and resilience to environmental change (Wang et al., 2021). Conclusively, even though nitrogen and phosphorus are the key drivers of such macronutrients on cyanobacterial bloom formation, availability and reaction of micronutrients are important, crucial regulatory role. Iron, sulfur, and trace metals are necessary to maintaining photosynthetic efficiency, assistance of nitrogen fixation, and enzyme activity. The limitation or co-limitation of micronutrients may moderate bloom dynamics and determine which species dominates a bloom, composition and defines physiological traits of dominant cyanobacterial populations. The identity of the role of micronutrients is, therefore, critical in accurate understanding, simulation of bloom behavior, ecosystem responses, and whole system designs. Management approaches for addressing macro- and micronutrient imbalances in aquatic ecosystems.

4.5 External Nutrient Loading

External nutritive loading is arguably the most important anthropogenic forcing agent for cyanobacteria, bloom occurrence, and resistance in the freshwater and coastal milieus. The term refers into the absorption of nutrients, especially nitrogen (N) and phosphorus (P) from sources beyond the aquatic system mainly as a result of human activities. These nutrient inputs increase the availability of essential growth-limiting elements, tipping the ecological balance in favor of opportunistic organisms like bloom-forming cyanobacteria. The rapid urbanization, intensification of agriculture, and industrial expansion witnessed over the last century have led to a marked increase in external nutrient inputs, significantly altering nutrient budgets and stoichiometric balances in aquatic environments (Xu et al., 2021). Agricultural runoff remains the predominant source of external nutrient enrichment in many regions. The widespread application of synthetic fertilizers, animal manure, and nitrogen fixing crops contributes substantial quantities of nitrate, ammonium, and phosphate to surrounding water bodies via surface runoff, drainage systems, and leaching. Rainfall events, particularly during the early growing season, can lead to episodic surges in nutrient loading that initiate or intensify bloom formation. Additionally, poorly managed irrigation practices and excessive fertilizer application contribute to the buildup of "legacy nutrients" in soils, which can continue to leach into aquatic systems for decades. Nutrient-rich runoff is often exacerbated by soil erosion, which not only transports particulate phosphorus but also stimulates sedimentation processes that alter habitat conditions for aquatic biota (Bhardwaj et al., 2024). Municipal and industrial wastewater discharges also represent a significant source of external nutrients. In many developing countries, inadequate sewage treatment results in the direct release of nitrogen- and phosphorus-laden effluents into rivers, lakes, and coastal zones. Even in regions with advanced wastewater treatment infrastructure, conventional secondary treatment processes may not sufficiently remove dissolved nutrients, especially phosphorus, unless enhanced biological phosphorus removal or chemical precipitation steps are included. Industrial discharges, including those from food processing plants, textile manufacturing, and mining operations, can introduce both nutrients and trace metals, further complicating bloom dynamics and water quality outcomes. Urban storm water, carrying pet waste, organic litter, detergents, and residual fertilizers from residential lawns, acts as a non-point source of nutrients, particularly during heavy rainfall and flooding events. Atmospheric deposition, while often overlooked, is another pathway through which nitrogen—mainly in the forms of nitrate and ammonium—enters aquatic systems. Fossil fuel combustion, industrial emissions, and intensive livestock production release reactive nitrogen compounds into the atmosphere. These compounds eventually return to Earth through wet deposition (rain and snow) or dry deposition (dust and aerosols), contributing to nitrogen enrichment in surface waters. This form of nutrient input can be particularly important in remote or otherwise pristine systems that lack direct terrestrial inputs, where even modest atmospheric nitrogen loading can trigger or exacerbate bloom development (Nõges et al., 2024). The influence of external nutrient loading on cyanobacterial bloom dynamics is multifaceted. First, it provides the essential substrates required for rapid cell division, leading to increased biomass accumulation and bloom formation. Second, elevated nutrient concentrations—particularly phosphorus—can disrupt the competitive balance among phytoplankton, favoring cyanobacteria due to their slower sinking rates, nitrogen-fixation abilities, and tolerance to light limitation. Third, external inputs can modify the timing and duration of blooms. For instance, sustained nutrient loading throughout the growing season may support recurrent or prolonged bloom episodes, whereas a single large pulse of nutrients following a storm event may initiate a short but intense bloom. A myriad of case studies in the world peoples highlight the strong relationship between external loading and cyanobacterial notching outbreaks. For example, Lake Erie in

North America has experienced severe blooms of Microcystis caused by phosphorus loading from agricultural watersheds, particularly the Maumee River Basin (King et al., 2022). A similar situation has been occasioned by Lake Taihu in China and Lake Dianchi. cyanobacterial because of a combination of urban wastewater discharge, agricultural runoff, and internal nutrient cycling (Donghao et al., 2022). In Australia's Murray–Darling Basin, nitrogen-rich runoff following rains has been associated with widespread blooms of Anabaena circinalis and produces saxitoxins with neurotoxic effects. The interaction of external nutrient loading is a point of consideration and hydrological conditions. Maintenance of high levels of nutrients in times of high rainfall and runoff is augmented. Delivery of water column occurs with stratification which forms excellent conditions for water column, cyanobacterial growth. By contrast, in the case of drought or low flow conditions, we have less flushing. Rates can extend persistence of bloom by reducing dilution and increasing residence times. In both cases, nutrient enrichment is the facilitating effect that controls bloom magnitude, ecological consequences. Finally, the external nutrient loading is a main and in many cases controllable contributor to the emergence of the process and propagation of cyanobacterial blooms. Whether originating from agriculture, urban wastewater, discharge of industrial effluent, or atmospheric deposition, such nutrient sources cause eutrophication of aquatic systems and provide favorable conditions for harmful bloom-forming species. Actively and through policy, the external loading must be well-managed interventions, determination of land use, better agriculture practices and enhanced wastewater. While bloom risks are mitigated and ecosystem health restored through treatment technologies, it is a necessity. Knowledge of the pattern, timing, and chemical form of nutrient inputs gives a crucial view of the approach espoused in the article, basis for predicting bloom behavior, as well as site-specific control strategies.

4.6 Nutrient Limitation and Co-Limitation

Nutrient limitation is a major concept in aquatic ecology which has a major impact on the dominant phytoplankton taxa community composition, productivity, and succession in particular cyanobacteria (Rizhinashvili, 2022). It refers to a condition in which the deficiency of one or more essential nutrients limits the biological growth and functioning of an ecosystem. In many aquatic systems, the two most commonly limiting macronutrients are nitrogen and phosphorus. However, it is well established that under certain conditions, micronutrients such as iron, silica, or molybdenum can also become limiting or even detrimental to biological productivity. The particular nutrient that will limit growth may change from place to place depending on the trophic status of the river, the water body, land use within the environment, seasonal dynamics as well as internal nutrient cycling process (Kuczyńska-Kippen et al., 2024). Systems which are oligotrophic or mesotrophic phosphorus usually are the limiting nutrient, owing to its low solubility and high affinity to adsorption on the particles or precipitation metals. On the other hand, in eutrophic or hyper-eutrophic systems—especially those exposed to mixed contamination or nutrient pollution—there is often a wider variety of species whose productivity is enhanced by nutrient enrichment, such as in sewage-impacted waters or environments with intense phosphorus loading from anthropogenic sources. Under such conditions, nitrogen may become the limiting factor. This change produces a situation that has a competitive edge for nitrogen-fixing cyanobacteria. They benefit from their ability to utilize atmospheric nitrogen, thereby altering the structure and function of the phytoplankton community. For instance, if phosphorus is plentiful and nitrogen is limited, diazotrophic species, like Dolichospermum and Nodularia, increase considerably and will grow significantly likely to prevail, whereas in the situations that both nutrients are in abundance neither is likely to predominate. Such diazotrophic bloom formers such as Microcystis or Planktothrix common may crowd out the others because of their convenience as a competitor and their high rates of growth and higher uptake kinetics (Patiño et al., 2023). The ratio of nitrogen to phosphorus (N:P) becomes a practical indicator concerning the water column of potential nutrient limitation. Traditionally, the Redfield ratio (16:1 molar N:P) has been used as a benchmark. Deviations below this ratio often indicate nitrogen limitation, while values exceeding this threshold suggest phosphorus limitation. Empirical studies have proposed various thresholds; for instance, an N:P ratio below 10:1 often signals strong nitrogen limitation, whereas a ratio above 30:1 may point toward phosphorus limitation. However, these thresholds are not absolute, as different cyanobacterial species possess varying nutrient requirements and uptake capacities, leading to species-specific responses to nutrient supply ratios (Tarafdar et al., 2021). Beyond single-nutrient limitation, co-limitation—where two or more nutrients are simultaneously limiting—has emerged as a more realistic representation of nutrient dynamics in aquatic ecosystems. In many cases, phytoplankton growth may not respond to the addition of a single nutrient but may increase significantly when multiple nutrients are added together. For instance, the simultaneous addition of nitrogen and phosphorus may stimulate more robust growth than either nutrient alone, especially in systems experiencing seasonal or spatial variability in nutrient inputs. Similarly, co-limitation involving a macronutrient and a micronutrient, such as phosphorus and iron, has been documented in several lake systems, where the availability of iron determines the effectiveness of nitrogen fixation and thus indirectly controls the phosphorus demand (Rock & Collins, 2024). Cyanobacteria have evolved various strategies to cope with nutrient limitation and co-limitation. Many species produce high-affinity nutrient transport systems that allow them to scavenge nutrients at very low concentrations. Others form colonies or filaments that reduce sinking rates, increase surface-to-volume ratios, and create microenvironments favorable for nutrient retention. Some cyanobacteria also possess storage capabilities, such as polyphosphate granules or cyanophycin reserves, enabling them to buffer against periods of nutrient scarcity. In environments with fluctuating nutrient availability, these adaptations confer resilience and enhance their capacity to dominate over other phytoplankton groups (Bannon et al., 2022). Furthermore, nutrient limitation can influence bloom toxicity and ecological impacts. Studies have shown that toxin production, particularly microcystins, is often linked to nitrogen availability, with higher toxin levels associated with nitrogen-replete conditions. Under nitrogen-limiting scenarios, toxin synthesis may decrease, but this can be offset if nitrogen is supplied through fixation or organic nitrogen sources. Similarly, phosphorus limitation can change cellular stoichiometry and slow down growth rates while determining the competitive balance. Between species possibly elevating the phosphorus-acquiring ability species such as high alkaline phosphatase activity. There are many community structure shifts and consequence of nutrient limitation patterns. For instance, a high phosphorus and low nitrogen conditions lead to a tendency toward diazotrophic cyanobacterium when nutritional conditions are balanced. Inputs may increase diversity in the phytoplankton assemblages. Long-term monitoring of lakes and the changes in nutrient limitation regimes may cause regime shifts in reservoirs which has been demonstrated in the major phytoplankton communities, at times with sustained ecological impacts. Such changes are often hard to reverse once they get started, because positive feedback works for them, for example internal loading, light attenuation from dense blooms, and sediment nutrient regeneration (Khazaei et al., 2023). In total, nutrient limitation and co-limitation play the primary roles of drivers of cyanobacterial bloom, formation, intensity, and persistence. Learning the limiting nutrients, their functioning in order to combinations, and the under what environmental contexts is critical in coming up with reliable, prediction models and management interventions with efficiency. The complexity of nutrient interactions necessitates integration, as factors of external nutrient are considered, inputs and internal cycling processes, as well as species-specific physiological traits, ecosystem responses. Holistic assessment of nutrient limitation creates a route for holistic vision, restoring the eutrophic systems and reducing negative consequences of harmful cyanobacterial blooms.

4.7 Cyanobacterial Adaptations and Feedback Mechanisms

The physiological adaptations and feedback of the cyanobacteria are highly impressive. mechanisms that do not just allow them to survive in changing aquatic environments, but actually aggressively cause the development, maintenance, and recurrence of noxious algal blooms (HABs). The most ecologically important adaptation of some cyanobacterial species to be used to fix atmospheric nitrogen by a process known as biological nitrogen fixation. This characteristics is highly prevalent in diazotrophic cyanobacteria including Dolichospermum. Aphanizomenon, and Nodularia which harbor the enzyme nitrogenase that reduces inert N_2 gas to bioavailable ammonia (NH_3). Due to this sensitivity of nitrogenase to oxygen, these infected plants provide highly sensitive near ultraviolet light. Species tend to have specialized cells known as heterocysts which produce anaerobic countries and microenvironment in order to protect the enzyme and efficient nitrogen fixation. This adaptation occupies an especially beneficial position in nitrogen-depleted, phosphorus-rich conditions. Non-diazotrophic phytoplankton is nitrogen-limited. As a result, diazotrophic cyanobacteria frequently mask blooms for communities from systems with low ratios of N:P, modifying composition of communities and sustaining the whole biomass of the bloom by endogenous nitrogen contributions (Yadav et al., 2022). In addition to nitrogen obtaining, cyanobacteria are specialists in phosphorus scavenging, particularly under phosphorus limitation, which is a widespread character of many freshwater and stratified systems. Production of alkaline phosphatase is one important strategy. Enzyme that catalyses the removal of phosphate groups from organic compounds thus liberating inorganic. a phosphate readily assimilable. This enables the cyanobacteria to tap pools of available kinds of dissolved organic phosphorus (DOP) which are unavailable for most other phytoplankton. Besides, cyanobacteria produce high-affinity phosphate transporters and demonstrate luxury unnecessary amount of phosphate extraction from surroundings uptake behavior whereby they take in and store phosphate during times of over abundance in intracellular polyphosphate granules for later use during scarcity. These phosphorus acquisition strategies enhance their competitiveness, enabling sustained growth in environments with fluctuating nutrient availability (Xiao et al., 2022). Internal nutrient recycling is another critical factor contributing to the sustainability of cyanobacterial blooms. During bloom development and senescence, cyanobacterial cells undergo lysis, releasing previously assimilated nutrients back into the surrounding water. This regeneration of nitrogen, phosphorus, and other growth-limiting nutrients supports continued proliferation of the remaining population or new bloom initiation. The microbial decomposition of organic matter further contributes to nutrient recycling, while in stratified or hypoxic conditions, nutrient fluxes from sediments—especially phosphorus—can be accelerated due to the breakdown of metal-phosphate complexes under anoxic conditions. This internal loading effectively decouples bloom dynamics from immediate external nutrient inputs, allowing blooms to persist even when watershed management practices reduce nutrient inflow. Cyanobacterial blooms are also reinforced by ecological and physicochemical feedback loops that enhance bloom resilience and hinder ecosystem recovery. For instance, dense surface scums formed by buoyant cyanobacteria reduce light penetration into the water column, suppressing the growth of competing phytoplankton such as diatoms and green algae. This light shading effect alters primary productivity and can shift community dominance further in favor of cyanobacteria. Additionally, bloom decomposition can result in oxygen depletion (hypoxia or anoxia), especially in bottom waters, which promotes the release of nutrients from sediments—a feedback that sustains or reinvigorates the bloom. Some cyanobacteria also produce allelopathic compounds or toxins that inhibit the growth of other planktonic species and deter grazing by zooplankton, providing a chemical defense that increases bloom longevity and reduces biotic resistance (Lakshmikandan et al., 2024). These adaptations and feedback mechanisms together create a self-reinforcing cycle: cyanobacteria establish blooms by exploiting available nutrients through specialized uptake and fixation strategies; they then alter the environment—chemically and physically—in ways that favor their continued dominance. The combination of biological flexibility, nutrient recycling, and ecosystem engineering makes cyanobacterial blooms highly resilient to short-term mitigation efforts. Effective management of such blooms therefore requires not only controlling external nutrient inputs but also addressing internal loading, monitoring feedback dynamics, and understanding the ecological traits that underpin cyanobacterial success.

4.8 Ecological and Biogeochemical Implications

Cyanobacterial blooms exert profound ecological and biogeochemical impacts on aquatic ecosystems, reshaping food webs, disrupting biogeochemical cycles, and posing serious risks to biodiversity and public health. One of the most immediate and visible consequences of bloom formation is the alteration of aquatic food web dynamics. Dense surface scums formed by buoyant cyanobacteria reduce light penetration into the water column, impairing photosynthesis in submerged aquatic vegetation and non-bloom-forming phytoplankton. This leads to a decline in primary productivity by other autotrophs and disrupts the energy flow to higher trophic levels. Moreover, many cyanobacteria are filamentous, colonial, or produce mucilage that reduces their edibility, making them a poor food source for zooplankton grazers such as Daphnia and copepods. In some cases, cyanobacteria even release compounds that are toxic or deterrent to grazers, further disrupting the link between primary producers and herbivores. As a result, food webs become simplified and less efficient, often leading to decreased fish populations, especially those reliant on phytoplankton-based food chains during their larval stages (Kumari et al., 2022).

Another major ecological consequence of cyanobacterial blooms is the disruption of oxygen dynamics in aquatic systems. During the initial stages of bloom development, photosynthetic activity can lead to supersaturation of oxygen in surface waters. However, as the bloom collapses and biomass decomposes, microbial respiration increases dramatically, consuming dissolved oxygen and leading to hypoxic (low oxygen) or anoxic (no oxygen) conditions, especially in bottom waters of stratified systems. These oxygen-depleted zones, often referred to as "dead zones," can cause massive fish kills, disrupt benthic communities, and lead to the release of nutrients such as phosphorus and ammonium from sediments—a process known as internal loading. The resulting feedback loop perpetuates eutrophication and makes it more difficult to restore ecosystem balance even after external nutrient inputs are reduced (Lakshmikandan et al., 2024).

One of the most alarming consequences of cyanobacterial blooms is the production of a wide range of potent toxins, collectively referred to as cyanotoxins. These include hepatotoxins such as microcystins, neurotoxins like anatoxin-a and saxitoxins, and cytotoxins such as cylindrospermopsin. Toxin production varies by species, strain, and environmental conditions, but many bloom-forming genera—including Microcystis, Planktothrix, Dolichospermum, and Nodularia—are known producers of harmful metabolites. These toxins pose acute and chronic risks to aquatic organisms, causing liver damage, neurological effects, impaired growth, and mortality in fish, amphibians, and invertebrates. Zooplankton exposed to sublethal toxin concentrations may experience reduced reproduction and mobility, further weakening grazing pressure and allowing blooms to persist. In addition to ecological effects, cyanotoxins present serious public health threats when blooms occur in recreational waters or drinking water sources. Human exposure through ingestion, inhalation, or dermal contact can lead to symptoms ranging from skin irritation and gastrointestinal distress to liver failure and, in extreme cases, death. The 2014 water crisis in Toledo, Ohio, where microcystins contaminated the city's water supply, highlighted the vulnerability of drinking water systems to cyanotoxin contamination (Moreira et al., 2022).

Beyond immediate biological effects, cyanobacterial blooms have lasting implications for sediment–water nutrient fluxes and long-term ecosystem trajectories. During bloom decay, settling biomass contributes large amounts of organic material to sediments, altering sediment structure and redox conditions. Under anoxic conditions, bound phosphorus is released from iron-phosphate complexes, increasing the availability of phosphate in the overlying water and driving internal nutrient loading. Similarly, ammonium and other reduced nitrogen species accumulate under anaerobic conditions, and easily accessible for nutrients for subsequent bloom events. In the long run, this nutrient re-use from sediments has accumulated and is able to support the state of the chronic eutrophication; hence, the system cannot be able to regain its balance. Periodic dissolved oxygen depletion also realigns the communities found at the sea bottom toward more tolerant and less diverse ones assemblages, further weakening ecosystem resilience (Bargu et al., 2023). In short, the phenomenon of cyanobacterial blooms is not just isolated biotic event, but a deep one built into that of aquatic ecosystems' ecological and biogeochemical functioning. They initiate cascading effects across food webs, change oxygen and nutrient cycling and have serious consequences for the ecosystem's health and for human wellbeing. These changes often create feedbacks stimulating a bloom persistence and eutrophication, which are significant challenges for restoration and management. The impacts have to be addressed with an integrated approach knowledge of the dynamics of bloom, ecosystem interaction, and nutrient movement to inform c effective long-term interventions.

4.9 Management and Mitigation Strategies

Cyanobacterial blooms are arising, and remain, significant and intensifying threats to freshwater and coastal systems and ecosystems calling for the establishment and implication of efficient management and mitigation strategies. Being central to these efforts are nutrient reduction policies targeted at control of the main culprits of bloom formation—i.e., excessive nitrogen (N), phosphorus (P) inputs. These policies affect both point sources—for example, wastewater treatment plants and industrial discharges, and non-point sources, agricultural runoff and urban storm water. Regulatory frameworks most often come with fertilizer bans, rates, and schedules for application and requirements for buffer zones along water courses, and requirements for nutrient removal technologies in treatment center. In many developed regions, the requirement of total maximum daily load (TMDL) bounds on nitrogen and phosphorus has been a foundation stone of nutrient management. While phosphorus control has historically been given top priority, notwithstanding the fact that it has already limited role in freshwater systems, emerging research increasingly understands the need for nitrogen management especially because of its effect on toxin production and bloom persistence. Nutrient reduction programs that are successful take years to complete a commitment, enforcement mechanism, and stakeholder involvement to compliance and sustained impact (Summers & Ryder, 2023).

At the same time as measures of nutrient control, biomanipulation measures have been drawing attention as means for change of the biological structure in the aquatic ecosystems to minimize bloom susceptibility. These approaches typically involve modifying grazer populations, such as increasing densities of zooplankton species (e.g., Daphnia) that feed on cyanobacteria or managing fish communities to favor zooplanktivorous species that exert top-down control. For instance, reducing populations of planktivorous fish like carp or tilapia can allow grazer populations to recover and exert greater grazing pressure on phytoplankton, potentially suppressing cyanobacterial dominance. However, biomanipulation outcomes are often system-specific and depend on food web complexity, nutrient levels, and species interactions. In eutrophic systems where cyanobacteria are poor-quality food sources or produce toxins that deter grazers, the effectiveness of biomanipulation may be limited unless combined with other nutrient management actions (Summers & Ryder, 2023).

Beyond conventional approaches, a range of innovative remediation techniques has emerged to address both acute bloom events and long-term water quality improvement. One such technique is the application of Phoslock, a lanthanum-modified bentonite clay that binds free phosphate in the water column and forms stable complexes, thereby reducing phosphorus availability for phytoplankton uptake. This method has been successfully applied in various lake restoration projects and is particularly useful for controlling internal phosphorus loading from sediments. Algaecides, including copper-based compounds and hydrogen peroxide formulations, offer rapid bloom suppression but must be used cautiously due to potential non-target effects on aquatic organisms, risk of toxin release upon cell lysis, and the development of resistance in cyanobacterial populations (Haque, 2021). Nature-based solutions such as floating treatment wetlands represent another promising strategy. These systems use buoyant platforms planted with aquatic vegetation to intercept nutrients, provide habitat for microbial communities that facilitate nutrient transformation, and shade the water surface to reduce light availability for cyanobacterial growth. Floating wetlands are low-cost, visually appealing, and can be integrated into urban and rural water bodies for both remediation and aesthetic enhancement.

Their efficacy can be enhanced by selecting plant species with high nutrient uptake rates and by combining them with other treatment methods (Breil et al., 2022). Recent advances in genetic and biotechnological tools also offer exciting prospects for bloom control. These include gene editing techniques to engineer cyanobacteria with reduced toxin production or nutrient uptake efficiency, synthetic biology approaches to introduce nutrient-consuming microbial consortia, and targeted RNA-based methods to silence key bloom-related genes. While still largely in the research and development phase, such interventions hold potential for highly specific and environmentally sensitive control of cyanobacterial blooms. However, ethical considerations, ecological risks, and regulatory hurdles must be thoroughly evaluated before deployment (Anabtawi et al., 2024). A comparative overview of key bloom management strategies, their mechanisms, effectiveness, and limitations is presented in Table 4.2. Collectively, these strategies underscore the importance of a multi-faceted, integrated approach to bloom management. No single method is universally effective, and the success of mitigation efforts often depends on system characteristics, historical nutrient loads, climatic variability, and socio-political context. Combining nutrient reduction with biological and technological interventions offers the best chance of achieving long-term control of cyanobacterial blooms and restoring ecological balance in affected water bodies.

4.10 Future Perspectives

The future of cyanobacterial bloom management lies in adopting predictive, integrated, and ecosystem-based approaches. Nutrient modeling in connection with remote sensing technologies provides strong instruments for early detection and prediction of blooms. By simulating nutrient monitors and tracking surface markers, such as chlorophyll and turbidity, managers are in a position to makeup, predict bloom events, and take timeous interventions. Bloom dynamics will become more problematic with the additional rise in water temperatures caused by climate change, altered rainfall patterns, and prolonged stratification all of which favor cyanobacterial dominance. It is essential for understanding these interactions for development of resilient management strategies. Ecosystem management, which focuses on watershed level actions, restoration, and stakeholder involvement, offers an integrated model for long-term bloom control. Against this backdrop, stoichiometry can inform the refinement strategies including the manipulation of industries to embrace precision or precision-like strategies like adjusting Nuthin:Phosituation where N:P ratios or limiting trace nutrients to inhibit bloom-forming species and to provide boron and silver during storage for both asparagus crops, said to exist in ecological balance. At the end, the integration of scientific tools with flexible and system-wide features takes place. Cyanobacterial bloom management in a changing world will rely on approaches.

Table 4.2 Management and mitigation strategies for cyanobacterial blooms

Strategy	Mechanism	Effectiveness	Limitations
Nutrient reduction (N & P)	Limits primary drivers of eutrophication by reducing external nutrient inputs	High when sustained; foundational for long-term control	Requires long-term commitment, infrastructure, and stakeholder cooperation
Phoslock® (Lanthanum-modified Bentonite)	Binds free phosphate, forming insoluble complexes to reduce bioavailable phosphorus	Effective in reducing internal phosphorus loading; long-lasting effects	May impact non-target organisms; effectiveness varies with pH and anoxic conditions
Floating treatment wetlands (FTWs)	Utilize aquatic plants to uptake nutrients and provide habitat for microbial nutrient transformation	Promising for nutrient removal and algal bloom mitigation	Variable performance; effectiveness depends on plant species and environmental conditions
Biomanipulation	Alters food web dynamics by managing fish and zooplankton populations to control phytoplankton	Can reduce cyanobacterial biomass; enhances water quality	Outcomes are system-specific; may be less effective in systems with complex food webs
Algaecides (e.g., copper sulfate, hydrogen peroxide)	Chemically suppress cyanobacterial populations	Rapid bloom suppression; cost-effective	Potential non-target effects; risk of toxin release upon cell lysis; short-lived effects
Genetic and biotechnological tools	Employ gene editing and synthetic biology to reduce toxin production or nutrient uptake in cyanobacteria	Potential for targeted, environmentally sensitive control	Still in research phase; ethical and ecological considerations; regulatory hurdles

4.11 Conclusion

Cyanobacterial blooms are a growing global threat to aquatic ecosystems, driven primarily by nutrient enrichment, stoichiometric imbalances, and climate change. This chapter has explored how excess nitrogen and phosphorus, along with key micronutrients, fuel bloom formation and how deviations from the Redfield ratio influence species dominance and toxicity. Cyanobacteria's adaptive strategies such as nitrogen fixation, phosphorus scavenging, and internal nutrient recycling further reinforce their resilience and ecological impact. These blooms not only disrupt food webs and oxygen dynamics but also pose serious risks to human and animal health through toxin production and long-term ecosystem alteration. Effective bloom management requires more than just nutrient reduction it demands a deep understanding of the stoichiometric and ecological processes underpinning bloom dynamics. Integrating tools like nutrient modeling, remote sensing, and ecosystem-based restoration will be essential for accurate prediction and sustainable control. Moving forward, interdisciplinary collaboration among ecologists, chemists, hydrologists, policymakers, and local communities is vital to protect water quality, safeguard biodiversity, and build resilient aquatic ecosystems.

References

Amin, N., Jaiswal, M., & Kannaujiya, V. K. (2024). Effects of temperature on morphology, physiology, and metabolic profile of diazotrophic cyanobacteria inhabiting diverse habitats. *Plant Physiology and Biochemistry, 216*, Article 109186.

Anabtawi, H. M., Lee, W. H., Al-Anazi, A., et al. (2024). Advancements in biological strategies for controlling harmful algal blooms (HABs). *Water, 16*, 224.

Bannon, C., Rapp, I., & Bertrand, E. M. (2022). Community interaction co-limitation: Nutrient limitation in a marine microbial community context. *Frontiers in Microbiology, 13*, Article 846890.

Bargu, S., Hiatt, M., Maiti, K., et al. (2023). The future of cyanobacteria toxicity in estuaries undergoing pulsed nutrient inputs: A case study from Coastal Louisiana. *Water, 15*, 3816.

Bhardwaj, A., Singh, P., Gupta, N., Bhattacharjee, S., Srivastava, A., Parida, A., & Mishra, A. K. (2024). Cyanobacteria: a key player in nutrient cycling. In *Cyanobacteria* (pp. 579–596). Elsevier.

Breil, P., Pons, M. N., Armani, G., Amer, R., Pienaar, H., Oberholster, P., & Namour, P. (2022). Natural-based solutions for bioremediation in water environment. In N. S. El-Gendy (Ed.), *Sustainable solutions for environmental pollution: Air, water and soil reclamation* (1st edn, pp.1–93). Wiley.

Devlin, M., & Brodie, J. (2023). Nutrients and eutrophication. In A. Reichelt-Brushett (Ed.), *Marine pollution—monitoring, management and mitigation* (pp. 75–100). Springer Nature Switzerland.

Donghao, W., Fangfei, C., Jiaxin, H., et al. (2022). The declining cyanobacterial blooms in Lake Taihu (China) in 2021: The interplay of nutrients and meteorological determinants. *Ecological Indicators, 145*, Article 109590.

Erratt, K. J., & Freeman, E. C. (2024). Cyanobacteria in the Anthropocene: Synanthropism forged in an era of global change. *Water Environment Research, 96*, Article e11066. https://doi.org/10.1002/wer.11066

Feng, L., Wang, Y., Hou, X., et al. (2024). Harmful algal blooms in inland waters. *Nature Reviews Earth & Environment, 5*, 631–644.

Haque, S. E. (2021). How effective are existing phosphorus management strategies in mitigating surface water quality problems in the US? *Sustainability, 13*, 6565.

Inomura, K., Deutsch, C., Jahn, O., et al. (2022). Global patterns in marine organic matter stoichiometry driven by phytoplankton ecophysiology. *Nature Geoscience, 15*, 1034–1040.

Khazaei, B., Bravo, H. R., & Hamidi, S. A. (2023). Interactions between sediment processes and ecosystem responses in the Green Bay of Lake Michigan. *Wires Water, 10*, Article e1668. https://doi.org/10.1002/wat2.1668

Khoshru, B., Mitra, D., Nosratabad, A. F., et al. (2023). Enhancing manganese availability for plants through microbial potential: A sustainable approach for improving soil health and food security. *Bacteria, 2*, 129–141.

King, W. M., Curless, S. E., & Hood, J. M. (2022). River phosphorus cycling during high flow may constrain Lake Erie cyanobacteria blooms. *Water Research, 222*, Article 118845.

Klip, H. C. (2023). *Balance of power: Dissolved nitrogen-to-phosphorus ratios and phytoplankton growth rate determine the balance between bottom-up and top-down processes in planktonic food webs.* PhD Thesis, Universität Bremen.

Kramer, B. J., Jankowiak, J. G., Nanjappa, D., et al. (2022). Nitrogen and phosphorus significantly alter growth, nitrogen fixation, anatoxin-a content, and the transcriptome of the bloom-forming cyanobacterium. *Dolichospermum. Frontiers in Microbiology, 13*, Article 955032.

Kuczyńska-Kippen, N., Kozak, A., & Celewicz, S. (2024). Cyanobacteria respond to trophic status in shallow aquatic ecosystems. *Science of the Total Environment, 949*, Article 174932.

Kumari, N., Mishra, S., Häder, D.-P., & Sinha, R. P. (2022). Cyanobacterial blooms and their implications in the changing environment. *Advances in Environmental and Engineering Research, 3*, 1–41.

Lakshmikandan, M., Li, M., & Pan, B. (2024). Cyanobacterial blooms in environmental water: Causes and solutions. *Current Pollution Reports, 10*, 606–627.

Luong, H. A., Rohlfs, A. M., Facey, J. A., Colville, A., Davie, A. W., Pera, J. B., & Mitrovic, S. M. (2025). Influence of macronutrient and iron enrichment on phytoplankton productivity and community dynamics: An in situ microcosm study in a drinking water supply reservoir. *Inland Waters*, (just-accepted), 1–30. https://doi.org/10.1080/20442041.2025.2475662

Madany, P., Olsen, D., Makled, S. O., et al. (2025). Innovative multiplex qPCR method for rapid and reliable detection of microcystin-producing genes during harmful algal blooms: Insights from Utah Reservoirs. *Water Research, 277*, Article 123322.

Magana-Arachchi, D. N., Gunathilaka, H. M. S. A. T. (2025). Cyanotoxins: Industrial potential. In: M. N. V. Prasad, S. D. Tetali, C. Bennetau-Pelissero (Eds.), *Biotoxins* (pp. 89–120). Springer Nature Switzerland.

Manic, D. C., Redil, R. D., & Rodriguez, I. B. (2024). Trace metals in phytoplankton: Requirements, function, and composition in harmful algal blooms. *Sustainability, 16*, 4876.

Mioni, C., Kudela, R., Baxa, D., et al. (2011). Harmful cyanobacteria blooms and their toxins in Clear Lake and the Sacramento-San Joaquin Delta (California). *Delta (California), 10*, 058–150.

Molot, L. A., Schiff, S. L., Venkiteswaran, J. J., et al. (2021). Low sediment redox promotes cyanobacteria blooms across a trophic range: Implications for management. *Lake and Reservoir Management, 37*, 120–142. https://doi.org/10.1080/10402381.2020.1854400

Moreira, C., Vasconcelos, V., & Antunes, A. (2022). Cyanobacterial blooms: Current knowledge and new perspectives. *Earth, 3*, 127–135.

Nõges, T., McCarthy, M. J., Nõges, P., & Kumar, R. (2024). Nitrogen cycle and its effect on phytoplankton community structure. In *Role of green chemistry in ecosystem restoration to achieve environmental sustainability* (pp. 315–324). Elsevier.

Patiño, R., Christensen, V. G., Graham, J. L., et al. (2023). Toxic algae in inland waters of the conterminous United States—A review and synthesis. *Water, 15*, 2808.

Qiang, H. U., Yan, C., Rui, X. I. A., et al. (2024). Weakened hydrological oscillation period increased the frequency of river algal blooms. *Water Research, 255*, Article 121496.

Rizhinashvili, A. L. (2022). An outline of the theory of the functioning of aquatic ecosystems: Nutrient limitation. *Biology Bulletin Reviews, 12*, 596–608.

Rock, L. A., & Collins, S. M. (2024). A broad-scale look at nutrient limitation and a shift toward co-limitation in United States Lakes. *Environmental Science and Technology, 58*, 11482–11491. https://doi.org/10.1021/acs.est.4c03135

Summers, E. J., & Ryder, J. L. (2023). A critical review of operational strategies for the management of harmful algal blooms (HABs) in inland reservoirs. *Journal of Environmental Management, 330*, Article 117141.

Tarafdar, L., Kim, J. Y., Srichandan, S., et al. (2021). Responses of phytoplankton community structure and association to variability in environmental drivers in a tropical coastal lagoon. *Science of the Total Environment, 783*, Article 146873.

Villalobos, T., Suárez-Isla, B., & Garcia, C. (2025). Health and environmental impacts of cyanobacteria and cyanotoxins from freshwater to seawater. *Toxins, 17*, 126.

Wang, Y., Xu, C., Lin, Q., et al. (2024). Modeling of algal blooms: Advances, applications and prospects. *Ocean & Coastal Management, 255*, Article 107250.

Wang, Z., Akbar, S., Sun, Y., et al. (2021). Cyanobacterial dominance and succession: Factors, mechanisms, predictions, and managements. *Journal of Environmental Management, 297*, Article 113281.

Wei, N., Chen, A., Guo, X., et al. (2023). Changes in nitrogen metabolism of phosphorus-starved bloom-forming cyanobacterium Microcystis aeruginosa: Implications for nutrient management. *Science of the Total Environment, 903*, Article 166832.

Welbara, D. E., Gebre-Meskel, D. K., & Hailu, T. F. (2024). Morpho-functional traits of phytoplankton functional groups: A review. *Biologia, 79*, 1983–1998.

Xiao, M., Burford, M. A., Wood, S. A., Aubriot, L., Ibelings, B. W., Prentice, M. J., Galvanese, E. F., Harris, T. D., & Hamilton, D. P. (2022). Schindler's legacy: from eutrophic lakes to the phosphorus utilization strategies of cyanobacteria. *FEMS Microbiology Reviews, 46*(6), fuac029.

Xu, C., Wang, X., Zhang, F., Lao, Y., Liu, J., & Du, J. (2024). Potential linkages between submarine groundwater (fresh and saline) nutrient inputs and eutrophication in a coastal aquaculture bay. *Journal of Geophysical Research: Oceans, 129*(10), e2024JC021501. https://doi.org/10.1029/2024JC021501

Xu, H., McCarthy, M. J., Paerl, H. W., et al. (2021). Contributions of external nutrient loading and internal cycling to cyanobacterial bloom dynamics in Lake Taihu, China: Implications for nutrient management. *Limnology & Oceanography, 66*, 1492–1509. https://doi.org/10.1002/lno.11700

Yadav, M., George, N., & Dwibedi, V. (2024). *Trace element pollution in the aquatic environment: impacts on aquatic macrophytes*. Springer.

Yadav, P., Singh, R. P., Rana, S., et al. (2022). Mechanisms of stress tolerance in cyanobacteria under extreme conditions. *Stresses, 2*, 531–549.

Yan, Z., Kamanmalek, S., Alamdari, N., & Nikoo, M. R. (2024). Comprehensive insights into harmful algal blooms: A review of chemical, physical, biological, and climatological influencers with predictive modeling approaches. *Journal of Environmental Engineering, 150*, 03124002. https://doi.org/10.1061/JOEEDU.EEENG-7549

Chapter 5
Carbon Cycling and Sequestration in Cyanobacterial-Dominated Ecosystems

Marzia Batool Laila, Nasir Assad, Tooba, Barkat Ullah, Muhammad Nauman Khan, Maryam Bibi, Shah Fahad, and Sezai Ercişli

Abstract Cyanobacterial-dominated ecosystems play a crucial role in global carbon cycling and sequestration, influencing both atmospheric CO_2 levels and aquatic carbon dynamics. This chapter explores the mechanisms by which cyanobacteria contribute to carbon fixation, organic matter production, and long-term carbon storage. Cyanobacteria act as primary producers in freshwater and marine environments, utilizing photosynthesis to convert CO_2 into biomass, which fuels aquatic food webs and supports microbial decomposition pathways. Their ability to fix inorganic carbon through various carbon-concentrating mechanisms (CCMs) enhances their competitive advantage in nutrient-rich and low-CO_2 environments. The interplay between carbon, nitrogen, and phosphorus cycles in cyanobacterial blooms is examined, highlighting the role of organic carbon production in shaping microbial community dynamics and sediment carbon deposition. Additionally, the impact of harmful cyanobacterial blooms (HCBs) on carbon fluxes is discussed, particularly in relation to dissolved organic carbon (DOC) release, oxygen depletion, and greenhouse gas

M. B. Laila (✉) · N. Assad · Tooba
Institute of Chemistry, University of Sargodha, Sargodha, Pakistan
e-mail: marziabatoollaila@gmail.com

B. Ullah · M. N. Khan
Department of Botany, Islamia College Peshawar, Peshawar, Pakistan

M. N. Khan
Biology Laboratory, University Public School, University of Peshawar, Peshawar, Pakistan

M. Bibi
Department of Botany, University of Malakand, Chakdara, Lower Dir, Khyber Pakhtunkhwa, Pakistan

S. Fahad
Department of Agronomy, Abdul Wali Khan University, Mardan, Pakistan

S. Ercişli
Department of Horticulture, Agricultural Faculty, Ataturk University, Erzurum, Turkey

S. Fahad et al. (eds.), *Cyanobacterial Blooms: Ecology, Evolution and Biogeochemical Impacts*, https://doi.org/10.1007/978-3-032-06042-6_5

(GHG) emissions such as methane (CH_4) and nitrous oxide (N_2O). The sequestration potential of cyanobacterial-derived organic carbon in sediments and its contribution to long-term carbon storage is also addressed. Understanding the processes governing carbon cycling in cyanobacterial ecosystems is critical for predicting their response to climate change and anthropogenic disturbances. This chapter evaluates strategies for enhancing carbon sequestration in aquatic ecosystems, including bloom management, biogeochemical modeling, and nature-based solutions. By integrating ecological and biogeochemical perspectives, this chapter provides insights into the role of cyanobacteria in regulating carbon fluxes and maintaining ecosystem stability.

5.1 Introduction

The carbon cycle around the world is largely managed by aquatic environments which both release and store carbon. Among all the organisms that are involved in these processes, cyanobacteria are considered especially important (Tréguer et al., 2018). Old photosynthetic prokaryotes which are found today helped shape Earth's atmosphere and assisted the start of life that needs oxygen. Cyanobacteria can be found in modern water systems and they use photosynthesis to change carbon dioxide into food. Biological activity of diatoms influences both marine ecosystems and carbon storage in the deep (Armbrust, 2009; Flombaum et al., 2013).

Cyanobacteria are efficient at taking in inorganic carbon because of their features. The presence of carbon-concentrating mechanisms (CCMs) enables cyanobacteria to harvest carbon dioxide from the environment and move it close to the area where it will be fixed into organic compounds. CCMs assist RuBisCO by concentrating CO_2 in the enzyme's vicinity, helping cyanobacteria to outperform most other phytoplankton, mainly in waters with excess nutrients and stratified layers (Wilczek et al., 2011). Thus, cyanobacteria are frequently the most common organisms in water systems whenever environmental conditions favor their growth, especially when both temperature and nutrients are high (Price et al., 2008).

Cyanobacterial growth and accumulation during blooms are responsible for most of the primary production and organic matter on such bodies of water. Fixed carbon gets used by living organisms to make carbohydrates, proteins, lipids, and cellular gear and most can be later picked up by higher organisms or microbes as food. A main part of the organic matter is recycled deep within the ocean, but a smaller percentage sinks and adds to the sediment at the seafloor. Putting carbon into sediment is vital for carbon sequestration, because it removes CO_2 from floating in the ocean or the atmosphere and stores it over long geological periods. How this carbon ends up depends on its rate of deposition, the chemical nature of the organic material, and the sediments' oxygen level (Yin et al., 2025).

Carbon cycling in cyanobacterial environments is strongly related to the nitrogen and phosphorus biogeochemical cycles. The presence and ratio of essential minerals influence the growth of cyanobacteria, the types found, and how thick their blooms will be. Many cyanobacteria flourish in nutrient-filled waters, allowing them to overshadow other phytoplankton and trigger widespread blooms that influence the properties of the water source (Taranu et al., 2015). Certain cyanobacteria are capable of fixing nitrogen from the air, so they do well when nitrogen is scarce. Because they can both capture carbon and nitrogen, their environmentally friendly nature lets them survive in a variety of habitats. Cyanobacteria are important, because they affect the amount of food available in aquatic ecosystems and affect both the short-term and long-term storing of carbon in the water (Kramer et al., 2022).

An important environmental and chemical event linked to cyanobacteria is the presence of harmful blooms called harmful cyanobacterial blooms (HCBs). These blooms have a lot of toxin-producing species, harming water, local life, and people's health. The release of dissolved organic carbon (DOC) into the water can be high when runoff from HCBs takes place. DOC is easier for heterotrophic bacteria to process than comparative particulate organic carbon, so it can start a microbial respiration process and leave little oxygen for aquatic life in areas of limited mixing. Because of this lack of oxygen, aquatic ecosystems are damaged, fish and invertebrates are affected, and redox conditions in the sediment in charge of recycling nutrients and storing carbon become different (Bailey et al., 2020; Dell'Aquila, 2024).

Another important problem caused by less oxygen is that there are more methane and nitrous oxide produced. As the water lacks oxygen, different microbes become important and cause CH_4 and N_2O to be released from the bottom and water. Such gases are much more powerful than CO_2 when it comes to climate change and they also encourage further climate change ourselves (Sterinbach, 2024; Tang et al., 2023).

Even so, these benefits might not last long if carbon dioxide and oxygen are consumed more quickly and microbes break it down faster, under warming temperatures. Furthermore, floods and droughts can make sediments unstable and cause carbon that is not currently involved in any process to move which makes it hard to forecast how carbon will flow in aquatic settings (Japri et al., 2024).

Understanding how both carbon fixation and loss happen in cyanobacterial environments requires knowledge of the interactions among species, local water regimes, and the pressures created by people. Strategies for effective management should pay special attention to the issue of nutrient loading. Applying suitable farming techniques, treating wastewater and looking after watersheds help lower algal blooms and maintain the health of aquatic species. When wetland vegetation is restored and water regimes are improved, ecosystems can remove even more carbon from the atmosphere and cope better with changes in climate.

They are now using enhanced biogeochemical models to guess how carbon will behave in cyanobacterial systems when affected by changes in the environment. The models use information about production, respiration, sedimentation, and outgassing of greenhouse gases to simulate carbon flow and gauge how aquatic ecosystems may act as carbon sinks (Chapra et al., 2017).

The use of climate predictions, nutrient cycles, and microbe processes in models allows them to suggest and shape important decisions and strategies for tackling climate change. Nature-based projects that take advantage of the carbon capture of cyanobacteria and similar settings, known as blue carbon initiatives, can improve carbon storage as well as help the environment and its living creatures (Hilmi et al., 2021; Lin, 2017).

Cyanobacteria's ability to take carbon out of the atmosphere and also stress the environment makes their effects on lakes, rivers, and oceans more complicated. On the one hand, through photosynthesis and producing biomass, they help take large amounts of carbon from the air and place it into the ground. In contrast, their danger in patterning oxygen reduction and greenhouse gases, mainly occurring during harmful blooms, calls for close and well-suited investigations of their role in the environment. Ongoing observation and multiple studies are necessary to determine the points at which cyanobacteria's impact on carbon cycling and general ecosystems turns negative (Fazi et al., 2021).

Cyanobacteria play a key role in the process of carbon change in water bodies. With their ancient origin, ability to convert sunlight into energy, and responsive metabolism, cyanobacteria have shaped the atmosphere and continue to influence carbon cycles today. Since they are involved in making food for life, recycling nutrients, and getting rid of waste by sedimentation, cyanobacteria help manage CO_2 in the air and keep carbon stores safe over time (Hilmi et al., 2021). Still, making them contribute to toxic blooms, nutrient discharge, and greenhouse gases leads to serious challenges requiring joint efforts of research and management. Both the environment and climate can benefit from learning to manage cyanobacterial-dominated habitats.

5.2 Role of Cyanobacteria in Carbon Fixation

Cyanobacteria which are both ancient and environmentally successful have contributed greatly to Earth's ecology and air composition. Carbon fixation by these microbes creates the base of the aquatic food web and helps move nutrients by turning inorganic carbon into organic matter via oxygenic photosynthesis. It all starts with absorbing sunlight and turning that energy into food for the plant which allows carbon dioxide to be absorbed into the plant's body (Mitsch et al., 2013). Cyanobacteria rely on photosynthesis to both live and thrive and their large populations frequently in freshwater, estuarine, and marine habitats tend to alter how carbon moves through the environment.

The structure and function of photosynthesis in cyanobacteria are closely linked to what is found in land plants and algae, but the mechanism is optimized for living in aquatic habitats. The light-dependent reactions of cyanobacteria photosynthesis require the simultaneous action of Photosystem I and Photosystem II. These photosystems are part of the thylakoid membranes in cyanobacteria and use chlorophyll-a plus extra pigments, phycocyanin and allophycocyanin, that help them collect various ranges of light. Once light energy is taken in, electrons become excited and are sent through several carriers in the electron transport chain, meaning ATP and NADPH are produced. These molecules are now converted in the Calvin–Benson cycle to reduce carbon dioxide into vital carbohydrates and other substances that the cell needs.

Being packed with a well-organized photosynthetic system, cyanobacteria are very effective at using both light and carbon. In numerous species, thylakoid membranes are tightly stacked so they can capture more light and change it into energy. Moreover, cyanobacteria are well adapted to conditions without much light because they possess phycobilisomes which efficiently collect and process light even when the water is turbid. Due to these physical adaptations, they are very capable of growing in different areas like oligotrophic oceans as well as eutrophic lakes and ponds.

Although they are good at making food from light, cyanobacteria must overcome a biochemical difficulty related to the enzyme RuBisCO. Because RuBisCO binds both carbon dioxide and oxygen, its oxygenase activity causes photorespiration that decreases how efficient photosynthesis is. As oxygen and carbon dioxide move more slowly through water, this problem becomes worse and the carbon dioxide levels can vary a lot (Liu et al., 2021). Cyanobacteria have come up with a set of special traits that, when combined, form what is known as CCMs. Carbon-concentrating mechanisms (CCMs) increase the internal CO_2 concentration around RuBisCO, thereby reducing oxygen-inhibiting reactions. A complex set of transportation, inorganic carbon uptake and tiny carboxysome compartments form the cyanobacterial CCM. Enclosing RuBisCO and carbonic anhydrase within these punctate structures leads to the conversion of bicarbonate to CO_2, ensuring RuBisCO only interacts with preferred CO_2 and minimizes contact with oxygen. Grouping steps spatially in such a way makes the whole process more effective at carbon fixation, mainly when CO_2 is in limited supply (Cushman, 2001).

The main way inorganic carbon spreads in the ocean is first by its forms: CO_2 and bicarbonate (HCO_3^-). CYAs have various kinds of bicarbonate transporters, all with various energy needs, so they can adapt to a wide range of carbon levels in the outside environment. Under conditions where CO_2 levels are not sufficient, such transporters are critical for efficiently finding and taking up the available carbon in the environment. When in the cell, bicarbonate can't pass through the membrane and builds up in the cytoplasm. As carbonic anhydrase operates on bicarbonate in carboxysomes, it changes it into CO_2 and helps construct an internal store for carbon that is instantly used by the Calvin cycle (Gupta et al., 2020; Liu et al., 2021) (Fig. 5.1).

The ecological features of this strategy are many. Because of this, cyanobacteria are able to survive in waters with little CO_2—which is common in both productive and alkaline aquatic environments (Kim et al., 2024). Due to their efficient CCMs, cyanobacteria usually take over different regions in aquatic ecosystems when conditions require it. In addition, CCMs can adjust their use of energy depending on important environmental sensors such as light, pH, and the amount of carbon available. By adjusting their carbon-fixing systems to the environment, cyanobacteria can maintain productivity and continue to do well in different surroundings.

This advantage of CCMs is easily noticeable when biofilms are formed in eutrophic regions rich in phosphorus and nitrogen. In these circumstances, cyanobacteria grow very quickly and cause the water surface to become thick with blooms. These blooms which usually form a green–blue layer on water are a sign of fast carbon fixing by high levels of photosynthesis. When blooms occur, cyanobacteria

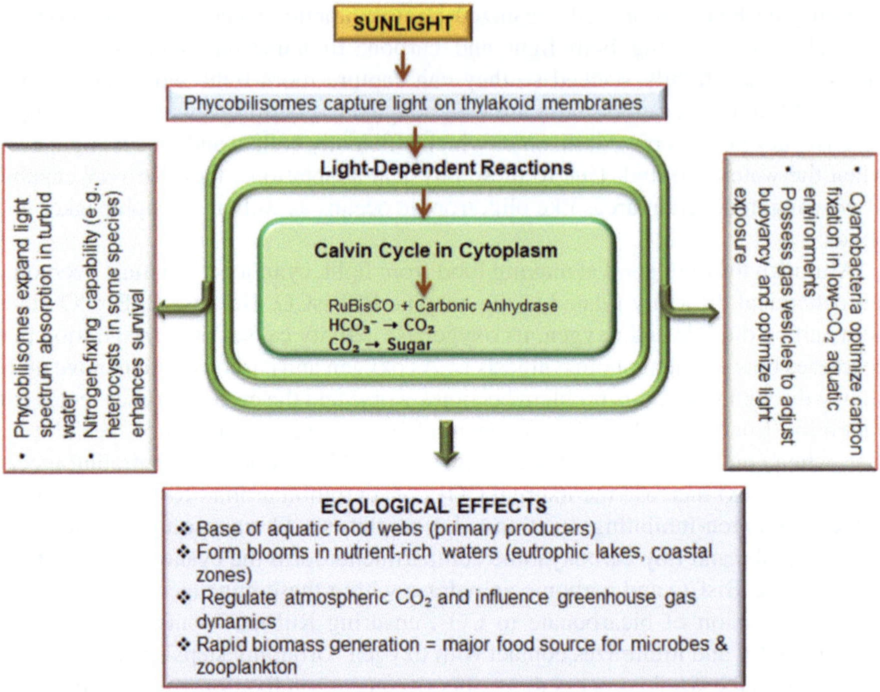

Fig. 5.1 Diagram showing how cyanobacteria use light-driven photosynthesis and carbon-concentrating mechanisms (CCMs) to fix CO_2 efficiently, form blooms, and influence aquatic ecosystems and global carbon cycles

are important in making new primary plant materials, taking in carbon and growing rapid amounts of biomass. This biomotion, in turn, provides food for microbes and zooplankton which matter greatly for the organization and running of aquatic food webs. In addition, cyanobacteria form blooms because of both their photosynthetic ability and their metabolic variation, as well as their interactions with other organisms. Several bloom-forming cyanobacteria use their gas vesicles to move vertically, better exposing themselves to both light and nutrients in the water. Light quality and intensity can influence the color in their pigments which allows them to continue promoting energy collection under more types of light. Also, some organisms have the ability to convert nitrogen from air into a form that allows them to sustain their development even when nitrate and ammonium are absent. Together with their functional CCMs, these adaptations enable cyanobacteria to lead in places that are always changing and rich with nutrients.

On a global scale, cyanobacteria shape carbon fixation in many ways. What they do together plays a part in managing regional greenhouse gases and regulating CO_2 levels above the Earth's surface.

Occurring in huge inland lakes and reservoirs, as well as along coastlines, blooms of cyanobacteria can lower CO_2 in the air for a while by turning it into matter that is buried in the sediment. Because bottom waters are oxygen poor, cyanobacteria store the carbon they create underwater, while other forms of life keep the carbon moving in the typical cycle (Ma & Wang, 2021).

Even so, this contribution involves some issues. Carbon movement into the soil by cyanobacteria can be disturbed by shifts in climate, pH, stream currents, and pollution from human activities. Besides, climate change means there are more risks for blooms, as unusual weather and increased heat can influence the blooms and how carbon is processed in the water. The increase in temperature often makes organisms require more energy, Phoenix Technical School, changes the balance of photosynthesis and respiration, and promotes the release of carbon in the water. In addition, with temperatures rising, greater stratification could keep surface blooms going for a longer period, limiting how nutrients and carbon are spread to the depths (Liu et al., 2021).

We can learn about carbon fixation by cyanobacteria by reviewing how they are organized in a cell and how they interact with their environment. It examines the results of modest changes such as CCMs, nitrogen fixers, and better light catching, on both living communities and whole ecosystems. Besides, knowing how changes in the environment impact cyanobacteria and how carbon is passed around the ecosystem over time is very important. So, researchers are able to predict how cyanobacteria will likely increase in the future and develop ways to handle their effects (Falkowski et al., 2008).

With efficient photosynthesis and unique carbon acquisition methods, cyanobacteria play a vital role in carbon fixation in aquatic environments. Microbiologists found it important that they created a process for capturing carbon, as it allowed them to work in areas without much carbon. Combined with an abundance of nutrients, staying light in water and turning sunlight into energy, cyanobacteria are very important for shaping sea life, carbon transfers, and stable environments. As a result, microbial carbon fixation is crucial for microbes, helps affect ecosystem activities, and contributes to the global balance of carbon. Table 5.1 summarizes the main factors affecting cyanobacterial carbon fixation, highlighting their physiological roles and impact on carbon cycling.

5.3 Organic Matter (OM) Production and Biomass Creation

Cyanobacteria contribute greatly to global carbon balancing as they transform carbon dioxide into organic matter important for aquatic life and influence carbon levels in the water. By means of oxygenic photosynthesis, cyanobacteria convert inorganic carbon from the atmosphere or from water into useful compounds for their physiology. CO_2 is captured at the start of the conversion when sunlight drives the Calvin–Benson cycle. With fixed carbon, cyanobacteria use biosynthesis to build carbohydrates, proteins, lipids, and nucleic acids, important parts of their body. Because

Table 5.1 Key parameters influencing cyanobacterial carbon fixation: mechanisms, environmental factors, and their roles in global carbon cycling

Parameter	Description	Impact on carbon fixation
Photosynthetic pigments	Chlorophyll-a, phycocyanin, allophycocyanin	Enhances light absorption in varied water conditions
Photosystems	Photosystems I and II located in thylakoid membranes	Drive light reactions, produce ATP and NADPH for CO_2 fixation
RuBisCO enzyme	Catalyzes the Calvin cycle; binds CO_2 and O_2	Inefficient due to oxygenase activity; improved via CCMs
Carbon-concentrating mechanisms (CCMs)	Systems including bicarbonate transporters, carboxysomes, carbonic anhydrase	Elevate local CO_2 for RuBisCO, reduce photorespiration
Environmental adaptability	Use of phycobilisomes, gas vesicles, buoyancy control	Thrive in both light-limited and nutrient-rich waters
Bicarbonate transporters	Multiple types with varying energy needs	Allow adaptation to low ambient CO_2 levels
Carbon sequestration	Fixation of CO_2 into biomass, sedimentation of dead cells	Long-term removal of CO_2 from atmosphere to sediments
Temperature sensitivity	Warming increases stratification and bloom duration	Enhances productivity but may destabilize carbon cycling
Nutrient interactions (N, P)	Nutrient levels influence bloom formation and fixation rates	Diazotrophic species fix N_2; high P boosts blooms
Global carbon role	Influence regional CO_2 balance via massive blooms and sequestration	Cyanobacteria act as carbon sinks or sources depending on bloom dynamics

they are primary producers, algae spark the movement of energy and nutrients in aquatic environments and feed many types of consumers, from microorganisms to fish. Cyanobacteria add both a large amount and quality of biomass to aquatic habitats. In certain lakes, rivers, and oceans, cyanobacteria can make up a big part of the phytoplankton population during blooms. All this biomass stores up carbon briefly, but it also supplies the local living things with fuel and organic matter. Both dissolved and solid organic materials are released by cyanobacteria at the end of their life cycles. Both sugars, amino acids, and organic acids belong to dissolved organic matter, whereas intact cells and their larger debris are classified as particulate organic matter. The organic substrates support the energy needs of heterotrophic microorganisms which turn the material into nutrients, returning it all to the system and producing CO_2 through breathing.

Cyanobacteria are important for controlling the structure and functioning of food webs in water environments. Because photoautotrophs use sunlight to produce their energy, they remain the foundation for energy transfer in ecosystems. Rotifers, copepods, and cladocerans in zooplankton species consume cyanobacteria, but how easily they can eat them and how much nutrition they get from cyanobacteria can depend on both species and the conditions during the bloom (Franklin, 2021). Certain cyanobacteria form chemicals that make them less palatable or even harmful to those eating them, sometimes disturbing the food web of the community. Even so, when they grow in a healthy environment or are part of a balanced diet, cyanobacteria help main food sources to increase in number and continue to move via food webs.

The flow of organic matter from cyanobacteria into microbial and detrital pathways is equally important. When cyanobacterial cells die or are lysed by viral infection, physical stress, or senescence, their biomass becomes available to a diverse community of bacteria and fungi. These decomposers break down complex organic molecules into simpler forms, a process that releases inorganic nutrients such as ammonium, phosphate, and CO_2. This mineralization of organic matter replenishes the nutrient pool, supporting new primary production and sustaining the microbial loop—a component of the ecosystem that recycles energy and matter at the microbial scale. The microbial loop enhances the efficiency of nutrient use in aquatic systems and contributes to the resilience and productivity of the ecosystem as a whole (Pina et al., 2011).

In the course of decomposition, some organic matter resists rapid breakdown and is transported to the bottom of water bodies through sedimentation. When cyanobacterial blooms settle on the floor of water bodies, organic matter is added to the sediment, most often in areas with light water movements and many blooms each year. The next step is up to a group of conditions in the environment, including temperature, oxygen, chemistry, and what type of sediment is involved. If conditions do not allow oxygen, decomposition greatly slows which helps organic carbon stay intact and last a long time. Eventually, buried carbon adds to the total sedimentary carbon that removes some CO_2 gas from the planet's cycle, thus reducing the level in the air (Kosten et al., 2012).

Long-term carbon sequestration by cyanobacterial-derived organic matter is affected by what biochemicals these organisms contain. These useful microbes produce polymers called peptidoglycan and exopolysaccharides which find it hard for microbes to break down. Such compounds, in addition to other resistant cell parts, tend to survive longer in sediments than in other environments. When detritus from blue-green algae is combined with certain mineral particles, it gets both physically protected and made chemically steady, reducing its chance of being broken down. The combination of biological production and geochemical stabilization makes it possible to estimate the amount of carbon stored and how well-preserved organic matter is in aquatic ecosystems (Lehmann & Kleber, 2015).

In areas with plenty of nutrients for cyanobacteria, the increase in organic matter tends to happen suddenly and only for a short period. As a result of lots of blooms, tons of biomass drift into the water and then into the sediment, encouraging the growth of microbes and the decomposition of nutrients. Although it supports the growth of many microorganisms and helps nutrients come back into circulation, it can also make the water at the bottom of water bodies hypoxic or anoxic, mainly in stratified systems. Oxygen shortages in the environment impair aerobic respiration and make anaerobic processes like denitrification, sulfate reduction, and methanogenesis important which can lead to both prettier and greenehouse gas releases.

Methane production is especially strong in areas lacking oxygen and adds to the overall carbon emission from aquatic systems into the atmosphere (Kosten et al., 2012).

In addition, cyanobacterial organic matter is important in places altered by humans, for example, reservoirs, aquaculture systems, and facilities that treat wastewater. Cyanobacteria can become major players in managing primary production which may have consequences for the amount and quality of organic material going to treatment or food production. To improve system results, save the water quality, and boost carbon retrieval, we must understand how biomass is formed and broken down.

Cyanobacteria perform the roles of producers and converters of organic matter in the water. Photosynthesis which transforms CO_2 into biomass in plants is important for the whole food web. How biomass is handled by grazing, microbes, or deposit in the bottom sediment decides the way carbon moves and how well it is kept in the ecosystem. Such processes ensure that cyanobacteria are important parts of aquatic carbon systems, taking part in nutrient cycling, shaping the lives of microbes and helping to remove carbon from the air (Bastviken et al., 2011).

5.4 Carbon-Nutrient Interactions

The movement of carbon in aquatic environments is linked with the supply and changes in essential nutrients, with nitrogen and phosphorus being the main examples. Being the leading primary producers in many waters, cyanobacteria influence the movement of elements by their metabolism, relationships with other organisms, and their tendency to form blooms. Photosynthesis does not happen separately from using nitrogen and phosphorus which are important for the making of proteins, nucleic acids, ATP, and phospholipids that build cell membranes. Changing one nutrient's availability may strongly affect the others which can then create changes in ecosystem productivity and stability.

Cell growth, division, and accumulation in cyanobacteria depend on having a proper supply of carbon, nitrogen, and phosphorus all at once. During photosynthesis, carbon dioxide undergoes change to organic matter, but this process lasts only when cells have adequate nitrogen and phosphorus. While nitrogen is needed for the assembly of amino acids and chlorophyll, phosphorus is used in energy molecules, genetic materials, and signals in the cell. Cyanobacterial biomass often matches the ratio of C:N:P in the local environment because this reflects its true need and the availability of its key elements. A mismatch in nutrient amounts in nature can restrain growth and change who lives in the community, influencing how much carbon is captured and the overall flow of nutrients.

Cyanobacteria have a unique capability to convert nitrogen gas (N_2) from the atmosphere into the compound ammonium (NH_4^+) for use by living organisms. Nitrogenase, the enzyme responsible for biological nitrogen fixation, enables diazotrophic cyanobacteria to grow in environments where combined nitrogen is scarce. This ability influences carbon–nitrogen coupling in cyanobacteria, as the cells can continue fixing carbon and growing without relying on external nitrogen sources—unlike many other organisms. Yet, nitrogen fixation takes a lot of energy, so the process requires a lot of ATP made from photosynthesis which requires more light and phosphorus (Bastviken et al., 2011). Because of the relationships between carbon assimilation and nitrogen fixation, there is a secure metabolic network in which the supply of phosphorus plays a key role.

Phosphorus is not found in our air like nitrogen is, so it mostly reaches aquatic environments through weathering, soil that washes into waterways, and things people do like fertilizer use and sewage treatment. In the water, phosphorus can be either inorganic or organic and orthophosphate is the form that living things can use most easily. Cyanobacteria battle phytoplankton and microbes over limited phosphorus, while the polyphosphate granules inside them allow them to be better fitted during changes in environmental factors. A number of cyanobacteria can make alkaline phosphatase enzymes that help release phosphorus from complex compounds, making it simpler for them to absorb this key nutrient when the environment does not contain much phosphorus.

It is during cyanobacterial blooms that the importance of connecting carbon, nitrogen, and phosphorus cycles is most clear. When light and temperature are right and water contains high levels of phosphorus, it's more likely that blooms will happen. As a result, cyanobacteria reproduce quickly, binding a lot of carbon and taking in a lot of nitrogen and phosphorus while doing so. Because of this sudden surge in primary production, the distribution of nutrients shifts and lowers nutrient availability for the rest of the community. The new community setup can limit other types of phytoplankton and allow cyanobacteria, especially nitrogen-fixing ones, to take over. As a cyanobacterial bloom grows and falls apart, the remnants are broken down, releasing nutrients into the water which may boot productivity up or add more nutrients than the system can handle. Microbial activities on dead cyanobacterial cells place organic carbon back into the dissolved state, while releasing ammonium and phosphate which can be used by the remaining cyanobacteria or other microbes. When corpses decompose fast and take place under low-oxygen situations, they can bring about hypoxic or anoxic zones, mostly at the bottom-water boundary (Bhardwaj et al., 2024). Changes in these conditions include the loss of nitrogen by denitrification and an increase in internal phosphorus loading caused by sediments. Internal recycling can create eutrophic waters and encourage events of algae blooms.

Cyanobacteria may transform nutrients and make them easier or harder for living things to use. As one instance, when bloom-forming cyanobacteria produce EPS, it can tie up dissolved nutrients and activate different uptake processes. Compounds and chemicals in the soil along with toxins and allelopathic agents can keep some organisms from thriving and influence the groups of microbes that break down nutrients. Cyanobacteria also form connections with small amounts of metals, mainly iron and manganese, that govern how phosphorus is attached and the redox processes in the cell. They reveal that cyanobacteria shape the main processes of nutrient cycling and biogeochemistry (Markou et al., 2014).

The rise of nutrients in water from farming, cities, and factories is the main reason for more cyanobacterial blooms, occurring more widely and with greater severity. Among all factors, excess phosphorus in the water is the main reason (Map Egypt Kahramanmaraş-Gaziantep-Dicle Valley Area) for bloom formation in lakes and ponds in freshwater systems. When combined with higher nitrogen levels and climate-related warming, these conditions become very welcoming to cyanobacteria. As a consequence, the natural link among carbon, nitrogen, and phosphorus cycles is more easily disturbed, upsetting ecosystem health, water quality, and the ability to store carbon (O'Neil et al., 2012; Paerl et al., 2011).

In both reservoirs and aquaculture ponds, sometimes changes to the nutrient mix are applied to either decrease the risk of bloom outbreaks or boost productivity. Getting a clear picture of the elemental coupling that cyanobacteria need is necessary to optimize the kind of intervention used. By cutting down on phosphorus, you can moderate algal blooms and keeping the nitrogen-to-phosphorus ratio within proper limits can deter blooms of nitrogen-fixing cyanobacteria. Furthermore, models that cover C:N:P can foresee bloom events and help create nutrient policies.

In short, the relationships among carbon, nitrogen, and phosphorus cycles are closely tied in ecosystems controlled by cyanobacteria. With help from their diverse metabolism, cyanobacteria bridge several nutrients and help determine the direction of biogeochemical changes. They can thrive where many other primary producers are lacking because of their ability to absorb and store carbon, nitrogen, and phosphorus. Because they absorb, utilize, and then return nutrients when their tissues decompose, bloom periods increase the amount of nutrients available to other organisms. This indicates that cyanobacteria are essential to elemental cycles and natural changes, as these processes have a significant impact on ecosystem health, productivity, and adaptability. Table 5.2 outlines the main aspects of carbon–nitrogen–phosphorus interactions in cyanobacteria, emphasizing their influence on carbon fixation and aquatic ecosystem processes.

5.5 Problems Caused by Harmful Cyanobacterial Blooms (HCBs)

Harmful cyanobacterial blooms (HCBs) pose a serious threat to aquatic ecosystems worldwide. They are characterized by a huge explosion in cyanobacteria, many of which produce substances that are toxic to humans and aquatic life, and are mainly brought on by nutrient overload and higher temperatures. In addition to being an annoyance, HCBs are crucial for greenhouse gas levels, ecosystem structure, water quality, and carbon sequestration. Their impacts cover various biogeochemical activities, mainly by increasing dissolved organic matter, disturbing oxygen, and emitting methane and nitrous oxide (Elser, 1999).

HCBs cause a huge increase in dissolved organic carbon being released into the environment. Polysaccharides, amino acids, proteins, and low molecular weight carbon substrates are among the organic compounds produced and released by cyanobacteria during bloom development. Exudation takes place by active

Table 5.2 Key aspects of carbon, nitrogen, and phosphorus interactions in cyanobacteria, highlighting their roles in metabolism, bloom dynamics, and ecosystem nutrient cycling

Aspect	Description	Impact on carbon fixation and ecosystem
Carbon fixation pathways	Utilization of CO_2 via the Calvin–Benson–Bassham (CBB) cycle and carbon-concentrating mechanisms (CCMs)	Efficient CO_2 fixation supports biomass production; however, energy-intensive CCMs may divert resources from nitrogen and phosphorus assimilation, potentially limiting growth under nutrient scarcity
Nitrogen fixation	Conversion of atmospheric N_2 to ammonium (NH_4^+) by nitrogenase enzymes	Enables growth in nitrogen-limited environments; however, the process is ATP-demanding and requires phosphorus for energy production, linking nitrogen fixation to phosphorus availability
Phosphorus acquisition	Uptake of inorganic phosphorus (orthophosphate) and organic phosphorus compounds	Essential for energy transfer (ATP), genetic material (DNA, RNA), and cellular signaling; phosphorus limitation can restrict growth and carbon fixation, particularly under eutrophic conditions
C:N:P stoichiometry	Elemental ratios in biomass, often approximated as 106:16:1 (Redfield ratio)	Deviations from this ratio can indicate nutrient imbalances, affecting primary production and ecosystem stability; for instance, phosphorus excess relative to nitrogen can lead to cyanobacterial blooms, altering carbon cycling dynamics
Bloom dynamics	Rapid proliferation of cyanobacteria under optimal light, temperature, and nutrient conditions	During blooms, cyanobacteria fix substantial amounts of carbon and nitrogen; upon senescence, decomposition releases nutrients back into the system, potentially leading to hypoxia and altering nutrient cycling, with implications for carbon sequestration and ecosystem health
Internal nutrient recycling	Release and reabsorption of nutrients during cell death and decomposition	Facilitates nutrient availability for other organisms; however, excessive recycling can exacerbate eutrophication, leading to shifts in community structure and function, and influencing carbon storage capacities
Environmental feedbacks	Interactions with light, temperature, and nutrient levels	Climate change and anthropogenic nutrient inputs can intensify cyanobacterial blooms, disrupting nutrient cycles and impacting carbon sequestration; understanding these feedbacks is crucial for managing aquatic ecosystems and mitigating climate change effects

metabolism and passive cell membrane rupture or spilling, mainly when a bloom ends. The DOC pool becomes a major and responsive feature of the carbon cycle, fueling the activity of bacteria and fungi. Microbes quickly use up DOC, causing respiration among bacteria and turning easily used carbon into more stable forms. Portions of the DOC may be oxidized or used again; still, a lot of it joins the microbial systems and alters nutrient cycling through stoichiometric coupling (O'Neil et al., 2012).

HCB-formed DOC is not the same as that made by other phytoplankton. Polymeric materials produced by cyanobacteria, including mucilage and other large molecules called EPS, often alter the water column's properties. Dispersants may be used to improve flocculation and form groups of particles, again affecting particle movement and sedimentation speed. At the same time, they can block light entering the water, decrease transparency, and stop photosynthesis in the water community below. Additionally, when dissolved organic carbon levels are increased, the result can be changes in pH, metal bonds, and oxidation state which can make other substances harder or easier for organisms to use. Among the main ecological impacts of HCBs is that they disrupt how water system oxygen is shared, mostly in water basins that do not mix well. At the start of a bloom, cyanobacteria use up so much oxygen that it can become oversaturated in surface waters. Yet, this does not last for long, and soon after the bloom ends, the oxygen quickly drops as the organisms decay. Breaking down cyanobacteria in the absence of light such as in the water's depths causes heterotrophic bacteria to consume masses of oxygen. Having lots of microbes in wastewater can quickly cause the water to become low in oxygen or completely oxygen free (Elser, 1999). Many of the impacts of HCBs, including killing aquatic organisms and altering community and ecosystem function, result in part from the oxygen depletion they cause.

Under conditions of hypoxia, fish, invertebrates, and aerobic microbes often experience stress or die. HCBs regularly lead to fish kills which may seriously harm the diversity of local species and damage fisheries. Lack of oxygen in an environment changes the chemical balance and directs microbes to break down material without oxygen such as denitrification, reducing sulfates and creating methane. Not only do these shifts alter how carbon and nitrogen move through the ecosystem, but they also have an effect on greenhouse gas emission. When organic matter degrades in sediments and under anoxic water, microbes emit methane (CH_4), but when nitrate is reduced, they produce nitrous oxide (N_2O) (Breitburg et al., 2018). Having a higher global warming potential than carbon dioxide, both gases help to cause climate change when they are emitted into the air.

In eutrophic lakes, reservoirs, and wetlands where there is thick organic sediment at the bottom, methane from cyanobacterial blooms is especially concerning. Cyanobacterial biomass becomes buried and in that process, methanogenic archaea digest the carbon, releasing methane, some of which might be broken down again by methanotrophs at the oxic-anoxic border and some is sent directly into the air by diffusion or bubble formation. The buoyancy created by gas vesicles in some cyanobacteria draws these microbes up, speeding up the discharge of methane. The way HCBs occur depends strongly on when and how intensely seasonal changes, water temperature layering, and water column mixing occur (Diaz & Rosenberg, 2008).

Production of nitrous oxide also rises after HCBs, mainly when organic matter that contains nitrogen decays under suboxic or anoxic conditions. Most N_2O is released because nitrification and denitrification interactions commonly allow some denitrification to occur which still leads to release of N_2O. In addition, cyanobacteria that carry out nitrogen fixation contribute further complexity, because new nitrogen they fix can boost microbial activity and N_2O production. Some cyanobacteria discharge nitrogen from within the cell when they burst which promotes an increase in ammonium and nitrate for nitrifying bacteria (Martikainen & Biochemistry, 2022).

The biogeochemical effects caused by HCBs influence both ecological health and the balance of carbon in the environment. Because DOC is released, oxygen is consumed, and greenhouse gases are emitted all at once, there is a loop that helps eutrophication and cuts back on how much carbon aquatic systems are able to store. Because of HCB, systems that should be storing carbon through organic burial tend to release carbon into the atmosphere, shrinking their usefulness in tackling climate change. Harmful algal blooms are becoming more common since actions by humans pollute streams, change the land, and cause global warming (Glibert et al., 2014; Van Spanning et al., 2005).

The negative effects of HCBs on the environment are usually long lasting and contribute to ongoing deterioration of essential water quality and the ability of ecosystems to recover. Frequent algal blooms can cause nutrient problems, shrink the variety of species, affect the ways predators and prey interact, and lower the usefulness of a healthy ecosystem (Bianchi et al., 2000). Cyanotoxins in water bodies can directly threaten human and animal health and this makes it harder to deal with the situation. HCB-caused damage costs a lot and is continuing to rise as bloom events happen more often and in more areas. To control the impacts of HCBs, it is important to link their role in carbon and nutrient cycling, oxygen behavior, and greenhouse gas emissions.

Approaches to control blooms need to pay attention to lowering main nutrient inputs, particularly of phosphorus and nitrogen, as well as possible climate changes involved. Working to repair wetlands, create riparian buffers, and update wastewater activities can cut down on nutrient loading. At the same time, checking DOC, oxygen, and gas levels can alert scientists and help guide actions to preserve the ecosystem and reduce climate feedbacks (Bastviken et al., 2011; Cole et al., 2007).

Basically, dangerous algae blooms are more than signs of nutrient buildup; they directly contribute to changes in ecosystems. By controlling the release of dissolved organic carbon, oxygen level, and greenhouse gas amounts, they transform the biochemistry of water bodies and add to changes happening worldwide. Managing and learning about the impacts that pollutants have are necessary for enhancing the health of waterways, protecting diverse species, and controlling climate change.

Even though the ocean is a sink for carbon, not all carbon ends up being trapped by sediments. Storing carbon in aquatic sediments is a major way to remove it from active biological and chemical cycles and keep it safe for long periods. The job of primary production and organic matter made by cyanobacteria greatly adds to this process. As photosynthetic species in many water bodies, cyanobacteria convert much carbon dioxide into organic matter. When they start to die, these compounds settle into the water and are included in sediments (Little et al., 2013; Mishra et al., 2020).

It is the start of sedimentary carbon absorption, a main way the world's carbon cycle protects the atmosphere from more greenhouse gases and helps maintain stable water habitats for years. When plants photosynthesize, they change inorganic carbon into their own biomass, starting the process toward sedimenting carbon. At times of bloom, cyanobacteria create plenty of organic material, with some ending up as suspended particles and some dissolved in water. As blooms are dying and the cells break apart, they form bigger groups or even bond to mineral areas which makes them sink to the lowest regions of the water. Sedimentation rate and extent can be influenced by water layering, turbulence, the type of organic material, and the presence of binding compounds such as extracellular polymeric substances. When cyanobacterial organic matter is caught in a mucilaginous coating or mixed with silt and clay, it often drops quickly below the surface where it is less likely to become remineralized.

After settling, organic carbon is controlled by different biological, chemical, and physical interactions. Microbes living at the bottom of oxygen-rich waters break down organic matter by breathing in oxygen and breathing out carbon dioxide. Consequently, part of the organic carbon is eventually recycled into both the water and air. In a lot of different water bodies, mainly when hypoxia, stratification, are increased productivity are present, the bottom waters can quickly become anoxic or suboxic. In these situations, aerobic breakdown cannot occur, so other anaerobic ways such as sulfate reduction, methanogenesis, or denitrification are promoted. Because they are not very good at digesting organic matter, many organics are left behind in the sedimentary record (Withers & Jarvie, 2008).

Sediment stability of organic carbon relies not only on oxygen absence but also on how the carbon and minerals interact. Many labile and recalcitrant parts are contained within cyanobacterial biomass. Bacteria and fungi break down labile compounds, but more stable substances such as lipids, pigments, cell wall polymers, and extracellular polysaccharides resist their effects. How fast and safely these compounds can be stored is affected by their molecular structure and the types of chemicals they contain. The presence of clay, iron oxides, and calcium carbonate minerals makes it easier for organic matter to be stabilized by sorption, co-precipitation, and the cover offered by small aggregates called microaggregates (Totsche et al., 2018).

The stabilization of carbon in several sedimentary settings occurs mainly through the process of forming organo-mineral complexes. The compounds in sea floor organic matter released by cyanobacteria could be tightly connected to clay particles or end up in fine sediments, reducing the ability of microbes and enzymes to access and use them. It reduces the organic carbon's speed of breakdown and lengthens its time inside the land. In a few cases, organic molecules entering the soil interact with reactive minerals such that further bonds emerge which decrease their accessibility to plants. These processes allow carbon to remain in sediments for over 10 thousand years, regardless of how long it is buried, what materials are deposited, or the prevailing environmental conditions (Engel et al., 2004).

There are two kinds of sedimentary carbon in this pool: particulate organic carbon (POC) and dissolved organic carbon (DOC), and both are trapped in the porewaters or attached to sediment particles. While sediments build up and are pressed through layers, the organic matter undergoes transformations resulting from the increase in pressure and compacting. During the slow change, humic substances and kerogen-like material can develop, both of which remain stable for a long time and hold a large portion of Earth's carbon. Burial in lakes and reservoirs is a major way that autochthonous carbon produced by cyanobacteria and primary producers is taken out of circulation. Sequestering carbon in sediment is crucial in marine systems because both coastal zones and continental shelves are highly productive and filled with sediments (Du et al., 2023).

Both environmental forces and actions by humans alter how sediments absorb carbon dioxide. The production, movement, and amount of organic matter in the environment can be affected by changes in land use, addition of nutrients, water flow, and climate. As an example, greater inputs of nutrients from agriculture and cities encourage cyanobacterial growth which could result in higher primary production and additional organic material settling in wetlands. Excessive levels of organic materials can use oxygen and foster the growth of greenhouse gases which cancel some of the positive effects from using organic fertilizer. Changes caused by climate change in temperature and water movement can affect sediment movement, microbial activities and levels of oxygen in deep ocean water which may change how much carbon ends up being buried or released.

With dams, dredging and covering sediments, people can change their structure and the way they trap carbon. With their long holding times and ability to trap sediment, reservoirs built by dams are ideal for putting carbon underground. It's possible that these formations emit methane because reactions happen without oxygen deeper down. It is important to know how much carbon is taken out of the air versus how much greenhouse gas is produced to really understand how these systems affect the climate. In addition, improving wetlands and riparian zones helps with sedimentation and storing carbon because plants there grow, trap suspended material, and secure sediments (Paerl & Otten, 2013).

In short, carbon from primary production by cyanobacteria is trapped in aquatic sediments, where it slowly turns into organic matter and is stored below. How efficiently organic carbon is turned into coal depends on the levels and quality of organic carbon produced, the place where it is deposited, and the changes it goes through inside the sediment. Due to their strong productivity and many blooms, cyanobacteria bring down a large amount of organic material to the bottom of eutrophic and stratified water bodies. Biomass kept under reduced oxygen and stabilized by mineral combination can endure for hundreds of years which cushion the atmosphere from rising levels of CO_2. While environmental stresses rise, caring for and improving sedimentary carbon sinks are vital for maintaining the carbon balance of water and supporting worldwide goals for climate control (Ma & Wang, 2021).

5.6 Climate Change and Anthropogenic Influences

Cyanobacterial ecosystems react swiftly to changes brought by climate change and human impacts. Higher temperatures and increased nutrient pollution have caused there to be more frequent, large cyanobacterial blooms that harm aquatic ecosystems and upset their biogeochemical cycles. Before, weather and nutrients helped these ecosystems stay regular, but now they are rapidly changing due to climate changes and human activity that adds more nutrients. It is important to know exactly how cyanobacteria respond to such shifts to better predict changes in ecosystems and to decrease the dangers posed by large blooms, the destruction of the environment, and unfamiliar carbon patterns (Kibuye et al., 2021).

Increased water temperature due to global warming changes the way cyanobacteria function, speed up their metabolism, and lead to new rivalries with other organisms. *Microcystis*, *Anabaena*, and *Dolichospermum*, as examples of bloom-forming vyanobacteria, are advantaged by certain traits in warmer conditions. For most cyanobacteria, the best growth temperature is higher than that used by many groups of phytoplankton like diatoms and green algae. Once surface water temperatures rise, the thermal layering of the water lasts longer, forming warm, rich-in-nutrients layers on top where cyanobacteria can thrive. Many bloom-forming cyanobacteria can use gas vesicles to float at the surface, unlike most phytoplankton that are usually carried downward when the water is too shallow. Such traits give them a better ability to absorb light, allowing them to dominate their environment further (Kibuye et al., 2021; Kumari et al., 2022).

The impact of eutrophication is not limited to increased biomass production; it also alters species interactions, food web structure, and biogeochemical cycling. As cyanobacteria dominate, they displace more nutritious phytoplankton taxa, leading to reduced food quality for grazers and disrupting the energy flow to higher trophic levels. Some cyanobacterial species produce toxins such as microcystins, anatoxins, and cylindrospermopsins, which can harm aquatic organisms, livestock, and human health. Moreover, these toxins increase ecological pressures, lower the diversity of species, and make developing more difficult.

Water is used for drinking water and for leisure purposes. However, when there are too many organic particles from bloom, the breakdown of dead fish causes microbes to use more oxygen, resulting in either hypoxia or anoxia at the bottom. Wave is a process that causes more phosphorus to come from sediments, resulting in further eutrophication results.

Anthropogenic influence can be found with chemical pollution, including the use of heavy metals and pharmaceuticals, as well as new contaminants related to microplastics. Various chemicals interact with cyanobacteria to different extents, sometimes completely shutting them down. Although they grow, pathogens often switch to resistant strains that change physically. Some presence of contaminants could stop nutrient uptake, disturb oxidative stress responses, or change internal cell work.

As a result, signaling can shape the way a community develops and the blooming process takes place. Furthermore, the mix of nutrients in water and chemical pollutants leads to new kinds of stress on aquatic life, that make it tough for aquatic ecosystems and cause challenges for those trying to manage them (Chorus et al., 2021).

The sensitivity of cyanobacterial ecosystems to light makes them particularly vulnerable to change, who is endangered by both changes in climate and pollution. A rise in the runoff of sediment from urban areas' low light can occur when turbidity and too many algae in water block light from reaching below the surface. Plants that are partly or fully underwater, as well as algae, are affected. But, a lot of cyanobacteria have adapted so they can live in cloudy and shaded areas because they adjust their coloring, how plants arrange themselves, and how much light they absorb. How well they are able to stay productive their dominance in disturbed or degraded areas comes from the fact that they are able to survive under light-limiting situations—places where other species find it hard to stay alive. Having the ability to adjust to light conditions allows cyanobacterial blooms to remain across a large variety environmental conditions.

In addition, climate change has indirect effects on the cycles and whereabouts of cyanobacterial species. With rising temperatures and longer times for plants to grow, blooms have expanded into higher latitudes and temperate parts of the world. Signs of cyanobacteria expanding toward the poles in several regions suggest that climate warming matters to their geographic distribution. Bloom-forming species placed into fresh ecosystems can cause different types of interactions with other organisms, competition for resources, and unpredictable changes in the ecosystem's function. In addition, harmful algal blooms appearing in drinking water supply and lakes used for recreation create new problems for public health, water treatment, and ecological management (Song et al., 2023).

Global climate change combining with local human pressure makes irreversible shifts in ecology more likely in systems dominated by cyanobacteria. When blooms happen more often, are stronger, and cover a wider area, the ability of water systems to stay healthy declines. Because of increased nutrient loading, a drop in biodiversity, and less food web support, it is not easy to recover ecosystems even though external stresses are removed. As a result, integrated watershed management needs to focus on dealing with climate issues and reducing nutrients. Pollution control measures should involve controlling fertilizer spread, improving wastewater management, restoring wetlands and riparian areas, and creating plans that lower runoff and erosion (Huisman et al., 2018).

Either global warming or pollution brought about by humans is enough to cause damage to ecosystems. How organisms are built, how they survive in their environment, and how they adjust to it both tell us about and cause changes to the environment. Because of a growing problem with both climate change and nutrient enrichment, cyanobacterial blooms are expected to last longer, occur more often, and cause more damage. We must work together to reduce nutrient inputs, track the effects of these contributions on ecosystems, and strengthen the ability of aquatic systems to resist problems. Learning about how cyanobacteria respond to changes in the environment helps us protect water quality, keep ecosystem services, and preserve the balance of global biogeochemical cycles at a time when ecological pressures are rising.

5.7 Management Strategies for Enhanced Sequestration

With climate change and nutrient pollution pushing cyanobacterial outbreaks higher, it's crucial that we come up with good ways to manage them. The presence of cyanobacteria in aquatic systems brings serious problems, but their ability to capture high amounts of carbon could offer an opportunity to improve carbon sequestration. Striking a good balance allows us to limit biomass increase and still make use of the support natural systems have for combating climate change. Bloom reduction, biogeochemical models, and methods that respect ecosystems allow the risks of cyanobacteria to be reduced without harming their capacity to store carbon for a long time. Managing blooms of cyanobacteria helps to hug the impact they leave in aquatic habitats and the chemicals they introduce. Bloom control can be organized into physical, chemical, and biological methods. Examples of these methods are taking out surface scums, using pumps to mix the water and putting oxygen into the bottom to prevent buildup of phosphorus. These ways may instantly help, but they usually need a lot of effort, are very expensive, and are difficult to use on a broad level. In other words, although chemical tactics such as using algaecide or aluminum sulfate to deal with blooms act quickly to solve the problem, they endanger non-bloom-causing organisms and are less sustainable for the environment. Moreover, these treatments mask the key factor leading to eutrophication, so they are often used again without getting to the source of the problem (Withers & Jarvie, 2008).

Using biologically based methods helps the environment and is sustainable. Examples are introducing animals that consume cyanobacteria, ensuring that they do not compete with in-bloom phytoplankton, and using plants or microbes that behave in ways that control cyanobacteria. For instance, some macrophytes and bacteria make secondary metabolites that reduce cyanobacterial growth but leave other aquatic life unaffected. Managing the food web by cutting down the activity of bottom-feeding fish is another approach to control bloom formation. It is important for biological control to succeed by closely watching local ecosystems, considering community structure and carrying out constant monitoring.

For effective management interventions, biogeochemical modeling highly contributes. The models replicate energy and nutrient movement in aquatic environments, bringing together data about hydrodynamics, sedimentation, microbiological activity, and how plants and algae start growing. Models are able to show how too many nutrients, changes in temperature, and water flow can influence cyanobacterial population sizes and the amount of carbon cycling. Resource managers can study the potential results of different measures and find the most suitable methods by considering feedback and variation in space and time changes in the models. Using a model, one can determine whether lowering phosphorus runoff by a specific percentage will help control algal blooms or if even more steps are required to decrease eutrophication (Hamilton et al., 2016).

Such models also offer new insights into how much carbon is sequestered in aquatic systems that largely contain cyanobacteria. Modelers use the understanding of fixed carbon's fate to determine the net amount of carbon found in every system. This information helps measure how much carbon lakes, reservoirs, wetlands, and estuaries hold globally. Also, when methane, nitrous oxide, and similar emissions are included in climate models, their influence on bloom events is better understood. Such ways of modeling provide basis for evidence-based decisions and encourage flexible practices in response to shifting environmental issues.

Although technology and models play a role, finding long-term answers for bloom control and increasing carbon sequestration depends on using more natural solutions. They place importance on conserving and repairing ecosystems that regulate the supply of nutrients, increase biodiversity, and keep sediments in place. Restoring wetlands is considered one of nature's most effective solutions. Wetlands help clean water by holding onto sediments and organics, as well as by giving a home to a variety of biological groups. When they intercept rich runoff, restored wetlands help reduce the amount of phosphorus and nitrogen entering nearby water sources which helps decrease conditions that lead to blooms of cyanobacteria. In addition, organic matter in wetlands is buried with little oxygen present, making them great places for long-term storage of carbon (Stroom & Kardinaal, 2016).

Besides, riparian buffer zones or vegetation planted close to rivers and lakes are important for regulating nutrients and supporting the stability of ecosystems. These types of buffers catch runoff from rain, help soak it into the soil, and stop sediment and nutrients from entering water. They protect water from the sun, also helping to regulate temperature and reduce light given to surface inhabitants. Adopting these two methods can lower the risks of both erosion and releasing nutrients into the water by farmers, ensuing bigger benefits for water ecosystems.

More often, cities address stormwater and nutrient-loading problems with wetlands, gardens that collect rain, and special pavement designed for moisture absorption. The approach is similar to how nature handles runoff and so less polluted water flows into areas around farms. Well-planned green spaces give extra benefits like a richer variety of living things in cities, more places for people to relax, and improved looks. By using these systems, watershed planning can combine city growth, environmental care, and climate efforts (Song et al., 2023).

Restoring aquatic ecosystems includes helping rebalance the connection between producers and consumers. Improving populations of fish that feed on cyanobacteria or upset the sediment bed has effects on the overall nutrient cycle and bloom events. Bringing back submerged aquatic vegetation may increase oxygen levels in the water, offer more space for animals to thrive, and use the same resources that cyanobacteria need. It is common that several approaches are needed to manage the varied conditions that create blooms and process carbon (Hudnell, 2008).

Overall, good management of cyanobacterial ecosystems for better carbon sequestration requires bringing together ecological, hydrological, and socio-economic factors. For nutrient reduction programs and ecology-related work to succeed, all stakeholders, including policymakers and scientists, must join forces to set and oversee actions. Teaching and informing the public help encourage people to use sustainable practices and support land and water management programs. Because climate change and nutrient pollution are going stronger, using new technology, environmental knowledge, and community action is necessary to guarantee aquatic ecosystems are secure and functional as natural carbon sinks (Ibelings et al., 2016). Table 5.3 outlines various management strategies aimed at enhancing carbon sequestration in cyanobacterial ecosystems, highlighting their potential impacts and considerations.

5.8 Conclusion

Cyanobacteria are central to aquatic carbon dynamics due to their high photosynthetic efficiency, ecological adaptability, and widespread distribution in freshwater and marine environments. Their role in fixing atmospheric carbon dioxide into organic matter positions them as key drivers of primary productivity and contributors to both short-term carbon cycling and long-term sequestration through sediment deposition. This chapter has explored the multiple dimensions of cyanobacterial influence, beginning with their specialized photosynthetic mechanisms and carbon-concentrating strategies that enhance carbon assimilation, particularly in low-CO_2 and nutrient-rich environments. Through the conversion of carbon into biomass, cyanobacteria form the base of aquatic food webs while also supporting microbial loops through the release of organic matter. Their interactions with nutrient cycles—particularly nitrogen and phosphorus—demonstrate a tightly coupled relationship where cyanobacterial activity not only responds to but actively modifies nutrient dynamics. The occurrence of blooms, especially harmful cyanobacterial blooms (HCBs), represents a turning point where their ecological function becomes problematic, often leading to oxygen depletion, disruption of aquatic life, and the release of greenhouse gases such as methane and nitrous oxide. Despite this, cyanobacteria remain critical to sedimentary carbon storage, particularly under anoxic conditions that slow decomposition and favor carbon stabilization. The chapter also reviewed how climate change and anthropogenic nutrient loading intensify cyanobacterial dominance, highlighting their sensitivity to warming, eutrophication, and hydrological shifts. In response, a range of management strategies was discussed—from bloom suppression techniques and predictive biogeochemical models to nature-based solutions that restore ecological balance and improve carbon sequestration. Together, these approaches underscore the need for integrated, ecosystem-based management that addresses both carbon dynamics and water quality concerns. Future directions should focus on developing high-resolution, climate-responsive models that can predict cyanobacterial behavior under varying nutrient regimes and environmental stressors. Long-term field studies are needed to quantify the permanence of sedimentary carbon in bloom-impacted systems and evaluate the trade-offs between productivity, emissions, and sequestration. Research into the genetic regulation of

Table 5.3 Management strategies for enhanced carbon sequestration in cyanobacterial ecosystems

Strategy	Description	Impact on carbon sequestration
Physical control measures	Techniques such as removing surface scums, using pumps to mix water, and adding oxygen to prevent phosphorus buildup	These methods provide immediate relief by reducing bloom severity, but they are resource-intensive and may not address underlying nutrient imbalances
Chemical control measures	Application of substances like algaecides or aluminum sulfate to manage blooms	While effective in the short term, chemical treatments can harm nontarget organisms and may lead to the recurrence of blooms if nutrient sources are not controlled
Biological control methods	Introduction of organisms or substances that naturally regulate cyanobacterial populations, such as macrophytes or specific bacteria	Sustainable and eco-friendly approach that can reduce bloom formation without harming other aquatic life; however, effectiveness depends on careful monitoring and ecosystem compatibility
Biogeochemical modeling	Use of models to simulate nutrient dynamics, water flow, and bloom development to inform management decisions	Provides valuable insights into the factors influencing bloom formation and carbon cycling, aiding in the development of targeted and effective management strategies
Wetland restoration	Rehabilitating wetlands to restore their natural functions, including nutrient filtration and carbon storage	Wetlands act as significant carbon sinks; restoring them enhances their capacity to sequester carbon and mitigate the impacts of cyanobacterial blooms by reducing nutrient influx
Riparian buffer zones	Establishing vegetated areas along water bodies to filter runoff and stabilize ecosystems	These zones help in reducing nutrient loading into water bodies, thereby decreasing the frequency and intensity of cyanobacterial blooms and promoting long-term carbon sequestration
Integrated management plans	Collaborative approaches involving stakeholders to develop and implement strategies for bloom management and carbon sequestration	Ensures a holistic approach that balances ecological health, economic considerations, and community involvement, leading to sustainable management outcomes

carbon and nutrient metabolism in cyanobacteria may also offer novel strategies for ecological engineering and biotechnological applications. By advancing our understanding of these organisms and managing their impacts wisely, cyanobacteria can be repositioned as allies in climate mitigation and ecological restoration, rather than as indicators of environmental degradation.

References

Armbrust, E. V. (2009). The life of diatoms in the world's oceans. *Nature, 459*(7244), 185–192.

Bailey, A., Meyer, L., Pettingell, N., Macie, M., & Korstad, J. (2020). Agricultural practices contributing to aquatic dead zones. In *Ecological and practical applications for sustainable agriculture* (pp. 373–393). Springer Singapore.

Bastviken, D., Tranvik, L. J., Downing, J. A., Crill, P. M., & Enrich-Prast, A. (2011). Freshwater methane emissions offset the continental carbon sink. *Science, 331*(6013), 50–50.

Bhardwaj, A., Singh, P., Gupta, N., Bhattacharjee, S., Srivastava, A., Parida, A., & Mishra, A. K. (2024). *Cyanobacteria: A key player in nutrient cycling* (pp. 579–596). Elsevier.

Bianchi, T. S., Engelhaupt, E., Westman, P., Andrén, T., Rolff, C., & Elmgren, R. (2000). Cyanobacterial blooms in the Baltic Sea: Natural or human-induced?. *Limnology and Oceanography, 45*(3), 716–726.

Breitburg, D., Levin, L. A., Oschlies, A., Grégoire, M., Chavez, F. P., Conley, D. J., Garçon, V., Gilbert, D., Gutiérrez, D., Isensee, K., & Jacinto, G. S. (2018). Declining oxygen in the global ocean and coastal waters. *Science, 359*(6371), eaam7240.

Chapra, S. C., Boehlert, B., Fant, C., Bierman Jr, V. J., Henderson, J., Mills, D., Mas, D. M., Rennels, L., Jantarasami, L., Martinich, J., & Strzepek, K. M. (2017). Climate change impacts on harmful algal blooms in US freshwaters: A screening-level assessment. *Environmental Science & Technology, 51*(16), 8933–8943.

Chorus, I., Fastner, J., & Welker, M. (2021). Cyanobacteria and cyanotoxins in a changing environment: Concepts, controversies, challenges. *Water, 13*(18), 2463.

Cole, J. J., Prairie, Y. T., Caraco, N. F., McDowell, W. H., Tranvik, L. J., Striegl, R. G., Duarte, C. M., Kortelainen, P., Downing, J. A., Middelburg, J. J., & Melack, J. (2007). Plumbing the global carbon cycle: Integrating inland waters into the terrestrial carbon budget. *Ecosystems, 10*(1), 172–185.

Cushman, J. C. (2001). Crassulacean acid metabolism. A plastic photosynthetic adaptation to arid environments. *Plant Physiology, 127*(4), 1439–1448.

Dell'Aquila, M. (2024). Characterizing water quality, nutrients, and dissolved oxygen dynamics in four Lake Ontario coastal wetlands.

Diaz, R. J., & Rosenberg, R. (2008). Spreading dead zones and consequences for marine ecosystems. *Science, 321*(5891), 926–929.

Du, J., Zhang, F., Du, J., Wang, Z., Ren, X., & Yao, Z. (2023). Source and sequestration of sediment organic carbon from different elevation zones in estuarine wetland, North China. *Science of the Total Environment, 859*, Article 160253.

Elser, J. J. (1999). The pathway to noxious cyanobacteria blooms in lakes: The food web as the final turn. *Freshwater Biology,42*(3), 537–543.

Engel, A., Delille, B., Jacquet, S., Riebesell, U., Rochelle-Newall, E., Terbrüggen, A., & Zondervan, I. (2004). Transparent exopolymer particles and dissolved organic carbon production by Emiliania huxleyi exposed to different CO_2 concentrations: A mesocosm experiment. *Aquatic Microbial Ecology, 34*(1), 93–104.

Falkowski, P. G., Fenchel, T., & Delong, E. F. (2008). The microbial engines that drive Earth's biogeochemical cycles. *Science, 320*(5879), 1034–1039.

Fazi, S., Amalfitano, S., Venturi, S., Pacini, N., Vazquez, E., Olaka, L.A., Tassi, F., Crognale, S., Herzsprung, P., Lechtenfeld, O. J., & Cabassi, J. (2021). High concentrations of dissolved biogenic methane associated with cyanobacterial blooms in East African lake surface water. *Communications Biology, 4*(1), 845.

Flombaum, P., Gallegos, J. L., Gordillo, R. A., Rincón, J., Zabala, L. L., Jiao, N., Karl, D. M., Li, W. K., Lomas, M. W., Veneziano, D., & Vera, C. S. (2013). Present and future global distributions of the marine Cyanobacteria Prochlorococcus and Synechococcus. *Proceedings of the National Academy of Sciences, 110*(24), 9824–9829.

Franklin, D. J. (2021). Examining the evidence for regulated and programmed cell death in cyanobacteria. How significant are different forms of cell death in cyanobacteria population dynamics?. *Frontiers in Microbiology, 12*, 633954.

Glibert, P. M., Icarus Allen, J., Artioli, Y., Beusen, A., Bouwman, L., Harle, J., Holmes, R., & Holt, J. (2014). Vulnerability of coastal ecosystems to changes in harmful algal bloom distribution in response to climate change: Projections based on model analysis. *Global Change Biology, 20*(12), 3845–3858.

Gupta, J. K., Rai, P., Jain, K. K., & Srivastava, S. (2020). Overexpression of bicarbonate transporters in the marine cyanobacterium Synechococcus sp. PCC 7002 increases growth rate and glycogen accumulation. *Biotechnology for Biofuels, 13*(1), 17.

Hamilton, D. P., Salmaso, N., & Paerl, H. W. (2016). Mitigating harmful cyanobacterial blooms: Strategies for control of nitrogen and phosphorus loads. *Aquatic Ecology, 50*, 351–366.

Hilmi, N., Chami, R., Sutherland, M. D., Hall-Spencer, J. M., Lebleu, L., Benitez, M. B., & Levin, L. A. (2021). The role of blue carbon in climate change mitigation and carbon stock conservation. *Frontiers in Climate, 3*, 2021.

Hudnell, H. K. (2008). Cyanobacterial harmful algal blooms: State of the science and research needs.

Huisman, J., Codd, G. A., Paerl, H. W., Ibelings, B. W., Verspagen, J. M., & Visser, P. M. (2018). Cyanobacterial blooms. *Nature Reviews Microbiology, 16*(8), 471–483.

Ibelings, B. W., Fastner, J., Bormans, M., & Visser, P. M. (2016). Cyanobacterial blooms. Ecology, prevention, mitigation and control: Editorial to a CYANOCOST Special Issue. *Aquatic Ecology,50*, 327–331.

Japri, D., Kasmuri, N., Zaini, N., Hamid, K. A., Nayono, S. E., & Mojiri, A. (2024). The efficacy of activated carbon from organic waste in treating wastewater effluent. *International Journal of Sustainable Construction Engineering,15*(1), 271–286.

Kibuye, F. A., Zamyadi, A., Wert, E. C. (2021). A critical review on operation and performance of source water control strategies for cyanobacterial blooms: Part II-mechanical and biological control methods. *Harmful Algae*, 109, 102119.

Kim, D. S., Moreno-Cabezuelo, J. Á., Schulz, E. N., Lea-Smith, D. J., & Sagaram, U. S. (2024). Recent advances in engineering fast-growing cyanobacterial species for enhanced CO2 fixation. *Frontiers in Climate, 6*, 1412232.

Kosten, S., Huszar, V. L., Bécares, E., Costa, L. S., van Donk, E., Hansson, L. A., Jeppesen, E., Kruk, C., Lacerot, G., Mazzeo, N., & De Meester, L. (2012). Warmer climates boost cyanobacterial dominance in shallow lakes. *Global Change Biology, 18*(1), 118–126.

Kramer, B. J., Jankowiak, J. G., Nanjappa, D., Harke, M. J., & Gobler, C. J., (2022). Nitrogen and phosphorus significantly alter growth, nitrogen fixation, anatoxin-a content, and the transcriptome of the bloom-forming cyanobacterium, Dolichospermum. *Frontiers in Microbiology, 13*, 955032.

Lehmann, J., & Kleber, M. (2015). The contentious nature of soil organic matter. *Nature*, 528(7580), 60–68.

Lin, S. (2017). *Climate change and algal blooms.* Michigan State University.

Little, S. H., Vance, D., Siddall, M., & Gasson, E. (2013). A modeling assessment of the role of reversible scavenging in controlling oceanic dissolved Cu and Zn distributions. *Global Biogeochemical Cycles, 27*(3), 780–791.

Liu, X.-Y., Hou, W.-T., Wang, L., Li, B., Chen, Y., Chen, Y., Jiang, Y.-L., & Zhou, C.-Z. (2021). Structures of cyanobacterial bicarbonate transporter SbtA and its complex with PII-like SbtB. *Cell Discovery, 7*(1), 63.

Ma, J., & Wang, P. (2021). Effects of rising atmospheric CO_2 levels on physiological response of cyanobacteria and cyanobacterial bloom development: A review. *Science of the Total Environment, 754*, 141889.

Markou, G., Vandamme, D., & Muylaert, K. (2014). Microalgal and cyanobacterial cultivation: The supply of nutrients. *Water Research, 65*, 186–202.

Martikainen, P. J. (2022). Heterotrophic nitrification—An eternal mystery in the nitrogen cycle. *Soil Biology and Biochemistry*, 168, 108611.

Mishra, R., Krishnamoorthy, P., Gangamma, S., Raut, A. A., & Kumar, H. (2020). Particulate matter (PM10) enhances RNA virus infection through modulation of innate immune responses. *Environmental Pollution, 266*, 115148.

Mishra, S., Kumari, N., Häder, D. P., & Sinha, R. P. (2022). Cyanobacterial blooms and their implications in the changing environment. *Advances in Environmental and Engineering Research, 3*(1), 1–41.

Mitsch, W. J., Bernal, B., Nahlik, A. M., Mander, Ü., Zhang, L., Anderson, C. J., Jørgensen, S. E., & Brix, H. (2013). Wetlands, carbon, and climate change. *Landscape Ecology, 28*, 583–597.

O'Neil, J. M., Davis, T. W., Burford, M. A., & Gobler, C. J. (2012). The rise of harmful cyanobacteria blooms: The potential roles of eutrophication and climate change. *Harmful Algae,14*, 313–334.

Paerl, H. W., Hall, N. S., & Calandrino, E. S. (2011). Controlling harmful cyanobacterial blooms in a world experiencing anthropogenic and climatic-induced change. *Science of the Total Environment, 409*(10), 1739–1745.

Paerl, H. W., & Otten, T. G. (2013). Blooms bite the hand that feeds them. *Sciences, 342*(6157), 433–434.

Pina, M., Bize, A., Forterre, P., & Prangishvili, D. (2011). The archeoviruses. *FEMS Microbiology, 35*(6), 1035–1054.

Price, G. D., Badger, M. R., Woodger, F. J., & Long, B. M. (2008). Advances in understanding the cyanobacterial CO_2-concentrating-mechanism (CCM): Functional components, Ci transporters, diversity, genetic regulation and prospects for engineering into plants. *Journal of Experimental Botany, 59*(7), 1441–1461.

Song, L., Jia, Y., Qin, B., Li, R., Carmichael, W. W., Gan, N., Xu, H., Shan, K., & Sukenik, A. (2023). Harmful cyanobacterial blooms: biological traits, mechanisms, risks, and control strategies. *Annual Review of Environment and Resources, 48*(1), 123–147.

Sterinbach, S. (2024). *Exploring rhizosphere nitrogen dynamics and losses: A comparative study of NUE in maize*. Iowa State University.

Stroom, J. M., & Kardinaal, W. E. A. (2016). How to combat cyanobacterial blooms: Strategy toward preventive lake restoration and reactive control measures. *Aquatic Ecology, 50*(3), 541–576.

Tang, Y., Li, T., Xu, Y., Ren, H., & Huang, H. (2023). Effects of electrical stimulation on purification of secondary effluent containing chlorophenols by denitrification biofilter. *Environmental Research, 216*, 114535.

Taranu, Z. E., Gregory-Eaves, I., Leavitt, P. R., Bunting, L., Buchaca, T., Catalan, J., Domaizon, I., Guilizzoni, P., Lami, A., McGowan, S., & Moorhouse, H. (2015). Acceleration of cyanobacterial dominance in north temperate-subarctic lakes during the Anthropocene. *Ecology letters, 18*(4), 375–384.

Totsche, K. U., Amelung, W., Gerzabek, M. H., Guggenberger, G., Klumpp, E., Knief, C., Lehndorff, E., Mikutta, R., Peth, S., Prechtel, A., & Ray, N. (2018). Microaggregates in soils. *Journal of Plant Nutrition and Soil Science, 181*(1), 104–136.

Tréguer, P., Bowler, C., Moriceau, B., Dutkiewicz, S., Gehlen, M., Aumont, O., Bittner, L., Dugdale, R., Finkel, Z., Iudicone, D., Jahn, O., Guidi, L., Lasbleiz, M., Leblanc, K., Levy, M., & Pondaven, P. (2018). Influence of diatom diversity on the ocean biological carbon pump. *Nature Geoscience, 11*(1), 27–37.

Van Spanning, R., Delgado, M., & Richardson, D. (2005). The nitrogen cycle: Denitrification and its relationship to N2 fixation. *Nitrogen fixation in agriculture, forestry, ecology, and the environment* (pp. 277–342). Springer.

Wilczek, A., Włoch, W., Iqbal, M., & Kojs, P. (2011). Position of rays and lateral deviation of vessel elements in the stem wood of some dicotyledonous species with storeyed, double-storeyed, and nonstoreyed cambia. *Botany, 89*(12), 849–860.

Withers, P. J. A., & Jarvie, H. P. (2008). Delivery and cycling of phosphorus in rivers: A review. *Science of the Total Environment, 400*(1–3), 379–395.

Yin, H., Bao, Y., Huang, T., Zhang, Y., Sun, T., Tao, P., Sun, Q., & Chen, K. (2025). Effects of cyanobacterial growth and decline on dissolved organic matter and endogenous nutrients release at the sediment–water interface. *Carbon Research*, *4*(1), 1–13.

Chapter 6
Nitrogen Fixation and Assimilation Pathways in Bloom-Forming Cyanobacteria

Nawab Ali, Muhammad Mehran Anjum, Jalal Bayar, Tehleem Akmal, Azam Ali Sher, Gul Roz Khan, Abdul Haq, Hayat Ali, and Rovaid Ali

Abstract Bloom-forming cyanobacteria play a pivotal role in aquatic nitrogen cycling due to their ability to fix atmospheric nitrogen and assimilate various combined nitrogen sources. This chapter explores the molecular, biochemical, and ecological dimensions of nitrogen fixation and assimilation in these organisms. We examine the structure and regulation of the nitrogenase enzyme complex, the genetic control of heterocyst differentiation, and adaptive strategies to overcome oxygen sensitivity. Additionally, the chapter discusses major nitrogen assimilation pathways, including nitrate/nitrite reduction, ammonium transport, and the GS-GOGAT cycle, alongside their regulation through signal transduction networks such as NtcA

N. Ali (✉) · J. Bayar
Biosystem and Agricultural Engineering, Michigan State University, East Lansing, USA
e-mail: alinawab@msu.edu

M. M. Anjum
Department of Agronomy, The University of Agriculture, Peshawar, Pakistan

T. Akmal
Department of Soil and Environmental Science, The University of Agriculture, Peshawar, Pakistan

A. A. Sher
Comparative Medicine and Integrative Biology Program, Michigan State University, East Lansing, USA

M. M. Anjum · G. R. Khan
Pakistan Council of Scientific and Industrial Research Laboratories Complex Peshawar, Peshawar, Pakistan

A. Haq
MaREI Research Center, University College Cork, Cork, Ireland

H. Ali
Department of Zoology, University of Swat, Swat, Pakistan

R. Ali
Department of Horticulture, The University of Agriculture, Peshawar, Pakistan

M. M. Anjum
Climate and Agriculture Division, Reckenholzstrasse AGRSOCPE, Zurich, Switzerland

S. Fahad et al. (eds.), *Cyanobacterial Blooms: Ecology, Evolution and Biogeochemical Impacts*, https://doi.org/10.1007/978-3-032-06042-6_6

and PII proteins. Emphasis is placed on physiological and molecular adaptations during bloom conditions, including nutrient sensing, diurnal metabolic coordination, nitrogen storage, and interactions with carbon and phosphorus metabolism. Understanding these integrated pathways is critical to predicting bloom dynamics and managing eutrophication in aquatic ecosystems.

6.1 Introduction

6.1.1 Overview of Cyanobacteria

Cyanobacteria, formerly known as "blue-green algae," are photosynthetic prokaryotes that have existed on Earth for approximately 3.5 billion years (Schopf, 1993; Shestakov & Karbysheva, 2017). Cyanobacteria constitute a phylogenetically ancient and ecologically significant group of Gram-negative prokaryotes that perform oxygenic photosynthesis. They are widely acknowledged for their role in the Great Oxidation Event that transformed Earth's atmosphere by introducing molecular oxygen, thereby paving the way for aerobic life. As primary producers, they play vital roles in aquatic and terrestrial ecosystems, contributing significantly to global carbon and nitrogen cycles (Soule & Garcia-Pichel, 2019), where other photosynthetic organisms are limited by light, nutrients, or stress factors.

Cyanobacteria are morphologically diverse, ranging from unicellular, colonial, and multicellular filamentous forms (Humbert et al., 2017). Their internal organization, while lacking membrane-bound organelles typical of eukaryotes, is highly specialized featuring thylakoid membranes, carboxysomes for carbon fixation, and phycobilisomes for harvesting light energy. Many filamentous cyanobacteria also develop specialized cell types such as heterocysts, which facilitate nitrogen fixation under aerobic conditions and akinetes, which are thick-walled resting spores that allow survival during unfavorable conditions (Sukenik et al., 2019).

Cyanobacteria are ubiquitous and resilient. They colonize a vast array of ecological niches, including freshwater and marine systems, deserts, polar habitats, thermal springs, and even symbiotic relationships within lichens and plant roots. Their ecological amplitude is enhanced by a versatile metabolism capable of photoautotrophy, mixotrophy, and even diazotrophy. They perform oxygenic photosynthesis and nitrogen fixation, playing a crucial role in carbon and nitrogen cycling across diverse environments, including extreme habitats such as deserts and polar regions (Kulasooriya, 2012).

Recent advances in genomics and molecular ecology have unveiled a staggering genetic diversity among cyanobacteria. Genomic plasticity, horizontal gene transfer, and large-scale genome rearrangements have facilitated their adaptation to diverse environmental gradients (Willis & Woodhouse, 2020). As model organisms in ecological genomics, cyanobacteria offer insights into microbial community structure, stress responses, and ecosystem resilience.

6.1.2 Ecological Significance of Bloom-Forming Cyanobacteria

Cyanobacterial blooms are dense, often visible accumulations of cells that occur under favorable environmental conditions, particularly in eutrophic or nutrient-enriched aquatic ecosystems. Cyanobacterial blooms are characterized by the accumulation of cyanobacterial biomass, leading to water discoloration and the formation of cell aggregations either at the surface (scums) or below the surface (deep chlorophyll maxima [DCM]) (Reinl et al., 2021; Harriss et al., 2024). These blooms are commonly dominated by genera such as *Microcystis*, *Dolichospermum*, *Aphanizomenon*, *Planktothrix*, and *Cylindrospermopsis*. The frequency, intensity, and duration of such blooms have increased globally, fueled by anthropogenic nutrient loading, altered hydrological cycles, and climate-driven changes in temperature, precipitation, and stratification.

Bloom formation is a complex, multifactorial phenomenon influenced by light availability, water column stability, nutrient ratios (N:P), and biotic interactions such as grazing and competition. Cyanobacteria possess several ecological and physiological traits that confer competitive advantages over other phytoplankton. These include:

- Buoyancy regulation using gas vehicles allowing migration to optimal light and nutrient zones
- Toxin production, which can deter grazers and affect competitors
- High-affinity nutrient transport systems, enabling uptake under limiting conditions
- Capacity to fix atmospheric nitrogen, particularly advantageous in nitrogen-depleted environments.

Many bloom-forming cyanobacteria also exhibit circadian rhythms and diel cycling in gene expression, which optimize their metabolic processes in sync with environmental light–dark cycles. This temporal regulation is crucial for managing the energetically expensive processes of photosynthesis and nitrogen fixation, which are often antagonistic due to oxygen sensitivity. Also, cyanobacteria possess unique physiological adaptations, including efficient light harvesting and the ability to fix nitrogen, which support their widespread success in a variety of aquatic environments (Bullerjahn & Post, 2014).

The ecological consequences of blooms are profound. They alter food web structures, reduce biodiversity, impair ecosystem services (e.g., fisheries, drinking water), and lead to hypoxic or anoxic zones through microbial decomposition of accumulated biomass. Moreover, the production of cyanotoxins including hepatotoxins (e.g., microcystins), neurotoxins (e.g., anatoxin-a), and dermatotoxins poses significant public health threats and economic burdens on water treatment and management. The resilience and recurrence of cyanobacterial blooms highlight the necessity of understanding the microbial and biochemical pathways underpinning their success, especially the central role of nitrogen fixation and assimilation.

6.1.3 Importance of Nitrogen Metabolism in Aquatic Ecosystems

Nitrogen is a vital nutrient required for the synthesis of amino acids, nucleotides, chlorophyll, and numerous coenzymes. In aquatic environments, nitrogen availability strongly influences primary productivity, species composition, and trophic dynamics. Despite its abundance in the atmosphere, molecular nitrogen (N_2) is inert and biologically unavailable to most organisms. Biological nitrogen fixation mediated exclusively by diazotrophs such as certain cyanobacteria plays a crucial role in converting N_2 into ammonia (NH_3), a form assimilable by living organisms (Marino & Howarth, 2022).

Cyanobacteria display exceptional versatility in nitrogen metabolism. They can utilize a wide spectrum of nitrogen sources, including:

- Ammonium (NH_4^+) energetically favorable and rapidly assimilated via the GS-GOGAT pathway
- Nitrate and Nitrite reduced to ammonium through assimilatory nitrate and nitrite reductases
- Urea and organic nitrogen compounds are processed via urease and amidase enzymes
- Atmospheric dinitrogen (N_2) fixed via the nitrogenase complex in diazotrophic strains.

Nitrogenases are broadly expressed in methanotrophic bacteria, enabling the coordination of methane oxidation and nitrogen fixation processes (Lee et al., 2017). The nitrogenase enzyme complex, composed of dinitrogenase (MoFe protein) and dinitrogenase reductase (Fe protein), is oxygen labile and energetically demanding, consuming at least 16 ATP per molecule of N_2 fixed. To overcome the challenge of oxygen sensitivity, filamentous cyanobacteria have evolved heterocysts specialized cells that create a microoxic environment through thickened cell walls, cessation of oxygenic photosynthesis, and upregulation of respiratory activity. Heterocyst-forming cyanobacteria such as *Anabaena*, *Nostoc*, and *Aphanizomenon* can spatially separate nitrogen fixation and photosynthesis, allowing both to occur concurrently in different cell types within the filament.

Non-heterocystous nitrogen-fixing cyanobacteria, such as *Trichodesmium*, temporally separate these processes, fixing nitrogen predominantly during early morning or nighttime when oxygen levels are lower. This reflects a high level of metabolic coordination and adaptation to fluctuating environmental conditions. In addition to nitrogen fixation, cyanobacteria employ several regulatory mechanisms to optimize nitrogen assimilation. These include:

- Global nitrogen regulators such as NtcA, which control transcription of genes involved in nitrogen uptake and assimilation
- Feedback inhibition of glutamine synthetase by end products like glutamine
- Two-component signaling systems that detect extracellular nitrogen levels and modulate gene expression accordingly.

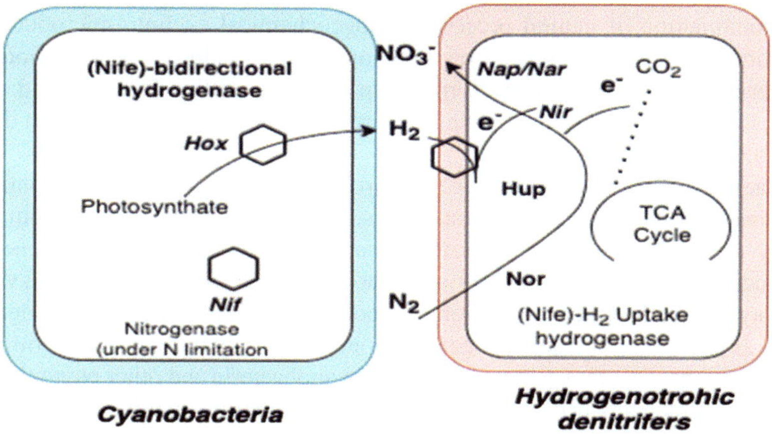

Fig. 6.1 Schematic representation of hydrogen metabolism in nitrogen-fixing bacteria

Cyanobacteria can fix atmospheric nitrogen into usable forms, which plants incorporate into organic molecules and transfer through the food chain (Bhardwaj et al., 2024). Cyanobacteria not only assimilate nitrogen for their own growth but also release bioavailable nitrogen into the environment through cell lysis, excretion, or sloughing of nitrogen-rich exopolymers. In this way, they act as both sources and sinks in the aquatic nitrogen cycle, influencing downstream processes such as denitrification and eutrophication.

The increasing prevalence of nitrogen-deficient conditions in eutrophic waters often due to phosphorus over enrichment favors the proliferation of nitrogen-fixing cyanobacteria. This shift in community structure exacerbates bloom formation and complicates management strategies aimed at nutrient reduction.

Moreover, climate change is projected to enhance nitrogen fixation rates in aquatic systems by promoting thermal stratification, extending growing seasons, and altering nutrient loading patterns. These environmental shifts could further amplify the dominance of bloom-forming diazotrophs in both freshwater and marine environments. Figure 6.1 shows the hydrogen metabolism pathways associated with nitrogen fixation and respiratory processes in diazotrophic bacteria.

6.2 Nitrogen Sources and Availability in Aquatic Environments

Nitrogen is one of the most essential macronutrients for all living organisms, particularly for primary producers such as phytoplankton and cyanobacteria in aquatic ecosystems. It plays a central role in the synthesis of amino acids, proteins, nucleic acids, and chlorophyll, making it indispensable for growth and reproduction. The availability, form, and cycling of nitrogen in aquatic environments are influenced by a combination of natural processes (biogeochemical cycling, microbial transformations) and anthropogenic pressures (agriculture, industry, urbanization). For cyanobacteria especially bloom-forming species, the ability to access and utilize different nitrogen forms is a defining feature that enables them to thrive in diverse and often fluctuating environments.

In aquatic ecosystems, the nitrogen cycle operates as a dynamic continuum involving multiple forms, sources, and sinks. Nitrogen enters the system from atmospheric deposition, terrestrial runoff, groundwater seepage, internal regeneration from sediments, and biological fixation. Once introduced, nitrogen undergoes various transformations mediated by microbial processes, including nitrification, denitrification, ammonification, and nitrogen fixation. These transformations determine the quantity and quality of nitrogen available to cyanobacteria and other phytoplankton, ultimately influencing ecosystem productivity, competition, and bloom dynamics.

6.2.1 *Inorganic Nitrogen Forms (NO_3^-, NH_4^+, N_2)*

Inorganic nitrogen in aquatic environments primarily exists in the forms of nitrate (NO_3^-), nitrite (NO_2^-), ammonium (NH_4^+), and atmospheric dinitrogen gas (N_2). The distribution and relative abundance of these forms are controlled by biological activity, redox potential, temperature, and hydrodynamics.

NO_3^- is the most common oxidized form of nitrogen in oxygen-rich surface waters. It is highly soluble, mobile, and persistent, particularly in systems with high external nitrogen loading such as agricultural runoff. The ability to assimilate nitrate is common among cyanobacteria (Flores et al., 2005), which reduce nitrate to nitrite using nitrate reductase, and subsequently to ammonium via nitrite reductase (NiR), before incorporating it into organic molecules. This two-step reduction is energetically demanding, requiring NADPH as a reductant, but is crucial when ammonium is scarce.

NH_4^+ is the most reduced and bioavailable form of nitrogen, and it is the preferred nitrogen source for most cyanobacteria due to its lower energy requirement for assimilation. Ammonium enters cyanobacterial cells through ammonium transporters (Amt family) and is assimilated into amino acids via the glutamine synthetase (GS) glutamate synthase (GOGAT) pathway. However, high concentrations of ammonium can be toxic, affecting pH balance and cellular metabolism. Moreover, the presence of ammonium can inhibit nitrate uptake and nitrogen fixation through feedback regulation.

Molecular nitrogen (N_2), although abundant in the atmosphere and dissolved in water, is inert and inaccessible to most aquatic organisms. Only certain prokaryotes, including many bloom-forming cyanobacteria, possess the enzyme nitrogenase that allows them to convert atmospheric N_2 into ammonia through biological nitrogen fixation. This process is particularly advantageous in nitrogen-depleted environments, allowing diazotrophic cyanobacteria (e.g. *Aphanizomenon*, *Dolichospermum*) to outcompete non-diazotrophs. However, nitrogen fixation is energy-intensive and highly sensitive to oxygen, necessitating specific cellular adaptations such as heterocyst formation or temporal separation of photosynthesis and nitrogen fixation.

The availability of these inorganic nitrogen forms is dynamic and fluctuates seasonally, daily, and spatially within the water column. Stratification, mixing events, upwelling, and biological uptake all influence nitrogen availability. Cyanobacteria, especially bloom-forming taxa, often exploit these fluctuations through physiological plasticity and regulatory mechanisms that allow them to rapidly switch between different nitrogen sources depending on availability and energetic constraints.

6.2.2 Anthropogenic Influences on Nitrogen Load

Human activities have drastically altered the nitrogen cycle over the past century, leading to significant increases in nitrogen inputs into freshwater and coastal ecosystems. Anthropogenic influences are now recognized as the dominant factor driving nitrogen enrichment and eutrophication in aquatic systems worldwide.

One of the primary sources is agricultural runoff, which introduces large quantities of nitrate and ammonium into water bodies from synthetic fertilizers and animal manure (Bijay-Singh & Craswell, 2021). The widespread use of nitrogen-based fertilizers in crop production often exceeds plant uptake capacity, leading to leaching and runoff, especially during rainfall or irrigation. These nitrogen inputs not only stimulate primary production but also promote the dominance of fast-growing and bloom-forming cyanobacteria, many of which are capable of withstanding extreme nutrient conditions.

Urbanization and wastewater discharge are also major contributors to nitrogen loading. Domestic sewage, stormwater runoff, and effluents from industrial facilities release significant amounts of ammonium, urea, and organic nitrogen compounds into rivers and lakes. Even advanced wastewater treatment facilities may not fully remove nitrogen, particularly in developing regions where infrastructure may be lacking or overloaded. In many urban lakes and reservoirs, nitrogen loading from point sources sustains perennial or recurrent cyanobacterial blooms.

Atmospheric deposition is another important pathway, especially in remote or forested watersheds. Combustion of fossil fuels (vehicular emissions, power plants) and intensive livestock production release nitrogen oxides and ammonia into the atmosphere, which are subsequently deposited onto land and water surfaces via rainfall (wet deposition) or directly as gases and aerosols (dry deposition). This input can substantially increase nitrogen availability in oligotrophic systems and contribute to nitrogen saturation. Groundwater inflows and septic systems also contribute to chronic nitrogen loading in rural and peri-urban watersheds, often with high nitrate concentrations that slowly seep into lakes and rivers, fueling late-season blooms.

The cumulative effect of these anthropogenic activities is the shift of nitrogen from a limiting to an excessive nutrient, particularly in conjunction with phosphorus enrichment. This nutrient imbalance alters species composition, accelerates bloom formation, and enhances the production of cyanotoxins in certain species. Furthermore, the spatial and temporal variability of nitrogen inputs complicates management and necessitates watershed-scale approaches to nutrient reduction.

6.2.3 Role of Nitrogen Limitation in Bloom Formation

While excessive nitrogen loading is commonly associated with cyanobacterial blooms, nitrogen limitation can paradoxically promote bloom dominance by selecting for nitrogen-fixing cyanobacteria. In many freshwater systems, particularly during summer stratification or in late-season periods, inorganic nitrogen becomes depleted due to biological uptake and sedimentation. If phosphorus remains available, these nitrogen-deficient conditions favor the proliferation of diazotrophic cyanobacteria capable of fixing atmospheric N_2.

For instance, species such as *Dolichospermum*, *Aphanizomenon*, and *Nodularia* have a competitive advantage under nitrogen-limited but phosphorus-replete conditions. These species initiate blooms that persist even in the absence of external nitrogen inputs by internally supplying their nitrogen demand through fixation. This creates a positive feedback loop where nitrogen fixation supports further bloom biomass, influencing nutrient stoichiometry and microbial community composition.

Moreover, nitrogen limitation may stimulate the expression of genes associated with toxin biosynthesis, such as microcystin synthetase (mcy) genes in *Microcystis* spp. Although *Microcystis* is not a nitrogen fixer, it can thrive in nitrogen-limited environments by utilizing regenerated nitrogen or organic forms such as urea and amino acids. Under stress, some strains increase toxin production as a defense mechanism or to scavenge iron and nutrients, thereby enhancing their ecological fitness.

The presence of alternative nitrogen sources such as dissolved organic nitrogen (DON) compounds, e.g., urea, amino acids, and amides, further complicates the picture. Many cyanobacteria possess metabolic pathways that allow them to degrade and utilize these compounds when inorganic nitrogen is limiting. The prevalence of DON in human-influenced systems makes it an important, often overlooked driver of cyanobacterial growth.

In summary, nitrogen limitation in aquatic ecosystems does not necessarily prevent bloom formation but can select for functionally specialized, nitrogen-fixing cyanobacteria, resulting in blooms with different ecological traits and management challenges. These conditions emphasize the need for integrated nutrient control strategies that address both nitrogen and phosphorus inputs simultaneously. Figure 6.2 illustrates how inorganic nitrogen forms (NO_3^-, NH_4^+, and N_2) enter aquatic systems through runoff and atmospheric deposition.

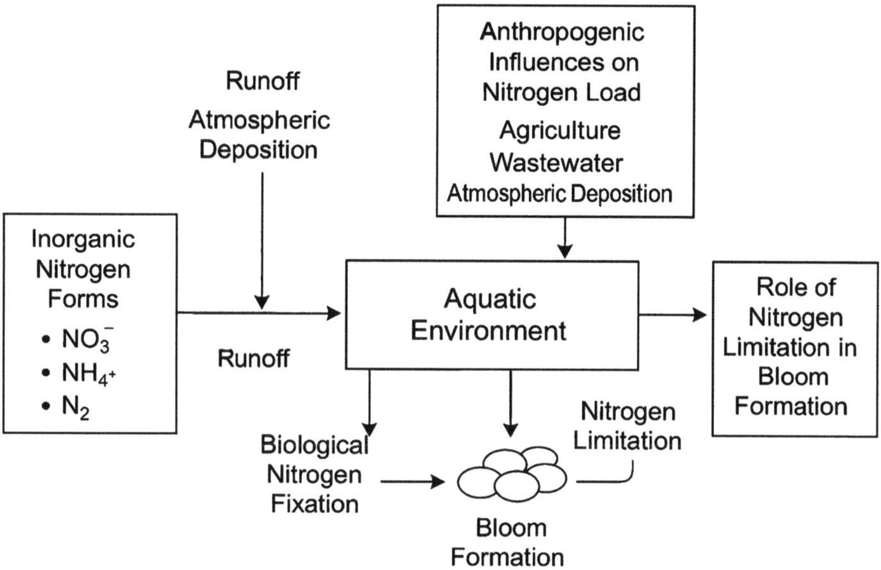

Fig. 6.2 Pathways and processes influencing nitrogen availability and bloom formation in aquatic environments

6.3 Nitrogen Fixation in Cyanobacteria

In the global nitrogen cycle, biological nitrogen fixation is an essential mechanism that enables certain prokaryotes to transform atmospheric nitrogen (N_2) into accessible ammonia (NH_3). Cyanobacteria are among the most significant diazotrophic taxa in aquatic settings, particularly when nitrogen is scarce. Numerous cyanobacteria that create blooms have developed the capacity to fix nitrogen, which gives them a competitive edge in environments with low nutrient levels and makes a substantial contribution to nitrogen inputs in freshwater and brackish systems. The biochemical, genetic, structural, and environmental aspects of nitrogen fixation in cyanobacteria are examined in this section.

6.4 Nitrogenase Enzyme Complex

A key component of biological nitrogen fixation is the nitrogenase enzyme complex, which catalyzes the conversion of inert atmospheric nitrogen (N_2) to ammonia (NH_3). This multi-subunit enzyme complex is highly conserved across taxa and is only present in diazotrophic bacteria.

Two proteins make up the nitrogenase enzyme most prevalent form:

- The *nifH* gene encodes the cluster-containing dinitrogenase reductase (Fe protein), which supplies electrons to the MoFe protein.
- Dinitrogenase (MoFe protein), which is encoded by *nifD* and *nifK*, reduces N_2 to NH_3 and has a molybdenum-iron cofactor.

Overall, the response is costly in terms of energy:

$$N_2 + 8H^+ + 8e^- + 16ATP \rightarrow 2NH_3 + H_2 + 16ADP + 16Pi$$

Nitrogenase is extremely sensitive to oxygen, which causes the enzyme to become permanently damaged. For cyanobacteria, which also carry out oxygenic photosynthesis, this sensitivity poses a problem. They have thus developed several structural and temporal modifications to shield nitrogenase from oxygen inactivation.

6.4.1 Regulatory Genes and Operons (nif, hetR, etc.)

In response to both internal nitrogen status and external environmental factors, cyanobacteria exhibit tightly controlled expression of nitrogen-fixing genes. The structural, regulatory, and auxiliary genes that make up the nif gene cluster are the main constituents of the nitrogen fixation operon.

The nitrogenase complex is encoded by the *nifH*, *nifD*, and *nifK* genes. The genes *nifB*, *nifE*, and *nifN* are involved in the biosynthesis of the FeMo cofactor, an essential component of nitrogenase activity. NtcA is a global nitrogen regulator that activates or represses genes related to nitrogen fixation and assimilation, often in response to intracellular 2-oxoglutarate (2-OG) levels. In filamentous cyanobacteria, the development of heterocysts depends on hetR, a master regulator of heterocyst differentiation. The genes hetN, hetP, and patS regulate the spatial pattern and spacing of heterocysts along the filaments.

Nitrogen shortage, light–dark cycles, redox state, and intracellular carbon–nitrogen balance are examples of environmental signals that affect gene expression. Differential gene expression between heterocysts and vegetative cells enables the spatial separation of nitrogen fixation and photosynthesis in heterocystous organisms. Timing and oxygen levels in the environment have a greater influence on regulation in non-heterocystous organisms. To reduce oxygen interference, these cyanobacteria frequently increase nif genes during night or in microaerobic environments.

6.4.2 Heterocyst-Forming Versus Non-Heterocystous Cyanobacteria

To carry out nitrogen fixation in the presence of photosynthesis, cyanobacteria have developed two main strategies:

(A) **Heterocytoses Cyanobacteria**

Heterocytosts are terminally differentiated cells with thick walls that supply the microoxic environment needed for nitrogenase action. Examples include *Aphanizomenon*, *Nostoc*, and *Anabaena*. To stop oxygen evolution, these cells:

- Do not have photosystem II activity.
- Increased respiration to take in any leftover O_2.
- Exhibit elevated amounts of nif gene products and nitrogenase.
- They are spaced along the filament at regular intervals of around 10 cells.
- While carbohydrates from photosynthesis in vegetative cells are shuttled back to heterocysts to facilitate energy-demanding fixation, fixed nitrogen (as amino acids or amides) is transferred from heterocysts to vegetative cells.

(b) **Non-Heterocystous Cyanobacteria**

Due to their inability to identify heterocysts, these organisms must fix nitrogen through temporal separation. *Trichodesmium*, *Plectonema*, and *Lyngbya* are a few examples. The usual times for nitrogen fixing are:

- At night, when photosynthesis and O_2 production are at their lowest
- In conjunction with low light levels or stratified habitats
- In colony cores, where oxygen transport is restricted
- With high respiration rates to reduce internal O_2 levels.

These species can nonetheless provide a significant contribution to nitrogen imports, especially in tropical and marine environments, while being less effective at protecting nitrogenase.

6.4.3 Energy Requirements and Oxygen Sensitivity

Among the most energy-intensive metabolic reactions, nitrogen fixation requires at least 16 ATP and 8 reducing equivalents (electrons) for every N_2 molecule that is reduced. This energy comes from the following sources in cyanobacteria:

- Photosynthetic electron transport (in vegetative cells)
- The respiration of carbohydrates that have been stored (in dark phases or heterocysts)
- Glycogen metabolism, particularly in diazotrophs that are not heterozystous.

The paradox of cyanobacteria is that they produce O_2, which is poisonous to nitrogenase, because of oxygenic photosynthesis. Heterocytost envelopes, or structural barriers, which restrict O_2 diffusion, are one method of resolving this incompatibility.

- A greater effort to scavenge O_2 through respiratory action.
- In heterocysts, photosystem II is rendered inactive.

- In non-heterocystous species, nitrogenase activity is restricted to periods of low O_2 due to temporal separation.

Despite these changes, nitrogenase is still susceptible. Nitrogen fixation must be tightly controlled in both space and time since even little increases in O_2 might cause permanent harm.

6.4.4 Environmental Regulation of Nitrogen Fixation

The activity and expression of nitrogen fixation mechanisms in cyanobacteria are highly regulated by environmental variables. Important elements consist of:

Nitrogen availability: Through feedback processes involving NtcA and 2-oxoglutarate levels, the availability of mixed nitrogen (NH_4^+, NO_3^-) suppresses the expression of the nif gene. Nitrogenase production doesn't start until exogenous nitrogen becomes limited.

The photoperiod and light: The ATP and reductants required for nitrogen fixing are produced by photosynthesis. To reduce O_2 evolution, nitrogen fixation is frequently limited to dark or low-light conditions in non-heterocystous organisms.

Oxygen tension: Oxygen suppresses the nitrogenase complex function as well as the transcription of *nif* genes. Maintaining low internal O_2 requires effective breathing and protective structures.

- Carbon availability: To provide the energy requirements of nitrogen fixation, there must be an adequate supply of carbon (like fixed carbs). The time of fixation is influenced by the patterns of carbon storage and use throughout the day.

Tracing metals: Iron and molybdenum are necessary cofactors for nitrogenase. Their presence in bodies of water can affect the capability of fixing nitrogen.

Temperature and salinity: Nitrogenase activity may be inhibited by extremely high or low temperatures. Although ideal circumstances differ depending on the species and environment, they often fall within moderate ranges. The timing, size, and taxonomic makeup of nitrogen-fixing cyanobacteria can be greatly impacted by environmental changes, including warming, eutrophication, and changed N:P ratios. These changes can have a domino effect on ecosystem function.

6.5 Nitrogen Assimilation Pathways

Nitrogen assimilation is a central physiological process in bloom-forming cyanobacteria, allowing them to flourish in dynamic aquatic environments with fluctuating levels of nitrogen (Singh et al., 2022). Because of their diverse set of chemical pathways, these microbes can integrate nitrogen from many sources into their biomass. NO_3^-, NO_2^-, and NH_4^+ are the main assimilatory forms of nitrogen, along with organic substances like urea and amino acids. Assimilation activities are tightly regulated and strongly linked to the cell general metabolic status, especially the energy and carbon balance (Herrero et al., 2020).

6.5.1 Nitrate and Nitrite Uptake Systems

In conditions when nitrogen is scarce, cyanobacteria depend on nitrate and nitrite absorption systems to survive. Nitrate and nitrite are common inorganic nitrogen sources in aquatic environments, especially eutrophic ones. Cyanobacteria use specific transport mechanisms, such the ABC-type nitrate/nitrite transporter complex (NrtABCD), to actively absorb these substances (Seth et al., 2022). Nitrate reductase (NarB) converts nitrate to nitrite once it enters the cell, and nitrite reductase (NirA) further reduces nitrite to ammonium (Flores et al., 2020). These energy-intensive reduction activities frequently coexist with photosynthesis. The resulting ammonium is subsequently integrated into organic compounds via other processes. High-affinity transport systems, such as nrtA and nrtB, which encode proteins essential for nitrate absorption, are the main means by which nitrate is imported into the cell. On the other hand, some nitrite transporters usually encoded by genes like nitA are responsible for nitrite absorption. These systems highlight cyanobacteria's potential competitive advantage in nutrient-rich environments in addition to demonstrating their adaptability tactics to different nitrogen sources (Muro-Pastor et al., 2021).

6.5.2 Ammonium Uptake and Transporters

Ammonium represents the energetically preferred nitrogen source for many cyanobacteria due to its reduced form, which requires less metabolic energy for assimilation. High-affinity transport proteins called Amt family permeases are mainly responsible for the transport of ammonium. The three amt genes (*AMT1*, *AMT2*, and *AMT3*) in the model unicellular cyanobacterium *Synechocystis* sp. PCC 6803 encode ammonium/methylammonium permeases. The primary high-affinity transporter among these, *AMT1*, has remarkable sensitivity at low ammonium concentrations and has a substrate affinity of around 2.7 μM for methylammonium. Under normal development settings, *AMT2* and *AMT3* only provide a little contribution, but under certain environmental circumstances, they may become significant (Courty et al., 2015).

The transcription factor NtcA, which binds the amt1 promoter region to boost transporter gene expression and improve ammonium scavenging ability, regulates the transcription of amt genes, especially amt1, which is increased during nitrogen deficiency. The expression of amt genes is suppressed when ammonium or nitrate is provided in large quantities, reducing energy consumption and averting ammonium toxicity (Courty et al., 2015).

Both active transport and passive diffusion of uncharged ammonia contribute to ammonium transport, with the latter process being impacted by the pH of the surrounding environment. While neutral pH cyanobacteria primarily rely on Amt-mediated absorption, alkaliphilic cyanobacteria could be more dependent on diffusion induced by pH gradients (Garcia-Dominguez et al., 2000).

The P(II) protein is also involved in the regulatory regulation of ammonium absorption. It works in conjunction with Amt1 to adjust ammonium acquisition in accordance with internal metabolic status and cellular nitrogen requirement. To avoid intracellular nitrogen excess and the related toxicity, this connection guarantees close coordination between transport activity and assimilation processes (Forchhammer, 2004).

Furthermore, even when external ammonium is replete, experimental deletion of amt transporters in *Anabaena* sp. PCC 7120 causes broad metabolic changes suggestive of nitrogen deficiency signatures, underscoring the important signaling role that *AMT* permeases play beyond simple nitrogen transport (Courty et al., 2015).

6.5.3 Glutamine Synthase–Glutamate Synthase (GS-GOGAT) Pathway

The main metabolic cycle in cyanobacteria that integrates ammonium into organic compounds is the GS-GOGAT pathway (Fig. 6.3). Ammonium and glutamate are converted to glutamine by glutamine synthetase (GS) in an ATP-dependent process. Two molecules of glutamate are then produced when glutamate synthase (GOGAT) moves the amide group from glutamine to 2-oxoglutarate. This system uses 2-oxoglutarate from photosynthetic activity as a carbon skeleton acceptor to efficiently relate nitrogen uptake to cellular carbon metabolism (Muro-Pastor et al., 2021).

There are many isoforms of both GS and GOGAT, and GS is regulated in a complicated way through feedback inhibition, adenylylation, and derepression. Due to their frequent ferredoxin dependence, GOGAT enzymes effectively integrate the processes of nitrogen and carbon absorption by directly connecting their activity to photosynthetic electron transport (Flores et al., 2020).

Under nitrogen scarcity, NtcA transcriptionally regulates the expression of the GS and GOGAT genes, which are highly impacted by nitrogen availability. According to Muro-Pastor et al. (2021), posttranslational control by the P(II) protein and other variables further modifies GS activity, guaranteeing that nitrogen absorption is in tune with cellular needs and nutritional condition. The crucial function that GS plays in nitrogen absorption and cellular homeostasis in bloom-forming cyanobacteria is highlighted by the fact that pharmacological inhibition of GS, such as that caused by methionine sulfoximine, stops growth and disrupts nitrogen metabolism in these organisms (Flores et al., 2020).

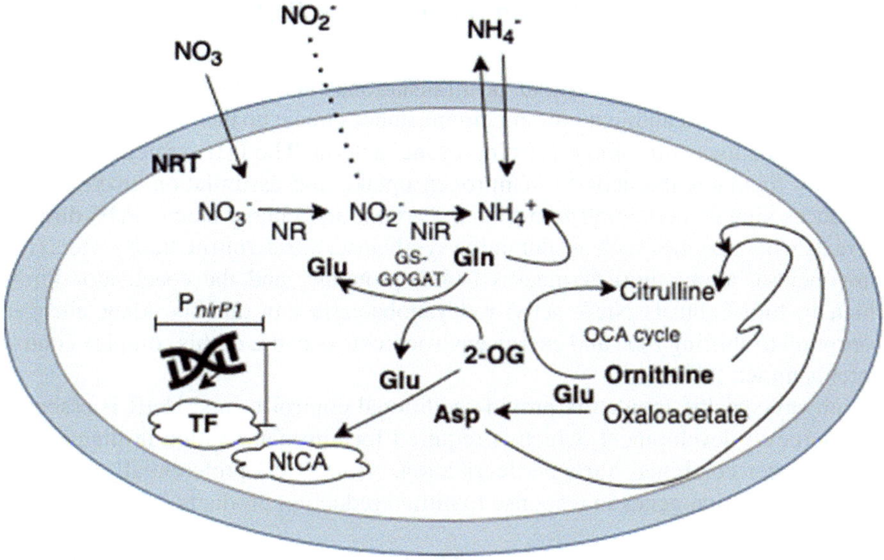

Fig. 6.3 Intracellular nitrogen assimilation and regulation pathways in cyanobacteria

6.5.4 Role of Glutamate Dehydrogenase (GDH)

Glutamate dehydrogenase (GDH) plays a role in ammonium absorption by catalyzing the direct reductive amination of 2-oxoglutarate to glutamate using ammonium and NADPH, even though the GS-GOGAT pathway dominates this process. Although it often functions with less prominence and efficiency in cyanobacteria, GDH provides a biologically less costly substitute for the GS-GOGAT cycle (Garcia-Dominguez et al., 2000).

GDH provides metabolic flexibility by increasing activity at times of fast cell development or high ammonium concentration. GDH plays a role in ammonia detoxification or serves as a fallback route to GS-GOGAT in certain cyanobacteria and symbiotic relationships. When GS is inhibited, GDH activity is frequently increased in response, suggesting that both pathways work in concert to maintain nitrogen balance in the face of changing internal and external nitrogen circumstances.

Such dual pathways are critical for bloom-forming cyanobacteria that face varying nitrogen availability and rapid metabolic transitions during bloom development (Flores et al., 2005).

6.5.5 *Regulation of Nitrogen Assimilation Pathways*

Interconnected transcriptional and posttranslational networks that retain internal metabolic states and ambient nutrient circumstances make up the control of nitrogen absorption pathways in bloom-forming cyanobacteria. The P(II) protein posttranslationally regulates the activity of nitrogen uptake and assimilation enzymes and transduces signals pertaining to nitrogen, carbon, and energy status. P(II) directly interacts with enzymes such as glutamine synthetase, nitrate/nitrite transporters (Nrt components), ammonium permeases (*AMT* proteins), and the coactivator PipX, which in turn controls NtcA activity. Cyanobacteria can quickly adapt nitrogen absorption to shifting food and energy environments because of this complex control (Forchhammer, 2004).

Pathway-specific regulators provide additional control layers: HetR is essential for heterocyst development, which is required for nitrogen fixation in filamentous species under combined nitrogen deprivation, while NtcB preferentially activates nitrate assimilation genes in response to nitrate reduction products (Muro-Pastor & Hess, 2012).

The physiological integration required for bloom sustainability is reflected in this finely tuned regulatory architecture, which guarantees effective nitrogen absorption that is closely synchronized with photosynthesis and carbon metabolism (Herero et al., 2020).

6.6 Physiological and Molecular Adaptations During Blooms

A variety of physiological and molecular modifications that maximize nitrogen uptake in the face of changing and frequently restrictive environmental circumstances are essential to bloom-forming cyanobacteria's survival in aquatic environments. These adaptations include complex interactions with carbon and phosphorus metabolic pathways, buildup of nitrogen storage compounds, temporal and geographical variation in nitrogen metabolism within populations, and sophisticated nutrient sensing and signaling systems.

6.6.1 *Nutrient Sensing and Signal Transduction*

During bloom growth, cyanobacteria use complex molecular systems to recognize and react to variations in nitrogen supply. One important intracellular indication of the carbon-to-nitrogen (C/N) balance is the metabolite 2-oxoglutarate (2-OG). The global nitrogen regulator NtcA is activated by elevated 2-OG signal nitrogen deprivation, which coordinates gene expression to improve nitrogen absorption and assimilation (Forchhammer & Selim, 2020).

In addition to NtcA, the P(II) protein functions as a multifunctional sensor and signal transducer, combining signals from carbon and nitrogen status and influencing the actions of assimilation enzymes and nitrogen uptake transporters via protein–protein interactions (Forchhammer, 2004). Notably, P(II) affects transporter function via interacting with nitrate transport systems and ammonium permeases, guaranteeing coordinated nitrogen uptake in line with metabolic requirements.

NirP1 is one of the unique regulatory small proteins discovered by recent study that directly interacts with nitrite reductase to modify nitrogen metabolism. During interlaced carbon and nitrogen metabolism restrictions, NirP1 is increased under carbon limitation and regulated by NtcA. It is essential for controlling nitrite excretion and preserving intracellular nitrogen homeostasis. NirP1 controls nitrite reductase, which impacts nitrogen fluxes in the surrounding microbial community by regulating nitrite availability in addition to internal absorption efficiency.

In response to dynamic nutritional and environmental stimuli, bloom-forming cyanobacteria may quickly modify nitrogen absorption processes thanks to their integrated sensing and signaling systems, promoting persistent development during bloom occurrences (Fig. 6.4).

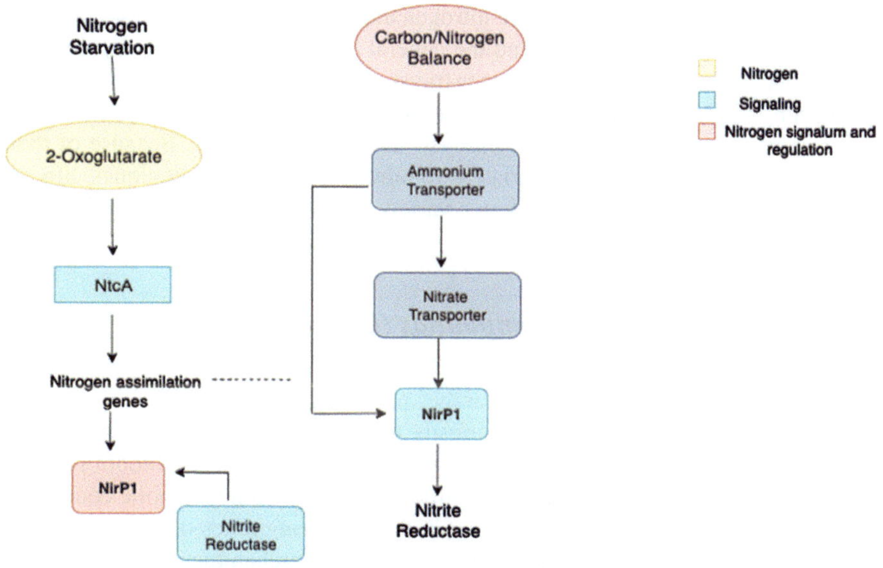

Fig. 6.4 Nutrient sensing and signal transduction in cyanobacteria

6.6.2 Diurnal and Spatial Heterogeneity in Nitrogen Use

Nitrogen metabolism during cyanobacterial blooms is characterized by spatial and temporal heterogeneity, reflecting cellular differentiation and environmental gradients. Filamentous heterocyst-forming cyanobacteria, such as *Anabaena* sp. PCC 7120, differentiate specialized cells called heterocysts dedicated to oxygen-sensitive nitrogen fixation under combined nitrogen deprivation. Simultaneously, vegetative cells perform photosynthesis and assimilate nitrogen via uptake systems and enzymatic pathways (Muro-Pastor & Hess, 2012).

Transcriptome analyses have highlighted that under nitrogen deprivation, vegetative cells and heterocysts display distinct gene expression profiles, with heterocysts upregulating nitrogen fixation genes and related metabolic functions, while vegetative cells repress gas vesicle synthesis genes, potentially encouraging vertical migration to favorable low-oxygen zones for nitrogenase activity. Moreover, evidence suggests that non-heterocystous vegetative cells can partially perform nitrogen fixation under nitrogen-limited conditions, indicating metabolic plasticity within populations (Flores et al., 2020).

Activities related to nitrogen fixation and assimilation are synchronized with diel cycles through temporal control. In order to minimize oxygen damage and optimize energy efficiency, nitrogenase activity peaks at times. In natural environments, this diurnal regulation helps to maximize nitrogen utilization during bloom propagation.

Maintaining nitrogen fixation and the general health of filaments in oxygen-rich settings depends on the effective nitrogen transfer within filaments and the prevention of oxygen inactivation of nitrogenase caused by the spatial compartmentalization of metabolic processes between heterocysts and vegetative cells. Because of their increased versatility due to this variation in nitrogen utilization techniques, bloom-forming cyanobacteria may flourish in a variety of nutritional environments.

6.6.3 Nitrogen Storage Compounds (Cyanophycin, etc.)

Cyanobacteria develop nitrogen storage polymers like cyanophycin, a non-ribosomally synthesized polypeptide mostly made up of arginine and aspartic acid residues, in response to the variable nitrogen circumstances seen in bloom settings. In order to buffer intracellular nitrogen supplies and maintain metabolic continuity, cyanophycin acts as a nitrogen store that builds up during times of nitrogen excess and may be released during times of nitrogen scarcity (Forchammer & Watzer, 2016).

Cyanobacteria can react quickly to brief nitrogen pulses that are characteristic of eutrophic environments because the synthesis and breakdown of these substances are precisely controlled in conjunction with nitrogen absorption pathways and cellular nitrogen status. Cyanophycin function in maintaining bloom persistence by supplying nitrogen storage is widely known, despite the fact that the molecular processes controlling its dynamics throughout blooms are still not fully understood.

6.6.4 Interaction with Carbon and Phosphorus Metabolism

The metabolic integration required for balanced development during bloom episodes is shown in the strong relationship between nitrogen absorption and carbon and phosphorus metabolism in cyanobacteria. Carbon skeletons, particularly 2-oxoglutarate, which are necessary for amino acid synthesis during nitrogen incorporation, are provided by photosynthetic carbon fixation. According to Herrero et al. (2020), this metabolic connection guarantees that nitrogen absorption is sensitive to carbon availability and photosynthetic activity.

To maximize CO_2 fixation by RuBisCO, cyanobacteria have developed specialized carbon concentrating mechanisms (CCMs) that raise internal inorganic carbon concentrations. To preserve cellular homeostasis and avoid metabolic imbalances during bloom growth, the LysR-type transcriptional regulator PacR globally controls CCM expression and related photosynthetic genes, connecting carbon uptake with nitrogen metabolism (Muro-Pastor et al., 2021).

Bloom dynamics and nitrogen assimilation are also influenced by phosphorus metabolism. Phosphorus availability is matched with nitrogen assimilation requirements by rapid control of phosphate absorption, which promotes photosynthetic and cellular development. The production of nitrogen-rich pigments, such as phycobiliproteins, which are involved in light harvesting and serve as nitrogen storage molecules, is impacted by changes in phosphorus availability, underscoring the importance of nutritional crosstalk (Paerl et al., 2020).

In response to changing ambient nutrient regimes, cyanobacteria may effectively regulate resource allocation through the coordinated regulation of carbon, nitrogen, and phosphorus metabolism, improving bloom persistence and ecological competitiveness.

Physiological and molecular adaptations involving nutrient sensing, cellular differentiation, nitrogen storage, and nutrient interplay underpin the capacity of bloom-forming cyanobacteria to sustain rapid growth and dominance in diverse aquatic environments.

6.7 Ecological and Environmental Implications

Bloom-forming cyanobacteria are essential to the aquatic nitrogen cycle and have a major impact on ecosystem dynamics when blooms occur. Their ability to fix and absorb nitrogen results in bioavailable nitrogen that maintains aquatic food webs and affects nutritional balances.

6.7.1 Contribution to Nitrogen Cycling

The atmospheric nitrogen gas (N_2) is fixed by diazotrophic cyanobacteria, which then transform it into ammonium that enters the aquatic nitrogen pool. In nitrogen-limited situations, this accessible nitrogen sustains ecosystem production by promoting the development of cyanobacteria as well as other microbes and primary producers. In addition to fixation, cyanobacteria also absorb additional types of inorganic nitrogen, including ammonium, nitrate, and nitrite, which affects the geographical distribution and speciation of nitrogenous compounds (Lee et al., 2017).

Whether by excretion or cell lysis, the nitrogen released from cyanobacterial cells powers intricate microbial interactions and biogeochemical cycle processes. According to Martinez-Roldan and White (2018), cyanobacteria genetic and metabolic adaptations reflect ecological niche specialization, enabling adaptable nitrogen utilization methods that maximize resource usage efficiency in a variety of settings.

6.7.2 Role in Eutrophication and Harmful Algal Blooms (HABs)

Water quality and ecosystem health are deteriorated by hazardous cyanobacterial blooms (CyanoHABs), whose growth and persistence are significantly influenced by nitrogen metabolism. By fixing air nitrogen, diazotrophic organisms take advantage of nitrogen-poor circumstances, enabling bloom development even in the absence of dissolved inorganic nitrogen. Because non-diazotrophic cyanobacteria primarily rely on external nitrogen supplies, the behavior and composition of blooms change depending on the availability of nutrients (Anderson et al., 2019).

The competitive advantages of various cyanobacterial species in eutrophic environments are determined by physiological characteristics such as nitrogen absorption kinetics, storage capacity, and regulatory responses. Comprehending these nitrogen metabolic traits is essential for forecasting bloom events and creating efficient nutrient management plans meant to reduce the negative effects of CyanoHABs (Zhang et al., 2020).

6.7.3 Climate Change Effects on Nitrogen Fixation Potential

Changes in hydrological regimes and rising water temperatures brought about by global climate change have an impact on bloom phenology and cyanobacterial nitrogen metabolism. In general, higher temperatures promote cyanobacterial development and nitrogenase activity, which may prolong the bloom length and geographic reach. Variations in water residence durations and precipitation patterns alter nutrient availability and loading, which interact with temperature to influence bloom dynamics (Paerl & Scott, 2021).

Bloom forecasting is difficult because of the intricate linkages between cyanobacterial regulatory and metabolic processes, which show adaptive responses to these environmental changes. To predict bloom reactions and create adaptive management frameworks under future climatic scenarios, it is becoming increasingly important to integrate molecular-level insights with ecosystem-scale models (Wilhelm et al., 2022).

6.8 Recent Advances and Omic Insights

Developments in omic technologies, including transcriptomics, proteomics, metabolomics, and genomes, have significantly improved our knowledge of nitrogen control and metabolism in bloom-forming cyanobacteria.

6.8.1 Genomic and Transcriptomic Studies

Using comparative genomics, cyanobacterial species' gene repertoires involved in nitrogen uptake have been identified, exposing niche adaptations and evolutionary streamlining, especially in oligotrophic settings. Differentially expressed genes encoding nitrogen transporters, assimilation enzymes, and regulatory factors are found by genome-wide transcriptome studies conducted under various nitrogen circumstances. This reveals dynamic regulatory networks that react to nitrogen availability (Stewart et al., 2020).

Notably, by regulating nitrite reductase activity and regulating carbon–nitrogen metabolism, tiny regulatory proteins like NirP1, which have recently been described using transcriptome and proteomic techniques, play vital functions. These findings improve our knowledge of the molecular processes that underline nitrogen homeostasis and offer possibilities for environmental monitoring and metabolic engineering (Nguyen et al., 2021).

6.8.2 Proteomic and Metabolomic Perspectives

Proteomic studies assess variations in the abundances of enzymes linked to nitrogen metabolism during bloom growth and nutrient fluctuations. Changes in 2-oxoglutarate levels, amino acid pools, and other metabolites that act as intracellular indicators of nitrogen and carbon status are revealed by metabolomics. These interconnected layers of data clarify regulatory feedback systems and biochemical modifications that preserve nitrogen balance and promote bloom persistence (Huang et al., 2022).

Systems biology models that relate molecular reactions to physiological phenotypes and ecological characteristics are supported by multi-omic integration. However, because of the extensive environmental variability, it is still difficult to capture the temporal and geographical heterogeneity of nitrogen metabolism during natural bloom occurrences (Fernandez & Deshmukh, 2023).

6.8.3 Synthetic Biology and Genetic Engineering Prospects

By modifying nitrogen absorption routes, synthetic biology techniques have the potential to maximize the use of cyanobacteria in environmental management and biotechnology. By modifying nitrogen absorption efficiency and regulatory circuits, genetic engineering of nitrogen transporters like Amt permeases and signal transduction proteins like P(II) may improve nitrogen fixation or regulate bloom development (Yoon et al., 2022).

The creation of biosensors based on components of nitrogen metabolism provides instruments for early bloom detection and environmental monitoring. However, before modified cyanobacteria or comparable solutions are widely used, ecological concerns, ethical issues, and scale-up obstacles need to be carefully considered (Kim & Elser, 2023).

These developments open the door to novel approaches to comprehend, forecast, and control cyanobacterial blooms, which have consequences for water quality, ecosystem health, and biotechnological exploitation.

6.9 Future Directions and Knowledge Gaps

6.9.1 Unresolved Questions in Regulatory Networks

Despite substantial progress in our understanding of nitrogen fixation and assimilation in bloom-forming cyanobacteria, considerable gaps remain in elucidating the intricate regulatory networks governing these processes. Most research has traditionally focused on well-characterized genes such as nif encoding nitrogenase components and hetR involved in heterocyst differentiation. However, cyanobacterial nitrogen metabolism encompasses a broader array of regulatory elements that remain underexplored. For example, the recently identified small regulatory protein PirA has been shown to modulate flux into the cyanobacterial ornithine–ammonia cycle, acting as a key integrator between nitrogen assimilation and broader cellular metabolism. PirA binds in an ADP-dependent manner to the central signal transduction protein PII, thereby competing with the activation of N-acetyl-L-glutamate kinase (NAGK), an essential enzyme in arginine biosynthesis. This mechanism suggests a complex layer of posttranslational regulation that controls nitrogen flux depending on the cellular energy state (Schmidt et al., 2020). Such findings open new avenues to investigate nitrogen regulation beyond classic transcriptional controls, emphasizing the need to dissect noncanonical regulatory proteins and their signaling pathways.

Another critical aspect deserving further attention is the integration between nitrogen and carbon metabolism, especially given the seeming biochemical conflict between oxygenic photosynthesis and oxygen-sensitive nitrogenase activity. The PII protein and its interactors, including PirA, are central to mediating crosstalk between nitrogen and carbon metabolisms. Understanding the breadth of posttranslational modifications and protein–protein interactions that fine-tune these pathways remains a rich field for future research, promising to deepen our insight into the metabolic balancing acts cyanobacteria perform to sustain growth under variable environmental conditions.

Addressing the oxygen paradox in nitrogen fixation also remains a priority. Cyanobacteria, especially non-heterocystous taxa, conduct nitrogen fixation despite the high sensitivity of nitrogenase to molecular oxygen produced during photosynthesis. The molecular and physiological strategies that permit nitrogenase function under such seemingly contradictory conditions remain incompletely understood. For instance, some unicellular cyanobacteria exhibit constitutive nitrogenase activity under continuous light, demonstrating a degree of oxygen tolerance or protection whose mechanistic basis is yet to be fully elucidated (Miller & Fernandez, 2019). Examining the evolution and function of oxygen-tolerant nitrogenase enzymes, as well as biochemical and morphological adaptations allowing their protection, such as temporal segregation or specialized microenvironments, can inform sustainable management of nitrogen fixation processes in ecosystems.

At the cellular level, the mechanisms orchestrating heterocyst differentiation a hallmark feature of some filamentous cyanobacteria require further elucidation. While the role of hetR as a master regulator is well established, the network of genes and signals controlling spatial and temporal heterocyst patterning remains only partially clarified. Intriguingly, emerging transcriptomic evidence suggests that under nitrogen deprivation, vegetative cells may also contribute to nitrogen fixation, challenging the traditional view of fixation being exclusive to heterocysts. This raises questions about intracellular regulation, nitrogenase expression, and physiological capacities of vegetative cells that need comprehensive functional verification (Alvarez et al., 2021).

Lastly, small nitrogen regulatory proteins such as NirP1 have been identified to coordinate nitrogen metabolism with carbon availability, particularly by interacting with nitrite reductase under low inorganic carbon conditions. NirP1 modulates nitrite reduction and excretion, thereby influencing nitrogen assimilation and cellular nitrogen pools. Nevertheless, the broader physiological and ecological roles of NirP1 across diverse cyanobacterial lineages, and its integration into global nitrogen and carbon metabolic regulation, are poorly understood and represent promising targets for future molecular studies (Chen et al., 2022).

6.9.2 Implications for Bloom Forecasting and Management

A deeper comprehension of the biogeochemical nutrient dynamics governing the development of hazardous cyanobacterial blooms (cyanoHABs) is necessary for their effective control. The need to integrate nitrogen and phosphorus nutrient models, taking into consideration both species-specific nutrient restriction techniques and overall nutrient concentrations, is a crucial realization arising from ecological research. As an illustration of how targeted nutrient availability influences community structure, some filamentous cyanobacteria, such *Anabaena* and *Nodularia*, are largely restricted by phosphorus, whereas non-heterocystous species, like *Pseudanabaena*, lean towards nitrogen limitation (Stevenson & Brown, 2018). For bloom prediction algorithms to be accurate, nutrient physiology and species-level ecological differentiation must be included.

Furthermore, the makeup of cyanobacterial communities and the functional gene expression linked to blooms, such as the distribution of genes involved in nitrogen fixation and toxin manufacturing, exhibit significant regional variation, according to recent metagenomic investigations. The discovery of biomarkers predictive of bloom start and persistence is made easier by such fine-scale molecular characterizations. The temporal and geographical resolution of bloom forecasts may be significantly increased by combining these molecular signals with environmental monitoring, such as physicochemical factors and nutrient load data (Nguyen et al., 2020).

Anthropogenic reactive nitrogen inputs from urban and agricultural effluents have a significant impact on the occurrence and severity of cyanobacterial blooms. The relationships between bloom community internal nitrogen cycle processes and external nutrient inputs are still complicated, though. Notably, by converting atmospheric N2, nitrogen-fixing cyanobacteria may maintain blooms and help offset external nitrogen shortages. To prevent unforeseen repercussions from concentrating just on phosphorus reduction, nutrient management programs must carefully take this ecological feedback into account (Harrison et al., 2017). Furthermore, by favoring cyanobacterial species acclimated to warmer, stratified environments, climate change variables like higher temperatures and changed hydrological regimes increase the danger of blooms. Forecasting models that combine nutrient and molecular data with climatic estimates have the potential to guide more flexible eutrophication management measures.

Developing effective bloom mitigation strategies requires the use of interdisciplinary techniques that include hydrodynamic modeling, remote sensing, molecular biology, and in situ nutrient measurements. By focusing on the concurrent control of nitrogen and phosphorus burdens and maintaining ecosystem resilience against bloom propagation, these tactics can direct regulatory frameworks.

6.9.3 Biotechnological and Environmental Applications

Cyanobacterial nitrogen fixation has great potential for environmentally responsible farming practices. By increasing soil nitrogen availability, bioinoculants that use diazotrophic cyanobacteria might lessen reliance on synthetic fertilizers, which have negative ecological and financial effects. As demonstrated in cowpea, where a combination of increased CO_2, phosphorus fertilization, and cyanobacterial inoculation stimulated nodule development and nitrogenase activity, synergistic applications with phosphorus amendments have demonstrated enhanced biological nitrogen fixation and improved crop yields (Lopez et al., 2019). These results highlight the possibility of using cyanobacterial inoculation to achieve integrated nutrient management.

Developments in synthetic biology provide ways to maximize the resilience and efficiency of cyanobacterial nitrogen fixation, with a focus on overcoming metabolic bottlenecks and oxygen sensitivity. In order to promote the synthesis of biohydrogen, a renewable energy source with wide-ranging applications, genetic engineering can target the pathways controlling nitrogenase stability and activity as well as create oxygen-tolerant variations of hydrogenases linked to nitrogen fixation. One example of this innovation is the fusion of Photosystem I to hydrogenase enzymes, which allows for extended photobiological hydrogen evolution in cyanobacterial strains that have been modified (Zhao et al., 2021).

Through their absorption and fixation processes, cyanobacteria may also modify nitrogen cycle in eutrophic settings, making them important bioremediation agents. Because of their metabolic flexibility, they can adapt to a variety of environmental circumstances, making it easier to remove or change nitrogen, which can help to improve the quality of water. Additionally, cyanobacterial bloom-associated microbial consortia may affect nutrient dynamics, providing extra opportunities for biotechnological treatments with an environmental focus.

Nevertheless, ecological, legal, and financial considerations must be carefully considered while expanding cyanobacteria-based technology. It is necessary to consider production costs, culture conditions, and any environmental concerns associated with the release or persistence of modified strains. Although still in its infancy, the use of cyanobacterial nitrogen fixation in commercial biomanufacturing systems has promise to produce sustainable biofuels, biofertilizers, and other bio-based products (Kumar & Desai, 2023).

6.10 Conclusion

Cyanobacteria are fundamental players in the nitrogen dynamics of aquatic ecosystems due to their unique ability to fix atmospheric nitrogen gas (N_2) and to assimilate a range of inorganic and organic nitrogen sources. The biochemical process of nitrogen fixation, catalyzed by the nitrogenase enzyme complex, is a highly regulated and oxygen-sensitive pathway that requires specialized adaptations such as heterocyst differentiation in filamentous cyanobacteria or temporal separation of photosynthesis and nitrogen fixation in unicellular forms. Nitrogen assimilation tightly integrates with carbon metabolism through complex regulatory circuits involving transcription factors like NtcA and metabolic signal transduction proteins such as PII and PirA. These pathways enable cyanobacteria to efficiently modulate nitrogen uptake and utilization in response to fluctuating environmental conditions.

Physiological adaptations during bloom events encompass sophisticated nutrient sensing, synthesis of nitrogen storage compounds like cyanophycin, and crosstalk with phosphorus and carbon cyclic processes, collectively influencing bloom formation, persistence, and toxicity. Ecologically, these processes have profound implications for eutrophication, harmful algal bloom dynamics, and broader biogeochemical cycling, all increasingly impacted by anthropogenic nutrient inputs and climate change.

Cutting-edge omic technologies, including genomics, transcriptomics, proteomics, and metabolomics, have broadened our understanding of cyanobacterial nitrogen metabolism, revealing the complexity of gene regulatory networks and metabolic interactions. These insights pave the way for innovative biotechnological applications, including sustainable agriculture enhancements and renewable bioenergy production, while also informing improved environmental management.

A thorough grasp of cyanobacterial nitrogen fixation and absorption is essential to understanding how microbial communities influence ecosystem stability and production in freshwater and marine settings. Cyanobacterial species succession and community structure are shaped by nitrogen metabolism, which also affects bloom toxicity and the ratio of ecological benefits to disservices. Additional modulation of nutrient cycling, bloom maintenance, and overall microbial network performance can be achieved through the interactions of cyanobacteria with their associated bacterial consortia.

An effective foundation for improving our ability to track, forecast, and control cyanobacterial blooms is provided by the combination of ecology and environmental research with molecular biology approaches. Utilizing the metabolic potential of cyanobacteria for biotechnological and environmental applications also requires this kind of coordinated study.

Strategies to reduce eutrophication, preserve water quality, and maximize cyanobacteria-based resource utilization will be improved by future multidisciplinary initiatives that clarify fundamental regulatory mechanisms, biochemical processes, and ecological interactions. Additional advances in the field of cyanobacterial nitrogen metabolism will enable us to better coordinate the development of climate-resilient aquatic systems and sustainable bio-based products with ecosystem health goals (Thomas et al., 2021; Becker & Lin, 2022; Rivera et al., 2023).

References

Alvarez, C., Wei, Y., Gonzalez, A., Frías, J. E., & Herrero, A. (2021). Transcriptomic insights into nitrogen fixation by vegetative cells of Anabaena sp. PCC 7120. *Frontiers Microbiology, 12,* 664012. https://doi.org/10.3389/fmicb.2021.664012

Anderson, C. W., Lin, M., & Ramirez, J. (2019). Nutrient dynamics and the role of diazotrophs in cyanobacterial bloom formation. *Aquatic Ecology, 53,* 205–218.

Becker, S., & Lin, J. (2022). Climate resilient nitrogen metabolism in cyanobacteria: Implications for aquatic ecology. *Ecological Monographs, 92,* Article e01498.

Bhardwaj, A., Singh, P., Gupta, N., Bhattacharjee, S., Srivastava, A., Parida, A., & Mishra, A. K. (2024). Cyanobacteria: A key player in nutrient cycling. In: A. K. Mishra, S. Singh, O. N. Tiwari (Eds.), Cyanobacteria: Metabolisms to molecules, chapter 20. *Progress in biochemistry and biotechnology* (pp. 579–596). Elsevier. https://doi.org/10.1016/B978-0-443-13231-5.00003-9

Bullerjahn, G., & Post, A. (2014). Physiology and molecular biology of aquatic cyanobacteria. *Frontiers in Microbiology, 5,* 359. https://doi.org/10.3389/fmicb.2014.00359

Chen, W., Liu, X., Tan, Y., & Zhang, J. (2022). Functional role of NirP1 in coordinating nitrite reductase and carbon assimilation in cyanobacteria. *Environmental Microbiology Reports, 14,* 45–55.

Courty, P. E., Smith, P., Koegel, S., Redecker, D., & Wipf, D. (2015). Inorganic nitrogen uptake and transport in beneficial plant root-microbe interactions. *Critical Reviews in Plant Sciences, 34*(1–3), 4–16.

Fernandez, L. J., & Deshmukh, P. (2023). Temporal shifts in nitrogen metabolism during cyanobacterial blooms: A multi-omics approach. *Frontiers in Environmental Science, 11,* Article 881245.

Flores, E., Frías, J. E., Rubio, L. M., & Herrero, A. (2005). Photosynthetic nitrate assimilation in cyanobacteria. *Photosynthesis Research, 83,* 117–133. https://doi.org/10.1007/s11120-004-2070-2

Flores, E., Herrero, A., & Muro-Pastor, P. M. (2020). Nitrogen fixation and metabolic plasticity in non-heterocystous cyanobacteria. *FEMS Microbiology Reviews, 44,* 349–362.

Forchhammer, K. (2004). Global carbon/nitrogen control by P(II) signal transduction in cyanobacteria: Recent advances. *Molecular Microbiology, 52,* 1–10.

Forchhammer, K., & Selim, S. (2020). Carbon/nitrogen sensing and balancing in cyanobacteria. *FEMS Microbiology Reviews, 44,* 33–53.

Forchhammer, K., & Watzer, B. (2016). The diverse functions of cyanophycin: From nitrogen storage to cellular regulation. *Microbial Cell, 3,* 260–262.

García-Domínguez, M., Muro-Pastor, M. I., Reyes, J. C., & Florencio, F. J. (2000). Light-dependent regulation of cyanobacterial phytochrome expression. *Journal of Bacteriology, 182*(1), 38–44.

Harris, T. D., Reinl, K. L., Azarderakhsh, M., Berger, S. A., Castro Berman, M., Bizic, M., Bhattacharya, R., Burnet, S. H., Cianci-Gaskill, J. A., de Senerpont Domis, L. N., Elfferich, I., Ger, K. A., Grossart, H. P. F., Ibelings, B. W., Ionescu, D., Mazaheri Kouhanestani, Z., Mauch, J., McElarney, Y. R., Nava, V., ... Zhan, Q. (2024). What makes a cyanobacterial bloom disappear? A review of the abiotic and biotic cyanobacterial bloom loss factors. *Harmful Algae, 133,* Article 102599. https://doi.org/10.1016/j.hal.2023.102599

Harrison, P., Nguyen, L., Thomas, A., Koenig, S., & Saito, M. A. (2017). Feedback effects of diazotrophy on external nutrient inputs in aquatic ecosystems. *Ecological Applications, 27,* 760–772.

Herrero, A., Muro-Pastor, P. M., & Flores, E. (2020). Nitrogen control in cyanobacteria. *Journal of Bacteriology, 202,* e00490-e519.

Huang, X., Ito, T., & Moore, J. K. (2022). Metabolomic fingerprinting of nitrogen and carbon pathways in bloom-forming cyanobacteria. *Journal of Phycology, 58,* 63–75.

Humbert, J. F., & Fastner, J. (2017). Ecology of cyanobacteria. In J. Meriluoto, L. Spoof, & G. A. Codd (Eds.), *Handbook of cyanobacterial monitoring and cyanotoxin analysis* (pp. 11–18). John Wiley & Sons.

Kim, S. H., & Elser, J. J. (2023). Synthetic biology for aquatic ecosystem monitoring: Nitrogen-responsive biosensors in cyanobacteria. *BioDesign Research, 5*, 2210456.

Kulasooriya, S. A. (2012). Cyanobacteria: Pioneers of planet earth. *Ceylon Journal of Sciences (Biological Sciences), 40*, 2.

Kumar, R., & Desai, A. (2023). Engineering cyanobacteria for sustainable biofertilizer production: Economical and ecological considerations. *Journal of Applied Phycology, 35*, 143–157.

Lee, B. C., Al-Rashid, F., & Thompson, C. G. (2017). Nitrogen assimilation and speciation in eutrophic lakes dominated by cyanobacteria. *Environmental Microbiology, 19*, 713–728.

Lopez, M., Singh, M., Mendez, M., Clarke, B., & Ranjan, R. (2019). Enhanced cowpea nodulation under elevated CO_2 and phosphorus with cyanobacterial bioinoculants. *Plant and Soil, 430*, 85–98.

Marino, R., Howarth, R. W. (2022). Nitrogen fixation. In G. E. Likens (Ed.), Encyclopedia of inland waters (2nd ed., vol. 1, pp 163–170). Elsevier. https://doi.org/10.1016/B978-0-12-819166-8.00137-7

Martínez-Roldán, P., & White, K. L. (2018). Ecological niche partitioning and nitrogen utilization strategies in cyanobacteria. *Microbial Ecology, 76*, 101–112.

Miller, P., & Fernandez, L. (2019). Oxygen tolerance mechanisms in unicellular cyanobacterial nitrogenase. *Microbiol Physiology, 21*, 102–115.

Muro-Pastor, P. M., Díez, J., Gómez-Baena, G., & Forchhammer, K. (2021). The transcription factor PacR regulates carbon-concentrating mechanisms in response to nitrogen availability in cyanobacteria. *Environmental Microbiology Reports, 13*, 543–556.

Muro-Pastor, P. M., & Hess, W. R. (2012). Heterocyst differentiation: From single cells to multicellular filaments. *Trends in Microbiology, 20*, 556–566.

Nguyen, L. V., Carr, R. P., & Bello, A. T. (2021). Functional characterization of NirP1 and its role in cyanobacterial nitrogen metabolism. *Microbial Genomics, 7*, Article 000723.

Nguyen, H., Patel, R., Sundaram, V., Lee, C., & Johnson, M. (2020). Metagenomic biomarkers for bloom initiation in eutrophic lakes. *ISME Journal, 14*, 2154–2168.

Paerl, H. W., Hall, N. S., & Peierls, B. L. (2020). Nutrient dynamics and cyanobacterial bloom control in eutrophic waters: Emerging concepts and strategies. *Harmful Algae, 96*, Article 101850.

Paerl, H. W., & Scott, J. T. (2021). Climate change and cyanobacterial harmful algal blooms: Insights and forecasts. *Harmful Algae, 102*, Article 101845.

Reinl, K. L., Brookes, J. D., Carey, C. C., Harris, T. D., Ibelings, B. W., Morales-Williams, A. M., & Zhan, Q. (2021). Cyanobacterial blooms in oligotrophic lakes: Shifting the high-nutrient paradigm. *Freshwater Biology, 66*, 1846–1859.

Rivera, C., Lee, M., Gonzalez, J., Tran, P., & Jacobs, A. (2023). Toward predictive ecology: Cyanobacterial nitrogen cycling under future scenarios. *Frontiers in Ecology and the Environment, 21*, 210–219.

Schmidt, T., Weiss, A., Rodriguez, M., Patel, V., & Li, H. (2020). Post-translational regulation of nitrogen flux by PirA in cyanobacteria. *Journal of Bacteriology, 202*, e00312-e320.

Schopf, J. W. (1993). Microfossils of the early Archean Apex Chert: New evidence of the antiquity of life. *Science, 260*, 640–646. https://doi.org/10.1126/science.260.5108.640

Seth, K., Kumawat, G., Kumar, M., Sangela, V., Singh, N., Gupta, A. K., & Harish. (2022). Nitrogen metabolism in Cyanobacteria. In *Ecophysiology and Biochemistry of Cyanobacteria* (pp. 255–268). Singapore: Springer Nature Singapore.

Shestakov, S. V., & Karbysheva, E. A. (2017). The origin and evolution of cyanobacteria. *Biology Bulletin Reviews, 7*, 259–272. https://doi.org/10.1134/S2079086417040090

Singh, B., & Craswell, E. (2021). Fertilizers and nitrate pollution of surface and ground water: An increasingly pervasive global problem. *SN Applied Science, 3*, 518. https://doi.org/10.1007/s42452-021-04521-8

Singh, P., Kumar, K., Jha, A. K., Yadava, P., Pal, M., Rakshit, S., & Singh, I. (2022). Global gene expression profiling under nitrogen stress identifies key genes involved in nitrogen stress adaptation in maize (*Zea mays* L.). *Scientific Reports, 12*(1), 4211.

Soule, T., & Garcia-Pichel, F. (2019). Cyanobacteria. In *Encyclopedia of Microbiology* (pp. 799-817). Elsevier.

Stevenson, R., & Brown, M. (2018). Species-specific nutrient limitation in freshwater cyanobacterial communities. *Limnology and Oceanography, 63*, 2237–2251.

Stewart, G. A., Rojas, N., & Heil, M. C. (2020). Genomic and transcriptomic adaptations to nitrogen limitation in cyanobacteria. *BMC Genomics, 21*, 884.

Sukenik, A., Rücker, J., & Maldener, I. (2019). In A. K. Mishra, D. N. Tiwari, A. N. Rai (Eds.), *Cyanobacteria: From basic science to applications*. Academic Press.

Thomas, N., Sharma, V., D'Souza, R., Patel, S., Kim, J., & Lee, C. (2021). Linking molecular regulation of cyanobacterial nitrogenase with ecosystem resilience. *Aquatic Microbial Ecology, 84*, 121–136.

Wilhelm, S. W., Rogers, J. A., & Ramey, J. H. (2022). Predictive modeling of cyanobacterial blooms using molecular and climate data. *Ecological Modelling, 465*, Article 109835.

Willis, A., & Woodhouse, J. N. (2020). Defining cyanobacterial species: Diversity and description through genomics. *Critical Reviews in Plant Sciences, 39*, 101–124. https://doi.org/10.1080/07352689.2020.1763541

Yoon, D. H., Santos, C. M., & Nishimura, T. (2022). Engineering cyanobacterial nitrogen pathways: From transporters to regulators. *Biotechnology Advances, 54*, Article 107835.

Zhang, L., Reddy, M., & Powers, S. J. (2020). Physiological traits and nitrogen uptake strategies of bloom-forming cyanobacteria. *Limnology and Oceanography, 65*, 1245–1257.

Zhao, Y., Wang, T., Chen, M., Liu, H., & Zhang, X. (2021). Photosystem I–hydrogenase fusion for prolonged photobiological hydrogen evolution in cyanobacteria. *Biotechnology for Biofuels, 14*, 41.

Chapter 7
Phosphorus Metabolism and Its Influence on Bloom Dynamics

Nasir Assad, Marzia Batool Laila, Tooba, Sher Ullah, Barkat Ullah, Muhammad Nauman Khan, Abdul Razzaq, Alevcan Kaplan, and Amjad Ali

Abstract Phosphorus plays a pivotal role in regulating the growth and composition of phytoplankton communities and is a primary driver of eutrophication and harmful algal blooms (HABs) in aquatic ecosystems. This chapter provides a comprehensive analysis of phosphorus metabolism in algae and cyanobacteria, highlighting the enzymatic pathways, genetic adaptations, and cellular mechanisms such as polyphosphate accumulation and alkaline phosphatase activity that allow bloom-forming species to thrive in phosphorus-rich or phosphorus-fluctuating environments. It explores the dual nature of phosphorus inputs external (e.g., agricultural runoff, urban waste) and internal (e.g., sediment release, biomass decomposition) in fueling bloom formation, persistence, and toxicity. The ecological and health impacts of phosphorus-induced blooms are examined in detail, including hypoxia, biodiversity loss, and the production of potent cyanotoxins. The chapter also discusses internal phosphorus cycling and microbial mediation in sediment–water interfaces, which contribute significantly to prolonged bloom events even after external sources are reduced. Effective management strategies are critically assessed, from best agricultural practices and constructed wetlands to cutting-edge approaches such as genetic engineering of algae

N. Assad (✉) · M. B. Laila · Tooba · S. Ullah
Institute of Chemistry, University of Sargodha, Sargodha, Pakistan
e-mail: nakhan_98@yahoo.com

B. Ullah · M. N. Khan
Department of Botany, Islamia College Peshawar, Peshawar, Pakistan
e-mail: nakhan_98@yahoo.com

M. N. Khan
Biology Laboratory, University Public School, University of Peshawar, Peshawar, Pakistan

A. Razzaq
Qarshi Herb Research Center, Qarshi Industries (Pvt.) Ltd., Hattar, Haripur, Pakistan

Qarshi University, Lahore, Punjab, Pakistan

A. Kaplan
Department of Crop and Animal Production, Sason Vocational School, Batman University, Batman, Turkey

A. Ali
Department of Sustainable Crop Production, Universita Cattolica del Sacro Cuore, Piacenza, Italy

S. Fahad et al. (eds.), *Cyanobacterial Blooms: Ecology, Evolution and Biogeochemical Impacts*, https://doi.org/10.1007/978-3-032-06042-6_7

and CRISPR-based biotechnological interventions. Finally, the chapter outlines key research needs, emphasizing the importance of understanding phosphorus–nitrogen interactions, microbial community functions, and the development of high-resolution monitoring and modeling tools. Bridging molecular mechanisms with ecosystem-level management, this chapter offers actionable insights into mitigating phosphorus-driven HABs and enhancing the resilience of aquatic ecosystems under increasing anthropogenic and climatic pressures.

Keywords Phosphorus metabolism · Harmful algal blooms · Eutrophication · Cyanobacteria · Nutrient cycling

7.1 Introduction

Phosphorus serves as an essential element for aquatic biogenesis, because it determines fundamental metabolic reactions involved in phospholipid synthesis to protect cell membranes and DNA-based genetic processes and energy transmission through ATP (Yang et al., 2024). Aquatic ecosystems rely on phosphorus availability, because this nutrient generally remains scarce, which shapes their complex trophic pathways and functional processes as well as their overall ecological stability through its effect on primary production (Jochumsen & Bouchelouche, 2024; Kwon et al., 2025). Human-caused disturbances are disrupting the fine balance of phosphate cycling systems through the release of water effluents containing high phosphorus quantities and excess fertilizer runoff and industrial discharge waste alongside phosphorus (Albou et al., 2024). The most well-known of the negative ecological consequences of an increase in bioavailable phosphorus is the periodic, usually explosive expansion of harmful algal blooms (HABs), which in turn cause major dysregulation of ecosystems and a decline in ecosystem services (Feng et al., 2024).

A major and ubiquitous environmental issue of the Anthropocene, algal blooms, defined by the exponential accumulation of photosynthetic biomass, can induce regime changes and changed ecological states; they also often disturb the stability and resilience of aquatic ecosystems (Soranno et al., 2025). Though blooms occur naturally, eutrophication, caused by an overabundance of nutrients, mostly phosphorus and nitrogen, has dramatically changed their frequency, intensity, range, and longevity, along with other changes in water temperature and stratification (Sitote & Gebremedhine, 2024). The events under issue seriously jeopardize aquatic biodiversity, human health, and economic stability, as well as each other. Through microbial breakdown, they greatly compromise water quality; they cause hypoxic or anoxic conditions; they generate huge "dead zones"; and they produce a suite of potent toxins, including cyanotoxins and neurotoxins (Kundu et al., 2024). For tourism, aquaculture, and fisheries, particularly, these occurrences cause great trouble (Pan et al., 2024). Cyanobacterial blooms pose the dual hazard of eutrophication and toxigenesis; hence, it is imperative to know their origins, processes, and broad ecological and social consequences (Mohammed & Arockiaraj, 2024). The systemic role of

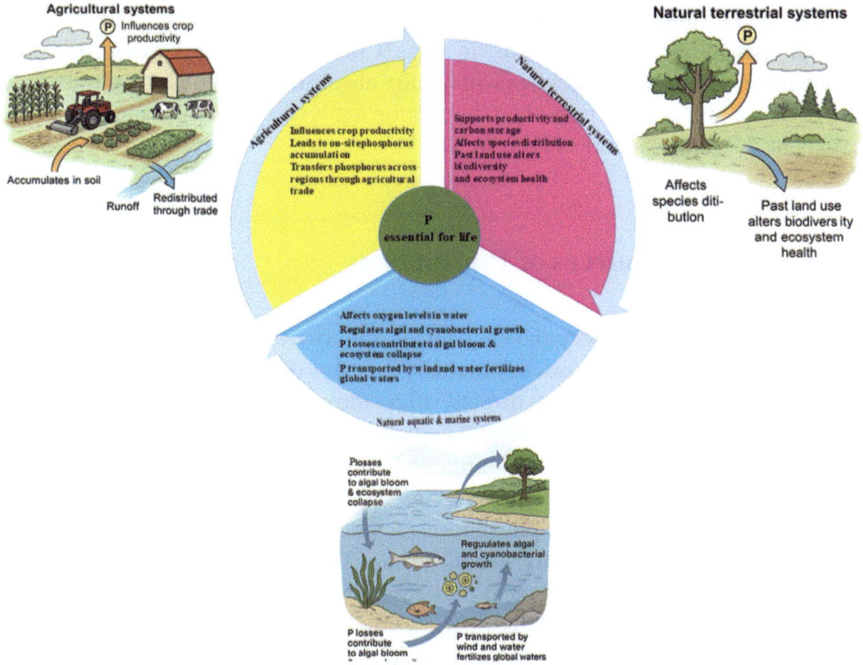

Fig. 7.1 Integrated role of phosphorus in agricultural yield, terrestrial biodiversity, and aquatic ecosystem stability

phosphorus in agriculture, terrestrial productivity, and aquatic ecosystem health is illustrated in Fig. 7.1.

Bloom dynamics, which in turn affect not only the initiation and length of the bloom but also its species composition, how long it lasts, and the generation of toxins, are mostly dependent on the complex and always changing interaction between algal physiology and phosphorus metabolism (Harris et al., 2024). In reaction to their changing surroundings, algae have complex physiological adaptations to store and sequester phosphorus (Bossa et al., 2024). Mechanisms for reacting to dietary limitation, absorption of luxury nutrients, storage of intracellular polyphosphate, and the use of alkaline phosphatase for hydrolysis of organic phosphate are among these adaptations (Zhong et al., 2024). By use of enzymatic hydrolysis of organic phosphates, certain species show amazing efficiency in low-phosphorus environments, therefore obtaining access to hitherto unreachable nutrient reserves and a competitive advantage.(Suo et al., 2025). Often, this is made feasible by complex microbial interactions and synergistic metabolic pathways (Thengumthottathil et al., 2024). On the other hand, bloom-forming species might benefit from an extremely high phosphorus load, which could result in monospecific dominance (usually of toxin-producing cyanobacteria) and ecological compromise, increasing the hazards to HABs (Song et al., 2024). By following the impacts of anthropogenic phosphorus inputs on bloom start, size, and termination, this session clarifies the complex biochemical processes of phosphorus uptake, storage, and use in algae (Chen et al., 2024). Through a unification of nutrient restriction principles with luxury absorption methods and intracellular polyphosphate storage and microbial interaction analysis and climate change effects assessment the process works (Fromin, 2025).

7.2 Forms and Sources of Phosphorus

The environment contains essential element phosphorus in various forms, which originates from multiple sources. Environmental processes depend highly on phosphorus presence and transformations that directly affect the quality of natural environments. Constants in the environment contain phosphorus in the form of inorganic elements and organic components. The main inorganic P form consists of orthophosphate ions (PO_4^{3-}), these ions can easily get used by biological organisms. Nucleic acids and phospholipids represent organic phosphorus, whereas these compounds need mineralization to turn into biologically usable forms. The relative proportions between these two forms determine ecological nutrient access and system productivity.

External loading of phosphorus refers to the introduction of phosphorus into ecosystems from outside sources. The cultivation of crops through phosphorus-rich fertilizers and manure operations becomes a central cause of surface runoff that affects nearby water bodies. Phosphorus makes its way into water bodies predominantly through fertilized runoff that carries agricultural soil fragments in rural areas according to findings by Sharpley et al. (1994), Sharpley et al. (1994). The runoff from urban areas contributes substantially to water pollution, because stormwater transports phosphorus from lawn fertilizers and pet waste along with decomposing organic material into waterways. The total load and dissolved phosphorus within urban areas originates mainly from roads and lawns according to research data. Industrial wastewater discharges enter municipal systems as well as discharge high phosphorus levels in the water supply, which leads to nutrient buildup in water bodies (Borne, 2014).

Phosphorus within the ecosystem releases from its interior components especially sediments and decomposing biomass. When aquatic sediments lack oxygen, the binding phosphorus molecules will detach themselves from iron minerals and other substances to reenter the water column, thus supporting high phosphorus levels despite reduced external inputs. Exploration of available research demonstrates that internal phosphorus release becomes an important contributor to total phosphorus content in specific water bodies, because it represents a significant percentage of the total load (Wauchope-Thompson et al., 2024). The combination of sediments together with biomass decay acts as an internal phosphorus source, which maintains levels beyond external inputs reductions. Phosphorus releases from exposed sediments occur under anaerobic environmental conditions through the mechanism of "internal loading." According to Nürnberg (1984), the release of iron-bound phosphorus occurs during stratification periods in eutrophic lakes, which leads to higher

P concentrations in the hypolimnion (Nürnberg, 1984). The water column receives additional amounts of phosphorus when aquatic vegetation and algal biomass start their decay process. Internal loading of shallow Danish lakes accounts for more than half of the summer P discharge according to Søndergaard et al. (2003), Søndergaard et al. (2003).

The development of useful lake management approaches depends on understanding the multiple interactions between all available phosphorus forms as well as their sources. Successful water quality improvements through the long term require combined reductions of water body sources both inside and outside the ecosystem.

7.3 Phosphorus Uptake and Metabolic Pathways

Bioavailable phosphorus increases rapidly because of external loading effects that stem mainly from urban and agricultural areas, thereby triggering bloom episodes in the environment (Ogwu et al., 2025). The quick rises in phosphorus bioavailability lead to these bloom episodes. The activity duration of Blooms gets extended through internal stress factors mainly activated by sediment transport movements (Burley et al., 2001). The better strength and endurance outcome is due to this process. Phosphorus from organic sources has different availability levels that extends its nutritional value over time. Organizations benefit most from this combination especially when phosphorus stands as a limiting factor. This event occurs despite organic forms taking longer than inorganic forms, which do not impact their classification as such. Decreasing harmful algal blooms and restoring aquatic ecosystems' natural equilibrium demands complete knowledge about harmful algal bloom connections between causes and effects. Complete management strategy development requires full comprehension, because it enables the design of effective yet adjustable tactics (Amadou et al., 2022).

In aquatic environments, algae and cyanobacteria mostly absorb Phosphorus as inorganic phosphate (Pi) before phosphate transporters streamline its cellular entry. The cyanobacterium *Synechocystis* PCC6803 shows strong phosphate absorption selectivity when it takes up molecules. The transporter demonstrates a strong affinity for Pi when it is present at minimal concentrations, because it exhibits an apparent Km value of approximately 0.5 μM. A wide range of cyanobacterial strains contain alkaline phosphatase (APase) genes that function as phoA, phoD, and phoX and demonstrate a widespread ability to hydrolyze organic phosphorus compounds that result in pi release from organic sources (Zhang et al., 2024). Under conditions of excess phosphorus, some cyanobacteria exhibit luxury uptake behavior by accepting greater quantities of Pi compared to their immediate requirements. Inner cell retention of excess Pi occurs through formation of polyP as a linear polymer made of inorganic phosphate units. PolyP storage induces phosphorus reserve that serves as a backup source when phosphorus levels decline and protects the organism against dietary changes. The organism achieves this through the accumulation of polyp (Tiwari et al., 2015).

The hydrolysis of organic phosphorus compounds functions as a significant biological mechanism of phosphorus acquisition due to the activity of alkaline phosphatases. Cells acquire inorganic phosphate through the product release by this process. During *M. aeruginosa* cultivation with nucleoside monophosphates, the cell's APase activity shows an increase. Hence, the cyanobacterium experiences higher growth together with increased phosphorus absorption. APase activity demonstrates a clear link between phosphorus resource availability and the health of the organism based on its physiological state. Research shows that different types of APase genes exist within a single cyanobacterium strain, which demonstrates genetic variability. The strain *Anabaena flos-aquae* displays two genes phoD and phoX simultaneously in its genetic structure. The germination of these genes happens during culturing conditions that involve AMP or lecithin organic phosphorus sources. Cyanobacteria demonstrate diverse enzymatic capabilities, because they can use numerous organic phosphorus sources found in their ecological habitat (Tazisong et al., 2015).

Numerous studies demonstrate that cyanobacteria use polyphosphate accumulation for phosphorus deficiency survival, because this method has stood the test of time. The cyanobacterium *Synechocystis* and other related strains accelerate their polyP granule production during phosphorus excess periods immediately following phosphate addition. Cyanobacteria maintain polyphosphate granules inside their cells as an intracellular substance for phosphorus storage to use during periods of nutrient scarcity. The accumulation of polyP alongside its degradation abilities varies based on the species type and responds to environmental conditions. The polyP accumulation in *Nodularia spumigena p*rimarily happens in vegetative cells; yet, heterocysts maintain low phosphorus amounts to showcase their functional specialization (Sun et al., 2024).

The control of polyphosphate metabolism occurs through enzymatic activity of polyphosphate kinases (PPKs) that perform phosphate polymerization functions. By elevating ppk gene expression levels in *Microcystis aeruginosa cells*, both phosphorus uptake ability and polyP synthesis capability increase. The importance of PPKs emerges from their role in storage processes that manage phosphorus. Algae together with cyanobacteria developed sophisticated mechanisms to acquire and store phosphorus, because they must survive during times of different nutrient availability. Through their absorption mechanisms combined with alkaline phosphatase enzyme activities and polyphosphate accumulation, these organisms successfully acquire available phosphorus to sustain biological processes as environmental limitations grow (Hildenbrand et al., 2023).

7.4 Phosphorus-Induced Bloom Formation

The occurrence of phosphorus-induced algal blooms, which are also commonly known as eutrophication events, poses a significant threat to the environment in aquatic ecosystems all over the world. One of the key factors that leads to the occurrence of these blooms is phosphorus, which is a nutrient that is considered to be a limiting factor in many freshwater and coastal marine ecosystems. This blooming phenomenon is mostly brought on by an increase in phosphorus levels. It is possible to attribute the initiation of these blooms as well as their intensity to a variety of factors that are interrelated with one another. Among these elements are nutritional thresholds, reactions that are peculiar to a species, and environmental conditions.

Some concentrations of phosphorus have the ability to operate as thresholds for the initiation of algal blooms, according to the results of the study that is now being conducted. The excessive application of phosphorus from agricultural and urban areas into Lake Winnipeg Canada resulted in widespread recurrent cyanobacterial blooms since the 1990s. Scientists use monitoring and modeling data to determine that elevated total phosphorus levels reaching 0.05 mg/L typically lead to bloom conditions (Bunting et al., 2016). Bloom incidents in Scotland's freshwater lochs are increasing because phosphorus concentrations are rising together with warming climate conditions which demonstrates how environmental factors affect nutrient thresholds (Tett &Edwards 2002). Under phosphorus enrichment Cyanobacteria demonstrate specific physiological traits that make them dominate as bloom-forming organisms. The species *M. aeruginosa* possesses luxury phosphorus uptake together with intracellular polyphosphate bodies that help it survive in short periods of phosphorus scarcity. The storage system they possess provides them a competitive edge because it grants dominance in fluctuating phosphorus environments and eliminates less-efficient phytoplankton from the community (Wan et al., 2019). Additionally, *Cylindrospermopsis raciborskii* increases its dominance by using sediment phosphorus while changing its nutrient acquisition pattern during low phosphorus conditions, thus reducing plankton diversity (Ryan et al., 2017).

The presence of phosphorus affects the two main characteristics of algal blooms through changes in their time span and toxicity levels. The growth of algae extends longer when phosphorus occurs abundantly, resulting in larger-scale persistent blooms. The excessive nutrient stimulates algae growth, because it is plentiful in water systems. Plentiful phosphorus creates conditions, which increase poisoning secretion by species that form algae blooms. The presence of this connection has now been identified. Extensive research on Ugandan freshwater lakes shows complete phosphorus levels increase both *Microcystis* abundance and dangerous microcystin toxin concentrations. The toxins released by these species can trigger water contamination and have negative effects on aquatic organisms and public health (Poste et al., 2013). Massive persistent toxin-producing *Microcystis* blooms have formed in Lake Taihu because of phosphorus enrichment caused by agricultural and industrial activities. The water supply security for millions of people faces threat through these incidents while extensive toxic substance removal procedures become necessary (Yue et al., 2014). Proper management and reduction of eutrophication require comprehensive knowledge of phosphorous effects on blooms as well as species-specific nutritional thresholds along with bloom traits. Precise knowledge about these factors plays an essential role for successful management techniques and response methods. To reach ecological objectives involving harmful algal bloom prevention and water quality protection and ecosystem health maintenance it is essential to establish methods which control phosphorus delivery into aquatic ecosystems (Table 7.1).

7.5 Internal Phosphorus Cycling During Blooms

Aquatic ecosystems experience eutrophication primarily because of internal phosphorus cycling in algal blooms, while this process strongly affects water quality together with biodiversity and harmful algal bloom (HAB) duration. Large-scale algal bloom collapses lead to the sinking of dying cells, which produce decomposed matter that distributes previously absorbed phosphorus into the water column. Dying *Microcystis* blooms in Lake Taihu, China, release 45% of their initial phosphorus content according to Han et al. (2022), which leads to repeated blooms in the water (Han et al., 2022). Different laboratory and field studies indicate that the complex process includes decaying algal biomass releasing phosphorus while sediment–water phosphorus exchange interacts with microbial functions that determine phosphorus availability (Chen et al., 2025).

After algal blooms end, phosphorus release from decomposing algal matter turns into an essential phosphorus source. During microbial decomposition of algal biomass, two types of phosphorus compounds are released: dissolved inorganic phosphorus (DIP) as orthophosphate and various forms of dissolved organic phosphorus (DOP). The decomposition process of organic matter produces organic phosphorus substances that migrate into sediment porewater. Microorganisms transform these compounds to create inorganic forms of soluble reactive phosphorus (SRP) available for algal uptake. The accumulation of dead algal material in research studies resulted in a peak phosphorus release from the sediments that exceeded 0.14 mg/L during a 40-day observation period (Chen et al., 2025). Research shows that rising phosphorus-cycling microorganism activity functions as the primary reason for this release because their metabolism increases with organic matter from decomposing algae.

Battle for phosphorus occurs at the sediment–water zone because it functions as a phosphorus source while serving as a phosphorus sink depending on environmental conditions. The water column receives phosphorus release from iron and manganese oxide compounds when anoxic conditions exist within eutrophic systems. The microbial reduction processes enable iron-reducing bacteria to convert Fe(III) to Fe(II) thus releasing phosphorus into solution. The increase of oxygen in the environment creates oxidation reactions that convert Fe(II) to Fe(III) thus triggering the sediment-based adsorption of phosphorus. The sediment–water interface experiences redox condition changes because of temperature fluctuations and organic carbon content variations

Table 7.1 Major cyanobacterial toxins, their producing genera, biological toxicity, and mechanisms of action

Toxins	Toxin-producing cyanobacterial	Biological toxicity	Possible mechanisms of action	Reference
Anatoxin-a(s)	*Anabaena*	Neurotoxic	Inhibition of ACh-esterase activity, hyper-excitability of nerve	Matsunaga et al. (1989)
Anatoxin-a	*Anabaena, Aphanizomenon, Cylindrospermum, Microcystis, Planktothrix, Raphidiopsis*	Neurotoxic	Depolarizing neuromuscular blocking	Christensen and Khan (2020)
Antillatoxin	*Lyngbya*	Neurotoxic	Blocking neuronal communication by binding to the voltage-gated Na^+ channels	(Li et al., 2001)
Aplysiatoxins	*Lyngbya, Schizothrix, Trichodesmium, Oscillatoria*	Dermatotoxic	Potent tumor promoters and protein kinase C activators	(FUJIKI et al., 1985)

(continued)

Table 7.1 (continued)

Toxins	Toxin-producing cyanobacterial	Biological toxicity	Possible mechanisms of action	Reference
Cylindrospermopsin	*Anabaena*, *Aphanizomenon*, *Cylindrospermopsis*, *Lyngbya*, *Oscillatoria* (*Planktothrix*), *Raphidiopsis*, *Umezakia*	Hepatotoxic, nephrotoxic, and cytotoxic	Irreversible inhibition of protein and glutathione synthesis, implicating cytochrome P-450, overexpression of DNA damage repair proteins	Neumann et al. (2007)
Cyanopetolin	*Microcystis*, *Planktothrix*	Neurotoxic activity	Transcriptional alterations of genes belonging to DNA damage and repair	Faltermann et al. (2014)
Homoanatoxin-a	*Anabaena*, *Oscillatoria* (*Planktothrix*), *Phormidium*, *Raphidiopsis*	Neurotoxic	Blockade of the neuromuscular transmission	Lilleheil et al. (1997)

along with microbial activities that influence phosphorus mobilization processes (Ebrahimi & Ojani, 2024).

The sediment's microbial communities act as main regulators of phosphorus availability throughout the sediment environment. The hydrogen sulfide produced by sulfate-reducing bacteria (SRB) allows it to react with Fe(II), leading to the formation of iron sulfides that may cause phosphorus release from sediments. The internal phosphorus cycles in sediments depend on polyphosphate-accumulating organisms (PAOs), because these organisms retain excess phosphorus when available for later release during nutrient-depleted periods. Multiple microorganism activities depend on temperature along with organic carbon levels and nutrient counts, since these factors affect their metabolic pathways leading to changes in phosphorus dynamics (Wang et al., 2024). The complete comprehension of these procedures enables effective management of nutrients in eutrophic water bodies. The continuous interdependence between algal blooms and sediment chemical composition together with microbial functioning demonstrates how complex the process of phosphorus recycling becomes for water quality.

7.6 Ecological and Health Consequences

Aquatic systems show an increasing occurrence of Harmful algal blooms (HABs) that produce major environmental and medical effects especially when phosphorus enrichment triggers these blooms. Blooms in the water system harm both environmental health and animal diversity. The proliferation of harmful algal blooms occurs, because water systems contain excessive amounts of phosphorus due to agricultural runoff and wastewater outflows. The Great Lakes experience escalating cyanobacterial blooms due to phosphorus contamination, which creates poor water conditions that boost waterborne disease risks (Michalak et al., 2013). The water body starts releasing excessive nutrients that result in dense algal development, which disrupts the health of aquatic organisms.

The main instant visual effect that cyanobacterial blooms produce is water quality degradation. The growth of cyanobacterial blooms releases microcystins that endanger aquatic life forms and humans. Microcystins cause liver damage that may develop into liver cancer in humans, thus creating a substantial health concern mostly affecting people who obtain water from nearby water sources for drinking and irrigation purposes (Carmichael, 2001). A major tragic incident in Brazil's history occurred in 1996, because a microcystin-contaminated water supply reservoir killed 56 citizens, illustrating the serious health threats that cyanobacterial toxins create for people. Health professionals face challenges during water treatment, because algal biomass toxins stay within the water even after decay. Traditional water filtration systems are unable to eliminate these toxic substances, leading to more health risks (Queiroz et al., 2007).

These blooms created by phosphorus pollution result in water environments becoming magnificently devoid of oxygen through the decay process until they reach hypoxic or anoxic states. Large amounts of deteriorating algae use up oxygen from the water, which suffocates aquatic organisms together with fish and crustaceans and oxygen-dependent marine life. Public waters in the U.S. Gulf of Mexico have developed death zones due to nutrient-fed algal blooms that produce low oxygen environments incompatible with marine life. Scientific records from Rabalais et al. (2002) detail the process through which oxygen depletion affects marine ecosystems from small invertebrates to large fish species, which results in total food web disruption. The phosphorus originating from agricultural runoff aggravates the oxygen loss problem by accelerating the fast growth of aquatic plant blooms (Rabalais et al., 2002).

Water quality degradation together with oxygen depletion represents only some of the effects that blooms create within aquatic environments. The development of blooms leads to persistent disruption of water biodiversity throughout their progression. The presence of phosphorus-induced cyanobacterial blooms in freshwater lakes results in the replacement of more efficient phytoplankton species like diatoms and green algae from their habitats. The distributional change in aquatic species creates a sequence of lower-level impacts that gradually depletes key food resources for aquatic organisms before affecting fish populations. The excessive amount of phosphorus in Lake Erie has resulted in major phytoplankton community transformations by promoting cyanobacterial growth while displacing different algae species causing broad ecosystem instability (Michalak et al., 2013).

The blooms of cyanobacteria stimulate increased phosphorus release from water column sediments, which aids in keeping the bloom going. Water column phosphorus metabolism through sediment to water transfers enhances the duration of bloom through an ongoing cycle that includes continued external nutrient inflow. In freshwater bodies, microbial processes that break down dead algal cells together with bacterial degradation activities release phosphorus compounds that maintain blooming conditions. Research indicates that when water bodies receive ongoing external nutrients and experience internal phosphorus cycling, both phenomena contribute to prolonged algal blooms beyond predicted lengths of time even though initial nutrient inputs decrease (Paerl & Paul, 2012).

Human health suffers serious damage as a side effect of these blooms. The production of cyanotoxins, such as microcystins and anatoxins, led to drinking water supply contamination, which resulted in the 2014 Toledo water crisis. Drinking water became unavailable for a certain period when Lake Erie showed elevated microcystin levels, which demonstrated how water systems remain exposed to HAB hazards (Mourelatos, 2014). These toxins, which can also bioaccumulate in aquatic food chains, pose significant health risks to consumers of contaminated fish and shellfish. Consumers who eat contaminated fish and shellfish face substantial health dangers, because the toxins tend to build up in aquatic food chains. The occurrence of phosphorus-induced blooms creates serious environmental impacts while negatively impacting human well-being. The toxic substances, along with oxygen depletion and biodiversity disruptions, emerged from these blooms as the main detrimental effects. Research findings will guide strategies to lower phosphorus pollution and enhance water purification methods for fighting the increasing worldwide spread of HABs.

7.7 Phosphorus Management and Bloom Mitigation

Aquatic ecosystems develop eutrophication together with harmful algal blooms (HABs) due to phosphorus accumulation levels exceeding the necessary amounts for plant growth. The environmental problems require careful management of phosphorus installations. This chapter discusses different P management approaches and HAB reduction methods currently used in aquatic settings based on recent research findings. The fundamental way to control phosphorus pollution is by implementing best management practices for nutrient reduction. Agricultural BMP systems were developed to decrease phosphorus discharges, which end up in water areas. The Alabama Cooperative Extension System presents pollinators and ground cover preservation with proper timing of fertilizer application and buffer zone implementation to minimize phosphorus discharge in agricultural settings (Sharpley et al., 2006). Such practices create barriers, which stop phosphorus from entering water systems, thus lowering the occurrence of eutrophication and HABs. The aquatic environment receives sustainable phosphorus removal through the implementation of bioremediation and nature-based solutions. Engineered CWs serve as wastewater treatment facilities by utilizing wetland vegetation and soils together with their respective microbial communities to perform natural wastewater treatment functions. The integration of plant growth-promoting rhizobacteria (PGPR) in CWs increases the phosphorus removal efficiency through improved plant development and enhanced phosphorus uptake capacity (Taylor & Robert, 2018).

The modern solutions to control phosphorus-driven algal growth stem from genetic and biotechnological developments. Genetic manipulation of microalgae allows them to better absorb phosphorus, which results in lower phosphorus availability for toxic algal species. Genetically modified microalgae demonstrate enhanced phosphorus assimilation functions according to research, which supports environmental phosphorus control efforts. The biotechnological advance of CRISPR/Cas9 gene editing enables researchers to create algal strains with improved phosphorus metabolism, which benefits bloom mitigation activities (Sarma et al., 2024). Multi-skilled phosphorus management and bloom reduction efforts become possible through the combined use of BMPs and bioremediation techniques and genetic engineering methods. When these methods become operational, they will create substantial reductions of phosphorus entering aquatic systems, which results in lowered chances for both eutrophic conditions and dangerous algal blooming. The sustainability of aquatic environments depends on ongoing research development focused on resolving phosphorus pollution challenges (Huser et al., 2016).

7.8 Future Directions and Research Needs

The metabolic processes of phosphorus work as a base mechanism that determines how algal blooms start and persist and develops toxicity. The understanding and management of phosphorus dynamics takes center stage now that aquatic environments experience rising human-made pressures alongside climate change impacts. Multiple essential research fields need development, because they will improve both our knowledge and control of phosphorus-driven bloom patterns (Chen et al., 2018; Chen MuSong et al., 2018).

The study of microalgae phosphorus uptake regulation along with storage mechanisms needs additional research attention. The polyphosphate (PolyP) storage mechanism enables microalgae biological organisms to accumulate phosphorus inside their cells. Microorganisms use various environmental and nutrient factors to control their PolyP synthesis pattern as well as their PolyP storage mechanics (Solovchenko et al., 2019a; 2019b). Scientists currently lack many details regarding the regulatory biological pathways controlling these cellular processes. The study of PolyP metabolic genetic and biochemical pathways will generate knowledge about phosphorus level adaptation mechanisms of microalgae alongside their effects on bloom phenomena. Studies demonstrate how specific microalgae species can use phosphoesters and phosphonates as organic phosphorus sources for growth, which demonstrates the diverse phosphorus acquisition practices found in aquatic systems (Solovchenko et al., 2024).

Research should focus on understanding how phosphorus interacts with other nutrients specifically with nitrogen to affect algal communities and the development of algal blooms (Zhao et al., 2023). The availability of nitrogen in freshwater ecosystems influences which algal species will become dominant even though phosphorus remains as the most restrictive nutrient. Research findings establish that elevated nitrogen quantities in the water promote abundant growth of *M. aeruginosa* and other nitrogen-fixing cyanobacteria, which subsequently create dangerous algal blooms. The growth of these species gets restrained when phosphorus becomes limited in the environment. Detailed nutrient management needs complete knowledge about how phosphorus and nitrogen both influence algal growth and bloom phenomena. More research needs to be done about microbial communities that transform phosphorus compounds (Mohammed & Arockiaraj, 2024). Microorganisms consisting of bacteria together with fungi take part in changing phosphorus compounds, thus influencing their accessibility to primary producers. The microbial process of organic phosphorus degradation within eutrophic systems evolves orthophosphate, which may stimulate algal multiplication. The study of microbial communities which perform phosphorus cycling functions will help create comprehensive models of phosphorus behavior in aquatic systems (Prabhu et al., 2019).

Advanced analytical methods open new paths for scientific investigation of the future. High-resolution mass spectrometry together with isotopic labeling enables researchers to track phosphorus pathways and transformations, which occur inside aquatic systems. The tools enable researchers to understand phosphorus movement

Table 7.2 Key research areas and critical questions for advancing phosphorus management and bloom mitigation

S. No.	Research area	Key questions/Needs
1	Internal loading	How can we better quantify and manage P release from sediments under varying oxygen and temperature conditions?
2	Bioavailability of P forms	Which forms of P (e.g., DOP, SRP) are most utilized by different cyanobacteria, and how does this affect bloom dynamics?
3	P-removal technologies	Develop and test sustainable in-lake and watershed technologies to reduce P availability, such as geo-engineering and bioremediation
4	Climate change impacts	How will warming temperatures and extreme weather events alter P cycling and cyanobacterial responses?
5	Modeling and forecasting	Improve predictive models that incorporate P dynamics, land use, hydrology, and cyanobacterial physiology
6	Policy and management integration	How can science better inform regulatory frameworks and stakeholder practices to reduce P loading and mitigate blooms?

between sources and sinks and clarify rates of transformation and sequestration, which becomes essential for building phosphorus dynamic models to forecast bloom occurrence. Understanding phosphorus metabolism and its bloom-enabling mechanisms better requires the solution of these scientific issues. The development of efficient management strategies to combat harmful algal bloom impacts on water quality requires this essential understanding. The advancement of this field of research requires integrated efforts from molecular biology with ecology and environmental science (Kruse et al., 2015) (Table 7.2).

7.9 Conclusion

Phosphorus metabolism is a central determinant of harmful algal bloom (HAB) dynamics in freshwater and marine ecosystems. This chapter has explored how both external nutrient inputs, such as agricultural runoff and urban discharge, and internal mechanisms, including sediment release and algal decay, elevate phosphorus concentrations and contribute to eutrophication. Cyanobacteria, through highly specialized adaptations like polyphosphate accumulation, alkaline phosphatase activity, and luxury uptake, gain dominance in phosphorus-rich environments, often leading to toxic bloom events. The ecological consequences of phosphorus-induced blooms are extensive. These include oxygen depletion, disruption of aquatic food webs, loss of biodiversity, and increased toxin production that directly threaten ecosystem stability and public health. Internal phosphorus cycling reinforces bloom persistence, creating a feedback loop that undermines remediation efforts even after external inputs are controlled. Addressing this complex issue requires an integrated management framework combining best agricultural practices, nature-based solutions, and modern biotechnological tools. Future research should focus on the interactions between phosphorus and other nutrients, the role of microbial communities in phosphorus transformation, and the influence of climate variability on bloom patterns. The development of predictive models and high-resolution monitoring systems will be essential for early warning and effective mitigation. A comprehensive understanding of phosphorus dynamics, rooted in both molecular and ecological perspectives, is critical for safeguarding water quality and ensuring the long-term sustainability of aquatic ecosystems.

References

Albou, E. M., Abdellaoui, M., Abdaoui, A., & Boughrous, A. A. (2024). Agricultural practices and their impact on aquatic ecosystems–a mini-review. *Ecological Engineering & Environmental Technology, 25*.

Amadou, I., Faucon, M.-P., & Houben, D. (2022). Role of soil minerals on organic phosphorus availability and phosphorus uptake by plants. *Geoderma, 428*, Article 116125.

Borne, K. E. (2014). Floating treatment wetland influences on the fate and removal performance of phosphorus in stormwater retention ponds. *Ecological Engineering, 69*, 76–82.

Bossa, R., Di Colandrea, M., Salbitani, G., & Carfagna, S. (2024). Phosphorous utilization in microalgae: Physiological aspects and applied implications. *Plants, 13*, 2127.

Bunting, L., Leavitt, P., Simpson, G., Wissel, B., Laird, K., Cumming, B., St. Amand, A., & Engstrom, D. (2016): Increased variability and sudden ecosystem state change in Lake Winnipeg, Canada, caused by 20th century agriculture. *Limnology and Oceanography, 61*, 2090–2107.

Burley, K. L., Prepas, E. E., & Chambers, P. A. (2001). Phosphorus release from sediments in hardwater eutrophic lakes: The effects of redox-sensitive and-insensitive chemical treatments. *Freshwater Biology, 46*, 1061–1074.

Carmichael, W. W. (2001). Health effects of toxin-producing cyanobacteria:"The CyanoHABs." *Human and Ecological Risk Assessment: An International Journal, 7*, 1393–1407.

Chen, M., Ding, S., Chen, X., Sun, Q., Fan, X., Lin, J., Ren, M., Yang, L., & Zhang, C. (2018). Mechanisms driving phosphorus release during algal blooms based on hourly changes in iron and phosphorus concentrations in sediments. *Water Research, 133*, 153–164.

Chen MuSong, C. M., Ding ShiMing, D. S., Chen Xiang, C. X., Sun Qin, S. Q., Fan XianFang, F. X., Lin Juan, L. J., Ren MingYi, R. M., Yang LiYuan, Y. L., & Zhang ChaoSheng, Z. C. (2018). Mechanisms driving phosphorus release during algal blooms based on hourly changes in iron and phosphorus concentrations in sediments.

Chen, W.-L., Zhang, M., Wang, J.-G., Huang, W.-J., Wu, Q., Zhu, X.-P., Li, N., Wu, Q., Guo, W., & Chen, J. (2024). Microbial mechanisms of C/N/S geochemical cycling during low-water-level sediment remediation in urban rivers. *Journal of Environmental Management, 359*, Article 120962.

Chen, Y., Li, D., Liu, S., Zhang, Y., Yan, X., Song, X., Li, Z., Li, B., Shan, S., & Zhu, Y. (2025). Long-term effects of dead algal deposition on sediment surfaces: Behavior of endogenous phosphorus release in sediments. *Water Research, 268*, Article 122742.

Christensen, V. G., & Khan, E. (2020). Freshwater neurotoxins and concerns for human, animal, and ecosystem health: A review of anatoxin-a and saxitoxin. *Science of the Total Environment, 736*, Article 139515.

Ebrahimi, E., & Ojani, M. R. (2024). *Phosphorus dynamics in soil-water-sediment environment, phosphorus in soils and plants*. IntechOpen

Faltermann, S., Zucchi, S., Kohler, E., Blom, J. F., Pernthaler, J., & Fent, K. (2014). Molecular effects of the cyanobacterial toxin cyanopeptolin (CP1020) occurring in algal blooms: Global transcriptome analysis in zebrafish embryos. *Aquatic Toxicology, 149*, 33–39.

Feng, L., Wang, Y., Hou, X., Qin, B., Kutser, T., Qu, F., Chen, N., Paerl, H. W., & Zheng, C. (2024). Harmful algal blooms in inland waters. *Nature Reviews Earth & Environment, 5*, 631–644.

Fromin, N. (2025). Impacts of soil storage on microbial parameters. *The Soil, 11*, 247–265.

Fujiki, H., Ikegami, K., Hakii, H., Suganuma, M., Yamaizumi, Z., Yamazato, K., Moore, R., & Sugimura, T. (1985). A blue-green alga from Okinawa contains aplysiatoxins, the third class of tumor promoters. *Japanese Journal of Cancer Research GANN, 76*, 257–259.

Han, C., Dai, Y., Sun, N., Wu, H., Tang, Y., & Dai, T. (2022). Algae bloom and decomposition changes the phosphorus cycle pattern in Taihu lake. *Water, 14*, 3607.

Harris, T. D., Reinl, K. L., Azarderakhsh, M., Berger, S. A., Berman, M. C., Bizic, M., Bhattacharya, R., Burnet, S. H., Cianci-Gaskill, J. A., & de Senerpont Domis, L. N. (2024). What makes a cyanobacterial bloom disappear? A review of the abiotic and biotic cyanobacterial bloom loss factors. *Harmful Algae, 133*, 102599.

Hildenbrand, J. C., Sprenger, G. A., Teleki, A., Takors, R., & Jendrossek, D. (2023). Polyphosphate kinases phosphorylate thiamine phosphates. *Microbial Physiology, 33*, 1–11.

Huser, B. J., Futter, M., Lee, J. T., & Perniel, M. (2016). In-lake measures for phosphorus control: The most feasible and cost-effective solution for long-term management of water quality in urban lakes. *Water Research, 97*, 142–152.

Jochumsen, M. R., & Bouchelouche, K. (2024). PSMA PET/CT for primary staging of prostate cancer-an updated overview. In *Seminars in nuclear medicine* (pp. 39–45). Elsevier.

Kruse, J., Abraham, M., Amelung, W., Baum, C., Bol, R., Kühn, O., Lewandowski, H., Niederberger, J., Oelmann, Y., & Rüger, C. (2015). Innovative methods in soil phosphorus research: A review. *Journal of Plant Nutrition and Soil Science, 178*, 43–88.

Kundu, D., Dutta, D., Joseph, A., Jana, A., Samanta, P., Bhakta, J. N., & Alreshidi, M. A. (2024). Safeguarding drinking water: A brief insight on characteristics, treatments and risk assessment of contamination. *Environmental Monitoring and Assessment, 196*, 180.

Kwon, J., Moghtader, A., Kang, C., Bibak Bejandi, Z., Shahjahan, S., Alzein, A., & Djalilian, A. R. (2025). Overview of dry eye disease for primary care physicians. *Medicina, 61*, 460.

Li, R., Carmichael, W. W., Brittain, S., Eaglesham, G. K., Shaw, G. R., Liu, Y., & Watanabe, M. M. (2001). First report of the cyanotoxins cylindrospermopsin and deoxycylindrospermopsin from *Raphidiopsis curvata* (Cyanobacteria). *Journal of Phycology, 37*, 1121–1126.

Lilleheil, G., Andersen, R. A., Skulberg, O. M., & Alexander, J. (1997). Effects of a homoanatoxin-a-containing extract from *Oscillatoria formosa* (Cyanophyceae/cyanobacteria) on neuromuscular transmission. *Toxicon, 35*, 1275–1289.

Matsunaga, S., Fusetani, N., Hashimoto, K., & Walchli, M. (1989). Theonellamide F. A novel antifungal bicyclic peptide from a marine sponge *Theonella* sp. *Journal of the American Chemical Society, 111*, 2582–2588.

Michalak, A. M., Anderson, E. J., Beletsky, D., Boland, S., Bosch, N. S., Bridgeman, T. B., Chaffin, J. D., Cho, K., Confesor, R., & Daloğlu, I. (2013). Record-setting algal bloom in Lake Erie caused by agricultural and meteorological trends consistent with expected future conditions. *Proceedings of the National Academy of Sciences, 110*, 6448–6452.

Mohammed, V., & Arockiaraj, J. (2024). Unveiling the trifecta of cyanobacterial quorum sensing: LuxI, LuxR and LuxS as the intricate machinery for harmful algal bloom formation in freshwater ecosystems. *Science of the Total Environment, 171644.*

Mourelatos, A. P. (2014). The pre-socratics: a collection of critical essays.

Neumann, C., Bain, P., & Shaw, G. (2007). Studies of the comparative in vitro toxicology of the cyanobacterial metabolite deoxycylindrospermopsin. *Journal of Toxicology and Environmental Health, Part A, 70*, 1679–1686.

Nürnberg, G. K. (1984). The prediction of internal phosphorus load in lakes with anoxic hypolimnia 1. *Limnology and Oceanography, 29*, 111–124.

Ogwu, M. C., Patterson, M. E., & Senchak, P. A. (2025). Phosphorus mining and bioavailability for plant acquisition: Environmental sustainability perspectives. *Environmental Monitoring and Assessment, 197*, 572.

Paerl, H. W., & Paul, V. J. (2012). Climate change: Links to global expansion of harmful cyanobacteria. *Water Research, 46*, 1349–1363.

Pan, Y., Zhang, S., & Zhang, M. (2024). The impact of entrepreneurship of farmers on agriculture and rural economic growth: Innovation-driven perspective. *Innovation and Green Development, 3*, Article 100093.

Poste, A., Hecky, R., & Guildford, S. (2013). Phosphorus enrichment and carbon depletion contribute to high Microcystis biomass and microcystin concentrations in Ugandan lakes. *Limnology and Oceanography, 58*, 1075–1088.

Prabhu, N., Borkar, S., & Garg, S. (2019). Phosphate solubilization by microorganisms: Overview, mechanisms, applications and advances. *Advances in Biological Science Research*, 161–176.

Queiroz, S. A., Figueiredo, G., Silva, J. A. I., Espasandin, A. C., Meirelles, S. L., & Oliveira, J. A. (2007). Estimativa de parâmetros genéticos da habilidade de permanência aos 48, 60 e 72 meses de idade em vacas da raça Caracu. *Revista Brasileira de Zootecnia, 36*, 1316–1323.

Rabalais, N. N., Turner, R. E., & Wiseman, W. J., Jr. (2002). Gulf of Mexico hypoxia, aka "The dead zone." *Annual Review of Ecology and Systematics, 33*, 235–263.

Ryan, C. N., Thomas, M. K., & Litchman, E. (2017). The effects of phosphorus and temperature on the competitive success of an invasive cyanobacterium. *Aquatic Ecology, 51*, 463–472.

Sarma, U., Hoque, M. E., Thekkangil, A., Venkatarayappa, N., & Rajagopal, S. (2024). Microalgae in removing heavy metals from wastewater–An advanced green technology for urban wastewater treatment. *Journal of Hazardous Materials Advances, 15*, Article 100444.

Sharpley, A. N., Chapra, S. C., Wedepohl, R., Sims, J., Daniel, T. C., & Reddy, K. R. (1994). Managing agricultural phosphorus for protection of surface waters: Issues and options. *Journal of Environmental Quality, 23*, 437–451.

Sharpley, A. N., Daniel, T., Gibson, G., Bundy, L., Cabrera, M., Sims, T., Stevens, R., Lemunyon, J., Kleinman, P., & Parry, R. (2006). *Best management practices to minimize agricultural phosphorus impacts on water quality. ARS-163.* USDA-ARS.

Sitote, Y. M., & Gebremedhine, M. G. (2024). *Comprehensive review of eutrophication in freshwater ecosystems: Causes, effects, assessment, and management strategies.* Preprints.

Solovchenko, A. E., Ismagulova, T. T., Lukyanov, A. A., Vasilieva, S. G., Konyukhov, I. V., Pogosyan, S. I., Lobakova, E. S., & Gorelova, O. A. (2019b). Luxury phosphorus uptake in microalgae. *Journal of Applied Phycology, 31*, 2755–2770.

Solovchenko, A., Khozin-Goldberg, I., Selyakh, I., Semenova, L., Ismagulova, T., Lukyanov, A., Mamedov, I., Vinogradova, E., Karpova, O., & Konyukhov, I. (2019a). Phosphorus starvation and luxury uptake in green microalgae revisited. *Algal Research, 43*, Article 101651.

Solovchenko, A., Plouviez, M., & Khozin-Goldberg, I. (2024). Getting grip on phosphorus: Potential of microalgae as a vehicle for sustainable usage of this macronutrient. *Plants, 13*, 1834.

Søndergaard, M., Jensen, J. P., & Jeppesen, E. (2003). Role of sediment and internal loading of phosphorus in shallow lakes. *Hydrobiologia, 506*, 135–145.

Song, Y., Wang, L., Chang, A, Fei, S., Fang, J., Liang, D., Qin, X., Ke, X., Li, Y., & Yan, C. (2024). Interannual monitoring of water environment and spatiotemporal dynamics of phytoplankton community and blooming risk in a tropical water source reservoir, China. *Applied Ecology & Environmental Research, 22.*

Soranno, P. A., Hanly, P. J., Webster, K. E., Wagner, T., McDonald, A., Shuvo, A., Schliep, E. M., Reinl, K. L., McCullough, I. M., & Tan, P.-N. (2025). Abrupt changes in algal biomass of thousands of US lakes are related to climate and are more likely in low-disturbance watersheds. *Proceedings of the National Academy of Sciences, 122*, Article e2416172122.

Sun, Q.-W., Gao, Y., Wang, J., Fu, F.-x, Yong, C.-W., Li, S.-Q., Huang, H.-L., Chen, W.-Z., Wang, X.-W., & Jiang, H.-B. (2024). Molecular mechanism of a coastal cyanobacterium Synechococcus sp. PCC 7002 adapting to changing phosphate concentrations. *Marine Life Science & Technology, 6*, 562–575.

Suo, Y., Li, T., von Sperber, C., Ge, L., Cao, C., Zhai, Z., Bu, Z., & Wang, M. (2025). Low molecular weight organic acids mobilize soil organic phosphorus for enzymatic hydrolysis in a temperate montane peatland. *Biogeochemistry, 168*, 1–15.

Taylor, L., & Robert, N. (2018). Phosphorus eutrophication and mitigation strategies. In *Phosphorus-recovery and recycling* (pp. 1–17). IntechOpen.

Tazisong, I. A., Senwo, Z. N., & He, Z. (2015). Phosphatase hydrolysis of organic phosphorus compounds. *Advances in Enzyme Research, 3*, 39.

Tett, P., & Edwards, V. (2002). Review of harmful algal blooms in Scottish coastal waters. *Scottish Environment Protection Agency Rep.*

Thengumthottathil, V., Ponnusamy, K., & Mohamed, S. N. (2024). Bioelectrochemical systems: Exploring microbial communities, interactions, and electron transfer. *Biochemical Engineering Journal*, 109442.

Tiwari, B., Singh, S., Kaushik, M. S., & Mishra, A. K. (2015). Regulation of organophosphate metabolism in cyanobacteria. A review. *Microbiology, 84*, 291–302.

Wan, L., Chen, X., Deng, Q., Yang, L., Li, X., Zhang, J., Song, C., Zhou, Y., & Cao, X. (2019). Phosphorus strategy in bloom-forming cyanobacteria (Dolichospermum and Microcystis) and its role in their succession. *Harmful Algae, 84*, 46–55.

Wang, F., Wang, J., Cao, T., Ji, X., Yan, J., Ding, S., & Chen, N. (2024). Seasonal hypoxia enhances sediment iron-bound phosphorus release in a subtropical river reservoir. *Science of the Total Environment, 936*, Article 173261.

Wauchope-Thompson, M. S., Baulch, H. M., & Cade-Menun, B. J. (2024). Internal phosphorus loading in a chain of eutrophic hardwater lakes in Saskatchewan, Canada. *Science of the Total Environment, 924*, Article 171493.

Yang, S.-Y., Lin, W.-Y., Hsiao, Y.-M., & Chiou, T.-J. (2024). Milestones in understanding transport, sensing, and signaling of the plant nutrient phosphorus. *The Plant Cell, 36*, 1504–1523.

Yue, T., Zhang, D., & Hu, C. (2014). Comparative studies on phosphate utilization of two bloom-forming Microcystis spp. (cyanobacteria) isolated from Lake Taihu (China). *Journal of Applied Phycology, 26*, 333–339.

Zhang, Q., Jia, L., Chen, Y., Yan, H., Chen, Q., Zhang, J., & Sun, H. (2024). Molecular mechanisms of the cyanobacterial response to different phosphorus sources. *Sustainability, 16*, 5642.

Zhao, Y., Wang, Z., Chen, M., Huang, X., & Luo, Z. (2023). Effects of nitrogen to phosphorus ratios on algal growth and arsenate metabolism by Microcystis aeruginosa with dissolved organic phosphorus and nitrate as nutrients. *Algal Research, 69*, Article 102922.

Zhong, J., Tang, L., Gao, M., Wang, S., & Wang, X. (2024). Beyond feast and famine: Cultivating hydrodynamic oxygenic photogranules with better performances under permanent feast regime. *Bioresource Technology, 401*, Article 130752.

Chapter 8
Trace Metal Interactions and Their Impact on Cyanobacterial Physiology

Nasir Assad, Muhammad Naeem-ul-Hassan, Arshad Ali, Barkat Ullah, Muhammad Nauman Khan, Hansa Gul, Iram Zahra, Sher Ullah, Majid Iqbal, and Alevcan Kaplan

Abstract Trace metals, although found in small amounts in aquatic environments, significantly influence the physiology of cyanobacteria and are crucial to the health and functionality of aquatic ecosystems. This chapter delves into the complex interactions between trace metals such as iron, copper, and zinc and cyanobacteria, emphasizing both their necessary roles and potential toxic effects depending on concentration and bioavailability. After discussing the vital roles of trace metals in processes like photosynthesis, nitrogen fixation, and cellular respiration, this chapter moves on to examine the mechanisms that cyanobacteria use to manage these metals. This includes uptake regulation and the balancing act necessary to avoid the detrimental impacts of metal toxicity, such as oxidative stress and growth inhibition. The adaptability and resilience of cyanobacterial metal homeostasis systems are highlighted, showcasing their ability to thrive in varying environmental conditions. The chapter further analyzes how environmental factors, such as pH fluctuations, the presence of

N. Assad (✉) · M. Naeem-ul-Hassan · A. Ali · I. Zahra · S. Ullah
Institute of Chemistry, University of Sargodha, Sargodha, Pakistan
e-mail: nakhan_98@yahoo.com

B. Ullah · M. N. Khan
Department of Botany, Islamia College Peshawar, Peshawar, Pakistan
e-mail: nakhan_98@yahoo.com

M. N. Khan
Biology Laboratory, University Public School, University of Peshawar, Peshawar, Pakistan

H. Gul
Department of Zoology, University of Mianwali, Mianwali, Punjab, Pakistan

M. Iqbal
Institute of Geographic Sciences and Natural Resources, Chinese Academy of Sciences, Beijing, China

University of Chinese Academy of Sciences, (UCAS), Beijing, China

A. Kaplan
Department of Crop and Animal Production, Sason Vocational School, Batman University, Batman, Turkey

S. Fahad et al. (eds.), *Cyanobacterial Blooms: Ecology, Evolution and Biogeochemical Impacts*, https://doi.org/10.1007/978-3-032-06042-6_8

competing ions, and human activities, affect trace metal availability and cyanobacterial responses. The implications of these interactions are explored, particularly how they influence cyanobacterial bloom dynamics, species competition, and toxin production, all of which have significant consequences for water quality and public health. Concluding with a synthesis of recent research findings, the chapter presents a nuanced understanding of trace metal interactions in cyanobacterial ecology. It also outlines promising research directions and management practices aimed at mitigating harmful blooms while supporting robust and diverse microbial communities in aquatic systems.

8.1 Introduction

The oldest and most important microorganisms on Earth are Cyanobacteria, which is also known as blue-green algae (Chankeshwara et al., 2025). The photosynthetic prokaryotic organisms dwell in many aquatic habitats and diverse terrestrial settings while actively participating in multiple biogeochemical cycling processes such as carbon and nitrogen cycle functions. Trillions of years of resilient survival and adaptable metabolic functions have allowed cyanobacteria to produce large quantities of oxygen, which helped atmosphere oxygenation during the Great Oxidation Event. Most primary production in marine and freshwater ecosystems depends heavily on these organisms. The success of cyanobacteria depends heavily on environmental conditions, but trace metals represent a specifically vital element for their growth. The essential cofactors, photosynthesis, and respiration depend upon along with nitrogen fixation and antioxidant defense require trace metals such as iron, copper, zinc and manganese, molybdenum and cobalt and nickel in small amounts (Lobus & Kulikovskiy, 2023).

The function of trace metals as enzyme components and protein structural parts leads to necessary changes in cell function along with metabolic rate optimization. The enzymatic functioning of cytochromes and ferredoxins in electron transport requires iron as a fundamental component and copper enables plastocyanin activity as well as zinc maintains proper structure for many transcription factors and enzymes. Photosystem II requires manganese as a vital component to establish numerous oxygen-evolving protein subunits, which facilitate water decomposition into oxygen molecules. Biological nitrogen fixation depends on molybdenum and vanadium even though these elements exist in relatively small amounts, since they enable proper nitrogenase enzyme operation. The survival and replication success of cyanobacteria depends fundamentally on trace metals according to these examples. The necessary connection between cyanobacteria and trace metals stands apart from basic requirements for survival. High concentrations of these metals together with bioavailable toxic forms lead to harmful effects against cyanobacterial cell structures, which results in stress-induced damage and growth restriction and often culminates in death (Leapaldt et al., 2025).

The double role of trace metals as nutrients while also being harmful substances creates intricate challenges for research on cyanobacterial cells. The trace metal toxicity and deficiency relationship depends on multiple factors such as metal speciation and pH conditions together with redox potential and chelating agent presence and ion interactions. The natural evolution of cyanobacteria led to the development of complex systems that help them tackle their difficulties. The uptake of necessary trace metals occurs through metal transport systems consisting of ATP-binding cassette (ABC) transporters together with major facilitator superfamily (MFS) proteins and divalent metal ion transporters, which efficiently select metals from their usually dispersed and variable environmental sources. After cells uptake metals, they are subject to metallothionein chelation as well as storage under intracellular compartments and metalloenzyme incorporation. Efflux systems along with antioxidant defenses become activated for metal excess reduction purposes under toxic conditions (Jenkins et al., 2023).

The regulatory approaches of cyanobacteria demonstrate their capacity to prosper under different environmental conditions that might be harsh for other organisms. The capability of cyanobacteria to regulate trace metals maintains their dominance across different ecosystems particularly in areas where nutrients are scarce or locations with environmental contamination. The availability of trace metal elements in their environment creates waves of physiological and metabolic changes throughout the cyanobacterial system. During iron-limited conditions, cyanobacteria alter photosynthetic structures but simultaneously turn on additional electron supply pathways so that they can continue functioning. Copper stress affects cyanobacterial gene activity, which shapes nitrogen metabolism and carbon fixation activities as well as their responses to oxidative stress situations. Trace metal dynamics show an inseparable connection with cyanobacterial functional biology, because they influence cells at the individual level while modifying population dynamics and shaping community structure and ecosystem processes (Singh et al., 2016).

Trace metals in aquatic ecosystems go through important changes in their accessibility and chemical forms due to rising anthropogenic operations such as mining and industrial outflows and agricultural drainage. The elevated levels of bioavailable metals generated by environmental changes create localized areas, which create new difficulties for microbial communities in the area. The fast growth rates and flexibility of cyanobacteria metabolism allow these bacteria to normally initiate first reactions when aquatic environments experience disturbances. The response by cyanobacteria can result in dangerous harmful cyanobacterial blooms (HCBs), which produce toxins that deplete oxygen and break down aquatic food chains. Certain trace metals both impact cyanobacterial toxin production and affect when blooms occur, thus demonstrating that available metal concentrations determine cellular metabolic rate and associated ecological consequences (Boota et al., 2024).

Knowledge about cyanobacterial interactions with trace metals is essential for scientists in both pure microbiology research and applied environmental research fields. Research on cyanobacteria trace metal interactions enables scientists to discover their biological adaptations as well as foresee harmful bloom risks while also identifying their potential uses in biotechnological applications like soil decontamination through biological remediation methods. Current research on cyanobacterial trace metal interactions has shown significant progress; yet, the full understanding remains incomplete. Studies that currently exist approach their investigations with few selected model strains tested in tightly controlled laboratory experiments without properly representing the natural ecosystems with their multiple interacting stress elements. Many scientists are actively exploring the genetic and molecular origin of metal uptake mechanisms while researching those pathways in nonmodel species that claim special ecological habitats.

Research gaps are currently addressed through interdisciplinary tactics, which combine molecular biology techniques together with environmental chemistry and systems biology methods alongside ecological modeling approaches. Through high-throughput omics methods such as transcriptomics proteomics and metabolomics, scientists have started to expose global cyanobacterial reactions to metal stress, thus identifying new genes and pathways important for maintaining metal equilibrium. The analysis of metal distribution through both metal imaging and speciation methods gives researchers detailed information about cellular distribution patterns along with chemical environment data. Synthetic biology together with genetic engineering techniques provide innovative approaches to enhance metal-processing capabilities of cyanobacteria while reducing sensitivity to metals, leading to future development of biosensing technologies alongside pollutant removal systems.

The goal of this chapter is to establish an extensive understanding of cyanobacteria's trace metal behavior concerning vital metal functions and toxic reactions together with cellular systems that support homeostasis and environmental regulators that govern these relationships. The study concentrates on metal ecological aspects for bloom behavior and interspecies competition and toxin manufacturing with a focus on innovative research strategies expected to advance our comprehension of this complex subject matter. To shape new research methods and create efficient protection strategies, this chapter combines recent findings with vital knowledge gaps about managing dual aquatic ecosystem threats from metals and cyanobacteria growth.

8.2 The Role of Trace Metals in Cyanobacterial Physiology

Trace metals function as vital components for cyanobacterial physiology, since they perform essential functions, which support multiple cellular activities. The metabolic processes of cyanobacteria depend on trace metals that function both as enzyme coenzymes as well as essential components of transport chains and essential elements for maintaining protein and membrane structure. Cyanobacterial growth together with metabolism and ecological function show strong sensitivity to trace metals, which bacteria need in extremely small amounts. Trace metals play essential roles for cyanobacteria where iron (Fe), copper (Cu), zinc (Zn), manganese (Mn), molybdenum (Mo), cobalt (Co), and nickel (Ni) belong to the most critical category. All metabolic balance functions and cellular existence depend upon the unique essential duties performed by these elements (Sukenik & Kaplan, 2021, Twining & Baines, 2013).

The trace metal Iron stands as the most extensively researched element in cyanobacteria, because photosynthesis and respiration heavily rely on it. The essential element serves as an integral part that enables electron transport in ferredoxins and iron–sulfur clusters and cytochromes present within photosystem I and photosystem II. Fe-SOD depends on iron to work as an enzyme, which detoxifies ROS during photosynthetic stress. The conversion of atmospheric nitrogen to ammonia through nitrogenase enzyme activity needs iron to be active in nitrogen-fixing cyanobacteria cells. The physiologic challenge of acquiring iron rises due to its existence as insoluble compounds mainly discovered in oxygenated water bodies. Cyanobacteria solved their iron procurement issue through the development of strong uptake mechanisms that utilize siderophores to capture environment-based iron molecules (Qiu et al., 2024).

Plastocyanin serves as a fundamental trace metal component, because it acts as an electron carrier that bridges cytochrome b6f with PSI. Plastocyanin enables photosynthetic electron flow maintenance when cyanobacteria experience iron limits, because it can replace cytochrome c6. The substitution process demonstrates how cyanobacteria adapt their metal utilization according to environmental resource availability. Multicopper oxidases that depend on copper act as cellular redox regulators as well as iron acquisition proteins. Copper must be carefully controlled, because free copper ions in high levels create conditions where they produce ROS through Fenton-like mechanisms, which harms proteins and damages lipids and DNA while leading to oxidative stress (Ueda et al., 2012).

Zinc operates both structurally and catalytically in numerous enzymes where it enables carbonic anhydrase to transform carbon dioxide and bicarbonate into efficient carbon fixation processes. DNA-binding proteins often contain zinc fingers along with other zinc-binding motifs, because these structures may serve a purpose in regulating transcription. The cellular zinc concentration in cyanobacteria functions in close connection to phosphate metabolism while two regulatory mechanisms of efflux systems together with metallothionein-like storage proteins control its homeostasis (Liu et al., 2004). Manganese serves a vital role in the WOC of PSII that enables the water molecule splitting process, which produces oxygen and protons and electrons for oxygenic photosynthesis. The PSII contains a Mn_4CaO_5 cluster, which exists with high frequency in all oxygen-producing organisms while sustaining ocean primary production. The antioxidative system benefits from manganese, because the element supports the cofactor role of Mn-dependent superoxide dismutase (Mn-SOD).

Photochemical tasks require the tiny amount of mineral molybdenum, which maintains nitrogen-based metabolic processes. Molybdenum functions as the central atomic element of molybdenum cofactor-containing enzymes, which include nitrate reductase and nitrogenase. One group of diazotrophic cyanobacteria requires molybdenum as its unique element to activate nitrogenase, yet these organisms can activate alternative nitrogenases using vanadium under situations without molybdenum.

URRED mutant cyanobacteria demonstrate how evolution creates survival strategies for metal scarcity between different ecological environments (Zerkle et al., 2011).

Nickel and cobalt exist in low concentrations but serve essential roles in running specific metabolic pathways. Vitamin B12 (cobalamin) contains cobalt as one of its components while some cyanobacteria do not synthesize the entire complex of B12 biosynthesis enzymes. Nickel functions as an essential component for the proper functioning of hydrogenase along with urease enzymes. Urease allows cyanobacteria to obtain nitrogen from urea while hydrogenases drive hydrogen metabolism, which is crucial for diazotrophic cyanobacteria during nitrogen fixation processes (Rodriguez & Ho, 2014) (Fig. 8.1).

The fundamental biological requirements of these trace metals demonstrate their essential functions in molding cyanobacterial physiological systems. Both excessive amounts and insufficient levels of any metal will disrupt photosynthesis and nitrogen fixation and result in oxidative harm and stunted growth. Cyanobacteria serve as critical primary producers and nitrogen fixers in aquatic food webs so their ecological health depends heavily on trace metal interactions. Cyanobacteria require optimal levels of internal metal contents to reach survival and ecological success.

The regulatory networks within cyanobacteria are sophisticated in their ability to detect metals as well as acquire and distribute them based on cellular requirements. Cyanobacteria regulate their metal handling components such as transporters along with chelators as well as storage proteins and metalloregulatory transcription factors in a cohesive manner. Mountain bacteria achieve survival by using these systems, which also help them dominate microbial competitors present in metal-constrained natural habitats. The natural process of trace metals regulating cellular metabolism

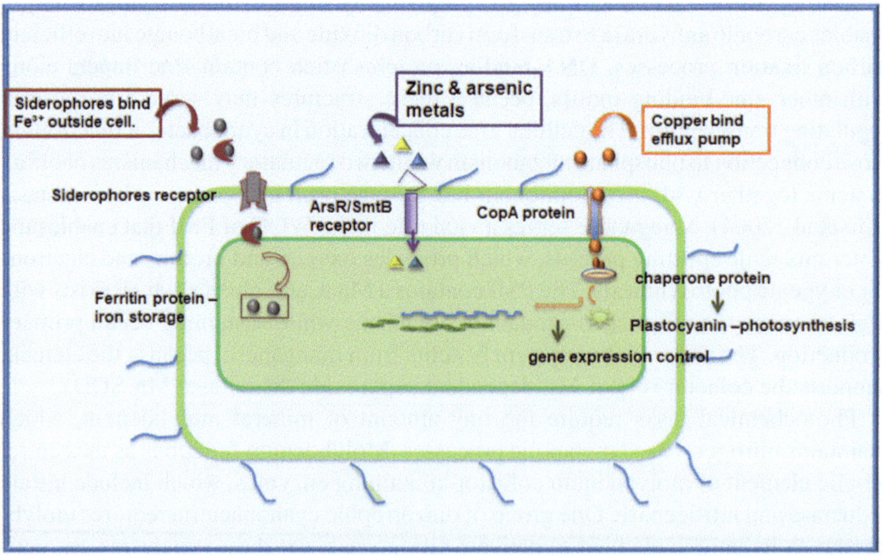

Fig. 8.1 Trace metal uptake and homeostasis in cyanobacteria

Table 8.1 Trace metals, and their essential functions and impacts

Trace metal	Essential functions	Toxic effects	References
Copper	Iron acquisition, electron carrier in plastocyanin	ROS production, disruption of ETC and photosynthesis	Ciani and Adessi (2023)
Zinc	Cellular structure, as cofactor for protein synthesis	Inhibit growth, affect protein synthesis	Abdeljelil et al.
Iron	Component of PSI & PSII, nitrogen fixation	ROS production, limit photosynthesis	Ipek et al. (2024)
Cadmium	No essential role	Damage to DNA, lipids, ROS production	Singh et al. (2018)
Nickel	Involved in hydrogenase enzymes formation, activity of urease	Inhibit enzymes activities	Mulrooney and Hausinger (2003)
Molybdenum	As a nitrate reductase, Nitrogenase	Affect nitrogen fixation	Navarro-Rodríguez et al. (2019)
Cobalt	As cofactor vitamin 12, Role in methylmalonyl-CoA mutase	Disrupts cellular respiration	Nies (1999)

influences larger-scale ecological events that include bloom development and shifts in community structure as well as relationships among planktonic organisms.

All aspects of cyanobacterial physiology require the fundamental role of trace metals in energy production and nutrient assimilation and stress response and gene regulation systems. The roles of trace metals for cyanobacteria exceed cellular functions by affecting aquatic environmental community dynamics and ecological function and biogeochemical cycling. The regulation and use of trace metals by cyanobacteria needs thorough study to determine how they will react to environmental changes and how to effectively control dangerous algal blooms (Table 8.1).

8.3 Mechanisms of Metal Uptake and Homeostasis in Cyanobacteria

Cyanobacteria have developed refined systems to acquire trace metals, which they transport and store and detoxify them for preserving homeostasis during changes in environmental conditions. The regulatory systems attain fine control, which enables them to strike a balance between the dual use and toxicity of metals iron zinc and copper. Cyanobacteria encounter considerable difficulties in metal acquisition within aquatic spaces because of limited bioavailability caused by pH shifts and redox potential variations and competitive ion interactions. The evolution of cyanobacteria has led to the development of both metal transporters with high acquisition affinity and multiple protective elements including metal-binding ligands along with intracellular chelators and regulatory circuitry between them.

Metal homeostasis begins with the absorption process that attracts metal ions from outside the extracellular area. The cell uses ATP-binding cassette transporters alongside membrane transport systems that exhibit high specificity toward individual metallic elements for extracellular metal intake. The iron uptake mechanisms of cyanobacteria utilize both Fe(III)-siderophore complexes and Fe(II) transporters with FeoB as a principal protein. The secretion of siderophores by cyanobacteria occurs under limited iron conditions to produce low-molecular-weight compounds that strongly bind with ferric iron, thus enabling its cellular uptake. The outer membrane receptors detect these siderophores, which then get transported into cells through TonB-dependent transporters by using the proton motive force as a power source. The next step of iron transport occurs through inner membrane permeases after the substance enters the periplasmic space. Fur (ferric uptake regulator) together with other metal-responsive transcription factors control iron uptake through mechanisms that either activate or repress gene expression according to intracellular iron levels. Under conditions of iron deficiency cyanobacteria use bacterioferritins to store excess iron inside cells while releasing it for healthy metabolic functions (Lis et al., 2015).

Specific transporters named CopA and P-type ATPases take charge of copper influx across cell membranes. The regulatory proteins CueR maintain copper homeostasis through their sensing mechanism, which activates either the copper efflux pumps or copper chaperones for expression control. Copper chaperones possess dual functions by transporting copper to plastocyanin and cytochrome oxidase enzymes but reducing the concentration of free copper ions, which causes oxidative damage (Giner-Lamia et al., 2012). The ZnuABC system functions as an ABC transporter to perform zinc transport in cyanobacteria while operating under zinc-deficient conditions. Glutamic acid and other proteins in the cytosol function as zinc dispersion channels for metalloenzymes and regulation means to balance internal zinc levels. The cell either uses CzcCBA transporters to export excess zinc or traps the substance with metallothionein-like proteins. Both groups of systems function together to maintain sufficient zinc supply for biochemical operations while protecting cells from toxic zinc levels.

The MntH transporter along with ABC-type Mn transporters help cyanobacteria absorb manganese when they encounter low manganese levels in their environment. The cyanobacterial organism utilizes manganese for dual physiological functions, because it serves both as an enzymatic cofactor while providing antioxidative protection that enhances superoxide dismutase activity. ACB transport proteins handle molybdenum uptake because these transporters possess some shared elements with sulfate transporters because of the close chemical structure of both anions (Eisenhut, 2019). Molybdate enters cyanobacteria cells and becomes part of the molybdenum cofactor (MoCo), which maintains the activity of enzymes responsible for nitrogen metabolism enzymes. The systems that control molybdenum uptake show strong correlations with nitrogen concentrations, because this element functions essentially for nitrogen assimilative pathways (Presta et al., 2015).

Cyanobacteria control total metal levels within their cells through transcriptional regulators, which respond to internal metal amounts and external environmental signals. The functionality of the ArsR/SmtB family of metalloregulatory proteins and MerR-type regulators allows them to bind DNA, thus enabling the transcription control of metal-responsive genes. The regulatory proteins modulate the activity of metabolism when metal binds to their active sites, which tolerates DNA operon regulation for metal uptake and metal release and storage (Baksh & Zamble, 2020). The maintenance of proper metal homeostasis represents an important ecological strategy that needs proper intracellular management to achieve effective competitiveness. Cyanobacteria establish their habitats primarily in metal-poor areas that other microorganisms also struggle to occupy. The dominance over other species in acquiring scarce metal resources creates substantial ecological advantages for these organisms especially when they form population blooms. Several species of cyanobacteria express allelopathic characteristics through the production of substances that create obstacles for competitors to gain metals or damage their metal regulatory processes.

Cyanobacteria adapt their physiological functions through developmental changes as metal concentrations vary in their environment. The absence of iron in the environment encourages various species to replace their iron-containing proteins with functionally equivalent copper protein alternatives so cytochrome c6 gets substituted with plastocyanin. The biological strategy of metalloprotein substitution represents "metalloprotein switching" enabling cyanobacteria to function normally when iron levels become too scarce for their system (Tengan & Moraes, 2017). Several antioxidant systems protect cyanobacteria against oxidative stress, which occurs due to metal overload. The defenses against reactive oxygen species consist of two categories: catalases, peroxidases and glutathione and thioredoxin systems. The synthesis of these antioxidant systems increases during periods when bacteria face metal stress, because they demonstrate the comprehensive integration between stress defense systems and regulatory mechanisms for metal homeostasis. The complex and effective mechanisms by which cyanobacteria obtain trace metals for distribution and regulation indicate their efficient adaptation to different aquatic environments throughout evolution. These systems help cyanobacteria to achieve survival under variable metal conditions and dominance in metal-variable regimes. Accurate predictions of cyanobacterial behavior toward anthropogenic metals require a full comprehension of their biological mechanisms combined with effective strategies for dealing with harmful cyanobacteria across aquatic environments (Latifi et al., 2009).

8.4 Metal–Metal Interactions and Competitive Dynamics

Trace metals within cyanobacteria produce conflicting reactions when present together, because these double or opposing effects substantially modify both survival and metabolic activity of the organism. The biological process enhancement occurs through synergistic interactions, which utilize molybdenum (Mo) and iron (Fe) as essential components for proper working of the MoFe nitrogenase enzyme during nitrogen fixation. Disruption of either metal's availability impairs nitrogen fixation and growth.

The relationship between metals results in antagonistic effects, because increased presence of one metal stops the uptake process and reduces functionality of other metals present. As Mn^{2+} concentration rises it blocks the cellular uptake of Fe^{2+} that leads to decreased activity of iron-dependent enzymes including cytochrome oxidase and photosystem I (PSI). Hydrodynamic conditions of excessive Cu^{2+} in the environment lead to the displacement of Zn^{2+} from superoxide dismutase enzymes causing breakdown of cellular antioxidant protection systems and raising oxidative stress levels (Fitsanakis et al. 2010; Shen et al. 2021).

The competitive uptake of metals has become difficult for cyanobacteria, since different metal species share equal chemical properties and occupy similar ionic spaces especially under conditions of coexisting multiple environmental metals. Cadmium (Cd^{2+}) along with Zinc (Zn^{2+}) shares transport protein Zrt/Irt-like proteins (ZIPs), thus resulting in zinc uptake inhibition under conditions of cadmium exposure. At the binding sites of P-type ATPase transporters like CopA both Cu^{2+} and Ag^{+} aggressively contend with each other while this affects copper homeostasis. Nonessential metals can win the competition for transport proteins by swapping essential metal cofactors out of enzymes, thus resulting in enzyme inactivation with potential ROS generation. The acquisition process of cyanobacteria depends on metal–metal interactions, because the bacteria must choose specific metals based on their availability and biological importance during metabolic operations. Due to metal resource competition cyanobacteria will experience metabolic disruptions along with stress responses that create difficulties for growth and cause cellular damage from oxidative stress (Lourembam et al., 2024).

8.5 Trace Metal Deficiency and Toxicity

The physical condition of lacking metals among cyanobacteria bacteria results in multiple negative effects for the growth cycle. The photosynthetic efficiency together with growth becomes impaired through iron deficiency, which blocks electron transport chain and cytochrome oxidase function. The reduction of zinc in the organism results in impaired superoxide dismutase (SOD) and carbonic anhydrase activity, which causes decreased oxidative stress control and slower growth and cell repair.

The condition of metal toxicity emerges when trace metals exceed maximum tolerable levels, thus triggering oxidative stress that simultaneously inhibits enzymes while damaging cellular structures. The three metal species lead (Pb^{2+}), cadmium (Cd^{2+}), and mercury (Hg^{2+}) exhibit high toxicity by creating reactive oxygen species (ROS), which initiate membrane lipid peroxidation and spark protein denaturation and rupture DNA strands. Oxidative damage intensifies cell injuries at the same time it sets off mutagenic reactions leading to both short-term and prolonged cellular function problems. The toxic substance copper (Cu^{2+}) makes proteins aggregate through its reaction with cysteine residues, which modifies their structure and functionality thereby disrupting cellular metabolic and signaling processes (Lajayer et al., 2017; Nowicka, 2022).

The direct interaction of toxic metals affects steady enzymatic processes apart from creating oxidative stress in the system. Cadmium interrupts the activity of essential enzymes carbonic anhydrase and RuBisCO while these enzymes are vital for carbon fixation and photosynthesis. The enzymes inhibition blocking carbon metabolism leads to decreased energy production and poor overall growth performance (Lis et al., 2015). The imbalancing of metal homeostasis in cyanobacteria caused by toxic metals leads to the improper binding of metalloproteins. The cellular activity and efficiency of enzymes with Mg^{2+} or Zn^{2+} cofactors decrease when cobalt (Co^{2+}) or nickel (Ni^{2+}) from the environment replaces these essential metal elements. The disruption in cellular functions such as signal transduction enzyme activity and cell division occurs because of metabolic imbalance.

Therapeutic relationships exist to determine the minimum quantities of metals until they become harmful. Among different cyanobacterial strains there exists a wide range of tolerance to metal toxicity due to unique detoxification capabilities even when the concentration climbs higher. The condition of metal toxicity emerges when trace metals exceed maximum tolerable levels, thus triggering oxidative stress that simultaneously inhibits enzymes while damaging cellular structures. The three metal species lead (Pb^{2+}), cadmium (Cd^{2+}), and mercury (Hg^{2+}) exhibit high toxicity by creating reactive oxygen species (ROS), which initiate membrane lipid peroxidation and spark protein denaturation and rupture DNA strands. Oxidative damage intensifies cell injuries at the same time it sets off mutagenic reactions leading to both short-term and prolonged cellular function problems. The toxic substance copper (Cu^{2+}) makes proteins aggregate through its reaction with cysteine residues, which modifies their structure and functionality (Singh et al., 2011).

8.6 Environmental Factors Influencing Metal Availability

The availability of metals in cyanobacteria depends greatly on three environmental factors including pH and redox potential together with organic matter complexation. In acidic areas, the solubility of copper and iron becomes elevated, because such conditions enable cyanobacteria to absorb these elements more easily. Under alkaline environments, metals tend to precipitate, thus diminishing their accessibility for biological usage. Environmental redox potential affects metal speciation, because iron and manganese along with other metals undergo redox reactions, which modify their availability and toxic levels. Metal availability to cyanobacteria depends on organic matter complexation, because it develops metal–organic compounds that simultaneously assist or prevent uptake of selected metals.

Human-caused activities that produce pollution along with agricultural runoff create major disruptions to metal availability rates in the environment. Certain metals including copper zinc and cadmium transfer from agricultural fields into water systems through runoff processes, thus changing the overall metal content of aquatic ecosystems. The excessive entry of metals into the environment creates toxic conditions that both damages the fragile metal ion equilibrium and causes metal toxicity. Urbanization plus industrial activities make pollution worse by bringing mercury and lead as well as other heavy metals into aquatic environments. The human-made pollution sources enhance total metal levels within ecological systems so that they create hazards for aquatic creatures while cyanobacteria face the challenge of adapting to these heightened metal concentrations (Gadd, 2009).

Biomolecules find substantially different metal availability levels in fresh versus saltwater environments. The lower ionic strength and distinctive chemical nature of freshwater environments increase metal availability, thus creating toxic conditions particularly in areas affected by industrial and agricultural pollution. The high salinity that characterizes marine systems decreases the availability of metals such as copper and zinc, thus making these metal elements inaccessible to cyanobacteria. A decrease in water salt content at estuarine locations results in elevated metal accessibility, thus increasing the potential for metal-related toxicity events. The differences in metal exposure between freshwater and marine environments result from multiple environmental conditions that control how metals become available and toxic to cyanobacteria (Midassi et al., 2020) (Fig. 8.2).

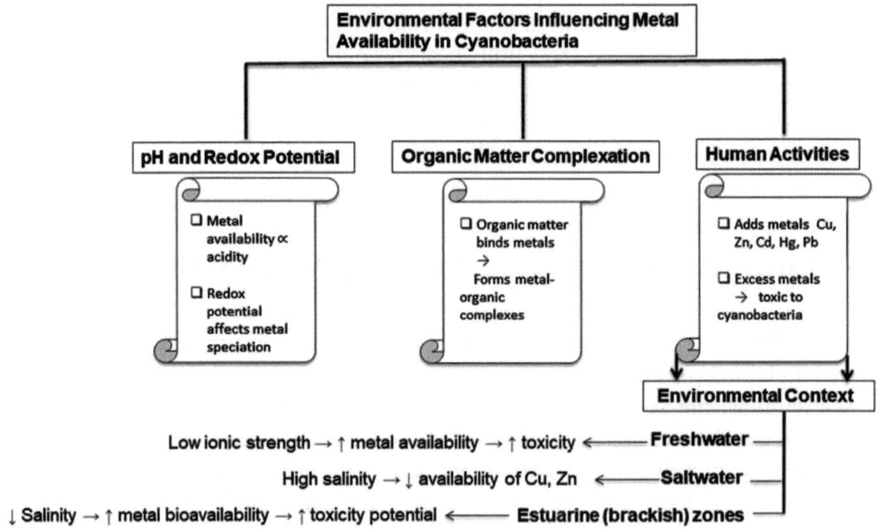

Fig. 8.2 Environmental factors influencing metal availability and their effects

8.7 Molecular and Biochemical Responses to Trace Metal Stress

Plants and animal species along with microorganisms face serious environmental problems from trace metals, because their metabolism becomes ill and doubtful when metals like copper zinc and iron appear in excessive amounts. The physiological need for these essential metals in enzyme activities and electron transport becomes detrimental to cellular health, since concentrations above baseline amount to ROS production, leading to oxidative damage. Organisms have several cellular mechanisms, which become activated to counter toxic effects of metals while sustaining homeostasis.

The cells utilize gene upregulation to enhance their ability to move and detoxify trace metals during periods of stress. The metal-response transcription factor MTF-1 controls the expression of both metallothioneins along with phytochelatins to capture and isolate metals from cells (Rubio et al., 2012). The production rate of phytochelatins increases when metal concentrations rise according to studies conducted on *Arabidopsis thaliana* and *Zea mays.* Under metal stress conditions, plants demonstrate greater expression of genes controlling stress signaling systems together with HSPs and chaperone genes that preserve cells from toxic metal-induced protein misfolding and aggregation events (Rumeau et al., 2007).

During trace metal stress, plants generate ROS including superoxide radicals ($O_2^{\cdot -}$), hydrogen peroxide (H_2O_2), and hydroxyl radicals ($^{\cdot}OH$) that inflict serious damage across cellular substances like lipids, proteins, and DNA. The excessive oxidative damage triggers organisms to enhance their antioxidant enzyme levels. When metal stress occurs, the major enzyme activation triggers superoxide dismutase (SOD) to convert superoxide radicals into hydrogen peroxide, which then peroxidases and catalase break down. Research shows that high zinc metal doses enhance SOD activity in the plants *Spinacia oleracea* and *Brassica juncea* along with the bacterial organism *Escherichia coli.* The presence of zinc as an oxidative stressor leads *E. coli* to enhance both SOD and peroxidase and catalase activities, which protect cells from ROS damage. The antioxidant molecule glutathione plays an essential part in regulating redox homeostasis. The essential plant defense mechanism of glutathione against ROS under metal stress requires the substance to maintain redox balance throughout metal-induced oxidative stress according to Ciani and Adessi (2023).

Metals that stress organisms cause profound modifications of metabolic products and proteins. A shift in the metabolomic profile occurs, because primary and secondary metabolites exhibit increased levels in the system. The stress induced by copper or cadmium exposure leads to amino acid accumulation of proline and polyamines, because these compounds protect cellular structures through osmoprotection and ROS scavenging. Secondary metabolites including phenolic compounds and flavonoids experience increased production, because they act as defense elements against stress by functioning as antioxidants while helping remove metal contaminants (Zheng et al., 2024).

The level of flavonoids rose in *Solanum lycopersicum* (tomato) following copper stress exposure and this increase demonstrated a protective mechanism against oxidative damage. The adaptive response to metal stress heavily depends on proteomic adaptations. Stress-related proteins and metal detoxifying proteins such as metallothioneins and chaperones show considerable increase in their expression levels. Organisms use synchronized molecular and biochemical pathways to deal with trace metal stress. Plants activate three main stress responses, which include increasing metal transporter levels and chelator amounts and activating antioxidant defense systems and modifying their metabolic and proteomic activities. Through adaptive mechanisms organisms acquire the necessary mechanisms to decrease metal toxicity and live in contaminated environments. Multiple studies demonstrate that all these stress responses function similarly in diverse species groups such as plants, animals and microorganisms for dealing with metal-induced oxidative stress and maintaining cellular equilibrium (Niu et al., 2023) (Table 8.2).

Table 8.2 List of abiotic stress tolerance in cyanobacteria with their references

Abiotic factors	Level	Impacts	References
Temperature	High	Disrupts enzyme function, induces heat shock proteins	Ismaiel and Piercey-Normore (2021)
	Low	Impairs membrane fluidity and protein synthesis	Gao et al. (2021)
Desiccation	–	Induces oxidative stress and membrane lipid peroxidation	Singh et al. (2013)
Chemicals	Heavy metals	Interferes with photosynthesis and membrane integrity	Srivastava et al. (2021)
	Pesticides	Causes ROS generation; upregulates SOD, catalase, phytochelatins, glutathione, flavonoids, chaperone	Kuritz et al. (1997)
Salt stress	Osmoticum	Induces water stress and affects turgor pressure	Apte et al. (1997)
	Salinity	Leads to ion imbalance and osmotic stress	Singh et al. (2022)
Nutrient deficiency	Essential nutrients	Limitation of macro/micronutrients affects growth and metabolism	Schwarz and Forchhammer (2005)
Light intensity	Light	Affects photosynthesis, pigment production, and oxidative balance	
Radition	Gamma	Oxidative stress and protein denaturation	Rai et al. (2013)
	UV	DNA, proteins; induces ROS	
Acidity	Low pH	Affects gene regulation, protein expression	

8.8 Future Directions and Research Gaps

The advancement of analytical technology has established fresh possibilities to study trace metal regulations within biological structures. Metalomics represents an important tool for comprehensive evaluation of cellular metal characteristics and species relationships in tissues and biological organisms due to its ability to explain emerging patterns of metal homeostasis. Advancements in modern imaging tools like synchrotron-based X-ray fluorescence microscopy and laser ablation inductively coupled plasma mass spectrometry (LA-ICP-MS) allow researchers to apprehend metal distributions at the cellular and subcellular resolutions. The innovative techniques offer once unprecedented in situ observations of metal spatial locations and possible functional duties, which establish fundamental knowledge for metal-based physiological processes (Perry et al., 2020).

The analysis of multiple metals in real-world complex environments requires further extensive research, because current progress falls short. Most existing research investigates metals as individual elements, yet organisms usually face numerous metal contaminants both in untainted and contaminated ecosystems. Exposure to multiple metals generates unpredicted combined impacts, because the effects may be either additive or synergistic or antagonistic. Approaches combining three different fields must take priority in metal stress research, so researchers can better understand the effects of multiple metal ions on biological systems. They need to study these effects from an ecological perspective alongside biochemical and molecular research methods.

Trace metal dynamics create implications, which affect both organismal stress responses as well as environmental transformations and major global processes. The environmental changes driven by climate change will modify the movement of metals and their susceptibility and dangerous levels throughout different ecological areas. Such variations affect the process of nutrient cycling and modify microbial populations and play a significant role in worldwide element cycle of carbon and nitrogen. Forecasting ecosystem resilience requires knowledge of how trace metal stress relates to major environmental alterations so scientists can create sustainable approaches for environmental restoration and resource management systems.

In summary, the study of trace metal stress has revealed a complex network of molecular, biochemical, and physiological adaptations that organisms use to maintain homeostasis under metal exposure. These responses include the activation of metal chelators, antioxidant defense systems, and shifts in metabolic and proteomic landscapes. However, key knowledge gaps remain, particularly regarding multi-metal interactions and their environmental context. As environmental and applied microbiology continue to intersect with climate science and molecular biology, understanding trace metal interactions will be crucial for addressing future challenges in environmental health, agriculture, and biotechnology.

References

Abdeljelil, N., Yahia, N. B. M., Landoulsi, A., Chatti, A., Wattiez, R., Van Houdt, R., Gillan, D. Cellular and molecular outcomes of pseudomonas putida Kt2440 exposure to aluminium. *Chemosphere.*

Apte, S., Fernandes, T., Iyer, V., & Alahari, A. (1997). Molecular basis of tolerance to salinity and drought stresses in photosynthetic nitrogen-fixing cyanobacteria. In *Plant molecular biology biotechnology* (pp. 258–268). Narosa Publications.

Baksh, K. A., & Zamble, D. B. (2020). Allosteric control of metal-responsive transcriptional regulators in bacteria. *Journal of Biological Chemistry, 295*, 1673–1684.

Boota, M. W., Xia, H., Qin, Y., Kakakhel, M. A., Yan, C., Weiran, L., & Xu, J. (2024). Distribution and bioaccumulation of trace elements in two Cyprinidae fish species in the Indus river, Pakistan, including the impact of hydraulic structure on macroinvertebrates' biodiversity. *Environmental Research, 252*, Article 118882.

Chankeshwara, N. N., Pant, M., & Pant, K. (2025). The cyanobacterial renaissance: Bioactive discoveries and future prospects. In K Sharma (Ed.), *Bio-prospecting of novel microbial bioactive compounds for sustainable development: Management of natural resources through microbial conversion into valuable products* (pp. 187–201). Springer Nature Switzerland.

Ciani, M., & Adessi, A. (2023). Cyanoremediation and phyconanotechnology: Cyanobacteria for metal biosorption toward a circular economy. *Frontiers in Microbiology, 14*, 1166612.

Eisenhut, M. (2019). Manganese homeostasis in cyanobacteria. *Plants, 9*, 18.

Fitsanakis, V. A., Zhang, N., Garcia, S., & Aschner, M. (2010). Manganese (Mn) and iron (Fe): Interdependency of transport and regulation. *Neurotoxicity Research, 18*, 124–131.

Gadd, G. M. (2009). Biosorption: Critical review of scientific rationale, environmental importance and significance for pollution treatment. *Journal of Chemical Technology Biotechnology: International Research in Process, Environmental Clean Technology, 84*, 13–28.

Gao, X., Zhu, Z., Xu, H., Liu, L., An, J., Ji, B., & Ye, S. (2021). Cold adaptation in drylands: Transcriptomic insights into cold-stressed Nostoc flagelliforme and characterization of a hypothetical gene with cold and nitrogen stress tolerance. *Environmental Microbiology, 23*, 713–727.

Giner-Lamia, J., López-Maury, L., Reyes, J. C., & Florencio, F. J. (2012). The CopRS two-component system is responsible for resistance to copper in the cyanobacterium Synechocystis sp. PCC 6803. *Plant Physiology, 159*, 1806–1818.

Ipek, Y., Ghosh, P., Mausbach, W. E., & Jeyasingh, P. D. (2024). Spatiotemporal variation in dissolved, bioavailable, and particulate elements and the abundance of harmful algae in grand lake. *ACS Est Water, 4*, 5492–5505.

Ismaiel, M., & Piercey-Normore, M. (2021). Gene transcription and antioxidants production in Arthrospira (Spirulina) platensis grown under temperature variation. *Journal of Applied Microbiology, 130*, 891–900.

Jenkins, J. A., Musgrove, M., & White, S. J. O. (2023). Outlining potential biomarkers of exposure and effect to critical minerals: Nutritionally essential trace elements and the rare earth elements. *Toxics, 11.*

Kuritz, T., Bocanera, L. V., & Rivera, N. S. (1997). Dechlorination of lindane by the cyanobacterium Anabaena sp. strain PCC7120 depends on the function of the NIR operon. *Journal of Bacteriology, 179*, 3368–3370.

Lajayer, B. A., Ghorbanpour, M., & Nikabadi, S. (2017). Heavy metals in contaminated environment: Destiny of secondary metabolite biosynthesis, oxidative status and phytoextraction in medicinal plants. *Ecotoxicology Environmental Safety, 145*, 377–390.

Latifi, A., Ruiz, M., & Zhang, C.-C. (2009). Oxidative stress in cyanobacteria. *FEMS Microbiology Reviews, 33*, 258–278.

Leapaldt, H., Ingalls, M., Soares, G., & House, C. H. (2025). Trace metal and phosphorus enrichments in cyanobacteria cells and cyanobacterial precipitated minerals. *Minerals, 15*, 378.

Lis, H., Kranzler, C., Keren, N., & Shaked, Y. (2015). A comparative study of iron uptake rates and mechanisms amongst marine and fresh water cyanobacteria: Prevalence of reductive iron uptake. *Life, 5*, 841–860.

Liu, T., Nakashima, S., Hirose, K., Shibasaka, M., Katsuhara, M., Ezaki, B., Giedroc, D. P., & Kasamo, K. (2004). A novel cyanobacterial SmtB/ArsR family repressor regulates the expression of a CPx-ATPase and a metallothionein in response to both Cu(I)/Ag(I) and Zn(II)/Cd(II). *Journal of Biological Chemistry, 279*, 17810–17818.

Lobus, N. V., & Kulikovskiy, M. S. (2023). The co-evolution aspects of the biogeochemical role of phytoplankton in aquatic ecosystems: A review. *Biology, 12.*

Lourembam, J., Haobam, B., Singh, K. B., Verma, S., & Rajan, J. P. (2024). The molecular insights of cyanobacterial bioremediations of heavy metals: The current and the future challenges. *Frontiers in Microbiology, 15*, 1450992.

Midassi, S., Bedoui, A., & Bensalah, N. (2020). Efficient degradation of chloroquine drug by electro-Fenton oxidation: Effects of operating conditions and degradation mechanism. *Chemosphere, 260*, Article 127558.

Mulrooney, S. B., & Hausinger, R. P. (2003). Nickel uptake and utilization by microorganisms. *FEMS Microbiology Reviews, 27*, 239–261.

Navarro-Rodríguez, M., Buesa, J. M., & Rubio, L. M. (2019). Genetic and biochemical analysis of the *Azotobacter vinelandii* molybdenum storage protein. *Frontiers in Microbiology, 10*, 579.

Nies, D. H. (1999). Microbial heavy-metal resistance. *Applied Microbiology and Biotechnology, 51*, 730–750.

Niu, L., Li, C., Wang, W., Zhang, J., Scali, M., Li, W., Liu, H., Tai, F., Hu, X., & Wu, X. (2023). Cadmium tolerance and hyperaccumulation in plants—A proteomic perspective of phytoremediation. *Ecotoxicology and Environmental Safety, 256*, Article 114882.

Nowicka, B. (2022). Heavy metal–induced stress in eukaryotic algae—Mechanisms of heavy metal toxicity and tolerance with particular emphasis on oxidative stress in exposed cells and the role of antioxidant response. *Environmental Science Pollution Research, 29*, 16860–16911.

Perry, W. J., Weiss, A., Van de Plas, R., Spraggins, J. M., Caprioli, R. M., & Skaar, E. P. (2020). Integrated molecular imaging technologies for investigation of metals in biological systems: A brief review. *Current Opinion in Chemical Biology, 55*, 127–135.

Presta, L., Fondi, M., Emiliani, G., & Fani, R. (2015). *Molybdenum cofactors and their role in the evolution of metabolic pathways.* Springer.

Qiu, G.-W., Zheng, W.-C., Yang, H.-M., Wang, Y.-Y., Qi, X., Huang, D., Dai, G.-Z., Shi, H., Price, N. M., & Qiu, B.-S. (2024). Phosphorus deficiency alleviates iron limitation in Synechocystis cyanobacteria through direct PhoB-mediated gene regulation. *Nature Communications, 15*, 4426.

Rai, S., Pandey, S., Shrivastava, A. K., Singh, P. K., Agrawal, C., & Rai, L. C. (2013). *Understanding the mechanisms of abiotic stress management in cyanobacteria with special reference to proteomics, 93.* CRC Press Boca Raton.

Rodriguez, I. B., & Ho, T.-Y. (2014). Diel nitrogen fixation pattern of Trichodesmium: The interactive control of light and Ni. *Scientific Reports, 4*, 4445.

Rubio, M. F., Fernandez, P. N. L., Alvarado, C. V., Panelo, L. C., Grecco, M. R., Colo, G. P., Martínez-Noel, G. A., Micenmacher, S. M., & Costas, M. A. (2012). Cyclin D1 is a NF-κB corepressor. *Biochimica et Biophysica Acta (BBA)—Molecular Cell Research, 1823*, 1119–1131.

Rumeau, D., Peltier, G., & Cournac, L. (2007). Chlororespiration and Cyclic Electron Flow around PSI during Photosynthesis and Plant Stress Response. *Plant & Cell, Environment, 30*, 1041–1051.

Schwarz, R., & Forchhammer, K. (2005). Acclimation of unicellular cyanobacteria to macronutrient deficiency: Emergence of a complex network of cellular responses. *Microbiology, 151*, 2503–2514.

Shen, L., Chen, R., Wang, J., Fan, L., Cui, L., Zhang, Y., Cheng, J., Wu, X., Li, J., & Zeng, W. (2021). Biosorption behavior and mechanism of cadmium from aqueous solutions by Synechocystis sp. PCC6803. *RSC Advances, 11*, 18637–18650.

Singh, H., Anurag, K., & Apte, S. K. (2013). High radiation and desiccation tolerance of nitrogen-fixing cultures of the cyanobacterium Anabaena sp. strain PCC 7120 emanates from genome/proteome repair capabilities. *Photosynthesis Research, 118*, 71–81.

Singh, J. S., Kumar, A., Rai, A. N., & Singh, D. P. (2016). Cyanobacteria: A precious bio-resource in agriculture, ecosystem, and environmental sustainability. *Frontiers in Microbiology, 7*, 529.

Singh, P. K., Wang, W., & Shrivastava, A. K. (2018). Cadmium-mediated morphological, biochemical and physiological tuning in three different Anabaena species. *Aquatic Toxicology (Amsterdam, Netherlands), 202*, 36–45.

Singh, R., Gautam, N., Mishra, A., & Gupta, R. (2011). Heavy metals and living systems: An overview. *Indian Journal of Pharmacology, 43.*

Singh, R. P., Yadav, P., Kujur, R., Pandey, K. D., & Gupta, R. K. (2022). Cyanobacteria and salinity stress tolerance. In *Cyanobacterial lifestyle and its applications in biotechnology* (pp. 253–280). Elsevier.

Srivastava, A., Biswas, S., Yadav, S., Kumar, S., Srivastava, V., & Mishra, Y. (2021). Acute cadmium toxicity and post-stress recovery: Insights into coordinated and integrated response/recovery strategies of Anabaena sp. PCC 7120. *Journal of Hazardous Materials, 411*, 124822.

Sukenik, A., & Kaplan, A. (2021). Cyanobacterial harmful algal blooms in aquatic ecosystems: A comprehensive outlook on current and emerging mitigation and control approaches. *Microorganisms, 9*, 1472.

Tengan, C. H., & Moraes, C. T. (2017). NO control of mitochondrial function in normal and transformed cells. *Biochimica et Biophysica Acta (BBA)—Bioenergetics, 1858*, 573–581.

Twining, B. S., & Baines, S. B. (2013). The trace metal composition of marine phytoplankton. *Annual Review of Marine Science, 5*, 191–215.

Ueda, T., Nomoto, N., Koga, M., Ogasa, H., Ogawa, Y., Matsumoto, M., Stampoulis, P., Sode, K., Terasawa, H., & Shimada, I. (2012). Structural basis of efficient electron transport between photosynthetic membrane proteins and plastocyanin in spinach revealed using nuclear magnetic resonance. *The Plant Cell, 24*, 4173–4186.

Zerkle, A. L., Scheiderich, K., Maresca, J. A., Liermann, L. J., & Brantley, S. L. (2011). Molybdenum isotope fractionation by cyanobacterial assimilation during nitrate utilization and N_2 fixation. *Geobiology, 9*, 94–106.

Zheng, X., Lin, H., Du, D., Li, G., Alam, O., Cheng, Z., Liu, X., Jiang, S., & Li, J. (2024). Remediation of heavy metals polluted soil environment: A critical review on biological approaches. *Ecotoxicology and Environmental Safety, 284*, Article 116883.

Chapter 9
The Agricultural NEXUS: How Farming Practices Amplify Nutrient Inputs and Harmful Algal Bloom Indices

Muhammad Mehran Anjum, Nawab Ali, Maaz Khan, Abdul Haq, Gul Roz Khan, Azam Ali Sher, and Jalal Bayar

Abstract Harmful algal blooms are becoming a major global environmental problem. These blooms, resulting primarily from nutrient over-enrichment in aquatic systems, cause depletion of oxygen, degradation of ecosystems, fish kills, and release of toxins harmful to humans and animals. The prime nutrients responsible for these events include nitrogen (N) and phosphorus (P), which are present in sufficient quantities in most fertilizer formulations used in agriculture. Such resource imbalances expressed in terms of nutrient runoff occur as global food production increases and such have since cumulated into perhaps the single most important element of eutrophication. This chapter addresses the intricacy of the connection between agricultural practices and harmful algal bloom formation. It covers the sources and pathways of nutrient inputs, the ecological mechanisms connecting such nutrients to bloom

M. M. Anjum (✉) · M. Khan
Department of Agronomy, The University of Agriculture, Peshawar, Pakistan
e-mail: mehrananjum@aup.edu.pk

M. Khan
e-mail: khanmaaz820@gmail.com

M. M. Anjum · G. R. Khan
Pakistan Council of Scientific and Industrial Research Laboratories, Peshawar, Pakistan
e-mail: gulrozkhan@aup.edu.pk

N. Ali · J. Bayar
Biosystem and Agricultural Engineering, Michigan State University, East Lansing, USA
e-mail: alinawab@msu.edu

J. Bayar
e-mail: bayarjal@msu.edu

A. Haq
MaREI Research Center, University College Cork-Ireland, Cork, Ireland
e-mail: haq_afridi88@yahoo.com

A. A. Sher
Comparative Medicine and Integrative Biology Program, Michigan State University, East Lansing, USA
e-mail: sherazam@msu.edu

S. Fahad et al. (eds.), *Cyanobacterial Blooms: Ecology, Evolution and Biogeochemical Impacts*, https://doi.org/10.1007/978-3-032-06042-6_9

formation, empirical observations from various ecosystems, and potential mitigation pathways. The goal is to provide a converging view of what the "Agricultural Nexus" between farming, nutrient loading, and degradation of water quality. This study, The Agricultural NEXUS: How Farming Practices Increase Nutrient Inputs and Harmful Algal Bloom Indices, also examines the important connection between agricultural land use and increased nutrient loading, particularly nitrogen and phosphorus, in aquatic systems. By combining data from watershed monitoring, remote sensing, and farm management records, we pinpoint key farming practices. These include synthetic fertilizer application, manure management, and tillage methods that lead to greater nutrient runoff. We present a HAB index framework that links these practices to bloom frequency, duration, and toxicity. Our findings show that heavily managed agricultural areas have much higher HAB indices, resulting from both diffuse and point-source nutrient inputs. The paper highlights the urgent need for integrated land–water management strategies, such as best management practices (BMPs), policy changes, and precision agriculture, to reduce the negative impacts of farming on water quality and ecosystem health.

9.1 Introduction

Agriculture is not just the pillar of food security worldwide but also a leading force in economic growth and rural livelihoods (de Oliveira Veras et al., 2021). Modern agriculture, particularly intensification and high-productivity agriculture, has spawned unintended environmental impacts. Among the most pressing of these is nutrient pollution, which plays an important role in the formation of harmful algal blooms (HABs) in water bodies (Hucke-Gaete et al., 2024).

Nitrogen and phosphorus are among the essential macronutrients that carry out irreplaceable functions in the growth of plants and animals within agricultural and natural ecosystems. These two constituent elements mainly determine food security and ecosystem functioning but are also relevant fertilization of climate stability and human health (Dimkpa et al., 2023; Penuelas et al., 2020). The extraction and utilization of N and P have become increasingly unsustainable due to increasing anthropogenic pressures (Bulacio Fischer et al., 2025; Witek-Krowiak et al., 2022; Zou et al., 2022). In plants, these elements are essential for photosynthesis, growth, energy transfer, and protein and nucleic synthesis (Wang et al., 2025). On the other hand, their availability in the environment and their geobiochemical cycles vary sharply. The N in the atmosphere must be converted to reactive forms by some biological or industrial fixation. The P is unique in having no other means to enter biological systems except mining of phosphate rock; it has no functional substitute in vital operations such as synthesis of DNA, functioning of membranes, and energy transfer (Vučić et al. 2021). On the whole, in terms of availability, N has an atmospheric overabundance, and P, by contrast, is mineralized from phosphate rock, the gradual renewal of which occurs with uplift of continental rocks (Guignard et al., 2017). Both N and P are usually among the co-limiting nutrients in terrestrial and aquatic ecosystems across the entire range of biomes (Guignard et al., 2017; Wang et al. 2025).

Overuse of nitrogen (N) and phosphorus (P) in agriculture, either through synthetic fertilizers or animal waste, results in an excess of nutrients in the soil (Yahaya et al., 2023). When rain occurs or irrigation is used, these nutrients are transported away, moving through runoff or leaching into streams, rivers, lakes, and even coastal waters (Alcamo, 2019; Mishra, 2023). In these water bodies, the surplus nutrients promote excessive algal growth, allowing conditions to be ideal for HAB formation. These flowers can release toxins, lower oxygen concentrations in the water, and displace aquatic food webs, threatening human health, aquatic organisms, and local economies that rely on fishery and tourism (Nath, 2023; Rajak et al., 2024).

The term agricultural NEXUS involves a complex interplay of agricultural interventions, nutrient inputs, and water quality results (Davidson et al., 2016). It shows that any environmental challenge requires an integrated approach where agricultural practices, nutrient management, water conservation, and policy considerations are set aside and considered together. This chapter describes how different sources and pathways of nutrient pollution from agricultural landscapes contribute to the development of HABs before discussing old and new approaches to mitigate these impacts (Alcamo, 2019; Liu et al., 2018; Tian et al., 2018).

The estimated number of people on the globe by 2050 is expected to be nearly 9.8 billion (Alexandratos & Bruinsma, 2012); intense agricultural production is therefore needed within such a time frame. Historically, this has been possible with the wide application of synthetic fertilizers and nutrient-rich pesticides to boost crop yield dramatically (Hayat et al., 2025). However, the high nutrient utilization has led to severe environmental consequences in terms of greenhouse gases, water shortages, eutrophication, and loss of biodiversity (Vasseghian et al., 2024). N pollution is multiple anthropogenic sources like synthetic fertilizers, fossil-fuel combustion, and livestock waste, with agriculture accounting for greater than 90% of all anthropogenic N and P inputs (Guignard et al., 2017; Hayat et al., 2025). Early in the twentieth century, the Haber–Bosch process changed agriculture because it led to the industrial-scale synthesis of ammonia out of atmospheric N (Gao et al., 2024a, b; Guignard et al., 2017). Whereas this process is today contributing more reactive N to ecosystems than natural biological N fixation, it disrupts the N cycles and exacerbates the environmental degradation.

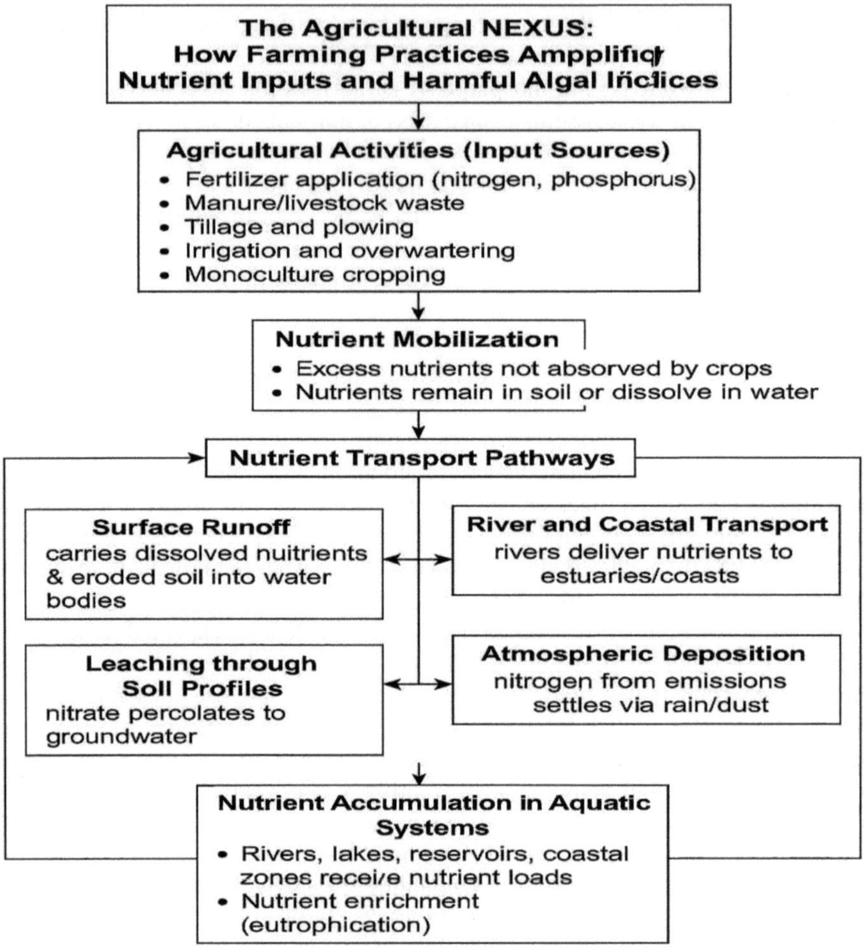

This flowchart diagram depicts the interrelated actions related to how agricultural practices contribute towards nutrient pollution and harmful algal blooms (HABs). It begins with important sources of agricultural inputs including fertilizer, animal waste, and irrigation. These inputs lead to the mobilization of nutrients, which include excess nitrogen and phosphorus that remain in the soil, or dissolve in water. The remaining nutrients travel through numerous transport pathways; they are either carried by surface runoff into water, leached through soil into waterways, transported via rivers to the coast, released into the air, and deposited into coastal zones. Each of these steps adds nutrients into the aquatic systems (rivers, lakes, coastal zones) and leads to nutrient accumulation, eutrophication, and algal blooms. The flowchart depicts and connects everything together in a clear sequential way, illustrating the agricultural nexus that contributed to w water pollution.

9.2 Agricultural Practices and Nutrient Inputs

Agricultural practices play a big role in adding nutrients to the environment, especially nitrogen and phosphorus. The widespread use of synthetic fertilizers to boost crop yields often results in excess nutrients that plants do not absorb. These nutrients can then leach into nearby water bodies through runoff or infiltration. Livestock operations add to this problem with manure, which, when not managed properly, introduces high levels of nutrients into the soil and waterways. Overfertilization, especially before rain, and conventional tillage disturb the soil. This increases erosion and the loss of nutrients. Irrigation can worsen the issue by moving dissolved nutrients beyond the root zone and into drainage systems. Additionally, monoculture systems can harm soil health, making it harder to retain nutrients. These combined factors lead to nutrient buildup in aquatic ecosystems. This buildup fuels harmful algal blooms, degrades water quality, and poses risks to human and environmental health. Sustainable land management and focused nutrient strategies are crucial to lessen these negative effects and restore ecological balance. Agricultural systems have evolved to maximize yield and efficiency, but many conventional practices inadvertently enhance the mobilization and transport of nutrients. Here, we explore the primary practices contributing to nutrient inputs in detail.

9.3 Fertilizer Use

9.3.1 The Role of Synthetic Fertilizers

The application of synthetic fertilizers including ammonium nitrate and urea and superphosphates serves to provide necessary nutrients for plant development. Through their manufacturing and application, farmers have been able to produce better crop yields which address the expanding worldwide population requirements (Giller et al., 2021). The uptake of nutrients by crops does not reach optimal levels during their growth period.

Synthetic fertilizers are essential in modern agriculture. They provide nutrients like nitrogen, phosphorus, and potassium, which help crops grow quickly and produce high yields. Although fertilizers have greatly increased food production around the world, their overuse has resulted in serious environmental problems (Tyagi et al., 2022). A maximum amount of fertilizers applied to the soil is not absorbed by different crops. Instead, it leaches into groundwater or runs off into surface waters, especially during rain. This nutrient runoff drives eutrophication, which leads to harmful algal blooms and oxygen-depleted "dead zones" in aquatic ecosystems. Moreover, relying on synthetic fertilizers over time can harm soil health by upsetting microbial balance and lowering organic matter. The production of these fertilizers also requires a lot of energy, leading to greenhouse gas emissions (Olowe, 2021; Sabina et al., 2025). While synthetic fertilizers are essential for meeting global food needs, their use needs careful management through precision agriculture and integrated nutrient management to reduce environmental damage.

9.3.2 Overapplication and Nutrient Imbalance

The application of fertilizers exceeds what crops can absorb when farmers make wrong nutrient requirement calculations or apply more fertilizers to avoid deficiencies. The excess nutrients become vulnerable to being carried away by rainwater or irrigation systems (Delgado et al., 2024). Overusing fertilizers is a common issue in intensive farming. Many farmers think that adding more nutrients will result in higher yields. However, this often creates nutrient imbalances in the soil, with nitrogen or phosphorus levels far above what crops require. This excess not only wastes resources but also raises the risk of nutrient runoff and leaching into water bodies, especially during irrigation or rainfall. Such imbalances can disturb soil chemistry, limit the uptake of other essential nutrients, and harm plant health (Ramamoorthy et al., 2024). Additionally, too many nutrients can promote the growth of harmful algal blooms in aquatic systems, leading to ecological damage and water quality problems. This imbalance may also affect microbial communities in the soil, reducing long-term fertility and resilience. To prevent these issues, effective soil testing and customized nutrient management plans (NMPs) are critical for maintaining crop productivity and environmental health (Thang et al., 2024; Shah & Wu, 2019).

9.3.3 Chemical Properties

Most synthetic fertilizers possess water solubility properties which enable them to dissolve easily when mixed with water. Although this trait provides immediate nutrient accessibility, it creates greater chances for fertilizers to enter groundwater through leaching or to be transported by surface runoff (Khan et al., 2018). The chemical properties of synthetic fertilizers are important for understanding how nutrients act in the soil and the surrounding environment. Most synthetic fertilizers have highly soluble forms of nitrogen, like nitrate or ammonium, and phosphorus, such as phosphate. These nutrients are readily available for plants to absorb (Yahaya et al., 2023). While this quick availability boosts plant growth, it also raises the risk of nutrient loss through leaching and runoff if not managed well. The high solubility allows nutrients to easily move past the root zone, especially in sandy or compacted soils. Using certain fertilizers continuously can change soil pH. This change can lead to increased acidity or alkalinity, which affects nutrient availability and microbial activity. Some compounds can also interact with soil minerals, causing either fixation or toxicity depending on their concentration (Michael, 2021). Understanding how these chemical behaviors work is crucial for using fertilizers effectively and reducing environmental harm. This shows the importance of balanced formulas and specific application strategies for each site.

9.3.4 Timing and Method of Application

The application of fertilizers before or during heavy rainfall events significantly increases the possibility of nutrient loss. The method of applying fertilizers through broadcasting instead of targeted approaches such as fertigation or banding produces greater fertilizer exposure on the soil surface which increases the chance of runoff losses (Shahena et al., 2021). The timing and method of applying fertilizer are key factors that affect nutrient efficiency and environmental impact. Applying fertilizers at the wrong time, such as before heavy rain or when plants aren't absorbing nutrients, can result in major nutrient losses through runoff or leaching. For instance, applying nutrients too early in the season might not match crop growth stages which result in the plants being unable to uptake the nutrients from the soil. This leads to lower uptake and higher pollution risks. Similarly, applying fertilizers to frozen or saturated soils can cause immediate nutrient wash-off. The method of application is also important. Spreading fertilizers on the surface can leave nutrients exposed. Incorporating them into the soil or placing them near roots can enhance absorption and cut down losses (Asadu et al., 2024). Methods like fertigation and controlled-release formulations enable more precise nutrient delivery based on crop needs. Using the right timing and methods not only boosts fertilizer efficiency and crop performance but also helps lessen environmental harm and promotes sustainable farming practices.

9.4 Organic and Slow-Release Fertilizers

9.4.1 Organic and Slow-Release Fertilizers

Nutrients in organic fertilizers such as composts and manure become available to plants at a gradual pace. Excess application of these materials together with adverse weather conditions can result in nutrient runoff (Verma et al., 2019). The industry faces challenges in adopting slow-release technology that protects nutrients through gradual matrix dissolution processes. Organic and slow-release fertilizers provide a more sustainable choice compared to regular synthetic options. They offer nutrients gradually and in a controlled way. These fertilizers come from natural sources like compost, manure, bone meal, or plant residues (Wu et al., 2024; Shaji et al., 2021). They improve soil health by boosting organic matter, increasing microbial activity, and enhancing soil structure. Slow-release fertilizers, including both organic and specially made synthetic types, release nutrients gradually. This process matches the rate at which crops absorb nutrients and lowers the risk of leaching and runoff. The steady release of nutrients ensures a more consistent supply, avoiding the extreme highs and lows that often come with quick-release fertilizers (Priya et al., 2024). Additionally, these fertilizers usually have a smaller environmental impact because they reduce the immediate release of excess nutrients into nearby ecosystems. While they may work more slowly and need careful management, their long-term benefits for soil fertility, water quality, and ecosystem health make them essential for sustainable farming.

9.4.2 Nutrient Synchronization

The main objective of slow-release fertilizers is to align nutrient delivery with the specific requirements of crops. The perfect matching process allows plants to absorb nutrients properly which decreases the chance of nutrient pollution in water bodies. Nutrient synchronization means matching the availability of nutrients in the soil with the needs of crops as they grow. This balance is important for increasing crop yield and reducing environmental losses (Fontaine et al., 2024). If nutrients are added too early or too late compared to plant demand, a lot of them can be lost through leaching, runoff, or evaporation instead of being taken up by the crop. By coordinating the release of nutrients with crop uptake, using techniques like split applications, slow-release fertilizers, or fertigation, plants get what they need at the right times, particularly during key growth stages. This approach makes nutrient use more efficient, cuts down on waste, and reduces the environmental effects of farming. Tools such as soil tests, leaf tissue analysis, and real-time nutrient sensors can help with timing and dosage (Swathy et al., 2024). Nutrient synchronization is an important concept in precision agriculture and is vital for sustainable nutrient management.

9.4.3 Environmental Impacts

The nutrient content in organic fertilizers varies extensively which creates difficulties for precise management despite their positive effects on soil structure and microbial activity. The environmental effects of poor nutrient management in agriculture are serious and widespread (Hartmann & Six, 2023). Overusing synthetic fertilizers and applying them at the wrong time often cause nutrient runoff into rivers, lakes, and coastal areas. This leads to eutrophication and harmful algal blooms. These blooms consume oxygen in the water and create "dead zones" where aquatic life cannot survive. Nutrients, especially nitrate, leach into groundwater and can contaminate drinking water sources. This poses serious health risks to humans. Additionally, producing and using too many nitrogen-based fertilizers increase greenhouse gas emissions, particularly nitrous oxide. This greatly contributes to climate change (Puga et al., 2020). Soil degradation, loss of biodiversity, and disruption of natural nutrient cycles are other results of unmanaged nutrient inputs. To reduce these effects, it is important to use integrated nutrient management practices that balance productivity with environmental care. This includes using resources efficiently, recycling organic materials, and protecting ecosystems.

9.5 Manure Management

Manure is an essential source of nutrients and organic matter, especially in mixed crop-livestock systems. However, the management of manure presents several challenges that can lead to nutrient losses (Ijaz et al., 2025). For agriculture to be sustainable, manure management is necessary. It entails properly handling, storing, treating, and applying animal waste in order to optimize its nutritional value and lower environmental risks. Manure is a valuable natural fertilizer because it contains high amounts of organic matter, phosphorus, and nitrogen. But improper management can result in damaging gas emissions like ammonia and methane, groundwater contamination, and nutrient runoff. In order to minimize nutrient losses, effective manure management integrates manure into the soil, uses storage systems that stop leaks and runoff, and schedules applications according to crop nutrient requirements (de Vries et al., 2023). Improved techniques, such as anaerobic digestion, stabilize nutrients for safer use and convert manure into biogas for energy.

9.6 Confined Animal Feeding Operations (CAFOs)

Large-scale livestock operations produce large volumes of manure. In CAFOs, the concentration of animals results in high nutrient loads in a relatively small area. Large farms called Confined Animal Feeding Operations, or CAFOs, keep hundreds or even thousands of animals, like cows, pigs, or chickens, in tight spaces to boost food production (MacDonald, 2018). This setup creates serious environmental problems, even though it helps satisfy the growing need for meat, milk, and eggs. These operations produce large amounts of wastewater and manure. If that waste is not managed correctly, it can seep into the ground or flow into nearby rivers, polluting the water with harmful levels of phosphorus and nitrogen. This pollution often leads to fish kills, algae blooms, and unsafe drinking water (Bhatt & Qayoom, 2021). CAFOs can also release unpleasant odors and harmful gases like methane and ammonia, which affect the climate and air quality.

9.6.1 Storage and Treatment

Manure is often stored in lagoons or pits until it can be applied to fields. The success of these storage facilities depends on their proper design together with ongoing maintenance. The storage facilities face significant nutrient spill risks when they leak or overflow especially during periods of heavy rain.

9.6.2 Regulatory Compliance

Many regions impose strict regulations about nutrient management on CAFOs. The enforcement of these regulations remains variable while technological solutions continue their development process.

9.7 Land Application and Direct Deposition

Manure serves as a common natural fertilizer which farmers use to enrich their agricultural fields. The practice of using manure requires proper management because excessive application causes soils to become oversaturated with nutrients. Land application and direct deposition are two common methods for returning manure and waste from livestock to the soil. In land application, farmers spread manure over fields to use it as a natural fertilizer (Zahoor & Mushtaq, 2023).

When done correctly, this practice recycles nutrients, improves soil health, and lowers the need for synthetic fertilizers. The nutrients, particularly nitrogen and phosphorus, can wash into neighboring rivers and lakes if too much manure is applied, if it is spread on frozen ground, or before heavy rain. Pollution and toxic algal blooms result from this. When cattle, in particular, graze near water sources, direct deposition takes place (Ijaz et al., 2025). Their waste may wind up in ponds or streams. Aquatic ecosystems can be harmed and water contaminated very quickly. Applying manure at the appropriate time, in the appropriate quantity, and away from delicate areas like waterways is essential to avoiding these problems. Utilizing buffer zones and managing grazing areas can help preserve water quality while utilizing the organic nutrients found in animal waste.

9.7.1 Application Timing

The same principles for synthetic fertilizer application apply to manure because heavy rainfall and frozen or saturated soils increase runoff possibilities.

9.7.2 Direct Deposition

Animals that graze in pastures receive unrestricted access to water bodies. The movement patterns of livestock enable direct deposition of their manure into waterways which causes increased nutrient levels and microbial contamination in local streams and ponds.

9.8 Tillage and Soil Disturbance

Soil structure, together with organic matter content and nutrient retention capacity, experiences direct impacts from tillage practices. The practice of traditional tillage through soil turning creates conditions that promote soil erosion and nutrient runoff. Turning and breaking up the soil, or tillage, have long been used to control weeds and prepare land for planting (Angon et al., 2023). However, aggressive or frequent tillage may be more detrimental than beneficial. It alters the soil's natural structure, increasing erosion susceptibility and decreasing its capacity to retain nutrients and water. Rain can wash away topsoil and carry nutrients like phosphorus and nitrogen into nearby rivers and lakes when the soil is loose and bare. This can lead to algal blooms and water pollution. Additionally, tillage can speed up the breakdown of organic matter and disrupt soil organisms that are vital for plant growth, reducing long-term fertility (Edem & Udo, 2017). Conversely, tillage reduction or conversion to no-till agriculture saves the soil, retains nutrients, and ensures improved water retention. It's an easy but effective change that makes agriculture more sustainable for the crops, the land, and the environment.

9.8.1 Soil Erosion and Nutrient Loss

- **Mechanisms of Erosion**: Tillage breaks down natural soil structure so wind and water can erode it more easily. When soil is eroded, attached nutrients are removed along with it, finding their way into nearby water bodies.
- **Loss of Soil Organic Matter**: Soil organic matter experiences reduction because of ongoing disturbances which harms both soil fertility and its ability to store water. This reduction further impairs the soil's ability to retain nutrients, making them more mobile and susceptible to leaching.

9.8.2 Conservation Tillage and No-Till Farming

9.8.2.1 Conservation Tillage

This method reduces the intensity of soil disturbance. By leaving crop residues on the field, conservation tillage protects the soil surface, reduces evaporation, and enhances water infiltration (Bekele, 2020). Agricultural techniques like conservation tillage and no-till farming aim to reduce soil disturbance and keep surface residue cover intact. This practice helps improve soil health and lessen environmental impacts. By leaving some crop residues on the soil's surface, conservation tillage systems make it easier for water to seep into the ground, reduce erosion, and maintain soil structure.

9.8.2.2 No-Tillage Practices

The complete avoidance of tillage operations allows the soil structure and organic matter content to remain intact. The combination of no-till farming with cover cropping demonstrates beneficial effects on reducing soil erosion and nutrient runoff while preserving soil health (Bezboruah et al., 2024). No-till farming eliminates mechanical soil turnover by directly seeding into undisturbed soil covered with residues from previous crops. These methods significantly lower the loss of nitrogen and phosphorus through erosion and runoff, which reduces nutrient overload in nearby water bodies. They also boost microbial activity, increase soil organic matter, and enhance carbon storage (Haque, 2021).

9.9 Irrigation and Drainage

Irrigation methods serve as essential elements within many agricultural frameworks yet these methods affect nutrient distribution throughout the system. Systems for drainage and irrigation are essential for agriculture productivity. They significantly affect environmental outcomes and nutrient transport. Poorly managed irrigation can lead to applying too much water (Li et al., 2009). In sandy or poorly structured soils, this can cause nutrients to leach into groundwater from below the root zone. This is particularly concerning because nitrogen moves easily as nitrate. On the other hand, inadequate drainage can lead to waterlogging. This affects soil microbes and reduces oxygen in the root zone. It may also increase the risk of denitrification and increase nitrous oxide emissions. Subsurface drainage systems, although effective in enhancing field conditions, could also rapidly transfer and dissolved the nutrients, particularly nitrate, to the surface water. In order to minimize the loss of nutrients, effective irrigation scheduling, precision water application (such as drip or sprinklers), and regulated drainage practices are essential (King et al., 2015). A combined management of irrigation and drainage is critical for enhancing the efficiency of nutrient use and reducing the environmental load of agricultural systems.

9.9.1 Leaching and Groundwater Contamination

- **Excess Irrigation**: When farmers apply water beyond soil retention capacity, nitrates move through the root zone and enter groundwater supplies. Groundwater contamination with nitrates presents a major public health threat which affects particularly areas that use well water.
- **Soil Texture and Structure**: Sandy soils, due to their high permeability, are particularly vulnerable to leaching. When water saturates soils containing more clay content, it may reduce movement speed although it still permits surface runoff.

9.10 Surface Runoff and Tile Drainage Systems

- **Flood Irrigation**: This traditional method of irrigation, where fields are flooded with water, often leads to significant nutrient losses as excess water carries dissolved nutrients away from the fields.
- **Tile Drainage**: In regions with poorly draining soils, farmers install tile drainage systems to remove excess water. While these systems improve field conditions, they also create a direct pathway for nutrients to enter water bodies, bypassing natural filtration processes.

9.11 Pathways of Nutrient Runoff to Water Bodies

Nutrient runoff from agricultural lands mainly happens through surface runoff and subsurface drainage. This process carries excess nitrogen and phosphorus into nearby water bodies. When farmers apply fertilizers too much or right before heavy rains, these nutrients wash off the fields and enter streams and rivers. Tile drainage systems, which are common in intensive farming areas, speed up the movement of soluble nutrients like nitrates into groundwater and connected waterways (Khan et al., 2018). Erosion also contributes by moving phosphorus attached to soil particles into aquatic systems. Poorly managed livestock operations worsen the problem by letting manure seep into nearby drainage channels. Irrigation practices can make nutrient leaching worse, especially in sandy or compacted soils that don't hold water well. These inputs build up in lakes, rivers, and coastal areas, creating conditions that are ideal for harmful algal blooms. As these blooms grow, they lower oxygen levels, disrupt aquatic food webs, and release toxins that can harm public health (Craswell, 2021).

Understanding these nutrient pathways is important for creating effective strategies to protect water quality. Nutrients from agricultural landscapes reach aquatic ecosystems through different pathways, each with unique traits and effects on water quality.

9.11.1 Surface Runoff

Surface runoff is a major transport mechanism for nutrients from agricultural lands into downstream waters, mainly causing eutrophication as well as harmful algal blooms (HABs). It refers to water flowing across land because saturation from rainfall or irrigation exceeds the soil infiltration capacity, and it also includes any nutrients, pesticides, and organic matter (Lan et al., 2024). In traditional farming systems where soils are bare or compact and have poor infiltration capacity, runoff risk is higher. We see runoff occurs with excess nutrients, primarily nitrogen and phosphorus that leach off landscapes. These nutrients enter streams/rivers/lakes easily because they move quickly off the agricultural field and I believe they are most often applied before heavy rains. Phosphorus is more often transported through erosion because it is chemically bound to particulate matter; for this reason, it is important to control soil loss on agricultural lands. Once nutrient laden water enters the water, they usually stimulate excessive algal growth that degrades the water quality of downstream waters and threatens other aquatic organisms in the ecosystem (Feng et al., 2024). So many variables contribute to surface runoff including soil type, slope, vegetation cover, and rainfall intensity. To mitigate runoff from agricultural fields, land managers can use some strategies including grassed buffer strips, contour farming practices, cover cropping, and reduced tillage. The abovementioned strategies are all aimed at slowing down water flow to allow water and nutrient to become trapped before leaving the agricultural field. Controlling surface runoff is needed to minimize HABs, maintain soil health, and support sustainable agricultural production.

- **Mechanisms and Dynamics**: Surface runoff happens when heavy rainfall surpasses the soil's ability to absorb water. This causes water—along with the nutrients it carries—to move quickly across the land into nearby streams and rivers.
- **Role of Topography and Land Cover**: The slope of the land, vegetation cover, and the presence of buffer strips or hedgerows all influence the volume and rate of runoff. Bare soils and steep slopes contribute to higher runoff rates, while well-vegetated areas can absorb and slow down water movement.

9.11.2 Leaching Through Soil Profiles

Leaching through soil profiles occurs when water soluble nutrients and especially NO3 are leached downward through soil profiles and into either the surface water or groundwater zones or possibly surface water. When we apply fertilizers or manure in excess of plant requirements, or when we apply those products and experience significant and/or excessive rainfall during or immediately after the application, all of the nutrients that are not used by the plant will continue to leach downward through the soil profile past the root zone of the plant (Craswell, 2021). Leaching with larger amounts of nitrate will be more consequential in sandy soils and soils that have poor structure and low water holding capacity and that leach readily. Nitrate is soluble and therefore mobile in water and can leach through porous soils and sediments to aquifers, wells, and drinking water sources, compared to phosphorus which can sorb strongly to soil particles. The nutrients that are leached may also re-emerge into streams, rivers, or lakes connected to the groundwater via groundwater discharge. If nutrient application practices are inconsistent or poorly timed with storms or prolonged periods of rainfall, this can eventually lead to nutrient loading to surface waters and the development of harmful algal blooms (HABs). Practices such as split fertilizer or manure applications, use of cover crops, and the planting of deep-rooted perennial vegetation will help reduce leachable nutrients from reaching the surface or groundwater (Nieder & Benbi, 2008). Understanding soil profile dynamics and the movement of leachable nutrients through the soil profile is important for developing sustainable nutrient management plans and maintaining water quality across time.

- **Process of Leaching**: Dissolved nutrients, specially nitrates, seep down through the soil profile. The leaching rate is affected by texture of soil, structure and organic matter content.
- **Groundwater Contamination**: These nutrients may be used as a source of drinking water once leached down and enter to the reservoir of ground water. A high significant health risk can be rise in infants like methemoglobinemia or blue baby syndrome due to leaching down of high nitrate in ground water.

9.11.3 Atmospheric Deposition

Atmospheric deposition is an often underestimated but important way nutrients, particularly nitrogen, are brought into aquatic ecosystems and can contribute to harmful algal blooms (HABs). Atmospheric deposition is the way of getting airborne nitrogen sources, like ammonia (NH_3) from livestock operations and nitrogen oxides (NO_x) from burning fossil fuels, onto the surface of land and water as a result of the precipitation "wet" deposition) or dust or gases (dry deposition). Once the nitrogen is deposited into the atmosphere, it quickly contributes to rivers, lakes, and coastal areas thus increasing the total nutrient pool that can stimulate algae growth (Zheng et al., 2019). In certain situations, atmospheric deposition accounted for 30% or more of nitrogen input into sensitive watersheds. Coastal and urban areas are particularly vulnerable due to a larger concentration of nitrogen emissions from industrialization and transportation activities. Atmospheric sources are more diffuse than conventional point sources of pollution, so solutions are not solely on the watershed but require a more systemic and broader approach to air quality management and regulations for point sources (Howarth, 2006). Managing atmospheric nitrogen involves integrating management activities across agriculture, energy, and transportation. Reduction atmospheric deposition of nitrogen is critical to assist in the soil input management of nutrient pollution and can limit the future occurrence of HABs.

- **Volatilization and Re-deposition**: Ammonia can volatilize into the atmosphere which is a byproduct of fertilizer application and livestock waste. This ammonia again recombines to the water bodies through precipitation in the form of acid rain further contributing to the nutrients loading.
- **Regional and Global Impacts**: Nutrient transportation via atmosphere in one area can decline the water quality in regions which is far away from the parent source. As a result of which, management efforts become complicated.

9.11.4 River and Coastal Transport

River and coastal transport facilitate the transfer of nutrients from agricultural fields to coastal waters where they cause harmful algal blooms (HABs). Obviously, rainfall and irrigation runoff carry nitrogen and phosphorus from agricultural fields to streams and rivers. These fields are situated in watersheds wherein the nitrogen and phosphorus are transported to lakes, estuaries, and coastal waters, sometimes hundreds of kilometers from the source of the nutrients (Paerl et al., 2018). Major rivers like the Mississippi, Yangtze, and Amazon can be considered nutrient highways, moving agricultural runoff to nutrient-sensitive coastal waters. In estuarine and coastal waters, accumulation of nutrients promotes algal blooms when warm and stagnant conditions prevail; tidal mixing, sediment resuspension, and freshwater discharge patterns determine the distance the nutrients are distributed and how long it remains suspended in the water column. Once the nutrients have settled out of the water column—specifically in deltas and bay systems—nutrients can accumulate to trigger seasonal hypoxia (dead zones), as we have seen in the Gulf of Mexico. If the goal of protecting coastal ecosystems is to be achieved, it is crucial that we manage nutrient loads upstream (Blann et al., 2009). Understanding river–coast linkages is fundamental to developing regional and regional nutrient reduction strategies and preventing large-scale degradation of aquatic communities.

- **Accumulation and Transport**: Nutrients are carried downstream and can accumulate in larger water bodies entering into the river system. These nutrients have the ability to form larger dead zones in coastal areas, where marine life is largely affected by oxygen depletion.
- **Complex Interaction**: Flow rates, sediment loads, and seasonal fluctuations can significantly affect nutrient movement through the river networks, which can also influence the timing and intensity of algal blooms in aquatic ecosystem downstream.

9.12 Nutrient Overload and Harmful Algal Blooms (HABs)

One of the main causes of harmful algal blooms (HABs) in freshwater and marine environments is too many nutrients, especially excess nitrogen (N) and phosphorus (P). In warmer water, these nutrients act like fertilizers for algal growth and lead to rapid growth of the algae in the water. While some algal blooms are consider harmless, some of them are considered very dangerous, in result which they produces harmful toxins like microcystins. These toxins can contaminate drinking water and kill aquatic life. Agricultural runoff is significant because fertilizers, manure, and wastewater increase nutrient levels beyond normal. After HABs occur, the oxygen levels in the water drop, creating dead zones or hypoxic zones that suffocate fish and other organisms (Chen et al., 2016). They also block sunlight, killing submerged plants that are vital for aquatic ecosystems. In coastal areas, these blooms impact fisheries, tourism, and public health. Satellite images and water quality monitoring frequently reveal a strong connection between nutrient spikes and bloom severity. Managing nutrient input from agriculture is crucial for controlling the frequency and intensity of HABs. Without intervention, these blooms are likely to worsen due to climate change and rising agricultural activity (Hamilton et al., 2016). The growth of harmful algal blooms as a result of nutrient of enrichment is well recorded. The fundamental mechanisms, ecological impacts, and real-world case studies of HABs are studied in this section.

9.12.1 The Science Behind Algal Blooms

Algal blooms happen when algae, especially phytoplankton and cyanobacteria, grow quickly because of good environmental conditions. The main causes are extra nutrients like nitrogen and phosphorus, warm temperatures, plenty of sunlight, and still or slow-moving water. These conditions let algae reproduce rapidly, creating thick mats or blooms that can change the color of water and lower oxygen levels (Mishra, 2023). In nutrient-rich areas, certain species, such as cyanobacteria, become dominant and can release harmful toxins. The biological process includes photosynthesis during the day and respiration at night, which can lead to significant changes in oxygen levels, putting stress on aquatic life. When algal cells die and break down, they further reduced oxygen and create a cycle of ecosystem decline. Today, monitoring uses satellite images, chlorophyll-a (CHL) measurements, and molecular methods to observe bloom formation and makeup. Understanding these processes helps scientists predict when blooms will occur and evaluate environmental risks. Good management needs to lower nutrient inputs and tackle the physical conditions that encourage bloom growth.

- **Limiting Nutrients and Ecological Stoichiometry**: Algae growth is mostly limit by a single nutrient in aquatic systems. Phosphorus is often the limiting nutrient in freshwater while nitrogen is the harmful factor in marine systems. The balance is disrupted when these limiting factors exceed from their normal level, which directly leads to imbalance growth of algal.

- **Species-Specific Responses**: Response of different algal species is altered towards nutrient enrichment. Certain species, especially cyanobacteria (blue-green algae), can outcompete other algae when nutrient availability ratios suit their growth. They have the potential to survive well to low light and limit oxygen availability. Toxins are released by many of these species which can harm both marine life and human health.
- **Interaction with Light and Temperature**: Algae growth is not the only factor by the nutrient availability. It also includes light intensity and water temperature, and even the presence of competing species also plays a significant role. Algal species that produce toxin also supported by warmer temperatures and greater availability of high light can synergistically promote the growth.

9.12.2 Eutrophication and Its Consequences

- **Eutrophication Process**: It is a process through which the water bodies become enriched with nutrients, leading to dense plant and algal growth. Initially, this may increase primary production, but as the system becomes overloaded, a number of disadvantages appear. These excess nutrients of nitrogen and phosphorus in the water speed up the growth of algae and aquatic plants. This disturbs the natural balance of ecosystems (Wurtsbaugh et al., 2019). As algal blooms grow, they block sunlight from reaching underwater plants. This hinders photosynthesis and kills beneficial vegetation. When the algae die, their decomposition uses up oxygen. This leads to low or no oxygen conditions, often called "dead zones." The lack of oxygen seriously affects fish, invertebrates, and other aquatic life, frequently resulting in mass die-offs. Eutrophication also changes food webs, reduces biodiversity, and harms water quality. In many instances, harmful algal blooms (HABs) develop, producing toxins that can harm human and animal health (Marques and Marques, 2020). Drinking water sources can become polluted, leading to economic losses in fisheries, recreation, and tourism. Long-term eutrophication can permanently alter aquatic ecosystems unless nutrient inputs are significantly reduced and managed.
- **Oxygen Depletion and Hypoxia**: During the decomposition of algal blooms, a huge amount of dissolved oxygen is consumed by bacteria, which often create a hypoxic condition (areas with very low oxygen). These hypoxic zones some time know as dead zones become inhospitable for most of the aquatic life which can cause mortality of fish and invertebrates.

9.12.3 Case Studies and Real-World Examples

Real-world case studies show the serious effects of nutrient runoff and harmful algal blooms (HABs) in different areas. One significant example is Lake Erie. Agricultural runoff from nearby farms has caused repeated toxic cyanobacterial blooms that endanger drinking water for millions. In 2014, a severe bloom led the city of Toledo, Ohio, to shut down its water supply for 3 days (Klasic, 2022). In a similar fashion, the Gulf of Mexico has one of the largest seasonal dead zones in the world, caused by nutrient-rich runoff from the Mississippi River Basin. In China, Lake Taihu has faced persistent algal blooms due to decades of fertilizer overuse, affecting both water supply and public health. In Australia, the Murray–Darling Basin has experienced toxic blooms because of droughts and nutrient pollution. These examples underline the urgent need for better nutrient management, improved agricultural practices, and international teamwork to tackle eutrophication and HABs globally.

- **Lake Erie**: Lake Erie has had many cyanobacterial blooms, making it one of the most researched examples of nutrient-driven HABs. Phosphorus discharge from intensive agriculture in the nearby watershed is the main offender. Drinking water sources have been impacted by the blooms, necessitating expensive water treatment measures.
- **Chesapeake Bay**: Decades of nutrient loading have left this ample estuary with enduring hypoxic zones. Fish population decreases and deteriorated habitat quality have been caused by the nitrogen inflow, which comes from both urban and agricultural runoff.
- **Gulf of Mexico**: One of the world's greatest dead zones is a result of the Mississippi River Basin's nutrient-rich waters. Seasonal hypoxia in the Gulf is caused by elevated nitrogen levels from extensive agricultural activities, which have a negative effect on the marine ecology and commercial fishing.

9.13 Mitigation Strategies and Sustainable Solutions

Addressing the agricultural NEXUS requires a multifaceted approach that combines best management practices (BMPs) improved nutrient management plans and regulatory frameworks. Mitigating nutrient runoff and harmful algal blooms (HABs) needs a mix of sustainable farming practices, policy changes, and restoring ecosystems (Xie et al., 2015). One important method is precision farming, which uses fertilizers based on the current needs of crops. This practice reduces the excess use of nutrients. Cover cropping and conservation tillage help stop soil erosion and nutrient loss during off-seasons. Installing vegetative buffer strips and created wetlands along field edges can filter out nutrients before they enter waterways. Nutrient management plans assist farmers in balancing the needs of their crops with environmental safety. On a larger scale, total maximum daily loads (TMDLs) set limits on nutrient inputs for polluted water bodies. Investing in green infrastructure and upgrades to wastewater treatment can also lessen nutrient loads from urban and agricultural sources (Areej et al. 2024; Harding, 2015). Education, incentives, and community-based watershed management are important for encouraging these practices. Achieving long-term sustainability requires approaches that integrate environmental health with farming productivity and resilience.

9.13.1 Best Management Practices (BMPs)

Best management practices (BMPs) are management techniques farmers and other land managers use to control the movement of nutrients and other water quality protection. The practices are designed to achieve a balance between farm productivity and water quality in the sense they attenuate the runoff of nitrogen, phosphorus, and sediment to nearby water bodies (Liu et al., 2017). The two (or more team members) are recognized to have successfully addressed the life elements at a level that causes an environmental benefit; this may include the following: Cover crops, buffer strips, conservation tillage, contour farming, and nutrient management planning. Structural practices such as grassed waterways, detention basins, and wetlands that collect and filter runoff are also BMPs. Adaptive management, farmer education, and ongoing monitoring are essential to optimizing BMP efficacy over time. As a cornerstone of sustainable agriculture, BMPs are vital instruments for maintaining water quality and soil health. BMPs are practical techniques or strategies designed to minimize nutrient losses from agricultural fields.

- **Buffer Strips and Riparian Zones**: These green spaces serve as organic filters between agricultural land and bodies of water. Before they reach streams, they absorb nutrients and hold onto sediments. Grass, trees, or wetlands created especially to reduce water flow can be found in buffer strips.
- **Cover Crops**: During off-season, growing cover crops helps prevent soil erosion, enhance soil organic matter, and add residual nutrients. It also improves microbial activity which can enhance soil fertility and limit nutrient leaching.
- **Controlled Drainage**: To enable plant uptake, this entails controlling drainage systems to keep water and nutrients on the field for extended periods of time. Existing tile drainage systems can incorporate controlled drainage to lessen the direct flow of nutrient-rich water into adjacent bodies of water.
- **Rotational Grazing and Reduced Stocking Densities**: Strategic grazing management in livestock enterprises avoids soil compaction, which can otherwise increase runoff, and lessens direct nutrient deposition on water bodies.

9.13.2 Nutrient Management Plans

Effective nutrient management is essential for optimizing fertilizer and manure application to meet crop needs while reducing excess nutrient loads. Nutrient management plans (NMPs) are strategies that farmers use to make the most of fertilizers while reducing harm to the environment. These plans focus on applying the right amount of nutrients at the right time, in the right place, and using the right method. This approach is known as the 4Rs of nutrient stewardship. NMPs depend on soil testing, crop nutrient needs, and local conditions to avoid applying too much nitrogen and phosphorus. By matching fertilizer use with what plants can take up, these plans help limit excess runoff into nearby water bodies. NMPs also take into account manure management, irrigation schedules, and crop rotations to improve how efficiently nutrients are used. In many areas, NMPs are required by law or offered through subsidies and conservation programs. When done well, they reduce the risk of eutrophication and harmful algal blooms (HABs) downstream. NMPs not only protect water quality but also boost farm profits by cutting down on waste. They serve as a practical, science-based tool for sustainable farming and caring for the environment.

9.13.3 Policy and Regulatory Frameworks

Government policies and regulations are crucial for enforcing best practices and incentivizing sustainable agriculture. Policy and regulatory frameworks are essential for managing nutrient pollution and preventing harmful algal blooms (HABs). Governments at local, national, and international levels have put in place various laws and guidelines to control agricultural runoff. In the United States, the Clean Water Act serves as the basis for regulating point and non-point-source pollution, which includes agricultural discharges. It is the kind of programs such as total maximum daily loads (TMDLs), which establish limits to nutrient inputs in impaired water bodies. Agencies like the Environmental Protection Agency (EPA) and the USDA Natural Resources Conservation Service (NRCS) help farmers with technical assistance, financial assistance, and incentives for conservation. An added notable example is the European Union's Nitrates Directive, under which member states are made to monitor and decrease agricultural nitrate pollution. Some areas have implemented nutrient trading schemes, which enable the flexibility to achieve pollution reduction goals. Enforceability and compliance remain issues despite these developments. This calls for more robust policies, enhanced monitoring, and increased farmer participation to reach long-term water quality objectives.

- **Clean Water Act (CWA) and Other Regulations**: Nutrient releases into aquatic bodies are regulated by environmental legislation in several nations. Organizations and farms are encouraged to implement nitrogen management techniques that lower runoff when these regulations are enforced.
- **Conservation Incentives and Subsidies**: Farmers who use BMPs may get financial assistance through programs like the Environmental Quality Incentives Program (EQIP) in the United States. The expenses of changing to more environmentally friendly activities are somewhat reduced by these incentives.

- **Nutrient Trading and Market-Based Approaches**: In certain areas, industry and farmers may earn credits for lowering nutrient runoff through nutrient trading schemes. Improved nutrient management can be economically compensated through the sale or exchange of these credits.

9.13.4 Innovative Technologies and Future Solutions

Emerging technologies are increasingly important in managing nutrient pollution and preventing harmful algal blooms (HABs). Precision agriculture employs GPS mapping, sensors, and unmanned aerial vehicles (drones) to precision-apply fertilizer, minimizing waste and runoff. Real-time water quality monitoring systems now identify nutrient spikes and algal toxins before they are at toxic levels. Bioreactors and denitrifying woodchip trenches placed at field borders remove excess nitrogen from drainage water. Scientists are also creating algae-harvesting technologies to eliminate blooms while recycling algae into biofuels or fertilizers. CRISPR gene editing and microbiome engineering hold future promise for modifying plant nutrient acquisition and lowering reliance on chemical fertilizers. Machine learning models and artificial intelligence are now being applied to forecast bloom occurrences with climate, land cover, and hydrology. Advancing technologies such as electrochemical treatment and phosphorus recovery technologies are looking to purify wastewater more effectively. These emerging technologies along with sustainable agriculture have a promise of a future where food production and clean water exist side by side. Advances in technology and novel approaches offer new avenues for reducing nutrient losses.

9.14 Future Perspective and Research Needs

In the upcoming decades, it is anticipated that the issues caused by nutrient contamination and HABs will get worse due to climate change and changing farming methods. Developing adaptive strategies will require continuous research and innovation. In the future, solutions to nutrient runoff and HABs will need a forward-looking, multidisciplinary framework involving science, technology, and policy. Future studies will need to examine how climate change by increasing temperatures, changing rainfall patterns, and extreme weather events amplifies nutrient loading and bloom behaviors. Increasingly, long-term watershed-scale research will be necessary that follows the path of nutrient transport from field to water. They need to investigate next-generation fertilizers, plant genetics, and microbial manipulation of the soil microbiome to decrease dependence on chemical inputs. Next-generation remote sensing, machine learning, and early warning systems can enhance HAB forecasting and response. In order to investigate farmer behavior, policy adoption, and community involvement, social science research will be equally important. Transdisciplinary collaboration among agronomists, ecologists, economics, and policymakers will require more funding. Lastly, the ability to apply science to fair and successful policy plans for global food and water security will be just as important to sustainable solutions as future discoveries.

9.14.1 Impact of Climate Change

- **Altered Prepetition Patterns**: Extreme weather events including prolonged droughts and heavy rainfall are predicted to occur more frequently as a result of climate change. While drought conditions could increase nutrients in smaller water bodies, increased rainfall events may exacerbate surface runoff.
- **Temperature Increases**: Warmer temperatures can encourage the growth of organisms that produce toxins and speed up the metabolism of algae. More frequent and severe HABs could result from the interaction of fertilizer loading and climate change.
- **Resilience and Adaptation**: In order to maintain the effectiveness of nutrient management techniques in an environment of changing weather patterns, future research must focus on creating varieties of crops and farming techniques that are resistant to shifting climatic conditions

9.14.2 The Promise of Regenerative Agriculture (RA)

Regenerative agriculture is a revolutionary method of farming that can not only "sustain" the planet but actually "restore" soil, biodiversity, and ecosystem function. Unlike conventional products, which may leach nutrients into runoff with HAB-producing consequences, Regen-A-GRO and Regen-A-Plant are focused on leaving nutrients in the ground. Practices like no-till farming, cover cropping, diverse crop rotations, and managed grazing improve soil structure and organic matter, which both reduce erosion and retain nutrients. These practices promote a good mix of soil life that provides its natural turnover of nutrients, minimizing the reliance on chemical fertilizers. Regenerative systems also enhance water infiltration—reducing storm runoff. Regenerative agriculture is a solution towards climate resiliency, sequestering carbon and sustaining food production in the long term. There is emerging evidence that it can decrease nutrient pollution and downstream HABs.

9.15 Conclusion and Recommendation

The agricultural NEXUS involves the interaction between farming practices, nutrient inputs, and water quality. This situation brings both challenges and opportunities. Modern agriculture relies heavily on synthetic fertilizers and intensive animal production, which has greatly increased nutrient loading. This sets the stage for harmful algal blooms that pose risks to aquatic ecosystems and human health. To tackle these issues, we need an integrated approach that includes on-farm practices, new technologies, policy changes, and community involvement. By using best management practices, precision nutrient management strategies, and exploring regenerative agricultural systems, stakeholders can cut down on nutrient runoff and lessen the effects of HABs. Additionally, using advancements in remote sensing, data analysis, and ecological research will be essential for adapting to a changing climate and ensuring long-term sustainability. As we progress, scientists, farmers, policymakers, and community members must work together to rethink farming practices and adopt solutions that safeguard our water resources. Through these collaborative efforts, we can create a future where agricultural productivity and environmental health support each other rather than conflict. Addressing these issues needs a combined effort that includes on-farm practices, new technologies, policy changes, and community involvement. By using the best management practices, precision nutrient management strategies, and exploring regenerative farming methods, stakeholders can cut down nutrient runoff and lessen the effects of harmful algal blooms (HABs). In addition, using improvements in remote sensing, data analysis, and ecological research will be essential for responding to a changing climate and ensuring long-term sustainability. As we move ahead, it is imperative for scientists, farmers, policymakers, and community members to work together in rethinking farming practices and embracing solutions that protect our water resources. Through these combined efforts, we can achieve a future where agricultural productivity and environmental health support each other.

To alleviate the effect of different agricultural practices on nutrient loading and harmful algal blooms (HABs), a multidimensional approach is necessary. The expansion and adaption of best management practices (BMP) such as precision nutrient management (PUM), and vegetative buffer zones and cover cropping could considerably reduce runoff. Using precision agriculture technologies, such as GPS-guided applicators and soil nutrient mapping, promotes more efficient use of inputs and less environmental impact. Policy changes should support conservation efforts with subsidies, nutrient trading programs, and performance-based payments. Improving manure and waste management, especially in areas with many livestock, is also key to reducing point-source nutrient pollution. At the same time, investing in real-time monitoring systems and a standardized HAB index can improve early warning capabilities and guide targeted actions. Public education and outreach programs for farmers are essential for promoting sustainable land use. Additionally, collaboration among agencies, scientists, and stakeholders at the watershed level will help ensure that mitigation strategies are both well-suited to local needs and backed by scientific knowledge.

References

Alcamo, J. (2019). Water quality and its interlinkages with the Sustainable Development Goals. *Current Opinion in Environmental Sustainability, 36*, 126–140.

Alexandratos, N., & Bruinsma, J. (2012). *World agriculture towards 2030/2050: The 2012 revision*

Angon, P. B., Anjum, N., Akter, M. M., et al. (2023). An overview of the impact of tillage and cropping systems on soil health in agricultural practices. *Advances in Agriculture, 1*, 8861216.

Areej, A., Usama, M., Zulfiqar, U., Sarwar, F., & Ashiq, A. (2024). Sustainable agriculture development: The role of biofertilizers in soil fertility and crop yield improvement. *Applied Agriculture Sciences, 2*(1), 1–5.

Asadu, C. O., Ezema, C. A., Ekwueme, B. N., Onu, C. E., Onoh, I. M., Adejoh, T., et al. (2024). Enhanced efficiency fertilizers: Overview of production methods, materials used, nutrients release mechanisms, benefits and considerations. *Environmental Pollution and Management.*

Bekele, D. (2020). The effect of tillage on soil moisture conservation: A review. *International Journal of Research Studies in Computer Science, 6*, 30–41.

Bezboruah, M., Sharma, S. K., Laxman, T., Ramesh, S., et al. (2024). Conservation tillage practices and their role in sustainable farming systems. *Journal of Experimental Agriculture International, 46*(9), 946–959.

Bhat, S. U., & Qayoom, U. (2021). Implications of sewage discharge on freshwater ecosystems. In *Sewage-recent advances, new perspectives and applications* (pp. 1–18). Intech Open.

Blann, K. L., Anderson, J. L., Sands, G. R., & Vondracek, B. (2009). Effects of agricultural drainage on aquatic ecosystems: A review. *Critical Reviews in Environmental Science and Technology, 39*(11), 909–1001.

Bulacio Fischer, P. T., Di Trapani, D., Laudicina, V. A., Muscarella, S. M., & Mannina, G. (2025). Nutrient recovery from zeolite and biochar columns: The case study of Marineo (Italy) wastewater treatment plant. *Water, 17*(6), 848.

Chen, L., Chen, J., Zhang, X., & Xie, P. (2016). A review of reproductive toxicity of microcystins. *Journal of Hazardous Materials, 301*, 381–399.

Craswell, E. (2021). Fertilizers and nitrate pollution of surface and ground water: An increasingly pervasive global problem. *SN Applied Sciences, 3*(4), 518.

Davidson, E. A., Nifong, R. L., Ferguson, R. B., Palm, C., Osmond, D. L., & Baron, J. S. (2016). Nutrients in the nexus. *Journal of Environmental Studies and Sciences, 6*, 25–38.

de Oliveira Veras, M., Parenti, E., & Neiva, S. D. S. (2021). Food security: Conceptual history and pillars. In *Zero hunger* (pp. 1–10). Springer.

de Vries, W., Kros, J., Voogd, J. C., & Ros, G. H. (2023). Integrated assessment of agricultural practices on large scale losses of ammonia, greenhouse gases, nutrients and heavy metals to air and water. *Science of the Total Environment, 857*, Article 159220.

Delgado, A., Quemada, M., Mateos, L., & Villalobos, F. J. (2024). Fertilization with phosphorus, potassium, and other nutrients. *Principles of Agronomy Sustainable Agriculture* (pp. 415–437). Springer International Publishing.

Dimkpa, C., Adzawla, W., Pandey, R., Atakora, W. K., Kouame, A. K., Jemo, M., & Bindraban, P. S. (2023). Fertilizers for food and nutrition security in sub-Saharan Africa: An overview of soil health implications. *Frontiers in Soil Science, 3*, 1123931.

Edem, I. D., & Udo-Inyang, U. C. (2017). Water-quality problems in the ecosystem as induced by mass fluxes of Phosphorus and Nitrogen. *Journal of Wetlands Biodiversity, 7*, 45–55.

Feng, L., Wang, Y., Hou, X., Qin, B., Kutser, T., Qu, F., et al. (2024). Harmful algal blooms in inland waters. *Nature Reviews Earth and Environment, 5*(9), 631–644.

Fontaine, S., Abbadie, L., Aubert, M., Barot, S., Bloor, J. M., et al. (2024). Plant–soil synchrony in nutrient cycles: Learning from ecosystems to design sustainable agrosystems. *Global Change Biology, 30*(1), Article e17034.

Gao, X., Gao, Y., Li, X., Zhang, C., Zeng, Q., et al. (2024a). Eleven-year canopy nitrogen addition enhances the uptake of phosphorus by plants and accelerates its depletion in soil. *Forests, 15*(3), 416.

Gao, Y., Lou, L., Liao, Y., Yao, H., Fang, J., & Liu, G. (2024b). Simultaneous nitrogen and phosphorus removal by immobilized bacterial particles of denitrifying phosphorus accumulating microorganisms and its application. *Biochemical Engineering Journal, 212*, Article 109495.

Giller, K. E., Delaune, T., Silva, J. V., Descheemaeker, K., Van De Ven, G., Schut, A. G., et al. (2021). The future of farming: Who will produce our food? *Food Security, 13*(5), 1073–1099.

Guignard, M. S., Leitch, A. R., Acquisti, C., Eizaguirre, C., Elser, J. J., & Hessen, D. O. (2017). Impacts of nitrogen and phosphorus: From genomes to natural ecosystems and agriculture. *Frontiers in Ecology and Evolution, 5*, 70.

Hamilton, D. P., Salmaso, N., & Paerl, H. W. (2016). Mitigating harmful cyanobacterial blooms: Strategies for control of nitrogen and phosphorus loads. *Aquatic Ecology, 50*, 351–366.

Haque, S. E. (2021). How effective are existing phosphorus management strategies in mitigating surface water quality problems in the US? *Sustainability, 13*(12), 6565.

Harding, W. R. (2015). *A feasibility evaluation of the total maximum daily (pollutant) load (TMDL) approach for managing eutrophication in South African dams: Report to the water research commission*. Water Research Commission.

Hartmann, M., & Six, J. (2023). Soil structure and microbiome functions in agroecosystems. *Nature Reviews Earth and Environment, 4*(1), 4–18.

Hayat, S., Li, P., Menhas, S., Liu, W., & Hayat, K. (2025) Exploring the nutrient nexus in environmental systems: Nitrogen and phosphorus cycling, removal, recovery, and management. *Environmental Research, 122162.*

Howarth, R. W. (2006). Atmospheric deposition and nitrogen pollution in coastal marine ecosystems. *Acid in the environment: Lessons learned and future prospects* (pp. 97–116). Springer.

Hucke-Gaete, R., Viddi, F. A., & Simeone, A. (2024). Marine mammals and seabirds of Chilean Patagonia: Focal species for the conservation of marine ecosystems. In Conservation in Chilean Patagonia: Assessing the state of knowledge, opportunities, and challenges (pp. 233–261).

Ijaz, M. U., Akbar, A., Eman, R., & Hayat, M. F. (2025). Mitigating nutrient pollution from livestock manure: Strategies for sustainable management. In *Agricultural nutrient pollution and climate change: Challenges and opportunities* (pp. 165–187). Springer Nature Switzerland.

Khan, M. N., Mobin, M., Abbas, Z. K., & Alamri, S. A. (2018). Fertilizers and their contaminants in soils, surface and groundwater. *Encyclopedia of the Anthropocene, 5*, 225–240.

King, K. W., Williams, M. R., Macrae, M. L., Fausey, N. R., Frankenberger, J., Smith, D. R., Kleinman, P. J., & Brown, L. C. (2015). Phosphorus transport in agricultural subsurface drainage: A review. *Journal of Environmental Quality, 44*(2), 467–485.

Klasic, M. R. (2022). *The greater the risk, the greater the reward: How environmental risk shapes policy response*. University of California.

Lan, J., Liu, P., Hu, X., & Zhu, S. (2024). Harmful algal blooms in eutrophic marine environments: Causes, monitoring, and treatment. *Water, 16*(17), 2525.

Li, S. X., Wang, Z. H., Malhi, S. S., Li, S. Q., Gao, Y. J., & Tian, X. H. (2009). Nutrient and water management effects on crop production, and nutrient and water use efficiency in dryland areas of China. *Advances in Agronomy, 102*, 223–265.

Liu, J., Hull, V., Godfray, H. C. J., Tilman, D., et al. (2018). Nexus approaches to global sustainable development. *Nature Sustainability, 1*(9), 466–476.

Liu, Y., Engel, B. A., Flanagan, D. C., Gitau, M. W., McMillan, S. K., & Chaubey, I. (2017). A review on effectiveness of best management practices in improving hydrology and water quality: Needs and opportunities. *Science of the Total Environment, 601*, 580–593.

MacDonald, J. M. (2018). CAFOs: Farm animals and industrialized livestock production. In Oxford research encyclopedia of environmental science.

Marques, L., & Marques, L. (2020) Collapse of biodiversity in the aquatic environment. *Capitalism and Environmental Collapse*, 275–301.

Michael, P. S. (2021). Role of organic fertilizers in the management of nutrient deficiency, acidity, and toxicity in acid soils—A review. *Journal of Global Agriculture and Ecology, 12*(3), 19–30.

Mishra, R. K. (2023). Fresh water availability and its global challenge. *British Journal of Multidisciplinary and Advanced Studies, 4*(3), 1–78.

Nath, N. (2023). A study on: Aquatic ecosystem, its features, functions and threats. *Jamshedpur Research Review, 9*(59), 57–65.

Nieder, R., & Benbi, D. K. (2008). Leaching losses and groundwater pollution. *Carbon and Nitrogen in the Terrestrial Environment*, 219–233.

Olowe, V. (2021). Africa 2100: How to feed Nigeria in 2100 with 800 million inhabitants. *Organic Agriculture, 11*(2), 199–208.

Paerl, H. W., Otten, T. G., & Kudela, R. (2018). Mitigating the expansion of harmful algal blooms across the freshwater-to-marine continuum.

Penuelas, J., Janssens, I. A., Ciais, P., Obersteiner, M., & Sardans, J. (2020). Anthropogenic global shifts in biospheric N and P concentrations and ratios and their impacts on biodiversity, ecosystem productivity, food security, and human health. *Global Change Biology, 26*(4), 1962–1985.

Priya, E., Sarkar, S., & Maji, P. K. (2024). A review on slow-release fertilizer: Nutrient release mechanism and agricultural sustainability. *Journal of Environmental Chemical Engineering*, 113211.

Puga, A. P., Grutzmacher, P., Cerri, C. E. P., Ribeirinho, V. S., & de Andrade, C. A. (2020). Biochar-based nitrogen fertilizers: Greenhouse gas emissions, use efficiency, and maize yield in tropical soils. *Science of the Total Environment, 704*, Article 135375.

Rajak, P., Ganguly, A., Nanda, S., Mandi, M., Ghanty, S., Das, K., et al. (2024). Toxic contaminants and their impacts on aquatic ecology and habitats. In Spatial modeling of environmental pollution and ecological risk (pp. 255–273). Woodhead Publishing.

Ramamoorthy, P., Sathiyamurthi, S., Pavithra, A., Sivasakthi, M., & Praveen, K. S. (2024). Impact of major nutrients fertilizer application on soil pollution and management measures. In Soil, water pollution and mitigation strategies: A spatial approach (pp. 301–314). Springer Nature Switzerland.

Sabina, R., Paul, J., Sharma, S., & Hussain, N. (2025). Synthetic nitrogen fertilizer pollution: Global concerns and sustainable mitigating approaches. In *Agricultural nutrient pollution and climate change: Challenges and opportunities* (pp. 57–101). Springer Nature Switzerland.

Shah, F., & Wu, W. (2019). Soil and crop management strategies to ensure higher crop productivity within sustainable environments. *Sustain, 11*(5), 1485.

Shahena, S., Rajan, M., Chandran, V., & Mathew, L. (2021). Conventional methods of fertilizer release. In *Controlled release fertilizers for sustainable agriculture* (pp. 1–24). Academic Press.

Shaji, H., Chandran, V., & Mathew, L. (2021). Organic fertilizers as a route to controlled release of nutrients. In *Controlled release fertilizers for sustainable agriculture* (pp. 231–245). Academic Press.

Swathy, R., Geethalakshmi, V., Pazhanivelan, S., Kannan, P., et al (2024). Real-time nitrogen monitoring and management to augment N use efficiency and ecosystem sustainability–A review. *Journal of Hazardous Materials Advances*, 100466.

Tian, H., Lu, C., Pan, S., Yang, J., Miao, R., Ren, W., Yu, Q., Fu, B., Jin, F. F., Lu, Y., & Melillo, J. (2018). Optimizing resource use efficiencies in the food–energy–water nexus for sustainable agriculture: From conceptual model to decision support system. *Current Opinion in Environmental Sustainability, 33*, 104–113.

Tyagi, J., Ahmad, S., & Malik, M. (2022). Nitrogenous fertilizers: Impact on environment sustainability, mitigation strategies, and challenges. *International Journal of Environmental Science and Technology, 19*(11), 11649–11672.

Van Thang, V., Nguyen, N. T. D., Nadagouda, M. N., Aminabhavi, T. M., Vasseghian, Y., & Joo, S. W. (2024). Effective removal of perfluorooctanoic acid from water using PVA@ UiO-66-NH2/ GO composite materials via adsorption. *Journal of Environmental Management, 368*, Article 122248.

Vasseghian, Y., Nadagouda, M. M., & Aminabhavi, T. M. (2024). Biochar-enhanced bioremediation of eutrophic waters impacted by algal blooms. *Journal of Environmental Management, 367*, Article 122044.

Verma, B. C., Pramanik, P., & Bhaduri, D. (2019). Organic fertilizers for sustainable soil and environmental management. In *Nutrient dynamics for sustainable crop production* (pp. 289–313). Springer Singapore.

Vucic, V., Suring, C., Harms, H., Muller, S., & Gunther, S. (2021). A framework for P-cycle assessment in wastewater treatment plants. *Science of the Total Environment, 760*, Article 143392.

Wang, L., Ju, C., Han, C., Yu, Z., Bai, M. Y., & Wang, C. (2025). The interaction of nutrient uptake with biotic and abiotic stresses in plants FA. *Journal of Integrative Plant Biology*.

Witek-Krowiak, A., Gorazda, K., Szopa, D., Trzaska, K., Moustakas, K., & Chojnacka, K. (2022). Phosphorus recovery from wastewater and bio-based waste: An overview. *Bioengineered, 13*(5), 13474–13506.

Wu, D., Zhang, Y., & Yu, X. (2024). Nitrogen and phosphorus metabolism together lead to a continuous increase in the environmental pollution risk in Minnan-Triangle cities. *Journal of Cleaner Production, 483*, Article 144274.

Wurtsbaugh, W. A., Paerl, H. W., & Dodds, W. K. (2019). Nutrients, eutrophication and harmful algal blooms along the freshwater to marine continuum. *Wiley Interdisciplinary Reviews: Water, 6*(5), Article e1373.

Xie, H., Chen, L., & Shen, Z. (2015). Assessment of agricultural best management practices using models: Current issues and future perspectives. *Water, 7*(3), 1088–1108.

Yahaya, S. M., Mahmud, A. A., Abdullahi, M., & Haruna, A. (2023). Recent advances in the chemistry of nitrogen, phosphorus and potassium as fertilizers in soil: A review. *Pedosphere, 33*(3), 385–406.

Zahoor, I., & Mushtaq, A. (2023). Water pollution from agricultural activities: A critical global review. *International Journal of Chemical and Biochemical Sciences, 23*(1), 164–176.

Zheng, T., Cao, H., Liu, W., Xu, J., Yan, Y., et al. (2019). Characteristics of atmospheric deposition during the period of algal bloom formation in urban water bodies. *Sustainability, 11*(6), 1703.

Zou, T., Zhang, X., & Davidson, E. A. (2022). Global trends of cropland phosphorus use and sustainability challenges. *Nature, 611*(7934), 81–87.

Chapter 10
Cyanobacterial Bloom Impacts on Aquatic Food Webs and Ecosystem Services

Waseem Ahmed Khattak, Muhammad Anas, Erdogan E. Hakki, and Shah Fahad

Abstract Cyanobacterial blooms are becoming increasingly frequent and intense globally, driven by nutrient over-enrichment, climate change, and anthropogenic disturbances to aquatic ecosystems. These harmful algal blooms (HABs) not only deteriorate water quality but also significantly disrupt aquatic food webs and threaten vital ecosystem services, including fisheries, potable water supply, biodiversity conservation, and recreational opportunities. This chapter provides a comprehensive examination of the ecological and socioeconomic impacts of cyanobacterial blooms, emphasizing their cascading effects across trophic levels and their role in destabilizing ecosystem resilience. Through ecological analysis and global case studies such as those from Lake Erie, Lake Taihu, the Baltic Sea, and the Murray Darling Basin, the chapter offers a holistic perspective on bloom dynamics and their broader environmental implications. The primary aim is to assess the influence of cyanobacterial blooms on aquatic food webs and ecosystem services. Objectives include investigating trophic disruptions, evaluating bloom-related ecosystem degradation in different regions, reviewing modern monitoring and assessment tools, and presenting integrated strategies for management and research. Key findings reveal that cyanobacterial blooms result in biodiversity loss, disrupted nutrient cycling, toxin transfer across food chains, oxygen depletion, and fisheries collapse. They impair provisioning, regulating, and cultural ecosystem services, with case studies

W. A. Khattak
Department of Plant Sciences, Faculty of Biological Sciences, Quaid-I-Azam University, Islamabad, Pakistan
e-mail: waseemkhattack@bs.qau.edu.pk

M. Anas · E. E. Hakki (✉)
Department of Soil Science and Plant Nutrition, Selçuk University. Campus, Konya, Turkey
e-mail: eehakki@selcuk.edu.tr

M. Anas
e-mail: muhammed.anas@selcuk.edu.tr

S. Fahad (✉)
Department of Agronomy, Abdul Wali Khan University Mardan, Mardan, Khyber Pakhtunkhwa, Pakistan
e-mail: shah_fahad80@yahoo.com

S. Fahad et al. (eds.), *Cyanobacterial Blooms: Ecology, Evolution and Biogeochemical Impacts*, https://doi.org/10.1007/978-3-032-06042-6_10

illustrating both regional variability and common ecological impacts. The chapter recommends future research focus on long-term ecological monitoring, multitrophic ecological modeling, climate-linked bloom forecasting, and the incorporation of ecosystem services into water governance. Adaptive, precautionary, and cross-sectoral management strategies are essential. In conclusion, cyanobacterial blooms represent a multifaceted environmental challenge. Mitigating their impacts requires an ecosystem-based approach that bridges scientific understanding, technological innovation, and policy integration to ensure the sustainability and resilience of aquatic ecosystems in a changing global climate.

Keywords Cyanobacterial blooms · Aquatic food webs · Ecosystem services · Trophic disruption · Climate change impacts

10.1 Introduction

10.1.1 Overview of Cyanobacterial Blooms

Cyanobacteria blooms, also referred to as blue-green algal blooms, have become an increasing environmental issue in freshwater systems and marine systems worldwide. A cyanobacterial bloom occurs when environmental conditions related to nutrient levels (especially nitrogen and phosphorous), thermally warmed water temperatures, prolonged sunlight exposure, and slow-moving hydrological conditions stimulate exaggerated growth of cyanobacteria (Borah et al., 2019). Cyanobacteria are naturally occurring components of freshwater ecosystems and indeed beneficial to aquatic ecosystems; however, cyanobacteria can also develop as harmful algal blooms (HABs) where excessive cyanobacterial biomass develops, and a range of secondary metabolites (toxic and nontoxic cyanotoxins) are excreted within or on the surface waters. Microcystins, anatoxins, saxitoxins, and cylindrospermopsin cyanotoxins represent some of the major cyanotoxins of concern to aquatic organisms, fauna, human health, and the overall ecological framework that aquatic habitats provide (Christensen & Khan, 2020). Recently, human-induced eutrophication, environmental climate change, and human impacts have happened over the last few decades, increasing the magnitude, frequency, and geographic distribution of cyanobacterial blooms in freshwater systems and marine systems worldwide, making cyanobacterial blooms a global problem in water resource management, biodiversity conservation, and food security (Barruffa et al., 2021).

Cyanobacteria are photosynthetic organisms like algae, but they are distinct from algae, because they belong to a group of prokaryotic microorganisms (Saber et al., 2022). Cyanobacteria have a variety of physiological traits that allow them to adapt, including buoyancy control through gas vesicles, nitrogen fixation, and defense against grazer activity, which provide them with an advantage over other species of phytoplankton. That is why they can outcompete the beneficial phytoplankton in times of nutrient enrichment and warm temperatures; they can invade water habitats and alter the natural tropic structure of the habitat. The blooms can contribute to hypoxia or even anoxia through the decomposition process, generate awful smells, reduce water clarity, and reduce the enforcement, aesthetic, and economic value of the waterbodies. Cyanobacterial blooms are not only restricted to lakes and ponds and are increasingly reported in estuarine and coastal marine systems and represent a pervasive ecological imbalance across aquatic habitats.

10.1.2 Scope and Importance of Aquatic Food Webs

Freshwater and marine food webs describe the complex networks or systems of trophic interactions and consist of the movement of energy and nutrients through the numerous biotic components in an ecosystem. Like terrestrial food webs, aquatic food webs consist of multiple trophic levels such as primary producers (e.g., phytoplankton, periphyton, aquatic macrophytes), primary consumers (e.g., zooplankton, benthic invertebrates), secondary consumers (e.g., small fish, macroinvertebrates), and tertiary or apex predators (e.g., large fish, aquatic birds, marine mammals). Aquatic food webs provide functions and services to ecosystems globally, so their integrity and resilience are critical for ecosystem functioning, biodiversity, and productivity (Oliver et al., 2015). Food webs orchestrate critical processes like nutrient cycling, energy transfer, regulation of fish and wildlife populations, and habitat stability. Disruption to even one of the trophic levels in the food web (e.g., in primary producer base) can cause cascading changes through each of the higher trophic levels, transforming the structure and dynamics of the whole food web (Shanafelt & Loreau, 2018).

Cyanobacterial blooms can profoundly change the structure of aquatic food webs by altering energy pathways, reducing food quality, and causing species shifts. For example, many cyanobacteria are relatively low in food quality, due to poor digestibility or low availability of essential fatty acids (fatty acids that organisms cannot synthesize on their own) and are an unsatisfactory food source for multiple primary consumers (e.g., zooplankton). Consequently, if conditions remain favorable for cyanobacteria, it is possible that zooplankton will decline in abundance or shift to species that can tolerate cyanobacteria (Ger et al., 2016). This translates to reduced recruitment and serious behavioral changes in fish community composition, through changes in food availability and quality for fish larvae and juvenile fish. Many detrimental effects of cyanobacteria in food webs are processed through cyanotoxin production. Cyanotoxins can poison or otherwise eviscerate organisms at every level of a food web. Cyanotoxins have the potential to bioaccumulate and biomagnify in organisms that consume cyanobacteria, exposing aquatic fauna, and potentially terrestrial predators, and even higher species like humans who consume contaminated aquatic organisms. Consuming cyanotoxins from contaminated aquatic species will cause direct harm, as well as chronic and possibly carcinogenic diseases. However, in addition, the shading effects of intense mat-forming cyanobacteria also exclude light penetration and reduces the amount of light available for growth of submerged aquatic vegetation (SAV) that are important habitat and food sources for many organisms, thereby placing extreme pressure on both biodiversity and ecosystem functioning and sustainability to mount a sustained response against ongoing degradation. If the destabilization and simplification of the food webs at cyanobacterial bloom conditions destroy or drastically reduce biodiversity, it could likely reduce or eliminate ALL ecosystem services (Paerl, 2018).

10.1.3 Ecosystem Services in Freshwater and Marine Systems

Ecosystem services are the benefitting aspects humans obtain from ecosystems (directly or indirectly), and aquatic systems are critical to sustaining life and livelihoods. Generally, services are classified as provisioning services (e.g. freshwater supply, fisheries, aquaculture, bioresources), regulating services (e.g., water purification, climate regulation, nutrient cycling), cultural services (e.g., recreation, aesthetic enjoyment, spiritual value), and supporting services (e.g., primary production, habitat provisioning, biodiversity maintenance). To sustain ecosystem services, it is essential, by sustaining healthy freshwater and marine ecosystems, to achieve ecological stability and equilibrium, socioeconomic development, and support human well-being (Yin et al., 2021). However, the occurrence and intensification of cyanobacterial blooms (and the resultant development of cyanotoxins) present a significant challenge to the provision of ecosystem services. From a provisioning services perspective, blooms directly reduce the water quality of drinking water supplies, which requires expensive treatment processes that either consume energy for removal of cyanotoxins and costly disposal of byproducts. Fish are killed due to environmental factors related to blooms or are rendered unusable from toxins accumulating in the edible tissues of fish (Samarajeewa, 2023). Aquaculture also suffers from reduced species composition by having fish killed or because aquaculturists choose not to purchase fish that are suffering and not through blame of aquaculturists. Regulating services are impaired due to environmental conditions that eliminate oxygen production (manufacturing) and reliance on decomposition of bloom biomass, which uses dissolved oxygen and creates anoxic conditions that are unfavorable for aerobic aquatic life.

Cultural services, such as recreational fishing, boating, swimming, and tourism, are negatively impacted during bloom events (Willis et al., 2018). Public health concerns, foul smells, and aesthetically displeasing water color can deter humans' desire to perform activities in a waterway and reduce the aesthetic and recreation value of the water body. Often, prolonged blooms have rendered lakes, beaches, and water-based tourism operations out of commission, resulting in economic losses to their communities. Also, there are public health concerns if humans dermally contact, inhale, or ingest water that contains cyanotoxins. Habitat loss and food scarcity also exist for aquatic fauna that rely on aquatic habitats for feeding, breeding, and nesting, demonstrating that aquatic ecosystems are profoundly affected. The supporting services of aquatic ecosystems are also heavily affected by cyanobacterial blooms. Bloom events will displace primary production by outcompeting beneficial phytoplankton, altering nutrient levels and sediment water interactions, and create nutrient loading feedback mechanisms under anoxic conditions (internal loading). A cycle also occurs when keystone species are lost, that creates an ecosystem increasingly vulnerable to repeated and heavy bloom events by removing self-regulatory capacity to maintain ecosystem stability (e.g., filter feeding bivalves, submerged aquatic plants) (Sherman, 2017).

10.2 Cyanobacteria and Trophic Dynamics

In aquatic ecosystems, cyanobacterial blooms can profoundly alter trophic functioning by altering primary productivity, disrupting consumer–resource relationships, and modifying the structure and functioning of aquatic food webs (Ullah et al., 2018). Despite being natural components of phytoplankton communities, cyanobacteria have a set of traits that enable them to dominate other phytoplankton and keep them suppressed when conditions are favorable. The interaction among the ecological strategies, physiological mechanisms, and chemical defenses of cyanobacteria defines how energy flows through aquatic ecosystems. Cyanobacterial blooms affect trophic functioning by outcompeting other primary producers, deterring zooplankton grazers, and disrupting fish recruitment (larval fish information), predator–prey interactions, and more. Cyanobacteria dominate and reprogram aquatic trophic position and relationships when they occur at the scale of food webs and ecosystems. Ultimately, these effects compound across the food web and ecosystem with consequences that include food web simplification (e.g., number of species), shifts in biomass (respiration vs primary productivity), and destabilization of ecosystem services that rely upon trophic functioning; aquatic ecosystems, for better or for worse, depend on their baseline interactions, producers, and consumers (Hammerschlag et al., 2019).

10.2.1 Primary Production and Competition with Phytoplankton

Cyanobacteria are potent primary producers capable of photosynthetic carbon fixation of light and CO_2 contributing measurable amounts of carbon to aquatic primary productivity, especially during bloom situations. However, while cyanobacteria number dominates phytoplankton communities, they also tend to negatively impact biodiversity and ecosystem functioning (Anas et al., 2025b). In contrast to eukaryotic phytoplankton like diatoms, green algae, and cryptomonads that represent nutritious and digestible food for higher trophic levels, cyanobacteria tend to have a lower nutritional quality: essential fatty acid content is low, as is predation potential that includes secondary metabolite production. Consequently, cyanobacteria will inherently contribute less to trophic transfer than primary producers that provide higher nutritional quality (Ger et al., 2016).

Nutrient-rich nutrient environments, particularly associated with high nitrogen and phosphorus loading, are one of the primary factors explaining the success of cyanobacteria (Anas et al., 2025a). A number of cyanobacteria genera (e.g., *Anabaena, Aphanizomenon*) can fix atmospheric nitrogen (N_2) by providing them with a competitive advantage in limitation conditions for nitrogen. Additionally, they have characteristics (e.g., colony size/large colonies, mucilaginous sheath, buoyancy through gas vesicles) that allow them to optimize light for growth and avoid sedimentation during favor light-limiting conditions, thereby facilitating survival and bloom formation in stratified water columns (Anas et al., 2024). Therefore, due to the ability of cyanobacteria taxa to displace multiple phytoplankton taxa, by creating a predominately cyanobacterial community composition, they have reduced phytoplankton diversity, altered successional dynamics, and delayed seasonal advantageous primary producer blooming. In these systems dominated by cyanobacteria, the base of the food-web shifted from quality, fast-growing phytoplankton taxa to less palatable, slow-growing, often toxic cyanobacterial, which significantly changes the flow and efficiency of energy through the food web (Moustaka-Gouni & Sommer, 2020).

10.2.2 Effects on Zooplankton Grazing and Community Structure

Zooplankton are a critical element in the movement of energy in aquatic food webs, acting as the bridge between primary producers and higher consumers like fish. However, cyanobacterial blooms disrupt grazing dynamics and community structure of zooplankton. Many cyanobacteria are either too large or morphologically complex (e.g., filamentous *Oscillatoria* or *Microcystis* colonies) to be effectively grazed by common filter-feeding zooplankton like *Daphnia* and copepods. Even when grazed, the poor nutritional quality and toxic properties of cyanobacteria may limit growth, reproduction, and survival of zooplankton. Microcystins, anatoxins, and other cyanotoxins have been shown to affect several zooplankton species in terms of feeding behavior, enzyme activity, and overall physiological health behavior (Chen et al., 2021).

Consequently, bloom conditions often result in loss of the larger-bodied efficient grazers like *Daphnia* and a successional shift toward smaller-bodied less efficient grazers like rotifers and nauplii, which can tolerate the poor quality and toxicity of the food (rather than getting sick or dying), but are less hospitable to effective trophic transfer. Loss of large-bodied grazer community structure results in reduced grazers that reduce grazing pressure and reinforce the loop and the ecosystem is susceptible to beeping and intensifying cyanobacterial blooms, in effect, making the ecological damage quasi-permanent. Furthermore, the loss of zooplankton population during bloom stress has been observed to cause valuable effects on the food web. Specifically, limited energy fixation for planktivorous fish as well as potential shifts in predator–prey interactions in the pelagic and littoral zones may occur (Mihalitsis et al., 2021).

The inhibition of the vertical migration patterns, as well as grazing efficiency and habitat availability for zooplankton due to hypoxia and low-light conditions below dense blooms will compound these effects. For example, in stratified lakes, cyanobacterial blooms can develop surface scums and stimulate hypoxic conditions at depth by consuming oxygen. These blooms strand zooplankton into less optimal habitat further reducing grazing efficiency and increasing risk of predation. The heightened vertical compressions reduce feeding efficiency, exposes and heightens predation capacity, and adds compounding pressures on zooplankton's regulatory role in phytoplankton community structure. In summary, cyanobacterial blooms not only reduce the zooplankton population through toxicity and/or palatability issues but also alter the functional and structural characteristics of the zooplankton community, lowering energy transfer to higher trophic levels (Ger et al., 2014).

10.2.3 Impact on Fish Populations and Recruitment

Cyanobacterial blooms have direct and indirect effects on fish populations. Fish are very dependent on planktonic organisms in their early life stages, and when cyanobacteria blooms dominate and displace the planktonic food base, fish may become food-limited, starve, or otherwise experience poor recruitment. Larval and juvenile fish require a highly nutritious and digestible zooplankton food source most often *Daphnia*, which are usually not abundant during bloom events (Nunn et al., 2012). For this reason, prolonged cyanobacterial dominance could severely limit fish recruitment, especially in fish populations exhibiting narrow feeding windows or strong reliance on prey types. In addition to nutritional effects, cyanobacterial blooms affect fish populations through deteriorated water quality, in large part through the initiation of hypoxic or anoxic conditions caused by bloom biomass decomposition. As bloom biomass decays, organic matter accumulation and microbial degradation are responsible for high dissolved oxygen demand, particularly in bottom waters, causing the development of hypoxic zones unsuitable for many fish. The loss of dissolved oxygen often can lead to severe fish kills often including benthic and demersal fishes. In stratified systems, deep water layers can become disconnected from oxygenated surface waters, limiting fish movement and habitat accessibility (O'Leary & Roberts, 2018).

Toxins produced by cyanobacteria can be directly toxic to fish. For example, the hepatotoxic contribution of Microcystins has been associated with liver damage, gill lesions, and immune suppression in fish (Li et al., 2023). Cyanotoxins can enter the fish food web through ingestion of contaminated food (the most common route), gill uptake, or via maternal transfer through fish eggs. In addition, sublethal levels of cyanobacterial toxins can result in reduced growth rates, behavioral changes, reproductive failure, and increased susceptibility to disease and parasites. In aquaculture systems, blooms of toxic cyanobacteria can reduce fish stock, diminish farm yield, or require costly management practices for mitigation. The trophic dynamics can also reform around these changes, resulting in new opportunities for growth of undesirable fish community members (Gårdmark et al., 2015). Blue-green algae dominance typically coincides with increased populations of omnivorous of planktivorous fish-like carp and tilapia that generally are tolerant to low-oxygen, high-turbidity waters. Furthermore, the disruption of sediment and other anthropogenic change can further catalyze bloom conditions from internal loading of nutrients and sediment resuspension. In contrast, fish such as salmonids or predatory cichlids that are piscivorous or strongly environmentally sensitive to such conditions might (in both growth and numbers) will subsequently decline due to habitat loss or food web collapse dependent on dramatic management interventions). Over time, this kind of degradation is often compounded by community changes that reinforces the degradation as resilience is slowly diminished, and fish populations are less capable of recovering after acute impacts (Fig. 10.1) (Pelletier et al., 2020).

Characteristic	Tolerant Fish (Carp, Tilapia)	Sensitive Fish (Salmonids, Cichlids)
Toxin Exposure	Ingestion of contaminated food	Gill uptake or maternal transfer
Direct Toxicity	Tolerant of low-oxygen, high-turbidity	Liver damage, gill lesions, immune suppression
Sub-lethal Effects	Increased populations	Reduced growth, behavioral changes, reproductive failure
Aquaculture Impact	Benefits from trophic dynamic changes	Reduced stock, diminished yield
Habitat Change	Tolerant of sediment disruption	Decline due to habitat loss
Community Resilience	Reinforces degradation	Less capable of recovering

Fig. 10.1 Contrasting fish responses to cyanobacterial blooms: A comparison of how tolerant (e.g., carp, tilapia) and sensitive (e.g., salmonids, cichlids) fish species are affected across ecological and aquaculture dimensions

10.2.4 Disruption of Trophic Cascades

Trophic cascades are top-down or bottom-up control effects creating effects that permeate through the food web by a disturbance at one trophic level. Cyanobacterial blooms can affect both types of trophic cascades. When cyanobacteria replace phytoplankton and limit zooplankton abundances, they establish an alternative base of the food web as a consequence of replacing organisms at the low trophic level (bottom-up) and affect the flow of energy and nutrients by weakening following steps of bottom-up cascades (Ullah et al., 2018). They also limit the influence of top-down productivity control by suppression of fish to indirectly regulate the abundances of phytoplankton through zooplankton: they attach themselves to the intermediate consumer, weakening the top-down control in the bottom up. Trophic cascades continually create regulatory feedback to influence distributions and abundance of organisms in food webs, but cyanobacterial blooms change that. The classic trophic cascade in the case of the productivity of fish is when predatory fish control the population of planktivorous fish, which creates opportunity for zooplankton to graze and limit abundance of phytoplankton. With limited grazing of zooplankton, and large abundances of inedible or toxic phytoplankton, individual or joint top-down control would not occur. Phytoplankton and zooplankton, as well as fish, may respond to bloom-related hypoxia; predator and prey species may not be affected evenly, and behavioral responses will likely alter interactions related to food web stability dynamics (Kazmi et al., 2022). To sum, in anthropogenically dominated systems, the established energetic flow is inefficient with cascading loss of biodiversity and an inconvenient persistently alternative stable state dominated by cyanobacteria. This disruption has significant implications for ecological restoration. Many types of aquatic ecosystem restoration efforts use some form of biomanipulation—increasing top predators or removing nutrient inputs, while systems that are experiencing one or several bloom impacts will have a broken trophic cascade already present and most likely will not produce any desired effects unless a full level of actions are aimed at re-establishing trophic linkages, restoring waters quality, and re-establishing ecosystem resilience. Understanding the mechanistic pathways through which cyanobacteria disrupt trophic cascades is important when considering management and rehabilitation actions (Hamilton et al., 2016).

10.3 Effects on Ecosystem Services

Cyanobacterial blooms have become one of the most detrimental ecological phenomena experienced by aquatic ecosystems world-wide. Their impacts are not only restricted to ecological factors such as water chemistry or community composition, but they directly impact the broad spectrum of ecosystem services human societies depend on as well. Ecosystem services are defined as the benefits humans obtain from ecosystems, and they are fundamentally important to environmental health, economic development, and social welfare. Ecosystem services are categorized into provisioning (e.g., food and water), regulating (e.g., climate, nutrient cycling), cultural (e.g., aesthetics, recreation), and supporting (e.g., biodiversity, ecosystem resilience). Cyanobacterial blooms caused by a combination of eutrophication, climate, and hydrological changes disrupt ecosystem services on more than one level. The persistence and severity of the blooms reduces water quality (and potential sources of drinking water), loss of biodiversity and potential collapse of fisheries, loss of recreational activity, and ultimately resilience of the ecosystems, with social, economic, and ecological repercussions (Sheergojri et al., 2023).

10.3.1 Water Quality Regulation

Cyanobacterial blooms have one of the most apparent and direct impacts on water quality, an important type of regulating service in aquatic ecosystems (Amorim & do Nascimento Moura, 2021). The bloom often forms dense surface scums that reduce light penetration, alter O_2 dynamics, and produce toxic compounds that affect the usability of the water. During bloom development, the photosynthetic capacity of cyanobacteria can greatly increase O_2 concentrations, which can be further elevated by the sun's energy; however, when blooms senesce, large amounts of cyanobacterial organic matter are decomposed by microbial activity. The aerobic decomposition of this organic matter consumes dissolved O_2 and can lead to hypoxia or anoxia in the bottom waters, affecting aerobic organisms, promoting anaerobic microbial processes, and releasing potentially harmful compounds such as ammonia and hydrogen sulfide (Boyd & Boyd, 2020). Cyanobacterial blooms also produce a suite of toxic metabolites known as cyanotoxins such as microcystins, cylindrospermopsin, anatoxins, and saxitoxins, which can enter drinking water supplies and threaten the health of animals and human users. Symptoms of toxicity can be related to damage to the liver (hepatoxins), neurotoxicity (neurotoxins), gastrointestinal illness, and in some instances, death. Many of these compounds will enter raw drinking water supplies and require expensive treatment technology (such as activated carbon filters, ozonation, or UV treatment) to eliminate, which provides a significant financial burden on municipal drinking water treatment facilities. Moreover, changes in pH, turbidity, and nutrient cycling caused by blooms can alter water chemistry and disrupt the balance of the ecosystem. The shift from external phosphorus loading to internal phosphorus loading under anoxic conditions, for example, can drive endogenous eutrophication and cyclic bloom formation; this degrading water quality is a feedback loop for the ecosystem. Thus, the regulatory role of freshwaters that provide clean, oxygenated, and nonharmful water is greatly diminished under cyanobacterial dominance (Sahu et al., 2024).

10.3.2 Biodiversity and Habitat Integrity

Cyanobacterial blooms have profoundly negative consequences for biodiversity by changing community composition, displacing native taxa, and reducing habitat complexity. The competitive advantage of these cyanobacteria over many other phytoplankton leads to decreases in species richness (otter diversity or primary producer diversity) that affect diversity of organisms at higher trophic levels (Terborgh, 2015). Many cyanobacterial species are harmful to food webs, because they are inedible or toxic, creating food voids for herbivorous zooplankton and filter feeders. Food voids can have cascading effects on consumers, which can lead to lower population sizes and trophic downgrading. The hypoxia and light limitation from blooms can hinder the survival of benthic organisms and submerged aquatic vegetation (SAV), both of which are critical habitats for fish and invertebrates and act as spawning grounds for many fish species. The growth of algal blooms reduces SAV, creates essential habitat, and limits niche diversity, which is essential for increasing habitat complexity and reducing local extinction risk (Meerhoff & de los Ángeles González-Sagrario, 2022). Prior to being displaced by algae or other vegetation, SAV creates more complex habitats, which can support a diverse group of species. The disappearance of a more complex benthic habitat as a response to algal blooms increases habitat homogeneity and reduces the habitat's suitability for many specialized species. Cyanobacterial blooms are particularly harmful for the aquatic organisms, amphibians, aquatic birds, and mammals that are reliant on these water bodies, as they can create local extinction events and permanent reductions in functional groups. Opportunistic bloom adapted organisms and generalists can expand their ranges into aquatic systems where valid habitats are reduced by algal blooms or other disturbance, making the food web homogeneous and invasive organisms replace native specialists, or can facilitate biotic homogenization where invasive or bloom-adapted organisms replace native specialists. Biodiversity loss not only harms ecosystem stability, it limits the possibilities of genetic and species-level resources for future adaptive capacity to respond to environmental change. Ultimately, the physical structure and function of aquatic habitats become less able to support diversified, resilient, and self-regulating biological communities (Svenning et al., 2019).

10.3.3 Fisheries and Food Provisioning

Aquatic ecosystems are important sources of provisioning services, mainly fish and shellfish which provide a source of protein, income, and livelihoods for millions of people worldwide. Cyanobacterial blooms impair these services by reduced the abundance of fish present, changing species composition, and possibly contaminating edible aquatic organisms with cyanobacterial toxins. The depletion of zooplankton populations during blooms removes available prey for juvenile fish, which slows down growth rates and reduces overall survival. In both wild and aquaculture environments, hypoxic conditions created by blooms can kill fish, along with hypoxia-sensitive species, and important commercial fish. In addition, some cyanobacteria bioaccumulate and translocate toxins through aquatic food webs. For example, microcystins have been documented in the tissues of fish, crustaceans, and shellfish. The consumption of contaminated fish and seafood can pose public health risks. Additionally, fishing bans can cause large—scale economic losses, and generally lower consumer confidence in their choice of local food. Such economic losses can be especially damaging in developing nations where entire communities rely on inland fisheries or aquaculture as a protein source. Blooms in aquaculture systems can also have a devastating effect due to increased operational management costs to manage water quality, increased incidence of fish disease, and losing stock. Intensive fish farming operations in nutrient-rich waters may compound bloom formation through nutrient runoff and inefficient waste management. Therefore, the interaction of cyanobacteria and food provisioning is a biological and socio-economic issue related to food security, income generation, and public health (Shukla et al., 2021).

10.3.4 Cultural and Recreational Services

Cyanobacterial blooms greatly affect a culturally and recreationally valuable service provided by aquatic ecosystems. Cyanobacterial blooms can reduce recreational services such as swimming, boating, angling, and eco-tourism when a bloom causes aesthetic concerns, foul taste and/or odors, or a health hazard. Cyanobacteria can discolor water, form unsightly scums, and release volatile organic compounds that smell like decaying organic matter or rotten eggs, resulting in people disengaging with the site and reducing the appeal of bodies of water for tourism or recreation (McGenity et al., 2020). Most commonly the health authority will take the measure to issue a health advisory or temporarily close beaches and bodies of water when cyanotoxin concentrations exceed a safe threshold. Contact with cyanobacteria in contaminated water can result in skin rashes, eye irritation, respiratory issues, and occasionally profound toxicological health effects due to inhalation or ingestion of toxins. The loss of recreational experiences decreases the quality of life of local communities and can be detrimental to the economy for businesses reliant on tourism and water-based activities. Many aquatic ecosystems also possess cultural, historic, or spiritual relevance for indigenous and local communities. Traditionally, water bodies are prominently represented in rituals, folklore, and religious practices. When cyanobacterial blooms occur, they destroy these important cultural elements, displacing communities from their culture, and lessen the socio-cultural value of ecosystems. It is hard to attach value to the loss of intangible benefits, but this loss impacts the overall decreased value of ecosystem services (Small et al., 2017).

10.3.5 Ecosystem Resilience and Functional Redundancy

Ecosystem resilience involves the capacity of an ecosystem to tolerate disturbance, its ability to reorganize and continue fulfilling its functions and structure. Cyanobacterial blooms threaten ecosystem resilience because they diminish the number of species, simplify food webs, and disturb the feedback mechanisms that regulate ecological functions (Paerl, 2018). With loss of biodiversity comes loss of functional redundancy, the presence of several species that perform these ecological functions. Without functional redundancy, ecosystems are increasingly vulnerable to stresses when they form and less likely to recover from perturbations such as pollution, climate change, and invasions. The feedback loops associated with cyanobacterial blooms may create ecological states or regimes that have long-lasting shifts, pushing systems toward alternate stable states dominated by cyanobacterial blooms. These regimes can be difficult to reverse, requiring long-term and coordinated management measures. Loss of functional groups that provide ecological functions like submerged vegetation, large zooplankton grazers, or filter-feeding bivalves further limits the ecological systems ability to self-repair or suppress conditions favoring cyanobacterial bloom development. Climate change makes this issue increasingly serious as water temperatures are rising and precipitation patterns are changing to enhance nutrients flowing to water bodies and favor cyanobacterial dominance (Paerl & Paul, 2012). Higher temperatures enhance the growing season for cyanobacteria, limit vertical mixing of the water column, and create stratified conditions that can contribute to bloom formation. Under these conditions, and recognizing the increasing frequency of cyanobacterial blooms, it is important to maintain resilience of ecosystems as a meaningful ecological goal and as part of climate adaptation efforts. Taking action to promote biodiversity, restore hydrographs, and control nutrients at the watershed scale are important steps to protect aquatic ecosystems in the future from the ongoing threat of cyanobacterial blooms (Fig. 10.2) (Hamilton et al., 2016).

10.4 Case Studies

Cyanobacterial blooms occur worldwide and exist in all locations and climates on the planet. Their commonality across freshwater lakes, coastal seas, and river basins illustrates the prevalence of eutrophication and the global issues of anthropogenic impacts and climate change. Case studies in various locations across the planet provide real-life examples of significant causes, consequences, and management responses to harmful cyanobacterial blooms. Through the lens of significant examples in the context of Lake Erie (North America), lake Taihu (China), the Baltic Sea (Northern Europe), and the Murray–Darling Basin (Australia), we can observe the socio-ecological nature of bloom formation, the interruption of the aquatic food web and associated ecosystem services, and possible management pathways.

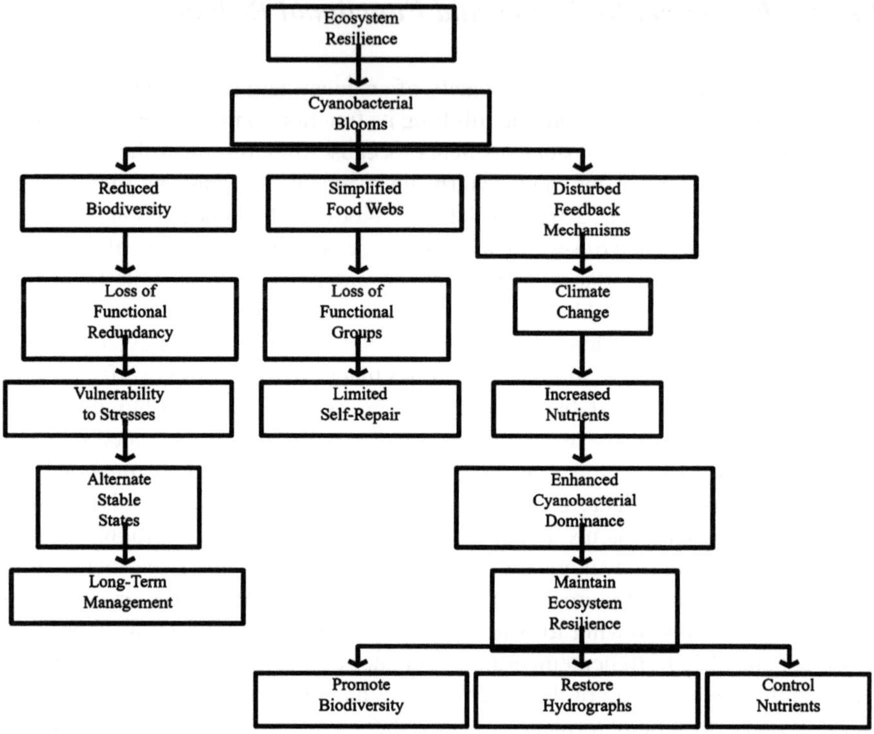

Fig. 10.2 Impact of cyanobacterial blooms on ecosystem resilience

10.4.1 Lake Erie (North America)

Lake Erie, one of the Great Lakes of North America, has become synonymous with harmful cyanobacterial blooms in temperate zones. Following the success of nutrient management indicated through declining phosphorus input, Lake Erie enjoyed a spike in blooms, particularly microcystin-rich blooms during the past two decades. This shift was largely attributed to excessive nonpoint agricultural runoff, chiefly from row crop agriculture within the Maumee River watershed, which supplied large amounts of dissolved reactive phosphorus to western Lake Erie. The acute impacts of *Microcystis aeruginosa*, a toxic cyanobacterium creating microcystins, impacted much more than the ecology of Lake Erie. One notable bloom in 2014 led to widespread contamination of Toledo, Ohio, municipal drinking water, impacting ~500,000 individuals, with toxin levels exceeding recommended safe thresholds for human consumption, triggering the temporary cessation of the municipal drinking water system and sparking national attention surrounding drinking water security. Biologically, the food web in Lake Erie has been destabilized. With reductions in edible phytoplankton and the increase in inedible or toxic cyanobacteria, zooplankton communities have shifted, and prey available to fishes has diminished (Briland,

2018). For example, economically valuable species (hydrographic distributions, growth conditions, etc.) such as walleye and yellow perch rely on food web dynamics to recruit and grow. Beyond ecological devastation, the economic impacts have been significant, resulting in negative impacts on tourism, fishing, and lakeside property values. Implementation of binational (Canada–U.S.) nutrient reduction targets for Lake Erie has included many components, including the Great Lakes Water Quality Agreement (GLWQA) that considers blooms and nutrient reduction. However, bloom trends are persistent, owing to complex watershed land-use patterns and continuing climate variability (Weber et al., 2020).

10.4.2 Lake Taihu (China)

Lake Taihu, the third largest freshwater lake in China, located in the densely populated and industrialized Yangtze River Delta, is one of the most intensively studied cases of cyanobacterial bloom crises in Asia. Since the 1990s, Microcystis-dominated blooms have occurred annually as a consequence of urbanization, industrial discharge, agricultural nutrient loading, and climate-induced shifts in hydrology and temperature. A severe bloom event in 2007 resulted in a water supply crisis in the city of Wuxi (over 2 million residents), as the bloom contaminated the intake water with cyanotoxins and organic material. This crisis caused widespread panic and a shortage of bottled water, made national headlines, and resulted in governmental intervention. The bloom event led to large-scale funding for lake restoration and monitoring activities (Reich & Lake, 2015).

The blooms in Lake Taihu have ecologically shifted the lake from a macrophyte-dominated system to a phytoplankton dominated system, altering the habitat value of the lake and chemical clarity. The lake experiences frequent anoxic events, resulting in fish kills and loss of biodiversity within both planktonic communities and benthic communities. Once in a cyanobacterial dominated state, the lake has faced extenuating difficulty to return to a lower nutrient state because of internal loading of phosphorus from the sediments that maintains these bloom cycles. To address the problem, China undertook the Taihu Basin Water Pollution Prevention Plan committing billions of yuan to reduce pollution inputs into the water body, improve sewage treatment, and facilitate ecological restoration moving toward dredging and wetland reconstruction. While some progress is being made, persistent blooms suggest further long-term integrated watershed management, increased agricultural controls, and continued public engagement in reduction of pollutant inputs may be needed (Wang et al., 2016).

10.4.3 Baltic Sea (Northern Europe)

The Baltic Sea is one of the largest brackish water bodies in the world, and its waters experience significant eutrophication as a result of urban waste and agricultural runoff from the surrounding countries—including Sweden, Finland, Poland, Germany, and Russia. The semienclosed form of the sea, in combination with low nutrient exchange, leads to nutrient accumulation and prolonged blooms. Cyanobacterial blooms in the Baltic Sea are often dominated by nitrogen-fixing species such as *Nodularia spumigena* and *Aphanizomenon flos-aquae*, which proliferate when other nitrogen forms are depleted, and phosphorus concentrations are high, and commonly occur during summer stratification. Visible from satellite imagery, these blooms occupy significant portions of the sea (Munkes et al., 2021).

These blooms lead to large dead zones, which are oxygen-depleted zones at the sea floor, created from degrading organic material and limited vertical mixing. These dead zones have contributed to loss of habitat for benthic fauna and substantial declines of many commercially important fish species, such as cod, that rely on oxygen from the bottom-water for spawning. Further, Nodularia toxins have raised concern for the safety of seafood and the health of the ecosystem overall. From a management perspective, the Helsinki Commission (HELCOM) has contributed towards coordination of multi-national action via the Baltic Sea Action Plan to limit nutrient inputs through improved agricultural practices, municipal wastewater upgrades and international collaboration. While some progress has been made, particularly in limiting point source nutrient inputs, nutrients internal to the system in the sediments and historic agricultural pressures continue to promote bloom events, highlighting long time scales required for recovery of ecosystem health (Wurtsbaugh et al., 2019).

10.4.4 Murray–Darling Basin (Australia)

The Murray–Darling Basin, which contains several of the most prominent river valley systems in southeastern Australia, has experienced recurring cyanobacterial bloom events that disrupted water management, ecosystem functioning, and agricultural production (Wurtsbaugh et al., 2019). One of the more infamous blooms reported occurred in 1991 along a 1000 km stretch of the Darling River that is frequently cited as one of the longest cyanobacterial (and noncyanobacterial) events on record worldwide. The Murray-Darling Basin is uniquely susceptible to blooms due to the arid climate, altered hydrology in relation to irrigation practices, and intermittent droughts, which provide relatively ideal bloom conditions of slow-moving, warm, nutrient-rich waterways. Blooms in this region typically consist primarily of *Anabaena circinalis* (now *Dolichospermum*), which produces neurotoxins and hepatotoxins that are potentially harmful to human beings, livestock, and wildlife health. The ecological impacts of cyanoHABs include widespread fish kills, the degradation of waterholes vital to biodiversity, and the destruction of habitat critical to aquatic invertebrates and waterbirds. Existing socioeconomic impacts include limited supply of water for towns and cities, livestock poisoning, loss of recreational water opportunities, and fisheries decline, as well as substantial costs associated with monitoring and management. In dealing with cyanoHABs, Australia has developed the National Cyanobacterial Risk Management Framework and regional early warning systems using remote sensing and field monitoring (Qiu et al., 2025). Additionally, water agencies are altering flow regimes to reduce the potential for blooms to develop to a manageable size. Nevertheless, increasing water demand, continuing land degradation, and variable climate conditions remain significant impediments to the goal of sustainably managing or controlling blooms in the Murray-Darling basin (Beavis et al., 2023).

10.5 Mechanisms of Impact

Cyanobacterial blooms can have complex and frequently synergistic feedback on freshwater ecosystems, influencing both biotic and abiotic processes. The ecological impacts of blooms operate through complex ecological processes that occur at different spatial and temporal scales. While the visible effects of blooms, such as water discoloration and rotten odors, may seize our immediate attention, they are caused by indirect processes triggered by blooms: burial and biological transfer of toxins, decreased dissolved oxygen levels, disruption of nutrient recycling, and reduced light penetration, each of which can ultimately destabilize aquatic food webs, diminish benefits and services of ecosystems, and decrease ecological resilience. Understanding these ecological processes will better enable us to predict the outcomes of blooms, evaluate the vulnerability of ecosystems to changing conditions, and design specific management responses to mitigate their negative impacts (Weißhuhn et al., 2018).

10.5.1 Toxin Transfer Through Food Chains

The development of cyanobacterial blooms has a very pernicious effect: the development and transference of toxins through aquatic food webs. Cyanobacteria can produce a variety of bioactive compounds (including microcystins, anatoxins, saxitoxins, and cylindrospermopsins) that can affect nearly all levels of a trophic system, and their toxins differ in chemical characteristics and physiological targets: they can be classified as hepatotoxins (affecting liver tissues), neurotoxins (affecting neural function), or cytotoxins (affecting cellular metabolism). Primary consumers such as zooplankton may be the first trophic level to take up these toxins, although they may be indirectly manufactured toxins through a detrital food chain. The uptake of the toxins can be assorted as zooplankton consumption and use of solid organic matter to either graze (direct uptake) or through detritus (indirect uptake). Even at low doses, toxins can impair zooplankton reproduction, feeding, and survival rates. This alters zooplankton population densities and alters community composition. Instead of a balanced assemblage of zooplankton composed of large and small taxa, possibly dominated by functioning functionally similar species. The shift in the tolerant or smaller species results in lower efficiencies of energy transfer to higher consumers (Eddy et al., 2021).

As the contaminated zooplankton and filter feeders (like mussels and clams) are consumed by fish, birds, or humans, the toxins are bioaccumulated and in some circumstances biomagnified. Many cyanotoxins do not biomagnify into fish. However, bioaccumulation into tissues is still a health concern for higher trophic organisms. Fish can bioaccumulate microcystins in their livers, kidneys, and muscles, and can pose a risk to piscivorous birds and mammals. In aquaculture and wild fisheries, products contaminated with microcystins can lead to economic losses from product recalls, consumption advisories, and lost confidence in the market. Additionally, microcystin sublethal effects may contribute to lowered fitness of organisms, making them more susceptible to disease, predation, and other environmental stressors. This affects food web integrity and the regulatory feedbacks of food webs, potentially facilitating cyanobacterial dominance (Ullah et al., 2018).

10.5.2 Oxygen Depletion and Anoxic Events

A fundamental ecological process that occurs with cyanobacterial blooms is oxygen depletion that happens during and especially after a bloom period when the remains decay (Huisman et al., 2018). In the beginning, cyanobacteria are able to increase dissolved oxygen in surface waters through photosynthesis; however, as a bloom remains longer and biomass begins to decompose, microbial respiration increases and uses large amounts of oxygen to generate hypoxic (low oxygen) or anoxic (zero oxygen) conditions. These conditions may intensify in stratified water bodies when warm surface waters are isolated from cooler bottom waters. The decomposition process quantifies mainly in the hypolimnion, because of poor mixing there is little replenishing of low levels of oxygen. In addition, the hypolimnion often develops anoxic zones near the sediments, with some form of stress on aquatic organisms, there are often massive fish kills and significant collapse of benthic invertebrate communities. Very sensitive species are lost or displaced, while more tolerant organisms (e.g., some anaerobic bacteria, or air-breathing fish) may survive, but cannot replace the ecosystem functions already expressed in the original community. Anoxic conditions favor the release of reduced substances such as ammonia, hydrogen sulfide, and methane, which are toxic and degrade water quality (Akhtar et al., 2021). With increasing anoxia, the metabolic pathways of microbial communities can change and shifts in biogeochemical cycling will take place. Hypoxia is recognized as an ongoing ecological crisis, especially within estuarine systems, as it often leads to the emergence of "dead zones," decimating coastal fisheries and affecting regional economies. The secondary effects of diminished oxygen will extend beyond just mortality. Anoxic and/or sublethal hypoxia can alter foraging and mating behavior, growth, egg development, and larval success in fish. For mobile species, increasing oxygen depletion pushes them into less suitable, more crowded habitable space and increases their stress as well as interspecies competition. Ecosystem-wide, increasing oxygen depletion will turn dynamic, productive systems into biogenically impoverished systems with weakened trophic interactions (Chakraborty & Chakraborty, 2021).

10.5.3 Altered Nutrient Cycling and Internal Loading

Cyanobacterial blooms are tremendously disruptive and can alter nutrient cycling, specifically nitrogen (N) and phosphorus (P) dynamics, which both drive productivity in aquatic environments. During bloom events, the processes of nutrient uptake, storage and release change fundamentally, in the water column and in sediment water column interactions (Zhu et al., 2013). A key issue to consider is internal nutrient loading especially phosphorus that sustain and increase bloom events. Many times, phosphorus is continually released from sediments and used by bloom events. During stratified or eutrophic regimes, sediment–water column anoxic conditions facilitate the release of phosphorus from sediments facilitated by a microbial reduction of iron oxides to which phosphate is bound. These internally sourced nutrients can sustain and extend bloom persistence remaining after external input reductions (Wurtsbaugh et al., 2019). Cyanobacteria are also known to actively act in changing the N cycling. Genus that can fix nitrogen like *Aphanizomenon, Dolichospermum,* and *Nodularia* transform atmospheric nitrogen (N_2) into the form of bioavailability ammonium and escape nitrogen limitation, which also play a role in the development of biomass. This process can alter the N:P ratio in which cyanobacteria tend to thrive far much better than other phytoplankton that rely on fixed nitrogen sources (Cottingham et al., 2015). In addition, blooms can hinder nitrification and denitrification, leading to changes in the pathway of nitrogen transformation and the accumulation or imbalance of nitrogen. Such changes will enhance eutrophic conditions, in which additional growth of blooms and ecological impoverishment are possible. Moreover, changing nutrient cycling can cause infancy of the biogeochemical buffering systems like wetlands and lawns on the coastal regions that genes up nutrient flows on a regular basis (Fig. 10.3). The reestablishment of natural nutrient cycling not only considers how to control the external inputs to restore the cycle but also emphasizes how to clean up sediment legacies of nutrients, which is costly and technically demanding venture (Tully & Ryals, 2017).

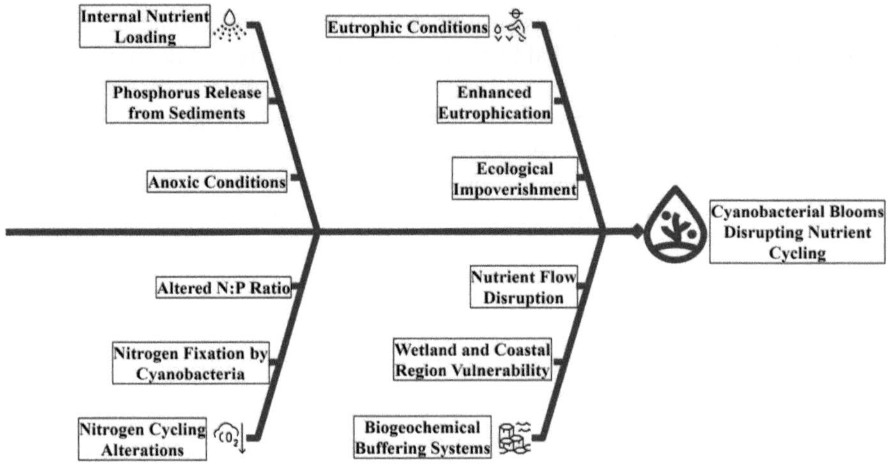

Fig. 10.3 Disruptive nutrient cycling by cyanobacterial blooms

10.5.4 Light Attenuation and Submerged Plant Loss

The other relevant pathway via which cyanobacterial bloom affects water ecosystems is the attenuation of light. Hypereutrophic concentrations of cyanobacteria on the surface of the water greatly inhibit transmittance of light in the water column, which causes shaded environments that suppress submerged aquatic vegetation (SAV) (Zhang et al., 2022). These plants are important in keeping the water clear, stabilizing the sediments as well as cycling the nutrients, and it also offers shelter to fish and other invertebrates. When lights are limiting, photosynthesis in submerged macrophytes will decrease resulting into slowed growth, reproduction, and death. Not only that will decrease habitat complexity but also eliminate one of the greatest competitors of phytoplankton leading to an even more tilted advantage to cyanobacteria. The reduction of SAV may result in regime shift of a clear-water and macrophyte-dominated condition into a turbid and phytoplankton-dominated condition—a change that is almost impossible to restore with the naked techniques. In addition, removal of macrophyte cover results in elevated resuspension of sediments, since plants with roots, which help to stabilize the sediments, are eliminated. This resuspension brings more nutrients to the water column and further decreases the light penetration, as well as enhances turbidity, which forms a positive feedback loop, leading to dominance of bloom. The effects of this loss of submerged plants are of ecological proportions such as the food web. Attached Fish species, which spawn or seek refuge on vegetation, are deprived of vital spawning or refuge habitat, and the invertebrate community becomes more generalist or tolerant (Gebrekiros, 2016). This causes collapse of indigenous biodiversity and growth of invasive, or bloom adapted species in most systems.

10.6 Monitoring and Assessment Tools

The increasing occurrence and intensity of cyanobacterial blooms in the world requires the need to develop or implement effective and integrated monitoring and assessment tools in order to get to know more about their impact, which can inform the management of such blooms and subsequently evaluate the success of such management or mitigation practices. Conventional water quality surveillance, based on a concentration measure of nutrients, chlorophyll-a, or cyanotoxins, can present important though limited information on the rich ecological interactions and on the wider ecosystem service outcomes of blooms. Thus, it is essential to have an ecosystem-based monitoring framework, which will use biological and chemical as well as socio-economic indicators. This chapter argues about the main tools and techniques that can improve the measurement of the effects of blood, especially the use of food web bio indicators, analysis of the stable isotopes, and ecosystem services valuation approaches (Alp & Cucherousset, 2022).

10.6.1 Food Web Bioindicators

The sensitivity of food web bioindicators increases its form of an early warning system as well as its investigative nature as a long-term diagnostic tool of determining the ecological variation. These bioindicators measure changes in the composition, abundance, physiology, or behavior of the different trophic levels and can give indications of food web integrity, the movement of energy in the ecosystem, and the health of the ecosystem. On the lowest trophic levels changes in phytoplankton communities, for example, the replacement of diverse assemblages of eukaryotic phytoplankton to blooms of cyanobacteria can be observed in terms of pigments (e.g., phycocyanin, chlorophyll-a) or molecular techniques (e.g., 16S rRNA gene sequence analysis) (Y. Li et al., 2022). Such transformations are indicators of eutrophication and disappearance of primary producers' diversity. Zooplankton, especially species like *Daphnia*, are good indicators of cyanobacterial pressure, since they have direct grazing behavior on their phytoplankton. While these losses may occur in their biomass, reproduction, or reduction toward smaller-sized taxa (e.g., rotifers), all are strong indicators of food quality degradation of the bloom and exposure to toxins. These effects can be measured with size structure, fecundity rates, and gut content analyses. Stress of bloom is bioindicated by fish populations, especially the planktivory and juvenile levels. Food web disturbance is denoted by changes in fish growth rates, condition indices, or diet composition (e.g., low-quality zooplankton or detritus-based feeding). Moreover, long-term exposure of fish to cyanotoxins can also be measured, for example, through liver histopathology or bioaccumulated toxin concentrations, and used as indicators of an ecosystem-scale. Other living things, like bivalves, water insects, and birds, are also effective in this capacity as bioindicators. Filter feeding bivalves bioaccumulated toxins providing information on tropic transfer and bird dying cases related to intoxication with cyanotoxins show high exposure in a top predator. Assessing the effect of the blooms using several taxa of bioindicators will provide a more detailed knowledge of the effects of the blooms at each of the trophic levels and the process of ecological prediction (Amorim & do Nascimento Moura, 2021).

10.6.2 Stable Isotope Analysis

Stable isotope analysis (SIA) has become an effective ecological tool that offers an inference into nutrient supply, trophic relation, and energy flow in aquatic conditions that are impacted by cyanobacterial blooms (McCue et al., 2020). Elements of the food webs will often be characterized by isotopic ratios of elements like carbon (13 C/ 12 C) and nitrogen (15 N/14 N) whereby the relative abundance of different isotopes will provide a natural tracer of material flow through the food webs and measure the role of cyanobacteria in impacting ecological structure and ecological activity. The presence of carbon isotopes (13 C) can be used to separate the sources of the primary production. Cyanobacteria can also exhibit quite different signature values of 13 C relative to other phytoplanktons, especially when they differ in mechanisms of carbon uptake (e.g., the use of bicarbonate) (Piatka et al., 2022). Through the study of 13 C in the consumers like zooplankton or fish, researchers can determine the portion of the diet of cyanobacteria. Greater levels of food web compression and lower energy transfer efficiency are usually represented by an increased dependence on cyanobacteria. Nitrogen isotopes (delta-15 N) can be used to explore food chain and to detect sources of nitrogen. Cyanobacteria that fixes N are likely to result in 915N smaller isotopic values (near 0 p increment) because of the uptake of atmospheric N_2 gas, whereas anthropogenic N (e.g., fertilizer or sewage) has larger 915N signatures. Consumer δ 15 N shifts can therefore show the nitrogen pathways changes and bloom-induced changes in the nutrient cycles. SIA can also help evaluate the alteration of Trophic position of organism as a bloom occurs. An example is that fish this feeding on detritus, or the so-called associations of food webs with cyanobacteria may portray values lower in the 815N than would be anticipated, demonstrating an altering to basal or degraded sources of energy. Besides, SIA may be used in combination with analyzing toxins to associate exposure levels with dietary sources and trophic pathways. Keen new isotope techniques such as compound-specific isotope analysis (CSIA) and dual-isotope mapping also support more resolution and enable the identification of subtle changes in ecosystem structure. SIA, as a tool that does not require any invasive processes, can be of significant value when a long-term monitoring as well as ecological impact study of cyanobacterial blooms should be conducted (Zahir et al., 2024).

10.6.3 Ecosystem Service Valuation Methods

Cyanobacterial blooms do not only lead to ecological damages but also they affect the provision of ecosystem services such as water purification, food, recreation, and aesthetic benefits. Because it is important to measure these effects in totality, valuation methods of ecosystem services are becoming widely used in the environment assessment and development of policies (Brouwer et al., 2013). The methods used to measure economic value are market-based valuation (direct financial loss of poorer services, e.g., fishing loss, water treatment costs) and nonmarket valuation implying indirect (intangible) losses, including health risks, biodiversity losses, or cultural displacement. Contingent valuation, travel cost method, and willingness-to-pay (WTP) surveys are techniques to determine the values attached by people to clean water or free of bloom environment. An example is that economic losses in the order of millions of dollars have been quantified in drinking water, tourism, and house values disruptions caused by blooms in Lake Erie and Lake Taihu, respectively. Instead, ecological evaluation is centered on biophysical measures of service delivery. These could be decompositions of fish biomass (provisioning), the disappearance of submerged vegetation (habitat support), or a decrease in water clarity (regulating services) (Benbow et al., 2020). These indicators are then combined in composites, that is, standardized as indices, as the Ecological Integrity Index (EII) or Water Quality Index (WQI), to provide comparative measures among sites or over time. To assess the changes in services with bloom scenarios spatially and policy relevant, tools, such as Integrated Valuation of Ecosystem Services and Trade-offs (InVEST) and the System of Environmental-Economic Accounting (SEEA), are emerging. These instruments aid in cost–benefit appraisals of administration interventions and aid in the choice of activities that present most ecological and societal return on investment activities. Furthermore, the participatory methods of assessment will involve local stakeholders in the process of defining the valued services, perceived impacts, and tolerable trade-offs so that valuation can contribute to equitable decision-making, which is based on community priorities. By expressing ecological changes in economic and social language, methods of valuing ecosystem services fill an informational gap between science and policy, thus enhancing an argument to actively manage use of bloom (Wainger & Mazzotta, 2011).

10.7 Management Implications

Alternatively, ecological valuation considers biophysical measures of service delivery. These could be decompositions of fish biomass (provisioning), the disappearance of submerged vegetation (habitat support), or a decrease in water clarity (regulating services). These indicators are then combined in composites, that is, standardized as indices, as the Ecological Integrity Index (EII) or Water Quality Index (WQI), to provide comparative measures among sites or over time (Yan et al.,

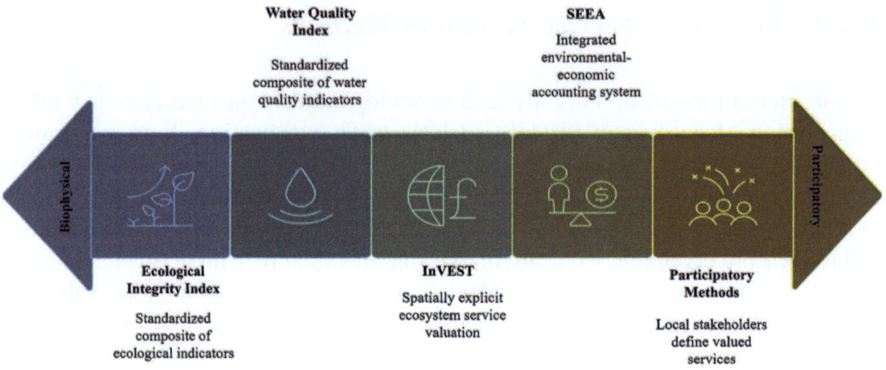

Fig. 10.4 Ecological valuation methods range from biophysical to participatory

2022). To assess the changes in services with bloom scenarios spatially and policy relevant tools, such as Integrated Valuation of Ecosystem Services and Trade-offs (InVEST) and the System of Environmental-Economic Accounting (SEEA), are emerging. These instruments aid in cost–benefit appraisals of administration interventions and aid in the choice of activities that present most ecological and societal return on investment activities. Furthermore, the participatory methods of assessment will involve local stakeholders in the process of defining the valued services, perceived impacts, and tolerable trade-offs so that valuation can contribute to equitable decision-making, which is based on community priorities (Asah & Blahna, 2020). By expressing ecological changes in economic and social languages, methods of valuing ecosystem services fill an informational gap between science and policy, thus enhancing an argument to actively manage use of bloom (Fig. 10.4).

10.7.1 Protecting Food Web Integrity

Preserving food web structure and functioning is one of the primary objectives of governing cyanobacterial blooms, because food webs are the main channel of energy flux and material cycling within aquatic systems and food webs form the basis of most ecosystem services. Blooms interfere with these networks by production of toxins, competition with phytoplankton, suppressing zooplanktons, and degradation of the habitat. Preserving food web integrity or its restoration takes the form of both the ecological and regulatory aspects of intervention. It all revolves around nutrient management. Lessening outside phosphorus and nitrogen recycling agricultural, urban drift and wastewater treatment plants is the way to curtail primary productivity and stratify the development of bloom (Wurtsbaugh et al., 2019). Nonetheless, strategies aiming at management of specific nutrients, as the management of dissolved reactive phosphorus (DRP) and stimulating more balanced N:P ratios, can be used to curb situations leading to precipitous advantage of cyanobacteria over other common and palatable phytoplankton. Another food web management tool is the biomanipulation. This would be by altering abundance of key species to bring back the ecological balance. As an example, grazing pressure by large-bodied zooplankton (e.g., *Daphnia*) can be intensified through fish stock management or decreasing planktivorous fish and in that way constrain phytoplankton bloom formation. Likewise, habitat structure can be enhanced by reintroducing or preserving submerged aquatic vegetation, which allows increasing the trophic diversity. Diverse food webs because of habitat restoration such as wetland rehabilitation, riparian buffer establishment, and sediment controls, nursery grounds, and stabilization of benthic communities, and water clarity (Geist & Hawkins, 2016). These measures also allow restoring the native flora and making it more resilient to another bloom. To safeguard the integrity of food webs, there is also a need to have monitor and early alarm. With bioindicators like the composition of zooplanktons, fish recruiting levels, and bioaccumulation of toxins, the managers will be able to recognize the alteration of food webs early and hence prevent catastrophic changes before it is too late (Stankovic & Stankovic, 2013).

10.7.2 Integrating Ecosystem Services in Water Governance

A common failure in traditional water governance is the lack of appreciation of ecosystem services in planning, policy, and economic evaluation. The occurrence of cyanobacterial blooms is a grim reminder that aquatic ecosystems do not only represent ecological systems, but they are also the very pinnacle of livelihoods, citizen health, cultural distinctive, and economic growth. The incorporation of ecosystem services into the governance structures is guaranteed that the additional benefit of healthy waters bodies will be appreciated and will survive. This will involve cross-sector coordination among water managers, government of states, communities, and farmers. Instruments like ecosystem service analysis, benefit costing, and multicriteria decision analysis (MCDA) are useful in determining trade-offs, investing priorities, and policy synchronization with the interests and needs of the society. Water policies will need to factor in not only the so-called provisioning services (e.g., drinking water, fisheries) but what are described as nonmarket services (e.g., recreation, biodiversity, climate regulation). As an example, cost–benefit analysis should incorporate the cost of the economic value of recreational losses during the bloom events to justify upstream nutrient reduction, wetland restoration, or the application of green infrastructure. Farmers, industries, or even municipalities can be rewarded by Payment for Ecosystem Services (PES) schemes to provide land and water practices that support the health of ecosystems. Nutrient runoff reduction/riparian zone conservation compensatory programs in the states that lack water bodies (such as the United States and the EU), there has been some promise in compensatory programs that pay landowners to minimize nutrient runoff or preserve riparian areas, helping to offset some pay off the risk of boom development and maintain sustainable economies of small farms. Also, equity issues should be included in governance structures. Indigenous communities, artisanal fishers, or low-income families are some of the most devastated communities due to bloom-related losses, which must be involved in the decision-making process. Participatory governance encourages trust, fosters local management, and improves legitimacy of management policies on water. Finally, the incorporation of ecosystem service frameworks to water governance allows to include the complete range of ecology, economic, and social values into the decision-making process, which makes the management strategies more sustainable, inclusive, and resilient (Jiménez et al., 2020).

10.7.3 Adaptive and Precautionary Strategies

The complexity, uncertainty, and evolving nature of cyanobacterial bloom dynamics demand a precautionary and adaptive approach to management. Such approaches emphasize learning-by-doing, flexibility, and proactive risk avoidance in the face of incomplete knowledge and rapidly changing environmental conditions. Adaptive management involves setting clear ecological goals, implementing interventions, monitoring outcomes, and adjusting strategies based on feedback. For instance, a watershed might implement phosphorus reduction measures, monitor bloom intensity and species shifts, and revise its strategy if internal loading or climate trends negate initial gains. This iterative process promotes evidence-based learning and long-term effectiveness. In the context of blooms, adaptive strategies may include Rotating aquatic vegetation zones to improve light penetration and prevent sediment resuspension. Implementing real-time monitoring systems using satellites, drones, and in situ sensors to track bloom formation and guide rapid response. Using decision-support models (e.g., hydrodynamic or biogeochemical models) to simulate bloom scenarios and evaluate policy options (Uusitalo et al., 2022).

Precautionary principle is the complement to adaptivity as it focuses more on preventing the possible serious or irreversible harm even though overall errors in scientific analyses are incomplete. That is particularly applicable in the case of bloom-related public health hazards (contamination of drinking water, accumulation of toxins in fisheries). Swift action can be taken by the authorities under precautionary focus on limiting water use or restrictions to access fishing because of indicators of early bloom or risk assessments. Also, strategies on climate adaption should be incorporated in management of bloom. Increasing temperatures, changes in rainfall, and unprecedented weather patterns promote the prevalence of cyanobacteria. Various mechanisms can minimize exposure to bloom in the future: creating floodplains that will absorb a surge of nutrients, expanding the capacity of reservoirs to mix thermally to dissolve nutrient loads, designing climate-resilient systems. Policy coherence also plays a main role. Due to the risk of a conflict between the goals, climate, agriculture, and biodiversity goals should be coordinated with water quality goals (e.g., to make sure that a spur to intensive production of biofuels crops does not increase the flow of nutrients into the ground). The aspect of interdisciplinary planning, cross-jurisdictional cooperation, scenario forecasting is required to gain synergistic benefits (Frusher et al., 2014). Supportive in any circumstance, adaptive governance becomes enhanced in the presence of public involvement and open communication. Community education of the risks of blooms, the trade-offs between bloom management and the individual contributions to pool likely behavior in communities (e.g., application of fertilizers, access of pets to bloom water) (Newton & Melaram, 2023).

10.8 Future Directions and Research Needs

A greater frequency, time, and spatial extent of cyanobacterial blooms emphasize the necessity to increase gaps in understanding, monitoring, and management reactions. Although much has been learnt about the proximate causes and ecological consequences of blooms, there are still many missing links between these short-term processes involving blooms and long-term food web processes, structural ecosystem losses, and social-ecological resilience. Multifactorial dynamics of blooms, including nutrient enrichment, climate changes, alterations in hydrology, and biotic interactions, require a multidisciplinary, prospective research approach. In this chapter, the three interrelated priorities of research are identified: (1) Ecological surveillance, (2) Multitrophic research integration, and (3) Counteracting the effects of climate change. The guidelines give a guideline on how to predict, prevent, and control any disruptions caused by blooms in water ecosystems while conserving the ecosystem services and biodiversity (Reid et al., 2019).

10.8.1 Long-Term Ecological Monitoring

The key to sustainable management of cyanobacterial bloom is healthy long-term ecological monitoring (LTEM) programs that have the capacity to track trends, assess effectiveness of interventions, and have enhanced predictive potential. Most of the existing monitoring systems are both temporally and geographically disaggregated and focused either on short water quality parameters or in a geographic scope. Future research should give significance to standardized datasets that cover long durations and incorporate biological, chemical as well as physical variables. The major items that should be considered are as follows: high-frequency remote sensing and satellite sensing data to determine the extent and intensity of the bloom. Noncyanobacterial competitors, grazers, and predators Time-series of plankton community composition, including noncyanobacterial competitors, grazers and predators. Indicators of ecosystem services, for example, fish yields, clarity of rivers and streams and recreation. Sediment and nutrient flux measurements to see the important aspects of internal loading and historic effects. The data that is long standing is required for the identification of delayed effects, feedback dynamics, and nonlinear ecological limits that can be missed by taking short-term studies. As an example, changes in the diversity of zooplankton after repeated exposure to blooms may not be realized unless on a decadal scale. In addition, the LTEM systems ought to be participatory and adaptive. The use of citizen science, community-based monitoring, and Indigenous knowledge will enable access to data-rich, spatially rich, and generally make people more involved. There are technologies that can increase monitoring at a reduced cost, which includes automated buoy sensors, unmanned aerial vehicles (UAVs), and eDNA sampling. Regional and global monitoring systems to coordinate measurement networks are important to share data and consensus on approach, and to allow cross-comparative analyses on freshwater and marine systems across climate gradients.

10.8.2 Integrating Multitrophic Approaches

The bloom of cyanobacteria is commonly examined through the primary production and nutrient processes, yet such incidents cause the ripple effect across all levels of trophic structure. It is a requirement that multitrophic research frameworks be developed, which would enable us to see the larger ecological impacts of the blooms and how to develop effective ecosystem-based management practices. In future, the following aspects should be given more importance: Trophic relationships including the impacts of cyanotoxin concentrations in zooplanktons on fish recruitment and health. Most behavioral and physiological adaptations of aquatic organisms to bloom stress, for example, changes in feeding, migration, and reproduction patterns. Food webs of microbes, the interaction of heterotrophic bacteria, and protozoa with cyanobacterial detritus and toxins. Experimental and modeling work that attempts to approximate whole-ecosystem conditions, by including fish and invertebrates as well as macrophytes and microbial communities, is crucial to forecast cascades and to measure resilience or weakness of various systems. The strategies are also able to identify key or functional groups of species that lessen or increase the impacts of the bloom. There is also the increasing pressure to know about functional redundancy and trophic compensation in bloom-affected systems. As an example, will there be replacements in the form of other species of grazers that will make up for the loss of *Daphnia*? Is the processing of bloom biomass by the benthic detritivores more efficient as opposed to pelagic pathways? The questions are essential in both the restoration of ecosystems and the formulation of biomanipulation. Nonlethal mechanisms through which multitrophic interaction can have effects over organismal fitness or growth and reproductive success should also be evaluated, allowing evaluation of nonlethal mechanism though exposure to sublethal toxins and shifts in food quality. The incorporation of ecotoxicology, physiology, and behavioral ecology will offer a more detailed overview of bloom effects than only they are instantaneous deaths.

10.8.3 Climate Change Considerations

Climatic change is quickly shifting the background surrounding the occurrence of cyanobacterial blooms and its establishment. An increase in the temperature of air and water, changes in precipitation, more runoff of storm water, and longer stratification of water columns, all contribute to bloom favorable conditions. As a result, climate change is not only an accelerator of the bloom frequency but also a transitioning effect, since it shifts the species balance, nutrient turnover, and ecosystem services. In the future studies, the projections about climate change must be clearly incorporated into the modeling of blooms, assessment of risks, and planning the management. The most important course are the Downscaled climate-bloom models to evaluate local bloom dynamics to regional variation in temperature and precipitation. Thermal tolerance and metabolic plasticity of Cyanobacteria that form blooms compared with other groups of phytoplankton. Investigation on compound strcssors, for example, the interacting warming, acidification, and nutrient enrichment on food webs. The knowledge on how the extreme events, such as droughts, floods, and hurricanes, transform the delivery of nutrients, internal loading, and hydrological residence times. In addition, there are some possibilities that climate change may change the geographical coverage, seasonal bloom, and extend such risks of cyanobacteria into an area where they have never been based or may alter the bloom phenology. All that requires advance alarm systems and seasonal predictions taking into consideration weather dynamics and long-terms climate pattern trends. Studies ought to assess also the manner by which modifications in land use, re-allocation of water, or re-engineering of reservoirs, among others, could significantly influence the risks of blooms inadvertently. As an example, enhanced runoff with irrigation or thermal stratification in deeper impoundments may confront even the severity of blooms unless taken care of beforehand. And finally, climate science, ecology, and social sciences need to be connected in interdisciplinary research so that the climate adaptation strategies are ecologically sound as well as socially just. The use of community-based resilience planning in the response of climate-bloom would be important in the future.

10.9 Conclusion

It is an important global ecological problem that cyanobacterial blooms have become a significant threat to the life of globally distributed aquatic ecosystems with serious impacts on food web processes, water quality, biodiversity, and the provision of countless ecosystem services. This chapter has discussed the complexity of the phenomenon of cyanobacterial bloom-starting with the ecological causes of the phenomenon and then entering the systemic consequences of the bloom on trophic relationships, ecosystem functioning, and human health. These flowers interfere with the aquatic food web by outgrowing other phytoplanktons, releasing toxic substances, inhibiting zooplanktons, and interfering with predator–prey relation. Consequently, primary productivity is inefficiently passed onto food chain to reduce fish populations and diversity in the food chain. At the same time, ecosystem services as the provision of drinking water, fisheries, recreation, and culture value are seriously affected. Break in nutrient cycle, emergence of hypoxic areas, and the detrimental ocean loss of submerged vegetation further contribute to ecological degradation and impair the resilience of the system. Tools of monitoring and assessment-tools aiming at comprehensive evaluations in a form of bioindicators, stable isotope analysis, and valued ecosystem services provide critical information on the dynamics of the blooms and decision-making. There is need to move toward ecosystem-based management where food web integrity, inclusion of ecosystem services in governance, and adaptive, precautionary measures will be provided. Moving forward, the monitoring of the ecology and multitrophic studies and climate-resistant planisphere beginner sayings will help to curb impending bloom menaces. With the rapidly increasing pace of climate change and the continuing anthropogenic pressures, interdisciplinary approach to cooperation, innovation, and active administration have never been as important. A multidimensional interpretation of the problem of cyanobacterial blooms and cascade consequences allows stakeholders and scientists to collaborate in protecting aquatic ecosystems and guaranteeing the delivery of the resulting services to the future generation.

References

Akhtar, N., Syakir Ishak, M. I., Bhawani, S. A., & Umar, K. (2021). Various natural and anthropogenic factors responsible for water quality degradation: A review. *Water, 13*(19), 2660.

Alp, M., & Cucherousset, J. (2022). Food webs speak of human impact: Using stable isotope-based tools to measure ecological consequences of environmental change. *Food Webs, 30*, Article e00218.

Amorim, C. A., & do Nascimento Moura, A. (2021). Ecological impacts of freshwater algal blooms on water quality, plankton biodiversity, structure, and ecosystem functioning. *Science of the Total Environment, 758*, 143605.

Anas, M., Falak, A., Hassan, S., Khattak, W. A., Saleem, M. H., Khan, K. A., Khalid, A., & Fahad, S. (2025a). Microbial interactions and bacterial responses to metal stress in plants: Mechanisms, adaptations, and applications for sustainable agriculture. *Journal of Crop Health, 77*(1), 36.

Anas, M., Khalid, A., Saleem, M. H., Ali Khan, K., Ahmed Khattak, W., & Fahad, S. (2025b). Symbiotic synergy: Unveiling plant-microbe interactions in stress adaptation. *Journal of Crop Health, 77*(1), 1–21.

Anas, M., Khattak, W. A., Sarfraz, W., Khalid, N., Naeem, N., Farid, M., Saleem, M. H., Khan, K. A., Khalid, A., & Ahmed, T. (2024). EDTA-enhanced phytoextraction: Leveraging Ricinus communis for the decontamination of Pb and Cd in agricultural soils. *Plant Growth Regulation*, 1–14.

Asah, S. T., & Blahna, D. J. (2020). Involving stakeholders' knowledge in co-designing social valuations of biodiversity and ecosystem services: Implications for decision-making. *Ecosystems, 23*, 324–337.

Barruffa, A. S., Sposito, V., & Faggian, R. (2021). Climate change and cyanobacteria harmful algae blooms: Adaptation practices for developing countries. *Marine and Freshwater Research, 72*(12), 1722–1734.

Beavis, S. G., Wong, V. N., Mosley, L. M., Baldwin, D. S., Latimer, J. O., Lane, P., & Lal, A. (2023). Water quality risks in the Murray-Darling basin. *Australasian Journal of Water Resources, 27*(1), 85–102.

Benbow, M. E., Receveur, J. P., & Lamberti, G. A. (2020). Death and decomposition in aquatic ecosystems. *Frontiers in Ecology and Evolution, 8*, 17.

Borah, D. K., Ahmadisharaf, E., Padmanabhan, G., Imen, S., & Mohamoud, Y. M. (2019). Watershed models for development and implementation of total maximum daily loads. *Journal of Hydrologic Engineering, 24*(1), 03118001.

Boyd, C. E., & Boyd, C. E. (2020). Microorganisms and water quality. *Water Quality: An Introduction*, 233–267.

Briland, R. (2018). *Evaluating the causes and consequences of ecosystem change in Lake Erie: From plankton to fish.*

Brouwer, R., Brander, L., Kuik, O., Papyrakis, E., & Bateman, I. (2013). A synthesis of approaches to assess and value ecosystem services in the EU in the context of TEEB. *VU University Amsterdam.*

Chakraborty, S. K., & Chakraborty, S. K. (2021). Trophic interactions and biogeochemical cycles in river ecosystem. *Riverine Ecology Volume 1: Eco-Functionality of the Physical Environment of Rivers*, 167–234.

Chen, L., Giesy, J. P., Adamovsky, O., Svirčev, Z., Meriluoto, J., Codd, G. A., Mijovic, B., Shi, T., Tuo, X., & Li, S.-C. (2021). Challenges of using blooms of Microcystis spp. In animal feeds: A comprehensive review of nutritional, toxicological and microbial health evaluation. *Science of the Total Environment, 764*, 142319.

Christensen, V. G., & Khan, E. (2020). Freshwater neurotoxins and concerns for human, animal, and ecosystem health: A review of anatoxin-a and saxitoxin. *Science of the Total Environment, 736*, Article 139515.

Cottingham, K. L., Ewing, H. A., Greer, M. L., Carey, C. C., & Weathers, K. C. (2015). Cyanobacteria as biological drivers of lake nitrogen and phosphorus cycling. *Ecosphere, 6*(1), 1–19.

Eddy, T. D., Bernhardt, J. R., Blanchard, J. L., Cheung, W. W., Colléter, M., Du Pontavice, H., Fulton, E. A., Gascuel, D., Kearney, K. A., & Petrik, C. M. (2021). Energy flow through marine ecosystems: Confronting transfer efficiency. *Trends in Ecology & Evolution, 36*(1), 76–86.

Frusher, S. D., Hobday, A. J., Jennings, S. M., Creighton, C., D'Silva, D., Haward, M., Holbrook, N. J., Nursey-Bray, M., Pecl, G. T., & van Putten, E. I. (2014). The short history of research in a marine climate change hotspot: From anecdote to adaptation in south-east Australia. *Reviews in Fish Biology and Fisheries, 24*, 593–611.

Gårdmark, A., Casini, M., Huss, M., Van Leeuwen, A., Hjelm, J., Persson, L., & de Roos, A. M. (2015). Regime shifts in exploited marine food webs: Detecting mechanisms underlying alternative stable states using size-structured community dynamics theory. *Philosophical Transactions of the Royal Society B: Biological Sciences, 370*(1659), 20130262.

Gebrekiros, S. T. (2016). Factors affecting stream fish community composition and habitat suitability. *Journal of Aquaculture and Marine Biology, 4*(2), 00076.

Geist, J., & Hawkins, S. J. (2016). Habitat recovery and restoration in aquatic ecosystems: Current progress and future challenges. *Aquatic Conservation: Marine and Freshwater Ecosystems, 26*(5), 942–962.

Ger, K. A., Hansson, L., & Lürling, M. (2014). Understanding cyanobacteria-zooplankton interactions in a more eutrophic world. *Freshwater Biology, 59*(9), 1783–1798.

Ger, K. A., Urrutia-Cordero, P., Frost, P. C., Hansson, L.-A., Sarnelle, O., Wilson, A. E., & Lürling, M. (2016). The interaction between cyanobacteria and zooplankton in a more eutrophic world. *Harmful Algae, 54*, 128–144.

Hamilton, D. P., Salmaso, N., & Paerl, H. W. (2016). Mitigating harmful cyanobacterial blooms: Strategies for control of nitrogen and phosphorus loads. *Aquatic Ecology, 50*, 351–366.

Hammerschlag, N., Schmitz, O. J., Flecker, A. S., Lafferty, K. D., Sih, A., Atwood, T. B., Gallagher, A. J., Irschick, D. J., Skubel, R., & Cooke, S. J. (2019). Ecosystem function and services of aquatic predators in the Anthropocene. *Trends in Ecology & Evolution, 34*(4), 369–383.

Huisman, J., Codd, G. A., Paerl, H. W., Ibelings, B. W., Verspagen, J. M., & Visser, P. M. (2018). Cyanobacterial blooms. *Nature Reviews Microbiology, 16*(8), 471–483.

Jiménez, A., Saikia, P., Giné, R., Avello, P., Leten, J., Liss Lymer, B., Schneider, K., & Ward, R. (2020). Unpacking water governance: A framework for practitioners. *Water, 12*(3), 827.

Kazmi, S. S. U. H., Yapa, N., Karunarathna, S. C., & Suwannarach, N. (2022). Perceived intensification in harmful algal blooms is a wave of cumulative threat to the aquatic ecosystems. *Biology, 11*(6), 852.

Li, T., Fan, X., Cai, M., Jiang, Y., Wang, Y., He, P., Ni, J., Mo, A., Peng, C., & Liu, J. (2023). Advances in investigating microcystin-induced liver toxicity and underlying mechanisms. *Science of the Total Environment, 905*, Article 167167.

Li, Y., Geng, M., Yu, J., Du, Y., Xu, M., Zhang, W., Wang, J., Su, H., Wang, R., & Chen, F. (2022). Eutrophication decrease compositional dissimilarity in freshwater plankton communities. *Science of the Total Environment, 821*, Article 153434.

McCue, M. D., Javal, M., Clusella-Trullas, S., Le Roux, J. J., Jackson, M. C., Ellis, A. G., Richardson, D. M., Valentine, A. J., & Terblanche, J. S. (2020). Using stable isotope analysis to answer fundamental questions in invasion ecology: Progress and prospects. *Methods in Ecology and Evolution, 11*(2), 196–214.

McGenity, T. J., Gessesse, A., Hallsworth, J. E., Garcia Cela, E., Verheecke-Vaessen, C., Wang, F., Chavarría, M., Haggblom, M. M., Molin, S., & Danchin, A. (2020). Visualizing the invisible: Class excursions to ignite children's enthusiasm for microbes. *Microbial Biotechnology, 13*(4), 844–887.

Meerhoff, M., & de los Ángeles González-Sagrario, M. (2022). Habitat complexity in shallow lakes and ponds: Importance, threats, and potential for restoration. *Hydrobiologia, 849*(17), 3737–3760.

Mihalitsis, M., Hemingson, C. R., Goatley, C. H., & Bellwood, D. R. (2021). The role of fishes as food: A functional perspective on predator–prey interactions. *Functional Ecology, 35*(5), 1109–1119.

Moustaka-Gouni, M., & Sommer, U. (2020). Effects of harmful blooms of large-sized and colonial cyanobacteria on aquatic food webs. *Water, 12*(6), 1587.

Munkes, B., Löptien, U., & Dietze, H. (2021). Cyanobacteria blooms in the Baltic Sea: A review of models and facts. *Biogeosciences, 18*(7), 2347–2378.

Newton, A. R., & Melaram, R. (2023). Harmful algal blooms in agricultural irrigation: Risks, benefits, and management. *Frontiers in Water, 5*, 1325300.

Nunn, A. D., Tewson, L. H., & Cowx, I. G. (2012). The foraging ecology of larval and juvenile fishes. *Reviews in Fish Biology and Fisheries, 22*, 377–408.

O'Leary, B. C., & Roberts, C. M. (2018). Ecological connectivity across ocean depths: Implications for protected area design. *Global Ecology and Conservation, 15*, Article e00431.

Oliver, T. H., Heard, M. S., Isaac, N. J., Roy, D. B., Procter, D., Eigenbrod, F., Freckleton, R., Hector, A., Orme, C. D. L., & Petchey, O. L. (2015). Biodiversity and resilience of ecosystem functions. *Trends in Ecology & Evolution, 30*(11), 673–684.

Paerl, H. W. (2018). Mitigating toxic planktonic cyanobacterial blooms in aquatic ecosystems facing increasing anthropogenic and climatic pressures. *Toxins, 10*(2), 76.

Paerl, H. W., & Paul, V. J. (2012). Climate change: Links to global expansion of harmful cyanobacteria. *Water Research, 46*(5), 1349–1363.

Pelletier, M. C., Ebersole, J., Mulvaney, K., Rashleigh, B., Gutierrez, M. N., Chintala, M., Kuhn, A., Molina, M., Bagley, M., & Lane, C. (2020). Resilience of aquatic systems: Review and management implications. *Aquatic Sciences, 82*, 1–25.

Piatka, D. R., Frank, A. H., Köhler, I., Castiglione, K., van Geldern, R., & Barth, J. A. (2022). Balance of carbon species combined with stable isotope ratios show critical switch towards bicarbonate uptake during cyanobacteria blooms. *Science of the Total Environment, 807*, Article 151067.

Qiu, Y., Huang, J., Luo, J., Xiao, Q., Shen, M., Xiao, P., Peng, Z., Jiao, Y., & Duan, H. (2025). Monitoring, simulation and early warning of cyanobacterial harmful algal blooms: An upgraded framework for eutrophic lakes. *Environmental Research, 264*, Article 120296.

Reich, P., & Lake, P. S. (2015). Extreme hydrological events and the ecological restoration of flowing waters. *Freshwater Biology, 60*(12), 2639–2652.

Reid, A. J., Carlson, A. K., Creed, I. F., Eliason, E. J., Gell, P. A., Johnson, P. T., Kidd, K. A., MacCormack, T. J., Olden, J. D., & Ormerod, S. J. (2019). Emerging threats and persistent conservation challenges for freshwater biodiversity. *Biological Reviews, 94*(3), 849–873.

Saber, A. A., El-Refaey, A. A., Saber, H., Singh, P., van Vuuren, S. J., & Cantonati, M. (2022). Cyanoprokaryotes and algae: Classification and habitats. In *Handbook of Algal Biofuels* (pp. 1–38). Elsevier.

Sahu, A. K., Mir, S., Nayak, B., & Baitharu, I. (2024). Sustainable management of eutrophication and problems associated with the algal toxin in ponds and lakes of rural areas. In *Water resources management for rural development* (pp. 155–170). Elsevier.

Samarajeewa, U. (2023). Emerging challenges in maintaining marine food-fish availability and food safety. *Comprehensive Reviews in Food Science and Food Safety, 22*(6), 4734–4757.

Shanafelt, D. W., & Loreau, M. (2018). Stability trophic cascades in food chains. *Royal Society Open Science, 5*(11), Article 180995.

Sheergojri, I. A., Rashid, I., Aneaus, S., Rashid, I., Qureshi, A. A., & Rehman, I. (2023). Enhancing the social-ecological resilience of an urban lake for sustainable management. *Environment, Development and Sustainability*, 1–26.

Sherman, B. H. (2017). Multiple marine ecological disturbance assessments for latin american and caribbean large marine ecosystems. *Environmental Development, 22*, 129–142.

Shukla, K., Shukla, S., Upadhyay, D., Singh, V., Mishra, A., & Jindal, T. (2021). Socio-economic assessment of climate change impact on biodiversity and ecosystem services. *Climate Change and the Microbiome: Sustenance of the Ecosphere*, 661–694.

Small, N., Munday, M., & Durance, I. (2017). The challenge of valuing ecosystem services that have no material benefits. *Global Environmental Change, 44*, 57–67.

Stankovic, S., & Stankovic, A. R. (2013). Bioindicators of toxic metals. *Green Materials for Energy, Products and Depollution*, 151–228.

Svenning, J.-C., Munk, M., & Schweiger, A. (2019). Trophic rewilding: Ecological restoration of top-down trophic interactions to promote self-regulating biodiverse ecosystems. *Rewilding, 26*, 66.

Terborgh, J. W. (2015). Toward a trophic theory of species diversity. *Proceedings of the National Academy of Sciences, 112*(37), 11415–11422.

Tully, K., & Ryals, R. (2017). Nutrient cycling in agroecosystems: Balancing food and environmental objectives. *Agroecology and Sustainable Food Systems, 41*(7), 761–798.

Ullah, H., Nagelkerken, I., Goldenberg, S. U., & Fordham, D. A. (2018). Climate change could drive marine food web collapse through altered trophic flows and cyanobacterial proliferation. *PLoS Biology, 16*(1), Article e2003446.

Uusitalo, L., Blenckner, T., Puntila-Dodd, R., Skyttä, A., Jernberg, S., Voss, R., Müller-Karulis, B., Tomczak, M. T., Möllmann, C., & Peltonen, H. (2022). Integrating diverse model results into decision support for good environmental status and blue growth. *Science of the Total Environment, 806*, Article 150450.

Wainger, L., & Mazzotta, M. (2011). Realizing the potential of ecosystem services: A framework for relating ecological changes to economic benefits. *Environmental Management, 48*, 710–733.

Wang, G., Mang, S., Cai, H., Liu, S., Zhang, Z., Wang, L., & Innes, J. L. (2016). Integrated watershed management: Evolution, development and emerging trends. *Journal of Forestry Research, 27*, 967–994.

Weber, S. J., Mishra, D. R., Wilde, S. B., & Kramer, E. (2020). Risks for cyanobacterial harmful algal blooms due to land management and climate interactions. *Science of the Total Environment, 703*, Article 134608.

Weißhuhn, P., Müller, F., & Wiggering, H. (2018). Ecosystem vulnerability review: Proposal of an interdisciplinary ecosystem assessment approach. *Environmental Management, 61*, 904–915.

Willis, C., Papathanasopoulou, E., Russel, D., & Artioli, Y. (2018). Harmful algal blooms: The impacts on cultural ecosystem services and human well-being in a case study setting, Cornwall, UK. *Marine Policy, 97*, 232–238.

Wurtsbaugh, W. A., Paerl, H. W., & Dodds, W. K. (2019). Nutrients, eutrophication and harmful algal blooms along the freshwater to marine continuum. *Wiley Interdisciplinary Reviews: Water, 6*(5), Article e1373.

Yan, T., Shen, S.-L., & Zhou, A. (2022). Indices and models of surface water quality assessment: Review and perspectives. *Environmental Pollution, 308*, Article 119611.

Yin, C., Zhao, W., Cherubini, F., & Pereira, P. (2021). Integrate ecosystem services into socio-economic development to enhance achievement of sustainable development goals in the post-pandemic era. *Geography and Sustainability, 2*(1), 68–73.

Zahir, M., Su, Y., Shahzad, M. I., Ayub, G., Rehman, S. U., & Ijaz, J. (2024). A review on monitoring, forecasting, and early warning of harmful algal bloom. *Aquaculture*, 741351.

Zhang, M., Zhang, Y., Zhou, Y., Zhang, Y., Shi, K., & Jiang, C. (2022). Influence of cyanobacterial bloom accumulation and dissipation on underwater light attenuation in a large and shallow lake. *Environmental Science and Pollution Research, 29*(52), 79082–79094.

Zhu, M., Zhu, G., Zhao, L., Yao, X., Zhang, Y., Gao, G., & Qin, B. (2013). Influence of algal bloom degradation on nutrient release at the sediment–water interface in Lake Taihu, China. *Environmental Science and Pollution Research, 20*, 1803–1811.

Chapter 11
Climate Change and Its Effects on Cyanobacterial Bloom Frequency and Distribution

Muhammad Anas, Waseem Ahmed Khattak, Javed Iqbal, Banzeer Ahsan Abbasi, and Shah Fahad

Abstract Cyanobacterial blooms are increasingly frequent and severe in freshwater ecosystems worldwide, driven by multiple climate-change-related factors. Rising global temperatures, altered precipitation patterns, prolonged water stratification, and elevated atmospheric CO_2 have created conditions that strongly favor bloom-forming cyanobacteria. These changes have not only enhanced bloom frequency and duration but also facilitated the expansion of blooms into previously unaffected geographic areas, including oligotrophic lakes, high-altitude regions, and polar ecosystems. Cyanobacteria possess adaptive traits such as thermal tolerance, nitrogen fixation, and allelopathic capabilities that allow them to dominate over other phytoplankton groups under these changing conditions. This chapter explores the complex interplay between climatic variables and bloom dynamics, emphasizing physiological responses, ecological shifts, and the resulting risks to ecosystem health, biodiversity, and human well-being. The chapter also examines emerging monitoring technologies, such as remote sensing and predictive modeling, alongside mitigation strategies like nutrient load reduction, hydrological control, and adaptive water resource management. Public health risks from cyanotoxins, economic losses due

M. Anas
Department of Soil Science and Plant Nutrition, Selçuk University, Konya, Türkiye
e-mail: muhammed.anas@selcuk.edu.tr

W. A. Khattak
Department of Plant Sciences, Faculty of Biological Sciences, Quaid-i-Azam University, Islamabad, Pakistan
e-mail: waseemkhattack@bs.qau.edu.pk

J. Iqbal (✉)
Department of Botany, Bacha Khan University, Charsadda, Khyber Pakhtunkhwa, Pakistan
e-mail: javed89qau@gmail.com

B. A. Abbasi
Department of Botany, Rawalpindi Women University, Satellite Town, Rawalpindi, Pakistan

S. Fahad
Department of Agronomy, Abdul Wali Khan University Mardan, Mardan, Khyber Pakhtunkhwa, Pakistan

S. Fahad et al. (eds.), *Cyanobacterial Blooms: Ecology, Evolution and Biogeochemical Impacts*, https://doi.org/10.1007/978-3-032-06042-6_11

to water quality degradation, and the need for interdisciplinary, climate-resilient management approaches are critically discussed. As climate pressures mount, the understanding of cyanobacteria's response to environmental drivers becomes essential. Integrated strategies that combine science, community engagement, and policy are key to minimizing bloom impacts and ensuring the sustainability of freshwater ecosystems.

Keywords Cyanobacterial blooms · Climate change · Freshwater ecosystems · Harmful algal blooms (HABs) · Eutrophication

11.1 Introduction

Cyanobacteria, commonly called blue-green algae, are a large group of photosynthetic prokaryotes inhabiting nearly all aquatic ecosystems. They are important primary producers and essential components in biogeochemical cycling, but some cyanobacteria group can proliferate into high-density bloom forming an ecological and health alarm. Harmful cyanobacterial blooms (HCBs) produce toxins, such as microcystins and cylindrospermopsin, which can greatly impact aquatic biodiversity, drinking water safety, and public health. In the past few decades, cyanobacterial bloom distribution in space and time has increased considerably in frequency, extent, and intensity. This increase was identified alongside anthropogenic climate change, and associating the link between cyanobacterial blooms and synoptic changes in global environments is staggering—but makes sense. All of the aspects of climate change (warming, alterations in precipitation regimes, and elevations of atmospheric CO_2) taken together create a perfect "storm" for the formation and re-formation of blooms. This chapter describes the relationships between climate change and the bloom dynamics of cyanobacteria—also summarizing current understanding of environmental drivers, reactions of organisms, and the wider consequences of these biological interactions, roughly. The aim of this chapter is to gain as comprehensive an understanding of the mechanisms involved in expanding blooms in a warming world, and to advocate for integrated management solutions to associated risks.

11.2 Climate-Driven Environmental Changes

Climate change has significant impacts on freshwater ecosystems and is creating environmental conditions that increasingly favor cyanobacterial dominance. Several interconnected subclimate variables (e.g., increasing temperatures, changing precipitation, and increased atmospheric CO_2) are changing ecosystem dynamics, nutrient availability, and the stability of the water column, all of which subsequently affect cyanobacterial physiology and bloom potential directly or indirectly (Ma & Wang, 2021).

11.2.1 Rising Temperatures

During the past century, global surface temperatures have increased dramatically, generating thermal stress on aquatic ecosystems. Cyanobacteria have a greater capability to tolerate elevated temperature ranges than most eukaryotic phytoplankton. Many cyanobacterial bloom-forming genera (e.g., *Microcystis, Dolichospermum,* and *Planktothrix*) have comparatively optimal growth temperatures exceeding 25 °C, prevailing under warming climate scenarios (Kumari et al., 2022). High temperatures not only increase rates of metabolism/reproduction in cyanobacteria, but also their ability to outcompete other phytoplankton, especially when nutrients are additionally available. Warmer water sometimes has lower dissolved oxygen content that can lead to hypoxia or anoxia at the bottom, releasing nutrients from sediments (internal loading), particularly phosphorus, and driving eutrophication and bloom maintenance (Fig. 11.1) (Howarth et al., 2011).

11.2.2 Altered Precipitation and Hydrological Patterns

Climate change is also affecting the global hydrological cycles, which may result in more intense, less predictable rainfall events, more frequent droughts, and more varied river flows. Ultimately, these changes will affect the dynamics of cyanobacterial blooms differently based upon the local context. Intense rain and runoff create nutrient inputs from both urban and agricultural areas, which can create nutrient pulses in lakes and reservoirs, causing blooms (Coffey et al., 2019). In contrast, low flow/drought can inhibit water turnover and increase water retention time that enhances stability and stratification of the water column and thus favoring bloom development and limiting nutrient dilution. In addition, less flushing in dry seasons allows for populations of cyanobacteria to build up and dominate the system. Retention time, the interaction of pulsed nutrient inputs, and longer retention time create an overall regime highly favorable to bloom production, especially shallow or slow-moving water bodies (Uk et al., 2022).

11.2.3 Elevated Atmospheric CO_2 and Nutrient Interactions

The level of CO_2 in the atmosphere has surpassed 420 ppm and is continuing to rise, which is concerning for photosynthetic organisms in general (Kabir et al., 2023). Although cyanobacteria possess CCMs, which can help them compete in low CO_2 environments, the presence of an increasing levels of CO_2 can provide a lower energetic demand for carbon acquisition for numerous species, which may facilitate the future growth of bloom-forming cyanobacteria. Significantly, different species either

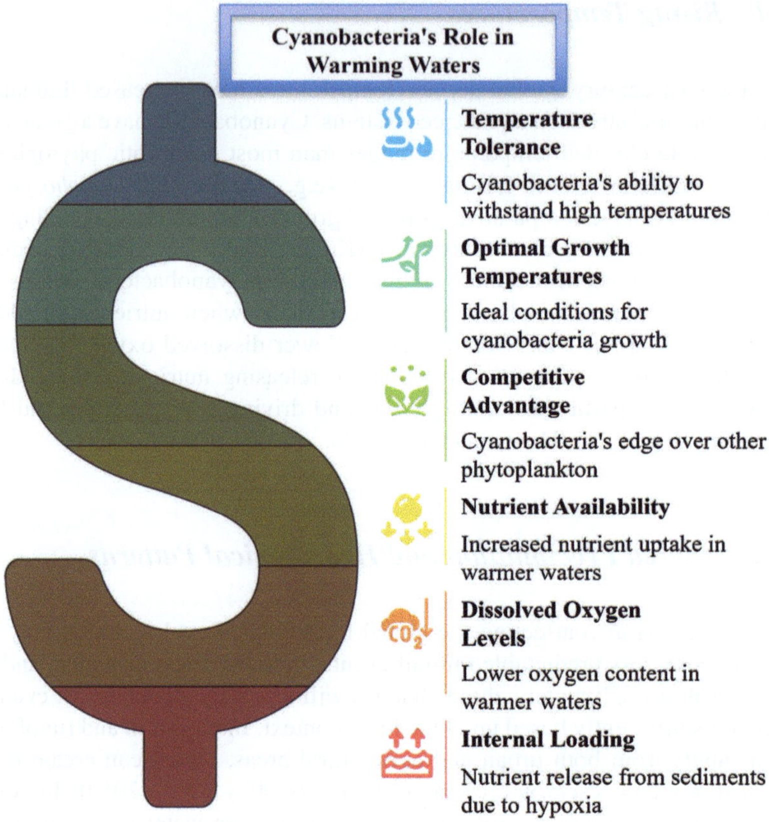

Fig. 11.1 Key factors contributing to cyanobacteria's dominance in warming aquatic environments, including thermal tolerance, enhanced nutrient availability, reduced oxygen levels, and internal loading—highlighting their ecological advantage in a changing climate

or both benefit from increased localized CO_2 concentrations. As an example, non-N_2-fixing genera such as *Microcystis* may benefit from direct CO_2 enrichment, whereas diazotrophic genera like *Dolichospermum* can act in low-nitrogen environments via atmospheric N-fixation, but only in environments with an abundance of phosphorus. Cyanobacteria can utilize nutrients in various and flexible ways to adequately express physiological functions in the face of an inevitable stoichiometric imbalance, particularly as climate change shifts nutrient fluxes and availability following precipitations (Nürnberg, 2024). Increased CO_2 also has an impact on water chemistry through its effects on pH and carbonate interactions, which may have community impacts by influencing community composition and cyanotoxin production. These chemical effects have an additional influence through the conditioning of thermal and hydrological regimes, which can also have influenced the dynamics of blooms in ways that are not fully understood yet (Paerl et al., 2011).

11.3 Cyanobacterial Responses to Climate Change

Cyanobacteria possess a collection of adaptive features that allow them to thrive under the environmental conditions amplified by climate change. Their physiological ability to switch traits, competitive traits, and ecological buffer traits promote their distinctiveness and superiority above other phytoplankton groups. As climate variables are changing, cyanobacteria are not only able to survive but are becoming dominant in freshwater systems, resulting in more frequent, longer duration, and extensive blooms (Carey et al., 2012).

11.3.1 Enhanced Growth and Bloom Frequency

With warm temperatures and increased light availability, many bloom-forming cyanobacteria enhance their cellular growth, producing faster population expansion. Cyanobacteria such as *Microcystis aeruginosa* and *Aphanizomenon flo-aquae* have high temperature optima and can proliferate rapidly during growing periods when water temperatures are over 25–30 °C, especially in stagnant or stratified waters (Burch et al., 2021). Additionally, warming during the spring has increased growing seasons, delaying fall cooling, increasing the amount of time available for development of blooms. This not only extends the length of the bloom but increases the likelihood that blooms will occur multiple times in a single season. In some regions, blooms now last into late autumn, which is a rare occurrence winding back through record histories of some of these blooms. It is a positive feedback loop driven by increases caused by elevated temperature, concentrations of CO_2, and nutrient addition, that cause blooms to occur frequently when once how via several natural processes productive ecosystems may have never bloomed. Once blooming happens, blooms tend to shade all phytoplankton competing for the same light in the water column, and while blooms are established they likely will persist with or without toxin release and allelopathic properties (Granéli et al., 2008).

11.3.2 Shifts in Bloom Timing and Duration

One observed consequence of climate change is phenological shifts in bloom timing. For example, both earlier blooms and later die-offs are increasingly accepted, altering the seasonality of freshwater ecosystems. These shifts can influence the structure of freshwater food webs, nutrient cycling, and the ecosystem's service provision. Warmer spring temperatures and earlier ice-off scenarios facilitate cyanobacterial growth earlier in the season, with decreased phytoplankton succession over time. Diatoms often dominate early season blooms, while cyanobacteria can dominate early season blooms, therefore disrupting migration timing and zooplankton recruitment potentially limiting grazing controls for subsequent freshwater spring bloom contributions. Additionally, prolonged blooming events will lead to hypoxia of bottom waters as decomposition of accumulated blooms increases oxygen consumption. This not only impractically reduces benthic habitat quality but also releases phosphorus from sediments potentially sustaining bloom conditions for extended periods (Wurtsbaugh et al., 2019).

11.3.3 Expansion into New Geographic Areas

Cyanobacterial blooms now extend beyond eutrophic and temperate freshwater systems. As climate continues to change, bloom-forming species are spreading into previously unaffected areas, including oligotrophic lakes, high-altitude water reservoirs, and into polar ecosystems entirely. Higher latitudes and altitudes that have warming waters have facilitated the poleward migration of many cyanobacterial taxa. For example, there have been increasing reports of *Dolichospermum* and *Nodularia* species in Arctic and sub-Arctic freshwater systems (Søgaard et al., 2021). In tropical and subtropical regions, warmer waters have exacerbated bloom severity, including in natural and anthropogenically created aquatic systems. In addition, invasive or non-native cyanobacterial strains that are being dispersed to different locations through ballast water, aquaculture, and migratory birds, can also quickly establish a foothold in newly affected areas if environmental, climatic, or nutrient conditions are conducive to their metabolic needs, thus displacing native phytoplankton communities; and potentially introducing new toxin risks (Table 11.1) (Smayda, 2019).

11.4 Ecological and Health Impacts

Under climate change, the proliferation of cyanobacterial blooms has serious implications for ecosystems and their stability, biodiversity, and public health. Blooms have the potential to modify entire food web structures, significantly increase nutrient loading to water bodies, and release toxins that have consequences for terrestrial and aquatic organisms. With an increase in bloom events occurring more frequently and affecting wider geographical areas, the impacts to ecosystem services and health are becoming more pronounced (Willis et al., 2018).

Table 11.1 Cyanobacterial responses to climate change, highlighting their physiological and ecological adaptations that enhance bloom formation, alter bloom timing, and facilitate expansion into new habitats

No	Response type	Description	Example species/notes	References
1	Elevated temperature tolerance	Thrive in water > 25–30 °C	*Microcystis aeruginosa, Aphanizomenon flos-aquae*	Paerl and Huisman (2008)
2	Accelerated cell division	Rapid growth in warm, stratified water	Competitive dominance over phytoplankton	Carey et al., (2012)
3	Extended growing season	Blooms begin earlier and last longer	Observed in temperate zones	Winder and Sommer (2012)
4	Multiple bloom events	More than one bloom per year	Shallow eutrophic lakes	O'Neil et al. (2012)
5	Increased CO_2 utilization	More efficient carbon fixation	Favors nondiazotrophs	Visser et al. (2016)
6	Positive feedback loop	Self-enhancing water conditions	Promotes bloom recurrence	Paerl and Huisman (2008)
7	Allelopathic suppression	Releases compounds to inhibit competitors	Stabilizes dominance	Huisman et al. (2018)
8	Shift in bloom onset	Blooms occur earlier in spring	Alters seasonal succession	Winder and Sommer (2012)
9	Delayed bloom decline	Blooms last into autumn	Prolonged exposure	Carey et al., (2012)
10	Disruption of food web timing	Zooplanktons lag behind bloom cycles	Less grazing control	Winder and Sommer (2012)
11	Prolonged hypoxia	Oxygen depletion lingers post bloom	Affects benthic fauna	Moss et al. (2011)
12	Sediment nutrient release	Anoxia mobilizes phosphorus	Feeds future blooms	Moss et al. (2011)
13	Expansion into oligotrophic systems	Blooms in previously nutrient-poor waters	Regime shift indicator	Reinl et al. (2021)
14	Colonization of high-altitude lakes	Appearance in previously too-cold regions	Elevation shift	Carey et al., (2012)
15	Spread into polar regions	Bloom presence in Arctic systems	*Dolichospermum, Nodularia*	Woodhouse et al. (2016)
16	Bloom intensification in tropics	Higher temperatures intensify blooms	Stronger, longer blooms	O'Neil et al. (2012)

(continued)

Table 11.1 (continued)

No	Response type	Description	Example species/notes	References
17	Invasive species spread	Introduced via ballast, birds, aquaculture	Displace native cyanobacteria	Carey et al., (2012)
18	New toxin profiles	Toxic strains emerge in new areas	Public health risks	Huisman et al. (2018)
19	Climate-nutrient synergy	Warming + eutrophication = stronger blooms	Synergistic driver	Moss et al. (2011)
20	Reduced seasonal ice cover	Earlier growth due to shorter ice duration	Extends active season	Carey et al., (2012)

11.4.1 Effects on Aquatic Ecosystems

Cyanobacterial blooms disrupt the natural operability of freshwater systems by taking a competitive advantage over other phytoplanktons, shading out submergent vegetation, and changing nutrient cycling. The presence of a bloom-forming species suppresses phytoplankton diversity and removes primary producers on which herbivorous zooplankton and fish larvae thrive (Moustaka-Gouni & Sommer, 2020). When dense blooms are subsequently decomposed, oxygen is consumed, which can lead to hypoxia or anoxia. This is a more severe risk in stratified lakes and reservoirs, particularly when there are layered bands to oxygen depletion, which is followed by rapid death of fishes, loss of benthic invertebrate diversity, and collapse of overall bottom-dwelling communities. Further, oxygen-depleted sediments release trapped phosphorus back into the system and together this leads to a feedback loop to the state of eutrophication. Some cyanobacteria are able to fix nitrogen from the atmosphere, increasing nitrogen sources in already impaired ecosystems and subsequently increasing cyanobacterial blooms. These changes unfavorably alter trophic connections, decrease transparency of the water, and transform the characteristics of habitats into low-resilience monocultures (Tiebel, 2020).

11.4.2 Cyanotoxin Risks to Public Health

Numerous cyanobacterial species manufacture harmful toxins—including microcystins, anatoxins, saxitoxins, and cylindrospermopsin—that threaten human and animal health. The generated toxins can be taken up by drinking water, and recreational waters (e.g., lakes, oceans, and swimming pools), or can be taken up through aquatic food chains. The conditions in which exposure may occur include ingestion, inhalation, or dermal contact through direct initiation (Shukla et al., 2024). Short-term acute health impacts could include modest skin irritation, gastrointestinal illness, or increased risk of liver damage and neurological damage. Long-term chronic exposure (with low concentrations) to cyanotoxins could lead to a higher risk of liver cancer, or other long-term chronic health impacts. Vulnerable populations such as children, the elderly, immunocompromised individuals, etc., may be particularly sensitive to continued exposures from cyanobacterial blooms (Chatterjee & More, 2023). These are public health issues known worldwide, but they may be worse in under-resourced areas where there may be no drinking water treatment plant, or the treatment plant may not be able to remove cyanotoxins from source water. The bloom also typically occurs in rural areas where people may rely on untreated or minimally treated surface water, that is, drinking or irrigation (Fig. 11.2) (Tariq & Mushtaq, 2023).

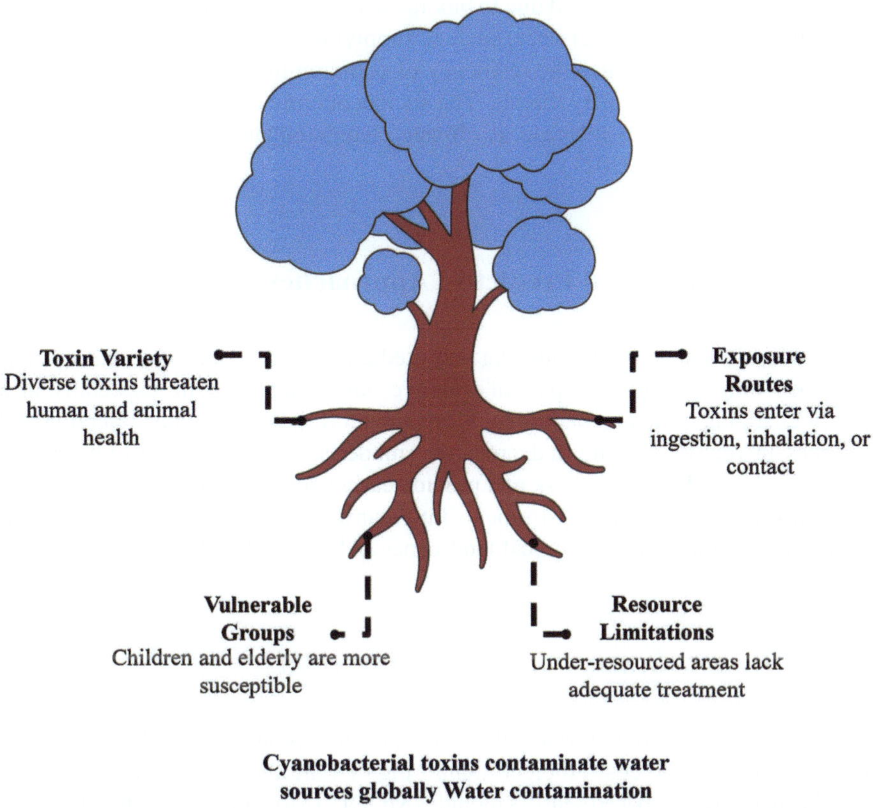

Fig. 11.2 Overview of the public health challenges posed by cyanobacterial toxins, highlighting their diversity, exposure pathways, impacts on vulnerable populations, and the disparities in treatment capacity underscoring the global issue of water contamination

11.4.3 Economic and Recreational Consequences

Apart from ecological and health-related consequences, cyanobacterial blooms have serious economic costs as they disrupt commercial and recreational fishery stocks, decrease tourism revenues, and increase water treatment costs. Dissolved organics and harmful algal blooms can not only ruin the aesthetics of water bodies, such as the presence of visible scums or smells, and dead fish, but they also limit recreational use activities such as swimming, boating, and fishing. Parties may be required to invest in costly monitoring systems, add-on or upgrade their treatments, or develop alternative water supplies for municipal purposes when blooms happen. The agricultural marketplace that uses surface water for irrigation or livestock water is also impacted if surface water is contaminated due to blooms in their supply system. The long-term degradation of ecosystem services, such as supplying clean water and maintaining biodiversity, can also reduce the resiliency of communities to environmental stressors, leading to food security threats. The socioeconomic burden of cyanobacterial blooms is very likely to increase as climate change enhances bloom drivers and impacts (Barruffa et al., 2021).

11.5 Monitoring and Predictive Approaches

Being able to manage the monitoring and prediction of cyanobacterial blooms has developed into an important part of water resource management, especially in a time of changing climatic conditions. More and more, real-time data collection and advances in remote sensing and predictive modeling are available to understand bloom characteristics, assess risk and provide early warnings, and develop intercession plans. Integrating these instruments into many ecosystems will be complicated by different climates and ecological landscapes (Albert et al., 2014).

11.5.1 Remote Sensing and Bloom Detection

Remote sensing has transformed the way the blooms are detected as the large volume, frequent, and cost-efficient monitoring of water bodies is possible with its help. Typical satellite-based sensors includes MODIS (Moderate Resolution Imaging Spectroradiometer), Sentinel-2, Landsat, which are commonly used to detect changes in the water color, chlorophyll-a level, phycocyanin signal, which are the major indicators of the presence of cyanobacteria (Obaid et al., 2021). These instruments give the possibility of monitoring the progression of bloom during a period. Use of high-resolution aerial imagery by drones is also coming in effect to allow gap-less surveillance of small or inaccessible water bodies in a more detailed manner. Also handheld fluorometers and in situ probes with pigment-specific sensors give real-time data to be used in conjunction with satellite data (Bertone et al., 2018). Remote sensing tools come with their limitations despite their usefulness. An issue that can impair accuracy is cloud cover, water turbidity, and challenge to differentiate between cyanobacteria and other algae. Furthermore, such technologies tend to reveal surface blooms, yet subsurface or benthic cyanobacteria are not revealed (Saleem et al., 2023).

11.5.2 Modeling Future Bloom Scenarios

The predictive models to identify the probability of a bloom occurrence as well as evaluate the risk of the occurrence in the future focusing on various climate and management conditions are important. Such models include empirical-statistical models of land-surface process through to sophisticated process-based simulations that combine physical, chemical, and biological processes (Sidle, 2021). CE-QUAL-W2, AQUATOX, and Bayesian networks are some of the models that are employed in simulating occurrences of bloom depending on various inputs such as temperature, nutrient concentrations, availability of light, and hydrodynamics. Introducing these projections of climate into such models assists in predicting the eventual changes in the frequency and intensity of bloom on an annual basis over the next few decades (Ralston & Moore, 2020). Some new models also add a machine learning part trained on historical bloom data so as to increase accuracy in the predictions. Nevertheless, difficulty in parameters, limited data, and complexity of the ecosystem commonly compromise the accuracy of the model application. Accordingly, model outputs must be considered quite carefully and should be combined with the field data and expert knowledge (Zhu et al., 2018).

11.6 Mitigation and Management Strategies

Traditional reactive strategies were able to deal with the outbreaks of cyanobacterial blooms but not with the rising intensity of blooms caused by climate change. It is now necessary to have an integrated and dynamic system of managing the mitigation both in terms of control of blooms as well as the factors that lead to blooms, namely, nutrient enrichment and climatic stressors. The strategies should be a mix of ecological solutions, engineering interventions, policy, and community responses as a way of preserving freshwater systems (Fig. 11.3) (Green et al., 2015).

11.6.1 Nutrient Input Control

The intervention that is most significant and highly recommended as the most appropriate in the long-term management of cyanobacterial blooms is mitigation of external

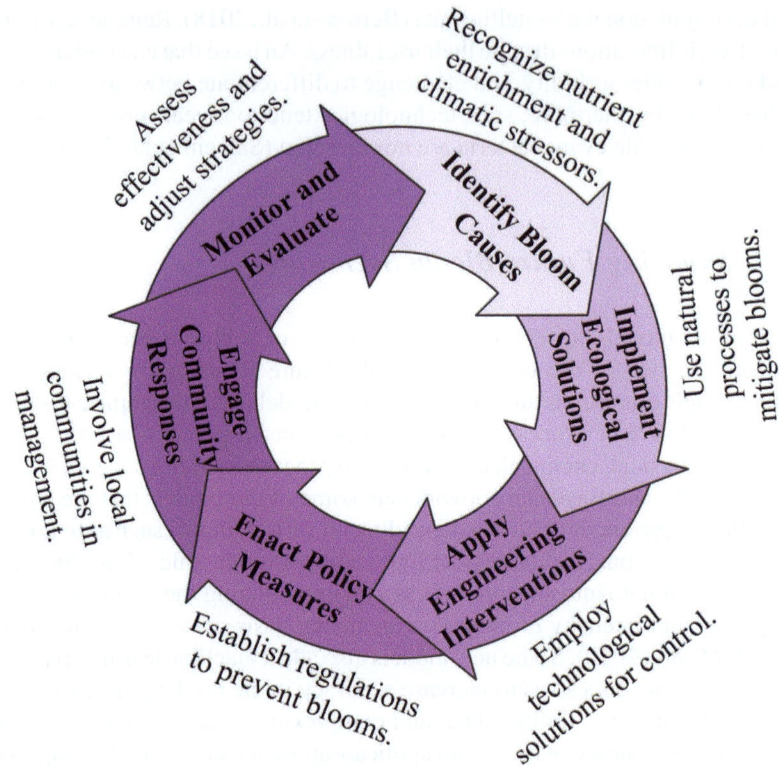

Fig. 11.3 Integrated cyanobacterial bloom management cycle

nutrient loading especially, phosphorus and nitrogen loads. The major sources of nutrient enrichment in fresh water are agricultural runoff, wastewater runoff, and storm water runoffs (Pericherla et al., 2020). Agricultural Best Management Practices (BMPs) that have the potential to substantially reduce nutrient losses in the waterways may include buffer strips, cover crops, manure/fertilizer restriction, and less cultivation. The risk of curbing eutrophication can also be played by upgrading wastewater treatment plant that helps in removing nutrients more effectively in them and by implementing stormwater retention basins in the urban parts (Yang & Lusk, 2018). Internal loading due to sediments is usually reduced by in-lake approaches like adding a strong phosphorus binder (e.g., alum or lanthanum-based compounds) to the lake. Nevertheless, they are taken as short-term solutions unless there is a combined effort to control nutrient in the watershed range (Lintern et al., 2020).

11.6.2 Adaptive Water Resource Management

One of the steps in climate-adaptive water management is to alter hydrological situation that predisposes the occurrence of blooms. Plans involve controlling the water flow, increasing the frequency of flushing as well as avoiding long stays of water. The bloom maintenance conditions created by physical factors, for example, selective withdrawal, destratification by aeration, or mechanical mixing and artificial circulation, can be disrupted by the technique used in managing reservoirs (Song, 2023). Internal release of phosphorus can also be reduced by controlling the blooms through manipulation of the water level in the shallow lakes due to exposure of the sediments to oxidation. But these should be well planned to prevent any adverse ecological effects, for example, destruction of homes of aquatic animals. Critical interventions can be made through risk-based infrastructures and early warning systems that combine the use of real-time surveillance with forecast models. Giving health-related advices, temporary closure of recreational locations of drinking water source change, etc., is important during bloom events (Coffey et al., 2019).

11.6.3 Community Engagement and Policy Implementation

Awareness by the people and the participation of stakeholders are vital to the process of good bloom management (West et al., 2021). Educational initiatives may use communities of the aspects and hazards of cyanobacterial blossoms and advocate ways of decreasing nutrient filth. On the policy front, the implementation of water quality guidelines, land-use policies, and incentives to sustainable farming should be done to guarantee long-term effects compliance. The more unified Water Resource Management (IWRM) approaches will synchronize the actions by sectors and the levels of governance, showing an integrated picture of a balance among water use, water conservation, and climate adaptation objectives (Ludwig et al., 2014). One of the most effective solutions to avoid the escalation of discretion problems caused by blooms in the next decades would be to integrate climate resilience into the national water policies, and add more investment into research activities, infrastructure, and capacity building (Table 11.2).

11.7 Conclusion

Cyanobacterial blooms are an expanding issue in the world with multidimensional effects of climate change. Warming temperature, change in precipitation patterns, long stratification, and increase in atmospheric CO_2 collectively provide a condition that comes closer to what bloom-forming cyanobacteria love. These climatic drivers do not only increase the rate of bloom as well as its geographical coverage but

Table 11.2 Key monitoring technologies and management strategies for predicting and mitigating cyanobacterial blooms under climate change conditions

No	Monitoring/management aspect	Application/function	References
1	MODIS satellite imagery	Detects surface water characteristics	Hu et al. (2010)
2	Sentinel-2 imagery	High-resolution temporal analysis	Ruescas et al. (2021)
3	Landsat data	Long-term landscape monitoring	Lunetta et al. (2015)
4	Drone-based aerial surveys	Detailed imagery for small systems	Teresi (2024)
5	Handheld fluorometers	On-site chlorophyll measurement	McQuaid and Zamyadi (2011)
6	In situ pigment sensors	Detect pigment-specific signals	Matthews et al. (2010)
7	Chlorophyll-a estimation	Estimates algal biomass	Binding et al. (2018)
8	Phycocyanin detection	Identifies cyanobacterial abundance	Simis et al. (2005)
9	Water turbidity assessment	Measures light penetration and clarity	Dogliotti et al. (2015)
10	Cloud cover correction	Improves image accuracy	Gao et al. (2020)
11	Surface bloom monitoring	Tracks bloom extent and timing	Binding et al. (2018)
12	Subsurface bloom limitation	Unable to detect deep-water blooms	Simis et al. (2005)
13	CE-QUAL-W2 modeling	Simulates physical–chemical interactions	Cole and Wells (2015)
14	AQUATOX simulations	Predict ecological risks from blooms	EPA (2020)
15	Bayesian network models	Model ecosystem probabilities	Hamilton et al. (2015)
16	Climate-integrated forecasting	Forecasts blooms under climate scenarios	Elliott (2012)
17	Machine learning predictions	Improve accuracy of bloom timing	Ahn et al. (2025)
18	Historical bloom data analysis	Trains models on bloom patterns	Stumpf et al. (2016)
19	Nutrient input reduction	Addresses root causes of eutrophication	Paerl et al. (2016)
20	Buffer strips in agriculture	Reduce nutrient runoff	Sharpley et al. (2006)
21	Cover cropping techniques	Enhance soil nutrient retention	Basche and DeLonge (2019)
22	Fertilizer management	Limit nitrogen and phosphorus loss	Withers et al. (2014)
23	Wastewater treatment upgrade	Reduces effluent nutrient discharge	Smith et al. (2017)

(continued)

Table 11.2 (continued)

No	Monitoring/management aspect	Application/function	References
24	Stormwater basin installation	Prevents pollutant wash-off	Hatt et al. (2009)
25	Alum-based phosphorus binding	Controls internal phosphorus loading	Welch and Cooke (1999)
26	Lanthanum-based treatment	Binds sediment phosphorus	Spears et al. (2013)
27	Water flow regulation	Modifies hydrology to prevent stratification	Olding et al. (2020)
28	Artificial destratification	Increases mixing and reduces bloom stability	Mitrovic et al. (2003)
29	Risk-based early warning systems	Enables proactive response	Davis et al. (2021)
30	Public education programs	Increase awareness and cooperation	Beversdorf et al. (2018)

The table outlines tools such as satellite remote sensing, modeling approaches, nutrient control methods, and adaptive water management practices

also lengthen the time of the bloom and introduce more chances of forming toxin, leading to ecological and human health hazards. The effects on the ecology are drastic: they vary from decreased biodiversity and ecosystem function to damage of food webs and chronic eutrophication. Meanwhile, there are both health and socioeconomic impacts, such as water pollution, medical costs, and the declining tourist and aquaculture industries, which need to be addressed with the utmost concern. Management of effective strategies implies the necessity of switching to proactive and holistic approaches to management. The watershed-scale nutrient reduction is the ongoing key aspect of mitigation of the bloom. This needs to be coupled with climate-adaptive water management, powerful monitoring and forecasting tools, and involvement of the stakeholders. The environment needs to be prepared for predictive models, remote sensing instruments, and early warning systems should; these must always be improved so that they can support climate variability and ecosystem complexity. In the future, there is a need to have interdisciplinary collaboration toward the understanding and response to the climate–cyanobacteria nexus. Future research projects should focus further on species-specific reaction to environmental change, the mechanical process that leads to regulation of toxins, and the socio-political systems that enable such interventions to be scaled. We can achieve resilient freshwater systems that can resist the twin forces of eutrophication and climate change by combining science, policy, and place-based action.

References

Ahn, J. M., Pyo, J. C., Lee, J., Abbas, A., & Kwon, D. H. (2025). Probabilistic machine learning-based phytoplankton abundance using hyperspectral remote sensing. *Geoscience & Remote Sensing*. https://doi.org/10.1080/15481603.2025.2484864

Albert, C., Aronson, J., Fürst, C., & Opdam, P. (2014). Integrating ecosystem services in landscape planning: Requirements, approaches, and impacts. *Landscape Ecology, 29*, 1277–1285.

Barruffa, A. S., Sposito, V., & Faggian, R. (2021). Climate change and cyanobacteria harmful algae blooms: Adaptation practices for developing countries. *Marine and Freshwater Research, 72*(12), 1722–1734.

Basche, A. D., & DeLonge, M. S. (2019). The impact of cover crops on water quality: A quantitative review. *Journal of Soil and Water Conservation, 74*(5), 562–574. https://doi.org/10.2489/jswc.74.5.562

Bertone, E., Burford, M. A., & Hamilton, D. P. (2018). Fluorescence probes for real-time remote cyanobacteria monitoring: A review of challenges and opportunities. *Water Research, 141*, 152–162.

Beversdorf, L. J., Rude, K., Weirich, C. A., Bartlett, S. L., Seaman, M., Kozik, C., et al. (2018). Integrating citizen science with professional monitoring improves bloom forecasting. *Environmental Management, 62*(2), 343–356. https://doi.org/10.1007/s00267-018-1049-0

Binding, C. E., Greenberg, T. A., & Bukata, R. P. (2018). The MERIS MCI and chlorophyll-a algorithms for estimating cyanobacterial biomass. *Remote Sensing of Environment, 204*, 450–465. https://doi.org/10.1016/j.rse.2017.10.022

Burch, M., Brookes, J., & Chorus, I. (2021). Assessing and controlling the risk of cyanobacterial blooms: Waterbody conditions. In *Toxic cyanobacteria in water* (pp. 505–562). CRC Press.

Carey, C. C., Ibelings, B. W., Hoffmann, E. P., Hamilton, D. P., & Brookes, J. D. (2012). Eco-physiological adaptations that favour freshwater cyanobacteria in a changing climate. *Water Research, 46*(5), 1394–1407. https://doi.org/10.1016/j.watres.2011.12.016

Chatterjee, S., & More, M. (2023). Cyanobacterial harmful algal bloom toxin microcystin and increased vibrio occurrence as climate-change-induced biological co-stressors: Exposure and disease outcomes via their interaction with gut–liver–brain axis. *Toxins, 15*(4), 289.

Coffey, R., Paul, M. J., Stamp, J., Hamilton, A., & Johnson, T. (2019). A review of water quality responses to air temperature and precipitation changes 2: Nutrients, algal blooms, sediment, pathogens. *JAWRA Journal of the American Water Resources Association, 55*(4), 844–868.

Cole, T. M., & Wells, S. A. (2015). *CE-QUAL-W2: A two-dimensional, laterally averaged, hydrodynamic and water quality model*. Version 4.0. U.S. Army Engineer Waterways Experiment Station.

Davis, T. W., Stumpf, R. P., & Wynne, T. T. (2021). Data-driven HAB early warning system for Lake Erie. *Journal of Great Lakes Research, 47*(3), 672–684. https://doi.org/10.1016/j.jglr.2021.01.008

Dogliotti, A. I., Ruddick, K. G., Nechad, B., Doxaran, D., & Knaeps, E. (2015). A single algorithm to retrieve turbidity from remotely-sensed data. *Remote Sensing of Environment, 156*, 157–168. https://doi.org/10.1016/j.rse.2014.09.020

Elliott, J. A. (2012). Is the future blue-green? Modelling climate-driven changes in phytoplankton population dynamics. *Environmental Modelling & Software, 32*, 147–155. https://doi.org/10.1016/j.envsoft.2011.11.017

EPA. (2020). *AQUATOX model for ecological risk assessment*. U.S. Environmental Protection Agency. https://www.epa.gov/aquatox

Gao, B., Yu, J., & Li, J. (2020). Cloud masking improvement for aquatic applications using Sentinel-2. *Remote Sensing Letters, 11*(5), 487–496. https://doi.org/10.1080/2150704X.2020.1735539

Granéli, E., Weberg, M., & Salomon, P. S. (2008). Harmful algal blooms of allelopathic microalgal species: The role of eutrophication. *Harmful Algae, 8*(1), 94–102.

Green, P. A., Vörösmarty, C. J., Harrison, I., Farrell, T., Sáenz, L., & Fekete, B. M. (2015). Freshwater ecosystem services supporting humans: Pivoting from water crisis to water solutions. *Global Environmental Change, 34*, 108–118.

Hamilton, D. P., Salmaso, N., & Paerl, H. W. (2015). Mitigating harmful cyanobacterial blooms: Strategies for control. *Harmful Algae, 54*, 213–222. https://doi.org/10.1016/j.hal.2015.01.002

Hatt, B. E., Fletcher, T. D., Walsh, C. J., & Taylor, S. L. (2009). Stormwater pollution control using constructed wetlands and retention basins. *Water Science & Technology, 60*(1), 285–292. https://doi.org/10.2166/wst.2009.343

Howarth, R., Chan, F., Conley, D. J., Garnier, J., Doney, S. C., Marino, R., & Billen, G. (2011). Coupled biogeochemical cycles: Eutrophication and hypoxia in temperate estuaries and coastal marine ecosystems. *Frontiers in Ecology and the Environment, 9*(1), 18–26.

Hu, C., Lee, Z., Ma, R., Yu, K., Li, D., & Shang, S. (2010). MODIS observations of cyanobacteria blooms in Taihu Lake, China. *Remote Sensing of Environment, 112*(11), 3898–3903. https://doi.org/10.1016/j.rse.2008.06.011

Huisman, J., Codd, G. A., Paerl, H. W., Ibelings, B. W., Verspagen, J. M. H., & Visser, P. M. (2018). Cyanobacterial blooms. *Nature Reviews Microbiology, 16*(8), 471–483. https://doi.org/10.1038/s41579-018-0040-1

Kabir, M., Habiba, U. E., Khan, W., Shah, A., Rahim, S., De los Rios-Escalante, P. R., Farooqi, Z.-U.-R., Ali, L., & Shafiq, M. (2023). Climate change due to increasing concentration of carbon dioxide and its impacts on environment in 21st century; a mini review. *Journal of King Saud University-Science, 35*(5), 102693.

Kumari, N., Mishra, S., Häder, D.-P., & Sinha, R. P. (2022). Cyanobacterial blooms and their implications in the changing environment. *Advances in Environmental and Engineering Research, 3*(1), 1–41.

Lintern, A., McPhillips, L., Winfrey, B., Duncan, J., & Grady, C. (2020). Best management practices for diffuse nutrient pollution: Wicked problems across urban and agricultural watersheds. *Environmental Science & Technology, 54*(15), 9159–9174.

Ludwig, F., van Slobbe, E., & Cofino, W. (2014). Climate change adaptation and integrated water resource management in the water sector. *Journal of Hydrology, 518*, 235–242.

Lunetta, R. S., Schaeffer, B. A., Stumpf, R. P., Keith, D., Jacobs, S. A., & Murphy, M. S. (2015). Evaluation of cyanobacterial detection methods using Landsat and hyperspectral data. *Remote Sensing of Environment, 157*, 196–209. https://doi.org/10.1016/j.rse.2014.04.034

Ma, J., & Wang, P. (2021). Effects of rising atmospheric CO2 levels on physiological response of cyanobacteria and cyanobacterial bloom development: A review. *Science of the Total Environment, 754*, 141889.

Matthews, M. W., Bernard, S., & Robertson, L. (2010). An algorithm for detecting cyanobacterial blooms in inland waters using MERIS. *Remote Sensing of Environment, 114*(3), 476–482. https://doi.org/10.1016/j.rse.2009.10.012

McQuaid, N., & Zamyadi, A. (2011). Performance of handheld fluorometers for on-site cyanobacteria detection. *Water Research, 45*(22), 7015–7024. https://doi.org/10.1016/j.watres.2011.10.003

Mitrovic, S. M., Oliver, R. L., & Rees, C. (2003). Mixing and cyanobacterial bloom management in a stratified lake. *Marine and Freshwater Research, 54*(8), 609–619. https://doi.org/10.1071/MF02163

Moss, B., Kosten, S., Meerhoff, M., Battarbee, R. W., Jeppesen, E., Mazzeo, N., et al. (2011). Allied attack: Climate change and eutrophication. *Inland Waters, 1*(2), 101–105. https://doi.org/10.5268/IW-1.2.359

Moustaka-Gouni, M., & Sommer, U. (2020). Effects of harmful blooms of large-sized and colonial cyanobacteria on aquatic food webs. *Water, 12*(6), 1587.

Nürnberg, G. K. (2024). *Lake functioning: Internal phosphorus loading, cyanobacteria, and climate change*. CRC Press.

O'Neil, J. M., Davis, T. W., Burford, M. A., & Gobler, C. J. (2012). The rise of harmful cyanobacteria blooms: The potential roles of eutrophication and climate change. *Harmful Algae, 14*, 313–334. https://doi.org/10.1016/j.hal.2011.10.027

Obaid, A., Ali, K., Abiye, T., & Adam, E. (2021). Assessing the utility of using current generation high-resolution satellites (Sentinel 2 and Landsat 8) to monitor large water supply dam in South Africa. *Remote Sensing Applications: Society and Environment, 22*, 100521.

Olding, D., Gilmore, B., & Thomas, R. (2020). Regulated flows and cyanobacterial bloom mitigation. *Journal of Environmental Management, 271*, 110973. https://doi.org/10.1016/j.jenvman.2020.110973

Paerl, H. W., Hall, N. S., & Calandrino, E. S. (2011). Controlling harmful cyanobacterial blooms in a world experiencing anthropogenic and climatic-induced change. *Science of the Total Environment, 409*(10), 1739–1745. https://doi.org/10.1016/j.scitotenv.2011.02.001

Paerl, H. W., & Huisman, J. (2008). Blooms like it hot. *Science, 320*(5872), 57. https://doi.org/10.1126/science.1155398

Paerl, H. W., Otten, T. G., & Kudela, R. (2016). Mitigating the expansion of harmful algal blooms. *Environmental Science & Technology, 50*(16), 8626–8629. https://doi.org/10.1021/acs.est.6b03491

Pericherla, S., Karnena, M. K., & Vara, S. (2020). A review on impacts of agricultural runoff on freshwater resources. *International Journal on Emerging Technologies, 11*, 829–833.

Ralston, D. K., & Moore, S. K. (2020). Modeling harmful algal blooms in a changing climate. *Harmful Algae, 91*, 101729.

Reinl, K. L., Brookes, J. D., Carey, C. C., Harris, T. D., Ibelings, B. W., Morales-Williams, A. M., et al. (2021). Cyanobacterial blooms in oligotrophic lakes: Shifting the high-nutrient paradigm. *Freshwater Biology, 66*(9), 1781–1799. https://doi.org/10.1111/fwb.13791

Saleem, F., Jiang, J. L., Atrache, R., Paschos, A., Edge, T. A., & Schellhorn, H. E. (2023). Cyanobacterial algal bloom monitoring: Molecular methods and technologies for freshwater ecosystems. *Microorganisms, 11*(4), 851.

Sharpley, A. N., Kleinman, P. J. A., Flaten, D. N., & Buda, A. R. (2006). Critical source area management of agricultural phosphorus. *Journal of Environmental Quality, 35*(5), 198–206. https://doi.org/10.2134/jeq2006.0057

Shukla, P., Srivastava, P., & Mishra, A. (2024). 8 Routes of exposure (inhalation, ingestion, dermal contact). *Heavy Metal Contamination in the Environment: Health Impacts and Potential Remediation Approaches, 114.*

Sidle, R. C. (2021). Strategies for smarter catchment hydrology models: Incorporating scaling and better process representation. *Geoscience Letters, 8*(1), 24.

Simis, S. G. H., Peters, S. W. M., & Gons, H. J. (2005). Remote sensing of the cyanobacterial pigment phycocyanin in turbid inland water. *Limnology and Oceanography, 50*(1), 237–245. https://doi.org/10.4319/lo.2005.50.1.0237

Smayda, T. (2019). Global epidemic of noxious phytoplankton blooms and food chain consequences in large ecosystems. In *Food chains, yields, models, and management of large marine ecosoystems* (pp. 275–307). Routledge.

Smith, V. H., Joye, S. B., & Howarth, R. W. (2017). Eutrophication of freshwater and marine ecosystems. *Limnology and Oceanography, 51*(1), 351–355. https://doi.org/10.4319/lo.2006.51.1_part_2.0351

Søgaard, D. H., Sorrell, B. K., Sejr, M. K., Andersen, P., Rysgaard, S., Hansen, P. J., Skyttä, A., Lemcke, S., & Lund-Hansen, L. C. (2021). An under-ice bloom of mixotrophic haptophytes in low nutrient and freshwater-influenced Arctic waters. *Scientific Reports, 11*(1), 2915.

Song, Y. (2023). Hydrodynamic impacts on algal blooms in reservoirs and bloom mitigation using reservoir operation strategies: A review. *Journal of Hydrology, 620*, 129375.

Spears, B. M., et al. (2013). A meta-analysis of lanthanum-modified bentonite clay (Phoslock) for controlling phosphorus in lakes. *Water Research, 47*(15), 5930–5942. https://doi.org/10.1016/j.watres.2013.07.024

Tariq, A., & Mushtaq, A. (2023). Untreated wastewater reasons and causes: A review of most affected areas and cities. *International Journal of Chemical and Biochemical Sciences, 23*(1), 121–143.

Teresi, J. (2024). Environmental public health surveillance of freshwater harmful algal blooms in Washington using drone technology. *University of Washington.* https://digital.lib.washington.edu

Tiebel, K. (2020). *The ability of pioneer tree species to mitigate the effects of site disturbance by fast and effective natural regeneration.*

Uk, S., Yang, H., Vouchlay, T., Ty, S., Sokly, S., Sophal, T., Chantha, O., & Chihiro, Y. (2022). Dynamics of phosphorus fractions and bioavailability in a large shallow tropical lake characterized by monotonal flood pulse in Southeast Asia. *Journal of Great Lakes Research, 48*(4), 944–960.

Visser, P. M., Verspagen, J. M. H., Sandrini, G., Stal, L. J., Matthijs, H. C. P., Davis, T. W., et al. (2016). How rising CO_2 and global warming may stimulate harmful cyanobacterial blooms. *Harmful Algae, 54*, 145–159. https://doi.org/10.1016/j.hal.2015.12.006

Welch, E. B., & Cooke, G. D. (1999). Effectiveness and longevity of phosphorus inactivation with alum. *Lake and Reservoir Management, 15*(1), 5–27. https://doi.org/10.1080/07438149909353954

West, J. J., Järnberg, L., Berdalet, E., & Cusack, C. (2021). Understanding and managing harmful algal bloom risks in a changing climate: Lessons from the European CoCliME project. *Frontiers in Climate, 3*, 636723.

Willis, C., Papathanasopoulou, E., Russel, D., & Artioli, Y. (2018). Harmful algal blooms: The impacts on cultural ecosystem services and human well-being in a case study setting, Cornwall, UK. *Marine Policy, 97*, 232–238.

Winder, M., & Sommer, U. (2012). Phytoplankton response to a changing climate. *Hydrobiologia, 698*, 5–16. https://doi.org/10.1007/s10750-012-1149-2

Withers, P. J. A., et al. (2014). Agricultural phosphorus management: A historical review. *Environmental Science & Policy, 42*, 129–140. https://doi.org/10.1016/j.envsci.2014.05.007

Woodhouse, J. N., Kinsela, A. S., Collins, R. N., Bowling, L. C., Honeyman, G. L., Holliday, J. K., et al. (2016). Microbial communities reflect temporal changes in cyanobacterial composition in a shallow ephemeral freshwater lake. *ISME Journal, 10*, 1337–1351. https://doi.org/10.1038/ismej.2015.213

Wurtsbaugh, W. A., Paerl, H. W., & Dodds, W. K. (2019). Nutrients, eutrophication and harmful algal blooms along the freshwater to marine continuum. *Wiley Interdisciplinary Reviews: Water, 6*(5), e1373.

Yang, Y.-Y., & Lusk, M. G. (2018). Nutrients in urban stormwater runoff: Current state of the science and potential mitigation options. *Current Pollution Reports, 4*, 112–127.

Zhu, A.-X., Miao, Y., Wang, R., Zhu, T., Deng, Y., Liu, J., Yang, L., Qin, C.-Z., & Hong, H. (2018). A comparative study of an expert knowledge-based model and two data-driven models for landslide susceptibility mapping. *CATENA, 166*, 317–327.

Chapter 12
Interactions Between Cyanobacteria and Other Microbial Communities During Blooms

Salma Noureen, Javed Iqbal, Banzeer Ahsan Abbasi, Tabassum Yaseen, Muhammad Anas, Ghulam Murtaza, Rashid Iqbal, Farishta Zarshan, Madiha Imtiaz, and Shah Fahad

Abstract Cyanobacterial blooms represent one of the most complex microbial phenomena in aquatic ecosystems primarily determined by nutrient eutrophication (over-enrichment). At the molecular level, eutrophication, characterized by elevated phosphorus and nitrogen concentrations, activates key regulatory pathways in cyanobacteria, including genes related to nutrient uptake, nitrogen fixation (e.g., nifH), carbon assimilation, and toxin production like mcy, cyr, sxt gene clusters.

The authors Salma Noureen and Javed Iqbal have contributed equally to this work.

S. Noureen · J. Iqbal (✉) · T. Yaseen · F. Zarshan · M. Imtiaz
Department of Botany, Bacha Khan University, Charsadda, Khyber Pakhtunkhwa, Pakistan
e-mail: javed89qau@gmail.com

T. Yaseen
e-mail: tabassumyaseen@bkuc.edu.pk

B. A. Abbasi
Department of Botany, Rawalpindi Women University, Rawalpindi, Pakistan

M. Anas
Department of Soil Science and Plant Nutrition, Selçuk University, Konya, Türkiye
e-mail: muhammed.anas@selcuk.edu.tr

G. Murtaza
School of Agriculture, Yunnan University, Kunming, Yunnan, China

R. Iqbal
Department of Agronomy, Faculty of Agriculture and Environment, The Islamia University of Bahawalpur, Bahawalpur, Pakistan

Department of Life Sciences, Western Caspian University, Baku, Azerbaijan

R. Iqbal
e-mail: rashid.iqbal@iub.edu.pk

S. Fahad
Department of Agronomy, Abdul Wali Khan University Mardan, Mardan, Khyber Pakhtunkhwa, Pakistan

S. Fahad et al. (eds.), *Cyanobacterial Blooms: Ecology, Evolution and Biogeochemical Impacts*, https://doi.org/10.1007/978-3-032-06042-6_12

These molecular responses enable rapid cellular proliferation under favorable environmental conditions such as high light intensity, warm temperatures, and particularly alkaline pH levels that often exceed 8.5 during bloom events due to intense photosynthetic activity and CO_2 depletion. The phylum Cyanobacteria comprises approximately 2700 species, classified into around 150 genera, such as *Microcystis, Dolichospermum (Anabaena), Planktothrix, Aphanizomenon,* and *Nodularia.* Metagenomic and single-cell sequencing shows that the true diversity is significantly underrated, with numerous of cryptic or uncultured species influencing bloom dynamics worldwide. These blooms are not cyanobacterial monocultures but complex consortia where cyanobacteria interact intricately with heterotrophic bacteria, viruses, archaea, and fungi. Mutualistic interactions facilitate bloom persistence and nutrient exchange, while antagonistic relationships, including viral lysis (via cyanophages) and bacterial allelopathy, may contribute to bloom senescence. Cyanotoxins (cyanobacterial toxins) such as anatoxinsmicrocystins and cylindrospermopsins can cause neurotoxic, hepatotoxic, and dermatotoxic effects in humans and animals. Treatment involves toxin adsorption, advanced filtration, and in some cases, medical detoxification. These methods are indeed reactive measures. Future bloom prevention strategies should be proactive and integrative, such as reducing nutrient loading through agricultural and wastewater reforms, deploying microbial biocontrol agents (e.g., algicidal bacteria), and employing synthetic biology to engineer microbial communities that suppress bloom-forming species.

Keywords Cyanobacteria · Microbial communities · Blooms · Interactions · Symbiotic relationships · Competition · Nutrient cycling · Toxin production · Ecological dynamics · Metagenomics

12.1 Introduction

The term Cyano Bloom refers to Cyanobacterial blooms, which are rapid proliferations of cyanobacteria in aquatic environments, often resulting in dense accumulations that can dominate the phytoplankton community (Lakshmikandan et al., 2024). These blooms are typically driven by a combination of environmental factors, including nutrient enrichment (particularly nitrogen and phosphorus), calm water conditions, elevated temperatures, and increased light availability (Lan et al., 2024). Anthropogenic activities, like wastewater discharge and agricultural runoff, contribute significantly to nutrient loading, supporting eutrophication and forming favorable conditions for bloom development (Li et al., 2025a, 2025b, 2025c). The characteristics of cyanobacterial blooms are complicated. They frequently appear as surface scums or mats, varying in color from blue-green to reddish-brown, and can generate unpleasant odors and taste issues, influencing recreational and drinking water uses (Thawabteh et al., 2023). Several cyanobacteria species produce cyanotoxins like asanatoxins, microcystins, and cylindrospermopsins, which pose risks to human health and aquatic organisms. These toxins can accumulate in the food web, leading to bioaccumulation and potential revelation through drinking water or consumption of contaminated aquatic organisms (Villalobos et al., 2025). The cyanobacterial blooms are significant ecological impacts. Bloom senescence leads to decomposition, consuming dissolved oxygen and causing anoxic or hypoxic conditions harmful to aquatic life (Kumari et al., 2022). This oxygen depletion can result in death of fishes and loss of biodiversity. Additionally, blooms can disturb trophic relations by suppressing other phytoplankton, and affecting zooplankton and fish populations, thus altering food web dynamics (Demertzioglou et al., 2022). Global warming worsens the occurrence and severity of cyanobacterial blooms, as rising global temperatures boost cyanobacterial growth rates and bloom persistence. Altered precipitation patterns and enhanced frequency of intense weather events can lead to increased nutrient runoff and changes in water column stratification, further encouraging bloom formation (Fig. 12.1) (Mohanty et al., 2025).

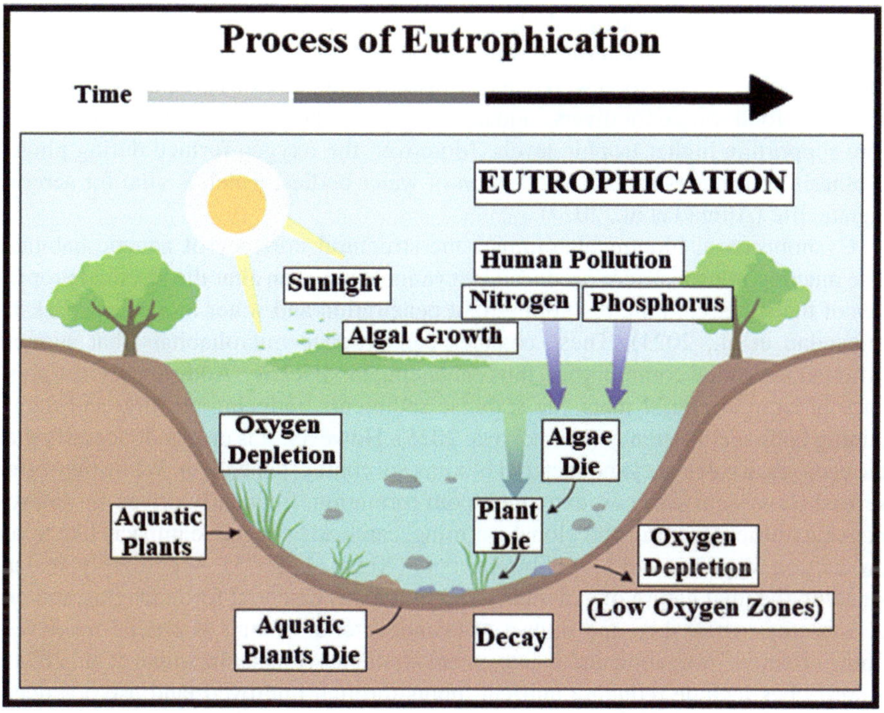

Fig. 12.1 Formation of algal bloom (eutrophication)

12.1.1 Importance of Cyanobacterial Booms in Aquatic Ecosystem

Cyanobacterial blooms, often perceived as detrimental due to their association with harmful algal blooms (HABs), also play significant roles in aquatic ecosystems. These blooms, primarily composed of cyanobacteria, are essential to diverse ecological processes such as nutrient cycling, primary production, and habitat structuring (Kumar et al., 2025). Several cyanobacterial blooms are primary roles is nitrogen fixation. Cyanobacteria possess the enzyme nitrogenase, facilitating them to convert atmospheric nitrogen (N_2) into ammonia (NH_3), a form available to other organisms. Nitrogen partial to aquatic environment that supporting the growth of other phytoplankton and maintains the productivity of the ecosystem (Saud et al., 2024). For instance, genera like *Gloeotrichia* do significantly influence phosphorus and nitrogen cycles in lakes, therefore affecting overall nutrient dynamics. In accumulation to nutrient cycling, cyanobacterial blooms contribute to primary production; through photosynthesis, cyanobacteria convert carbon dioxide into organic matter, forming the base of the aquatic food web (Chankeshwara et al., 2025). These organic matters serve as a food source for diverse aquatic organisms like fish and zooplankton, therefore supporting higher trophic levels. Moreover, the oxygen formed during photosynthesis contributes to the oxygenation of water bodies, which is vital for aerobic aquatic life (Ahmad et al., 2024).

Cyanobacterial blooms also impact the structural intricacy of aquatic habitats. The mucilaginous matrices produced by cyanobacteria can alter the physical properties of the water column, disturbing light penetration and water movement (Lakshmikandan et al., 2024). These changes can produce microhabitats that support different microbial communities, thus enhancing biodiversity. Additionally, the presence of cyanobacterial mats can stabilise sediments, inducing attrition and maintaining habitat consistency (Marečková, 2025). However, it is crucial to identify that the ecological roles of cyanobacterial blooms are context dependent. While they offer several ecosystem services, extreme bloom formation, frequently driven by anthropogenic nutrient loading and global warming, can lead to negative impacts like toxin production, hypoxia, and loss of biodiversity (Xia et al., 2025). Accordingly, understanding the dual nature of cyanobacterial blooms is essential for managing aquatic ecosystems sustainably, and impact of cyanobacterial blooms is crucial for developing effective mitigation and management strategies (Lakshmikandan et al., 2024). Approaches include reducing nutrient inputs through improved land use practices, enhancing water circulation to disturb bloom formation, and implementing premature warning systems for bloom detection (Hwang et al., 2024). Continued research is necessary to reveal the complex relations governing bloom dynamics and to inform policy decisions aimed at protecting public health and aquatic ecosystems (Shafik, 2025).

12.2 Overview of Microbial Interactions

12.2.1 Types of Interactions

Cyanobacterial blooms are intricate ecological phenomena involving complicated relationships between cyanobacteria and different microbial communities. These relations include commensal, antagonistic, and mutualistic, each playing an important role in shaping bloom dynamics and ecosystem functions (Fig. 12.2) (Du et al., 2022). Mutualistic associations are prominent during cyanobacterial blooms. Specific cyanobacteria genera, like *Richelia* and *Calothrix*, form symbiotic associations with diatoms like *Hemiaulus* and *Rhizosolenia* (Belmonte, 2025). In these relationships, cyanobacteria supply fixed nitrogen to their hosts, facilitating growth in nutrient-deprived environments, while benefiting from the organic carbon produced by the diatoms (Bhardwaj et al., 2024). Likewise, in the marine cyanobacterium *Trichodesmium*, associated heterotrophic bacteria contribute to nutrient acquisition like iron, enhancing the colony's metabolic capabilities (Zhong et al., 2025). Commensal relations also occur, where one organism benefits without affecting the other organism. For instance, cyanobacteria released organic compounds utilize bacteria during photosynthesis, supporting their growth without affecting the cyanobacterial population (Dey, 2024). Cyanophages, viruses that taint cyanobacteria, can promote bloom populations by lysing cyanobacterial cells, therefore changing community composition and nutrient cycling (Ain et al., 2024). Additionally, some bacteria generate allelopathic compounds that inhibit cyanobacterial growth, impacting bloom development and persistence (Cui et al., 2025). Understanding these microbial associations is important for comprehending the ecological roles of cyanobacterial blooms and their impacts on aquatic ecosystems. These relationships manipulate nutrient dynamics, primary production, and the overall structure of microbial communities during bloom phases (Chen et al., 2025).

12.2.2 Importance of Microbial Interactions in Shaping Ecosystem Dynamics

Interactions between Cyanobacteria and other microbial communities during bloom events are intricate and extensively affect aquatic ecosystem dynamics. Cyanobacterial blooms, often resulting from eutrophication, lead to significant moves in microbial community structures, disturbing both function and diversity (Yang et al., 2021a, 2021b). During bloom, cyanobacteria liberate various organic compounds, including toxins like microcystins, which can decrease the metabolic activities of co-occurring microorganisms, resulting in lowered microbial diversity (Manganelli et al., 2023). *Microcystis aeruginosa* can suppress the growth of other aquatic microorganisms through the production of such toxins, thereby facilitating its own dominance in

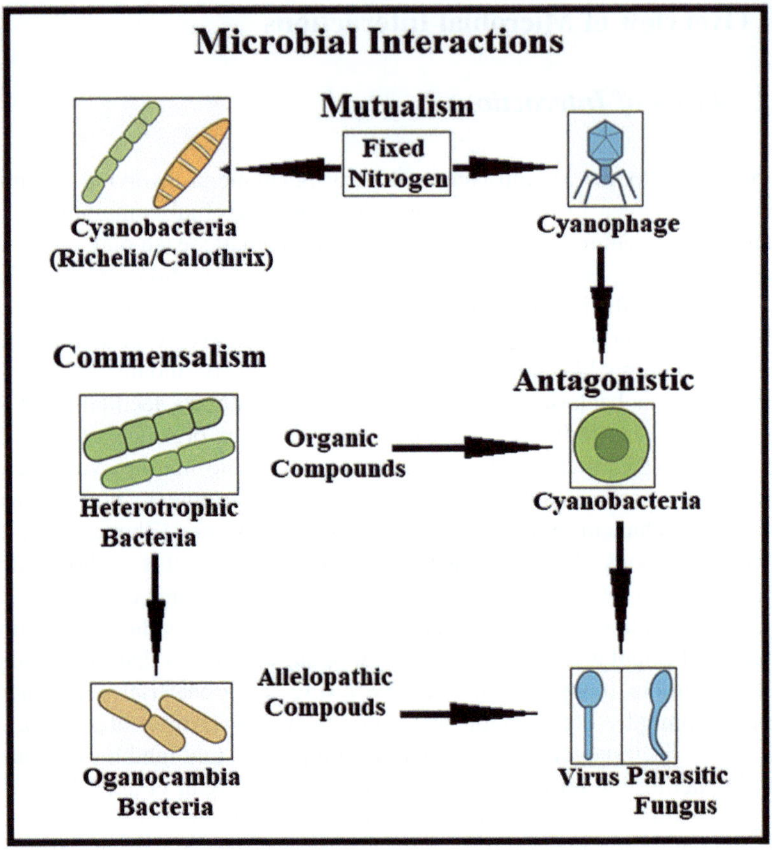

Fig. 12.2 Types of interactions between microbes

the ecosystem (Kang et al., 2024). Additionally, the decomposition of cyanobacterial biomass alters the nutrient composition, affecting heterotrophic bacteria's abundance and activity. These bacteria are important in nutrient cycling, decomposing organic matter and releasing nutrients rear into the ecosystem (Raza et al., 2023). However, the dominance of cyanobacteria can disturb these processes, leading to induces bacterial diversity and altered community structures (Du et al., 2022). Moreover, associations between cyanobacteria and heterotrophic bacteria can be mutualism. Heterotrophs can influence cyanobacteria organic compounds while providing crucial nutrients or remove detrimental substances, which in turn supports cyanobacterial growth and stability (Abate et al., 2024). Furthermore, parasitic relationships participate in bloom dynamics. For instance, parasitic and viruses fungi can infect cyanobacteria, leading to cell lysis and altering the microbial community structure. These relations can produce alternative energy pathways, increasing ecosystem resilience during bloom phases (Banerji et al., 2024).

12.2.3 Interactions Between Cyanobacteria and Other Microbial Communities

Cyanobacteria and other microbial communities' relationships during bloom events are multifaceted and complex, influencing bloom dynamics and ecosystem health. Cyanobacterial blooms, typically dominated by genera such as *Microcystis, Anabaena,* and *Planktothrix*, can considerably reform their structure and function of microbial communities in aquatic environments (Du et al., 2022). These blooms are not only a result of abiotic factors like nutrient availability and temperature but also biotic factors like co-occurring microorganisms. During bloom phases, cyanobacteria release different organic compounds, including toxins like microcystins, influencing surrounding microbial populations (Ren et al., 2023a, 2023b). For instance, *Microcystis's* microcystin production is linked to upregulated extracellular polysaccharide synthesis, enabling colony formation and potentially recruiting specific bacterial communities that can tolerate or degrade these compounds (Wang et al., 2025a, 2025b). These associated bacteria regularly possess genes for processing organic compounds, such as alkaline phosphatases and peptidases, assisting them to utilize cyanobacterial exudates and contribute to nutrient cycling (Li et al., 2025a, 2025b, 2025c). It is also noteworthy how different microbial communities evolve during cyanobacterial blooms. In eutrophic lakes, bloom onset can lead to induced in microbial diversity, with certain bacterial taxa dominating (Yang et al., 2021a, 2021b). As the bloom progresses and cyanobacterial biomass declines, often microbial diversity increases, with bacterial groups emerging that degrade cyanobacterial biomass (Croci et al., 2025). This dynamic succession highlights the role of microbial communities in both mitigating and supporting bloom phases. Furthermore, relations between cyanobacteria and heterotrophic bacteria can be antagonistic or symbiotic, affecting bloom dynamics through nutrient provides algicidal activity, or organic matter degradation (Abate et al., 2024). Environmental factors, such as nutrient concentrations and temperature, modulate these interactions. For instance, nitrogen limitation can alter nitrogen-fixing cyanobacteria: *Anabaena* to non-nitrogen-fixing genera like *Planktothrix* and *Microcystis.* These moves are frequently accompanied by alteration in the associated microbial communities, reflecting adaptations to the current environmental conditions (Mishra et al., 2021). Understanding the intricate relationships is crucial for predicting bloom occurrences and developing effective management strategies. Explaining complicated interactions key to predicting bloom dynamics and mitigating their adverse effects on aquatic ecosystems (Zhang et al., 2025).

12.3 Cyanobacterial Blooms

12.3.1 Causes of Cyanobacterial Blooms

Cyanobacterial blooms are complicated ecological phenomena affected by anthropogenic and natural factors. Leading among these is nutrient pollution, particularly the extreme input of nitrogen (N) and phosphorus (P) into aquatic systems. These nutrients, often originating from wastewater discharge, agricultural runoff, and urban stormwater, provide as primary growth-limiting parameters for cyanobacteria (Lakshmikandan et al., 2024). Elevated concentrations of N and P can disturb the stoichiometric balance of aquatic ecosystems, favoring cyanobacterial proliferation over other phytoplankton due to their efficient nutrient uptake and nitrogen fixing ability (Gao et al., 2025). Global changes further worsen the occurrence and intensity of cyanobacterial blooms. Growing global temperatures enhance thermal stratification in water bodies, creating stable, warm surface layers that are conducive to cyanobacterial growth. Some cyanobacteria possess gas vesicles for resilience, allowing and optimal light absorption (Chatterjee et al., 2023). Furthermore, improved temperatures also speed up metabolic rates, leading to faster growth. Changes in precipitation patterns and intense rainfall and droughts, all these contribute to nutrient loading through runoff, increased water residence, both of which favor bloom development (Wen et al., 2025).

Environmental parameters such as light availability, hydrodynamics, and water chemistry also play important roles in bloom dynamics. High light intensity can stimulate photosynthesis, but excessive irradiance may lead to photoinhibition in some phytoplanktons, whereas cyanobacteria often possess protective pigments effects to mitigate these (Simkin et al., 2022). Water chemistry factors like pH and dissolved oxygen levels influence nutrient solubility and bioavailability, consequently impacting cyanobacterial growth (Sun et al., 2025). Hydrodynamic conditions, such as reduced flow and mixing, can lead to the accumulation of cyanobacterial biomass in surface waters, forming dense mats that further reduces light penetration and oxygen dynamics. The interface between these factors underscores the complex nature of cyanobacterial bloom formation (Lakshmikandan et al., 2024). Anthropogenic nutrient improvement sets the stage for bloom potential, while climate induced alters in precipitation and temperature patterns modulate the regularity and severity of blooms (Paerl, 2023). Understanding these interconnected drivers is crucial for developing efficient mitigation and management strategies to address the growing challenge of cyanobacterial blooms (Fig. 12.3) (Barruffa et al., 2021).

12.3.2 Consequences of Cyanobacterial Blooms

Cyanobacterial blooms, or harmful algal blooms (HABs), pose enormous challenges to aquatic ecosystems, economies worldwide, and human health. These blooms create

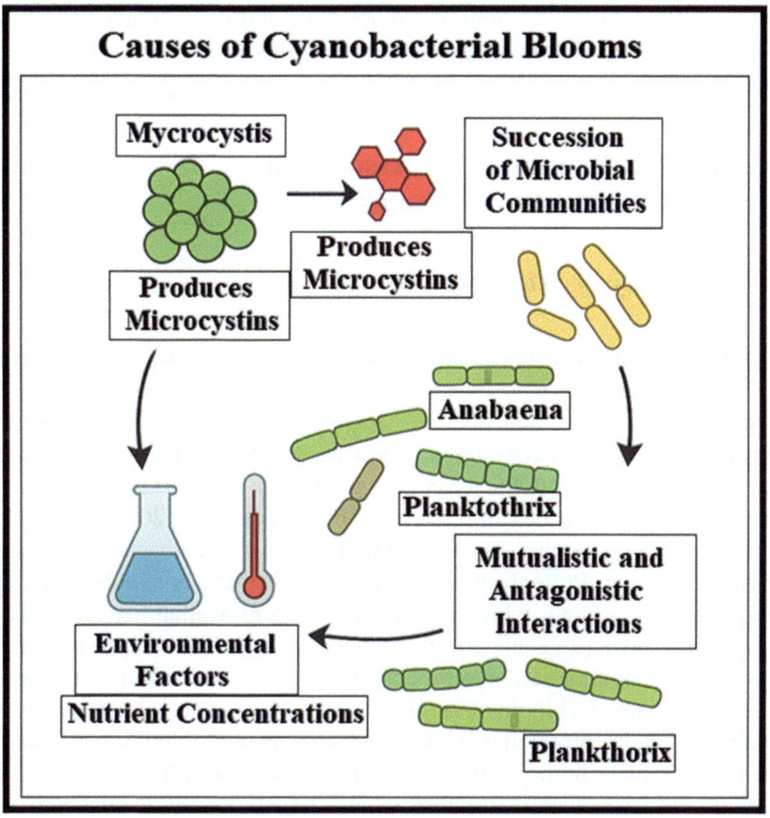

Fig. 12.3 Major causes of cyanobacterial blooms

a variety of cyanotoxins, such as microcystins, cylindrospermopsin, saxitoxins, and anatoxins, which contaminates drinking water resources and poses serious health risks to the general population (Massey et al., 2022). Exposure can occur through ingestion or inhalation, causing health effects ranging from gastroenteritis, liver damage, and neurological impairments to chronic conditions, including liver cancer and neurodegenerative diseases (Xiong et al., 2025). The (1996) Caruaru, Brazil, incident highlighted the risks, with 50 dialysis patients dying from microcystin-contaminated water. Ecologically, cyanobacterial blooms disrupt aquatic food webs by outcompeting and displacing native phytoplankton species, leading to loss biodiversity (Fotler, 2024). The decomposition of massive cyanobacterial biomass depletes oxygen levels, causing hypoxic conditions and death of fishes. Economically, cyanobacterial blooms impact water treatment facilities, face increased costs to remove toxins, and address taste and odor issues caused by compounds like geosmin and 2-methylisoborneol (Igwaran et al., 2024). Recreational activities like swimming, boating, and fishing are often restricted during bloom events, leading to significant losses in sightseeing revenue. Moreover, possessions values impacted water bodies can decline due to the diminished aesthetic and recreational appeal, further exacerbating economic burdens on local communities (Kutlu et al., 2025).

12.4 Microbial Communities Associated with Cyanobacterial Blooms

12.4.1 Diversity of Microorganisms

Cyanobacterial blooms involve dynamic relations among diverse microbial communities, including heterotrophic bacteria, viruses, and archaea, influencing bloom development, persistence, and decline, shaping the ecological and biogeochemical dynamics of aquatic ecosystems (Zhou et al., 2020). Heterotrophic bacteria, such as Proteobacteria and Bacteroidetes phyla, play important rolesin nutrient recycling and influencing bloom dynamics (Bhardwaj et al., 2024). Bacteroidetes degrade complex polysaccharides, while breakdown of cyanobacterial extracellular polymeric substances. Proteobacteria are involved in various metabolic processes, including nitrogen cycling, which can impact nutrient availability and thus affect cyanobacterial growth (Ye et al., 2020). Archaea are essential components of microbial communities in aquatic environments; only fewer Archaea are studied in the context of cyanobacterial blooms (Zhu et al., 2021). Methanogenic archaea contribute to carbon cycling by producing methane under anaerobic conditions. Halophilic archaea thrive well in high-salinity environments and can affect bloom dynamics in saline or brackish waters (Ibrahim et al., 2023). Research on the battle for resources and possible syntrophic linkages between archaea and cyanobacteria is still underway, underscoring the intricacy of microbial networks during blooms (Du et al., 2022).

Viruses, specifically cyanophages, are important regulators of cyanobacterial populations. Cyanophages contaminate cyanobacteria, leading to cell lysis and the release of cellular contents, which can suppress bloom intensity and support nutrient recycling. This process, known as the viral shunt, redirects organic matter from higher trophic levels rear into the microbial loop, affecting carbon and nutrient cycling within the ecosystem (Ain et al., 2024). Cyanophages can hold auxiliary metabolic genes (AMGs) that alter host metabolism, potentially increasing viral replication and affecting bloom dynamics (Grasso et al., 2022).

12.4.2 Role of Microorganisms in Boom Dynamics

Microorganisms play important role in the dynamics of cyanobacterial blooms, impacting nutrient cycling, organic matter degradation, and the overall growth and productivity of cyanobacteria (Bhardwaj et al., 2024). These relations are complicated and engage diverse microbial groups, including archaea, bacteria, and viruses, each contributing to the ecological balance within aquatic ecosystems. Heterotrophic bacteria are crucial to nutrient cycling during bloom phases (Pereira, 2025). They decay organic matter, releasing inorganic nutrients such as ammonium and phosphate, which are then available for uptake by cyanobacteria. This process affects sustaining bloom, duration and intensity (Yin et al., 2025). The microbial loop describing the recycling of dissolved organic carbon (DOC) through microbial communities highlights the importance of bacteria in maintaining nutrient accessibility (Wan et al., 2022). Archaea, like methanogens, contribute to carbon cycling by generating methane under anaerobic conditions frequently establish in stratified water bodies during bloom events. This methane production can affect greenhouse gas emissions and further influence the ecological dynamics of the aquatic environment (Fazi et al., 2021). Viruses, especially cyanophages, regulators of cyanobacterial populations, controlling bloom magnitude and facilitate the release of cellular contents back into the environment. This viral lysis contributes to the viral shunt, redirecting organic matter from higher trophic levels back into the microbial loop, therefore influencing carbon and nutrient cycling within the ecosystem (Grasso et al., 2022).The phycosphere, a region surrounding phytoplankton cells, is a hotspot for microbial interactions. Bacteria inhabiting the phycosphere feed on nutrients exuded by cyanobacteria, and in turn, can influence the productivity and cyanobacterial growth (Semple & Pandhal, 2025).

12.5 Mechanism of Interaction Between Cyanobacteria and Other Microorganisms

12.5.1 Nutrient Exchange and Competition

The interactions between cyanobacteria and other microbial communities during blooms are complex, involving concerning nutrient exchange and competition (Du et al., 2022). Cyanobacteria efficient nutrient uptake systems allowing them to flourish in environments where nutrients like nitrogen and phosphorus are limited (Bhardwaj et al., 2024). Heterotrophic bacteria play important role in nutrient cycling by decomposing organic matter, thus releasing inorganic nutrients back into the environment. This process not only supports cyanobacterial growth while competing for available nutrients, which influence bloom dynamics (Fig. 12.4) (Yin et al., 2025).

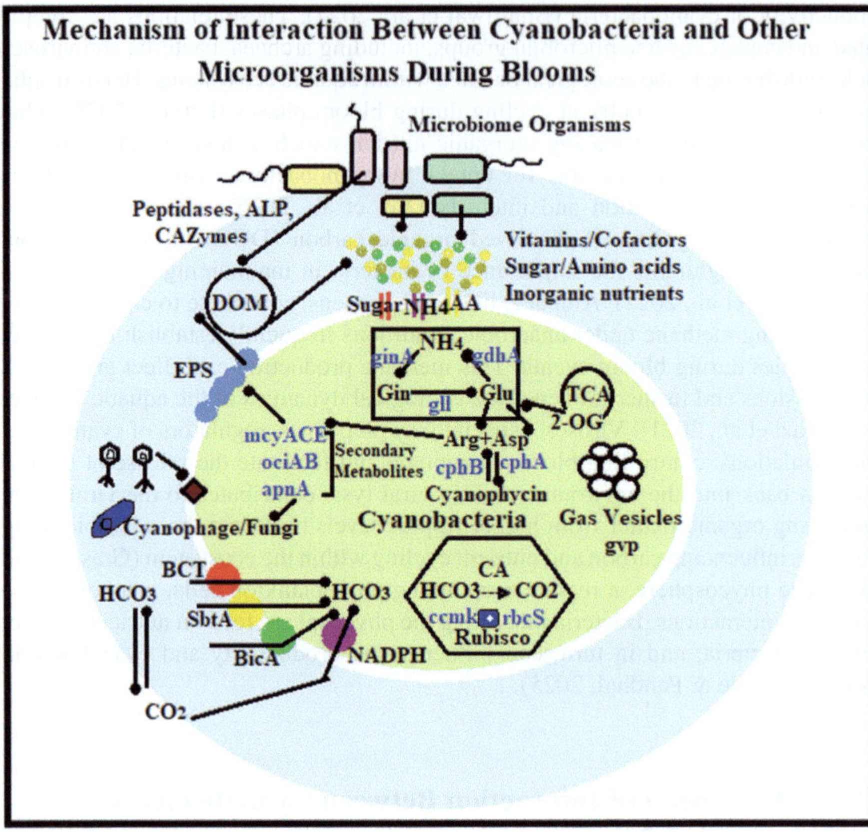

Fig. 12.4 Mechanism of interaction between microbes and cyanobacteria

12.5.2 Quorum Sensing and Signaling

Quorum sensing (QS) is intricate cell-to-cell communication mechanisms that facilitate bacteria to coordinate gene expression and behavior in response to population density. Central to this process are auto inducer molecules like N-acyl homoserine lactones (AHLs), which accumulate in the environment as bacterial populations grow (Shahni et al., 2025). Once a threshold concentration is reached, these molecules bind to specific receptors, triggering a cascade of gene regulatory events that alter bacterial physiology and behavior (Roemhild et al., 2022). In Gram-negative bacteria, AHLs are synthesized by enzymes like acyl-homoserine-lactone synthase and play a key role in regulating different phenotypes, such as biofilm formation, virulence factor production, and motility (Liu et al., 2022). In the bioluminescent bacterium *Aliivibriofischeri*, the addition of AHLs activates the lux operon, leading to light production as a classic example of QS-mediated gene regulation. Cyanobacteria, although traditionally associated with photosynthesis, also exhibit QS-like behaviors. They produce and respond to AHLs, influencing processes like biofilm development and secondary metabolite production (Coolahan & Whalen, 2025). The occurrence of AHLs in cyanobacterial communities suggests a role in coordinating activities that are beneficial at higher cell densities, such as the formation of protective biofilms and toxins production during bloom phases (Mohammed et al., 2024). Furthermore, QS also facilitates limited intra-species and even inter-kingdom signaling, where AHLs affecting of other microbial behavior species and eukaryotic hosts. This interplay can modulate microbial community structure and function, influencing nutrient cycling, pathogenicity, and ecosystem stability (Singh et al., 2022). Understanding the intricacies of QS and its role in microbial associations during cyanobacterial blooms is decisive. It offers insights into bloom regulation and potential mitigation strategies for harmful algal blooms on aquatic ecosystems and human health (Feng et al., 2024).

12.5.3 Allelopathy and Antagonistic Interactions

Cyanobacteria generate allelopathic relations that reduce competing microorganisms'growth and survival, thereby affecting microbial community dynamics during bloom phases (Wang et al., 2025a, 2025b). These allelochemicals, often-secondary metabolites, serve as chemical defenses, providing cyanobacteria with a competitive improvement in nutrient limited aquatic environments. Cyanobacterin is one of the prominent allelochemicals, produced by *Scytonemahofmanni*, which functions as a photosynthesis inhibitor that influences the synthesis of algicide and herbicide (Litchman & Tezanos Pinto, 2024). By worrying the photosynthetic mechanism of neighboring phytoplankton, cyanobacterin effectively suppresses potential competitors, facilitating the dominance of cyanobacterial populations during blooms. Allelopathic relations also surpass phytoplankton competition (Lakshmikandan et al., 2024). Cyanobacteria can influence bacterial communities by releasing compound that modulating their behavior. Certain cyanobacterial metabolites have shown to inhibit quorum sensing in bacteria, disturbing communication pathways crucial for biofilm formation and virulence. This interference can change microbial community structure, potentially dipping predation or competition pressures on cyanobacteria (Mohammed & Arockiaraj, 2024). The ecological implications of cyanobacterial allelopathy are profound. By chemically suppressing competitors and modulating microbial relations, cyanobacteria can create and maintain bloom conditions, frequently leading to loss of biodiversity and changed nutrient cycling within aquatic ecosystems (Sukenik & Kaplan, 2021). Understanding these allelopathic mechanisms can inform for developing strategies to forecast and mitigate the impacts of harmful cyanobacterial blooms (He et al., 2023).

12.5.4 Mutualistic Interactions

Cyanobacteria and heterotrophic bacteria engage in mutualistic relatioships, particularly limited nutrient conditions (Litchman & Tezanos Pinto, 2024). For example, in oligotrophic marine environments, cyanobacteria like *Prochlorococcus* and *Synechococcus connect* are in symbiotic relationship with heterotrophic bacteria, affecting nutrient cycling and enhancing community stability (Zhang et al., 2021). These relations involve the exchange of metabolic products, where cyanobacteria supply organic carbon through photosynthesis, and heterotrophic bacteria remineralize necessary nutrients such as phosphorus and nitrogen, making them available for cyanobacterial uptake. The mutualistic interaction is facilitated by nutrient availability (Zeng, 2024). For example, in co-culture systems, the addition of elevated nutrient concentrations led to the breakdown of mutualism between cyanobacteria and bacteria, suggesting the existence of a nutrient threshold that manages the balance between cooperation and competition (Tong et al., 2023). Heterotrophic bacteria contribute to cyanobacterial resilience by providing essential compounds vitamins (e.g., B12), enzymes, and protective agents against oxidative stress (Sing et al., 2023). In return, cyanobacteria support heterotrophic growth by releasing dissolved organic matter (DOM), which serves as a carbon and energy source. This reciprocal exchange improves the stability of the microbial consortium, allowing sustained carbon sequestration and primary productivity (Ataeian et al., 2022). Additionally, the dynamic nature of cyanobacteria and bacteria interactions is obvious in long-term co-cultivation studies, where relationships can shift from antagonism to mutualism. Such transitions are driven by alteration in community composition and the establishment of self-sufficient nutrient cycles, particularly nitrogen cycling processes like fixation and denitrification (Tong et al., 2023).

12.6 Molecular Study

12.6.1 Gene Expression Dynamics in Cyanobacteria During Blooms

Cyanobacteria like *Microcystis* exhibit distinct diurnal pattern in the gene expression during bloom events: daytime gene expression focuses on photosynthesis and energy metabolism genes, while nighttime expression shifts toward DNA replication, repair, and toxin biosynthesis pathways (Louer et al., 2020). But genes involved in secondary metabolite production, including polyketide synthase (PKS) and non-ribosomal peptide synthase (NRPS) clusters, do not show any diurnal pattern: these are actively transcribed throughout the day and night, demonstrating a continuous role in bloom maintenance and microbial interactions (Blyth et al., 2024).

12.6.2 Functional Interplay Between Cyanobacteria and Associated Microbial Communities

Cyanobacterial blooms are characterized by multifaceted interactions with co-occurring microbes, including Proteobacteria, Bacteroidetes, and Actinobacteria. These connections involve coordinated gene expression related to nutrient cycling such as nitrogen (N) and phosphorus (P) metabolism, and stress responses (Wang et al., 2021a, 2021b). Associated bacteria express genes for nitrogen assimilation like glutamine synthetase, and phosphorus acquisition like alkaline phosphatase, complementing cyanobacterial metabolic activities and supporting ecosystems stability and resilience of bloom (Xiong et al., 2023).

12.6.3 Role of Secondary Metabolites and Toxins in Microbial Interactions

Cyanobacteria generate secondary metabolites, including toxins like microcystins, which affecting microbial community structure and relations. These compounds can inhibit the growth of rival microorganisms and influencing nutrient cycling processes (Pei et al., 2020). Several bacteria possess gene clusters like mlrA-D enabling them to degrade microcystins, illustratinga dynamic interaction between toxin production and degradation in bloom communities (Wang et al., 2024a, 2024b).

12.6.4 Influence of Environmental Factors on Gene Expression and Community Dynamics

Environmental parameters, like nutrient availability, temperature, and light conditions, significantly impact gene expression patterns in both cyanobacteria and related microbes. For instance, phosphorus limitation triggers the expression of genes for alternative phosphorus acquisition, while changes in the temperature influence the composition and metabolic activities of the microbial community. These factors modulate the relations and functional dynamics in bloom ecosystems (Cordeiro-Araújo et al., 2022).

12.6.5 Metagenomic and Metatranscriptomic Insights into Bloom Microbiomes

Higher omics approaches, including metagenomics and metatranscriptomics, have showing the genetic and functional diversity of bloom-associated microbial communities. The studies expose that cyanobacterial blooms not solely dominated by cyanobacteria, however, entail a consortium of microorganisms with specialized functions (Xu et al., 2022). Metagenomic analyses have recognized symbiotic relations where heterotrophic bacteria contribute to nutrient recycling, and supporting cyanobacterial growth, emphasizing the significance of microbial interactions in bloom development and persistence (Yang et al., 2021a, 2021b).

12.7 Impacts of Interactions on Bloom Dynamics

12.7.1 Effects on Cyanobacterial Growth

Cyanobacterial blooms engage intricate relationships between cyanobacteria and surrounding microbial communities. These relationships can either stimulate or inhibit cyanobacterial growth, thereby influencing bloom dynamics (Du et al., 2022). Mutualistic relations between cyanobacteria and heterotrophic bacteria frequently play a key role in bloom dynamics. Heterotrophic bacteria can boost cyanobacterial growth by remineralizing nutrients such as nitrogen and phosphorus, making them more accessible to cyanobacteria. Additionally, some heterotrophic bacteria create vitamins and growth factors that are essential for cyanobacterial proliferation (Bhardwaj et al., 2024). Conversely, microbial relations can also inhibit cyanobacterial growth. Several bacteria produce antagonistic compounds in competitive interactions that suppress cyanobacterial populations. For example, the presence of sure bacterial taxa has connected with induced cyanobacterial abundance, suggesting inhibitory effects (Banerji et al., 2024). Moreover, microbial community composition moves during diverse bloom stages can lead to changes in the functional dynamics of the ecosystem, potentially limitation bloom persistence (Li et al., 2025a, 2025b, 2025c). The structural and functional succession of microbial communities during cyanobacterial blooms further underscores the affect of these associations. During bloom events, microbial diversity often decreases, with specific bacterial groups dominating and contributing to the degradation of cyanobacterial metabolites. In difference, post bloom phases may observer increased microbial diversity and the emergence of taxa involved in nutrient recycling and organic matter transformation (Fig. 12.5) (Zhou et al., 2020).

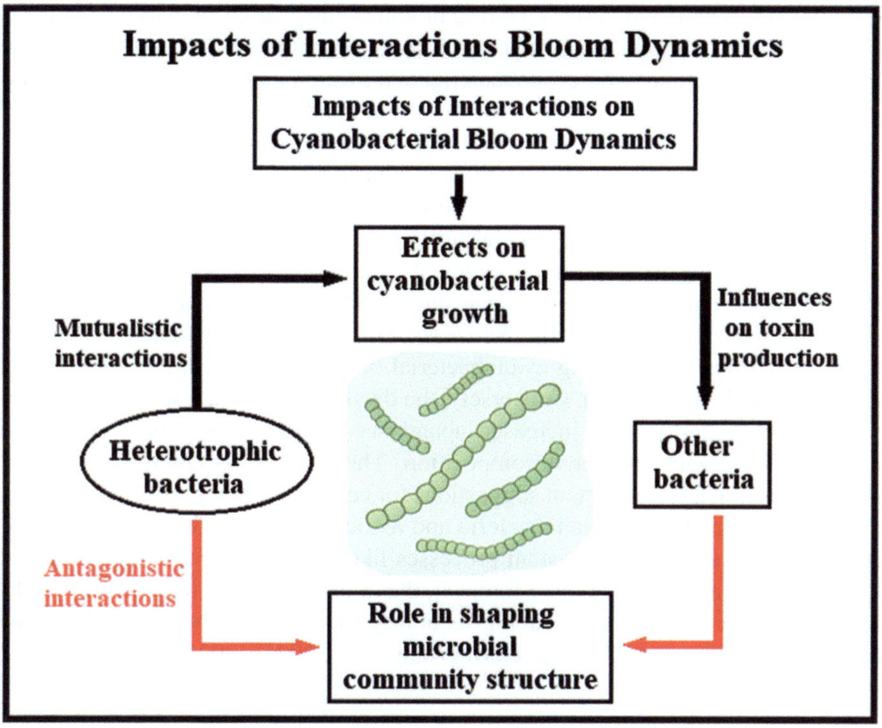

Fig. 12.5 Impacts on interactions on algal blooms

12.7.2 Influences on Toxin Production

Cyanobacterial blooms are multifaceted ecological phenomena influenced by complex interactions between cyanobacteria and surrounding microbial communities. These interactions can either stimulate or inhibit cyanobacterial growth, influencing blooms dynamics (Yu, 2025). Mutualistic associations between cyanobacteria and heterotrophic bacteria often play a pivotal role in bloom dynamics. Heterotrophic bacteria can increase cyanobacterial growth by remineralizing nutrients such as nitrogen and phosphorus, making them more accessible to cyanobacteria. For instance, certain bacteria associated with cyanobacteria promote nutrient cycling, thereby promoting bloom formation and stability (Zhang et al., 2021).

12.7.3 Role in Shaping Microbial Community Structure

Cyanobacterial blooms also influence microbial community structures, inducing diversity and altering composition, which impacts ecosystem functioning and resilience (Philippot et al., 2021). During bloom events, there is a marked decline in microbial diversity, particularly among bacterial communities. This decline is due to the dominance of specific cyanobacterial taxa, which outcompete other microorganisms for resources and change environmental conditions like oxygen levels and nutrient availability (Yan et al., 2025). For example, studies have revealed that cyanobacterial blooms can lead to decreased alpha diversity in bacterial communities, indicating a loss of microbial richness. The structural alteration in microbial communities is not uniform but displays spatial and temporal variabilities. Distinct microbial assemblages characterize different phases of a bloom like the beginning, peak, and decline (Arena et al., 2025). For instance, during the peak bloom event, certain bacterial genera like *Acinetobacter* and *Rhodobacter* are found in abundant, potentially due to their roles in degrading cyanobacterial biomass or resisting cyanobacterial toxins (Badshah et al., 2024). Conversely, the decline phase often sees a resurgence of microbial diversity, with increased abundance of bacteria involved in nutrient recycling and organic matter decomposition. These moves in microbial community composition have important suggestions for ecosystem functioning (Raza et al., 2023). The dominance of cyanobacteria and associated decline in microbial diversity can impair essential ecosystem processes like nutrient cycling, organic matter decomposition, and energy flow. Moreover, the microbial interactions can affect the resilience of the ecosystem, reducing its ability to recover from disturbances and maintain functional stability (Fig. 12.6) (Wang et al., 2025a, 2025b).

12.8 Implications for Bloom Management and Mitigation

12.8.1 Understanding Interactions to Predict Bloom Dynamics

Cyanobacterial blooms are dynamic ecosystems shaped by complicated interactions between cyanobacteria and co-occurring microbial communities (Krausfeldt et al., 2024). These connections significantly affect bloom formation, succession, and persistence, offering critical insights into bloom mitigation and management strategies. During bloom events, cyanobacteria are in mutualistic relationships with heterotrophic bacteria, particularly within the phycosphere and the microenvironment surrounding cyanobacterial cells (Litchman & Tezanos, 2024). These heterotrophs facilitate the remineralization of organic matter, converting complex compounds into bioavailable nutrients such as phosphate ammonium and, therefore supporting cyanobacterial growth in nutrient-limited conditions (Li et al., 2024). For example, in Harsha Lake, the transition from nitrogen-fixing *Anabaena* to non-nitrogen-fixing *Planktothrixwas* and *Microcystis* is connected to increased expression of genes involved in organic substrate processing by heterotrophic bacteria, indicating their role in alleviating nutrient limitations for cyanobacteria (Zhang et al., 2020).

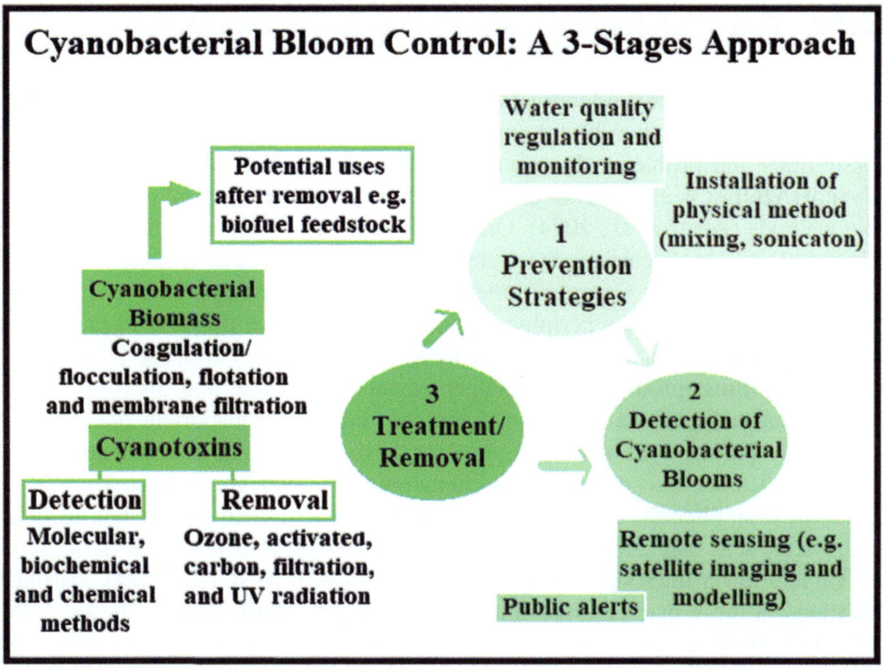

Fig. 12.6 Strategies to control cyanobacterial bloom

Biological and abiotic factors impact the composition and stability of microbial communities during blooms. Cyanobacteria and particle-associated bacteria show stable relationships, indicating that bacteria-bacteria interactions contribute to the survival of certain microbial assemblages throughout bloom (Zoccarato & Grossart, 2019). These communities are also significantly shaped by environmental factors such as water chemistry, temperature, and nutrient concentrations. Viral infections of cyanobacteria called cyanophages further alter bloom dynamics by lysing cyanobacterial cells and releasing nutrients and organic materials into the surrounding environment (Ain et al., 2024). By changing the availability of nutrients and encouraging the growth of other microbial taxa, this viral lysis affects the makeup and function of microbial communities in addition to controlling cyanobacterial numbers (Šulčius et al., 2025). It is crucial to comprehend these complex relationships in order to create bloom dynamics prediction models. The structure and function of microbial communities may be included in these models to improve bloom forecast accuracy and guide focused mitigation techniques (Wang et al., 2024a, 2024b).

12.8.2 Potential Applications of Microbial Manipulation

For bloom management and mitigation, the complex interactions between cyanobacteria and coexisting microbial communities during bloom occurrences offer both possibilities and problems. In order to prevent dangerous cyanobacterial blooms (CyanoHABs), it is essential to comprehend these interactions and design sustainable methods (Abate et al., 2024). One significant way that microbial communities might inhibit cyanobacteria is by allelopathy, which is the chemical suppression of one species by another. Complex collections of photoautotrophic and heterotrophic microorganisms known as periphyton biofilms have shown promise in preventing cyanobacterial blooms (Wang et al., 2025a, 2025b). Bloom collapse is the result of these biofilms' production of allelochemicals like indole and 3-oxo-α-ionone, which damage thylakoid membranes and stop electron transport in photosystem II, hence impairing cyanobacterial photosynthesis (Van et al., 2023). Allelopathic characteristics are also displayed by submerged macrophytes such as *Myriophyllumaquaticum.* Phenolic substances found in their aqueous extracts, including gallic acid, ellagic acid, and catechin, have been shown to suppress *Microcystis aeruginosa* growth and lower microcystin-LR concentrations (WIJESINGHE, 2024). Seasonal variations in these macrophytes' allelopathic capacity, with greater efficacy noted in the fall, indicate that timing is essential when using these biological controls. In addition to allelopathy, microbial manipulation presents encouraging approaches to bloom reduction (Lakshmikandan et al., 2024). Through direct cell-to-cell contact, some bacteria, such as *Streptomyces globisporus* strain G9, have algicidal action against *Microcystis aeruginosa,* which significantly lowers cyanobacterial populations. These relationships indicate their potential as biocontrol agents by reducing microcystin production and suppressing cyanobacterial development (Zhang et al., 2023). Viral infections of cyanobacteria called cyanophages lyse host cells, releasing dissolved organic materials into the microbial loop and controlling cyanobacterial populations (Ain et al., 2024).

12.8.3 Challenges and Limitations

Complex interactions between cyanobacteria and various microbial communities are a feature of cyanobacterial blooms. Bloom dynamics and ecosystem health are greatly influenced by these interactions, which include mutualistic, competitive, and antagonistic relationships (Grasso et al., 2022). Cyanobacterial aggregates are frequently colonized by heterotrophic bacteria, which aid in the remineralization of organic materials and the cycling of nutrients. These microorganisms can give cyanobacteria essential nutrients, which will promote their growth and bloom permanence (Bhardwaj et al., 2024). In contrast, there is a reciprocal relationship between cyanobacteria and heterotrophic bacteria, as the latter use the organic chemicals released by the former as substrates. However, environmental disturbances like temperature changes and nutrient inflow from runoff can upset the stability of these interactions and alter the shapes and functions of microbial communities (Yang et al., 2021a, 2021b). The regulation of toxin production is another aspect of microbial communities' function in bloom development. The expression of cyanobacterial toxin genes is influenced by certain bacteria, which may have an impact on bloom toxicity (Anabtawi et al., 2024). Furthermore, by altering gene expression linked to stress reactions and food intake, microbial interactions can impact the resilience of cyanobacterial communities (Pimentel et al., 2025). Even with improvements in our knowledge of these microbial relationships, there are still difficulties in extrapolating lab results to environmental environments. Simplified or axenic cultures, which are frequently employed in laboratory research, are not representative of the complexity of real microbial communities (Parikh et al., 2025). The need for in situ investigations that capture the dynamic nature of microbial interactions during blooms is highlighted by this contradiction, which restricts the application of laboratory results to field conditions. Biological controls have been investigated in cyanobacterial bloom management tactics, impacting the introduction of cyanophages and algicidal bacteria (Anabtawi et al., 2024). Although these methods appear promising, host specificity, the possibility of resistance, and unexpected effects on nontarget organisms limit their effectiveness. Furthermore, it is necessary to apply caution and conduct thorough ecological assessments, because the ecological effects of modifying microbial populations are not entirely understood (Ren et al., 2023a, 2023b).

12.9 Future Research Directions

12.9.1 Emerging Research Areas

Bloom dynamics and biogeochemical cycles are greatly influenced by the intricate interactions between cyanobacteria and related microbial populations, which define complex microbial ecosystems (Chen et al., 2025). Deeper understanding of these relationships is made possible by recent developments in metagenomics, metatranscriptomics, and metabolomics. According to metagenomic investigations, cyanobacterial blooms contain a variety of heterotrophic bacteria that support bloom stability and nutrient cycling, rather than just cyanobacteria (Butowski et al., 2025). For instance, research has shown that genes related to the storage of nutrients, the consumption of organic nitrogen and phosphorus, and defensive mechanisms against phages and oxidative stress increased during blooms (Krausfeldt et al., 2024). These functional pathways imply that microbial interactions are essential for enhancing cyanobacterial harmful algal blooms (cyanoHABs') resistance. Studies using metatranscriptomic analysis have provided more insight into the dynamic character of microbial communities during bloom occurrences (Zhu et al., 2021). Large-scale metatranscriptomic analyses conducted during cyanobacterial blooms in the Baltic

Sea have revealed alterations in gene expression linked to the cycling of carbon and nitrogen, indicating that cyanobacteria affect biogeochemical cycles by supplying nutrient pulses and have an impact on the makeup and activity of co-occurring microbial populations (Munkes et al., 2021). The symbiotic connections between cyanobacteria and heterotrophic bacteria are proved by metabolomic techniques (Teikari et al., 2025). It has been discovered that heterotrophic bacteria in communities dominated by *Microcystis* aid in the manufacture of vital vitamins like B12 and the breakdown of inhibitory chemicals, both of which are necessary for the growth of cyanobacteria (Croci et al., 2025). This cooperative relationship emphasizes how crucial microbial consortia are to preserving bloom stability. Moreover, it has been established that viruses, especially cyanophages, influence cyanobacterial populations (Ain et al., 2024). Through auxiliary metabolic genes (AMGs), cyanophages can affect host metabolism, influencing nutrition absorption and photosynthesis. These relationships have wider ramifications in addition to regulating cyanobacterial abundance (Bhattarai et al., 2025).

12.9.2 Opportunities for Interdisciplinary Research

The need for multidisciplinary study techniques is highlighted by the complex interactions that occur between cyanobacteria and other microbial communities during bloom occurrences. Integrating knowledge from environmental science, ecology, and microbiology is essential to understand the complex dynamics of these events (Idowu, 2024). According to recent research, cyanobacterial blooms are affected by intricate interactions with co-occurring microbial communities in addition to being the consequence of nutrient enrichment (Wang et al., 2021a, 2021b). For example, studies have demonstrated that when resources are plentiful during bloom occurrences, cyanobacteria can outcompete other microbes, which can have detrimental impacts on the local microbial population (Du et al., 2022). Additionally, more intricate and stable microbial networks are seen during the blooms' breakdown phase, indicating that microorganisms are working together to break down the cyanobacterial biomass. Traditionally, the goal of laboratory research has been to separate cyanobacteria in axenic cultures in order to comprehend their physiology (Chen et al., 2025). But the significance of the microbiome in natural environments is ignored by this method. The significance of microbial interactions in bloom dynamics is highlighted by co-culture experiments that have shown that the presence of heterotrophic bacteria can affect cyanobacterial gene expression and metabolic activity (Tong et al., 2023). Field research adds context-specific insights regarding bloom occurrences to laboratory findings. (West et al., 2021). Studies conducted in various freshwater environments, for instance, have shown that sediment-enriched cyanobacteria are essential for bloom development, with sediment microorganisms encouraging nutrient release that further stimulates cyanobacterial growth (Schreckinger, 2023). These findings highlight how important it is for bloom research to take into account both benthic and pelagic microbial ecosystems. A thorough knowledge of cyanobacterial blooms requires the integration of laboratory and field research (Astuti & Govindaraju, 2025). While field research offers ecological validity and captures the complexity of natural systems, laboratory experiments enable controlled variable manipulation and in-depth mechanistic examinations. By connecting these methods, scientists may create precise models to forecast bloom events and create efficient management plans (Gerhard et al., 2023).

12.9.3 Priorities for Future Studies

To understand the underlying mechanisms and create efficient management plans, a multidisciplinary research approach is necessary to address the intricate interactions that occur between cyanobacteria and other microorganisms during blooms (Abate et al., 2024). Examining the molecular and ecological mechanisms that govern these interactions should be the focus of future research. It is essential to comprehend the genetic, metabolic, and ecological elements that affect the behavior of cyanobacteria and their interactions with other microorganisms (Ain et al., 2024). This involves investigating the effects of environmental factors on the makeup and function of microbial communities, such as temperature, light, and nutrition availability. Technological developments in metabolomics, metatranscriptomics, and high-throughput sequencing provide hitherto unheard-of chances to examine the functional roles of distinct microbial taxa within bloom communities (Flörl et al., 2025). The mechanisms by which cyanobacteria and other microbes exchange nutrients, generate toxins, and affect one another's growth and survival can be uncovered by these tools. For instance, research has demonstrated that cyanobacteria can release compounds known as allelopathic compounds that inhibit the growth of competing phytoplankton, thereby increasing their own dominance during blooms (Patrick et al., 2025). On the other hand, certain bacteria can break down cyanotoxins, reducing the negative consequences of blooms. Simultaneously, creating efficient cyanobacterial bloom management plans necessitates a coordinated strategy that blends technological innovation with ecological knowledge (Anabtawi et al., 2024). The intricate and dynamic structure of bloom ecosystems has made traditional techniques, such as nutrient reduction and physical removal, less effective. Using biological control agents, such bacteriophages, which selectively target and lyse bloom-forming cyanobacteria to reduce their biomass and toxin production, is one of the emerging techniques (Grasso et al., 2022). Furthermore, bloom prediction and monitoring can be improved by the use of artificial intelligence and machine learning algorithms, enabling prompt and focused treatments. Additionally, addressing the complex issues raised by cyanobacterial blooms requires interdisciplinary cooperation (Zahir et al., 2024). The establishment of comprehensive management frameworks that consider ecological, economic, and social concerns can be facilitated by involving microbiologists, ecologists, environmental engineers, and policy makers. These kinds of partnerships can result in the creation of adaptive management strategies that are sensitive to shifting environmental circumstances and new scientific findings (Cannizzo et al., 2025).

12.10 Conclusion

Clarifying the complex interactions that control aquatic ecosystems requires an understanding of the relationships that cyanobacteria have with coexisting microbial populations during bloom occurrences. Intricate ecological phenomena are influenced by a variety of biotic interactions, such as competition, mutualism, and antagonism among bacteria, archaea, viruses, and cyanobacterial blooms, which also involve the expansion of a single taxonomic group, which is characterized by protists. These relationships influence the biochemical environment, community succession, and nutrient cycling, which in turn affects bloom initiation, persistence, and decline. The complex web of microbial interdependencies emphasizes that cyanobacterial dominance cannot be fully comprehended in isolation; rather, it results from the dynamic relationship within the larger microbiome, which is influenced by environmental factors like light, temperature, and nutrient availability. The need for integrated research methods that combine molecular, ecological, and computational techniques to analyze microbial interactions at various scales—from genes to ecosystems—is highlighted by this complexity. Such understandings are crucial for both practical applications and the advancement of basic ecological theory. The generation of toxins and oxygen depletion from cyanobacterial blooms present serious ecological and public health hazards, making the creation of prediction models that take into account the dynamics of microbial communities necessary. Strategies for managing blooms must take into account co-occurring microorganisms, whose actions may enhance or inhibit bloom development, in addition to cyanobacteria. Longitudinal and high-resolution investigations must be given priority in future study in order to document changes in microbial consortia and their functional characteristics over time during bloom cycles. Furthermore, developing techniques like metagenomics, metabolomics, and single-cell analysis will improve our comprehension of interactions between microbes and their environments. Designing more efficient, ecosystem-based management strategies to lessen the effects of blooms will be made easier by incorporating this data into ecosystem models. In the end, creating sustainable interventions to regulate cyanobacterial blooms while maintaining aquatic biodiversity and ecosystem services may be possible by integrating fundamental microbial ecology with practical environmental management.

References

Abate, R., Oon, Y. L., Oon, Y. S., Bi, Y., Mi, W., Song, G., &Gao, Y. (2024). Diverse interactions between bacteria and microalgae: A review for enhancing harmful algal bloom mitigation and biomass processing efficiency. *Heliyon.*

Ahmad, N. B., Jaafaru, M. S., Isa, Z., Abdulhamid, Y., Kakudi, R. A., Ugya, A. Y., & Meguellati, K. (2024). High pollution loads engineer oxygen dynamics, ecological niches, and pathogenicity shifts in freshwater environments. *Journal of Hazardous Materials Advances*, 100425.

Ain, Q. U., Wu, K., Wu, X., Bai, Q., Li, Q., Zhou, C. Z., & Wu, Q. (2024). Cyanophage-encoded auxiliary metabolic genes in modulating cyanobacterial metabolism and algal bloom dynamics. *Frontiers in Virology, 4*, 1461375.

Anabtawi, H. M., Lee, W. H., Al-Anazi, A., Mohamed, M. M., & Aly Hassan, A. (2024). Advancements in biological strategies for controlling harmful algal blooms (HABs). *Water, 16*(2), 224.

Arena, M., Pratolongo, P., Loisel, H., Tran, M. D., Jorge, D. S., & Delgado, A. L. (2025). Phenological patterns of coastal algal blooms in the optically complex waters of the northern Patagonian Continental Shelf: A remote sensing approach.

Astuti, W., & Govindaraju, R. (2025). Characterization and modeling of harmful algal blooms: A Review. *Journal of Hydraulic Engineering, 151*(2), 03125001.

Ataeian, M., Liu, Y., Kouris, A., Hawley, A. K., & Strous, M. (2022). Ecological interactions of cyanobacteria and heterotrophs enhances the robustness of cyanobacterial consortium for carbon sequestration. *Frontiers in Microbiology, 13*, 780346.

Badshah, S. L., Stirbet, A., Siddiquee, M., Govindjee, G., Kang, D. W., Bridgeman, T., & Seo, Y. (2024). Inhibition of CO_2 fixation as a potential target for the control of freshwater cyanobacterial harmful algal blooms. *ACS ES&T Water, 4*(8), 3309–3319.

Banerji, A., Brinkman, N. E., Davis, B., Franklin, A., Jahne, M., & Keely, S. P. (2024). Food webs and feedbacks: The untold ecological relevance of antimicrobial resistance as seen in harmful algal blooms. *Microorganisms, 12*(11), 2121.

Barruffa, A. S., Sposito, V., & Faggian, R. (2021). Climate change and cyanobacteria harmful algae blooms: Adaptation practices for developing countries. *Marine and Freshwater Research, 72*(12), 1722–1734.

Belmonte, G. (2025). Mediterranean sea biodiversity. In *Elements of Pelagos biology: With focus on the Mediterranean Sea* (pp. 97–146). Springer Nature Switzerland.

Bhardwaj, A., Singh, P., Gupta, N., Bhattacharjee, S., Srivastava, A., Parida, A., & Mishra, A. K. (2024). Cyanobacteria: A key player in nutrient cycling. In *Cyanobacteria* (pp. 579–596). Academic Press.

Bhattarai, B., Bhattacharjee, A. S., Coutinho, F. H., Li, H., Chadalavada, S., & Goel, R. (2025). Bacteriophages carry auxiliary metabolic genes related to energy, sulfur and phosphorus metabolism during a harmful algal bloom in a freshwater lake. *Chemosphere, 370*, 143819.

Blyth, H. R., Zarsav, A., Smith, D., Kanyuka, K., Hassall, K. L., Michaelson, L. V., et al. (2024). Feeling the Heat: Investigating the dual assault of *Zymoseptoria tritici* and heat stress on wheat (*Triticum aestivum*). In *32nd Fungal genetics conference.*

Butowski, C. F., Dixit, Y., Reis, M. M., & Mu, C. (2025). Metatranscriptomics for understanding the microbiome in food and nutrition science. *Metabolites, 15*(3), 185.

Cannizzo, Z. J., Hunter, K. L., Hutto, S., Selgrath, J. C., & Wenzel, L. (2025). Future-proofing the global system of marine protected areas: Integrating climate change into planning and management. *Marine Policy, 171*, 106420.

Chankeshwara, N. N., Pant, M., & Pant, K. (2025). The cyanobacterial renaissance: Bioactive discoveries and future prospects. In *Bio-prospecting of novel microbial bioactive compounds for sustainable development* (pp. 187–201). Springer.

Chatterjee, S., & More, M. (2023). Cyanobacterial harmful algal bloom toxin microcystin and increased vibrio occurrence as climate-change-induced biological co-stressors: Exposure and disease outcomes via their interaction with gut–liver–brain axis. *Toxins, 15*(4), 289.

Chen, J., Zhou, Y., Zhang, Y., Guo, Q., Zhang, S., & Ge, G. (2025). Succession of bacterial community during microbially driven cyanobacterial organic matter degradation and its relationship to water quality in Taihu Lake, China. *Aquatic Ecology, 59*(1), 125–144.

Coolahan, M., & Whalen, K. E. (2025). A review of quorum-sensing and its role in mediating interkingdom interactions in the ocean. *Communications Biology, 8*(1), 179.

Cordeiro-Araújo, M. K., Lorenzi, A. S., Chia, M. A., Mota, E. C., & do CarmoBittencourt-Oliveira, M. (2022). Insights into the impact of increasing temperature, light intensity, and UV-B exposure on the circadian rhythm of microcystin production and release, and the expression of mcy genes in the cyanobacterium *Microcystis aeruginosa*. *Journal of Applied Phycology*, 1–12.

Croci, C., de la Escalera, G. M., Kruk, C., Segura, A., Alvarez, S. D., & Piccini, C. (2025). Selective enrichment of active bacterial taxa in the Microcystis associated microbiome during colony growth. *PeerJ, 13*, e19149.

Cui, Z., Yang, C., Ma, L., Gu, X., Shen, X., Wan, B., et al. (2025). Floating-leaved and submerged macrophytes suppress filamentous cyanobacteria blooms and 2-MIB episodes in eutrophic shallow lakes. *Journal of Hazardous Materials, 492*, 138163.

Demertzioglou, M., Genitsaris, S., Mazaris, A. D., Kyparissis, A., Voutsa, D., Kozari, A., et al. (2022). A catastrophic change in a European protected wetland: From harmful phytoplankton blooms to fish and bird kill. *Environmental Pollution, 312*, 120038.

Dey, P. (2024). Good girl goes bad: Understanding how gut commensals cause disease. *Microbial Pathogenesis*, 106617.

Du, C., Li, G., Xia, R., Li, C., Zhu, Q., Li, X., et al. (2022). New insights into cyanobacterial blooms and the response of associated microbial communities in freshwater ecosystems. *Environmental Pollution, 309*, 119781.

Fazi, S., Amalfitano, S., Venturi, S., Pacini, N., Vazquez, E., Olaka, L. A., et al. (2021). High concentrations of dissolved biogenic methane associated with cyanobacterial blooms in East African lake surface water. *Communications Biology, 4*(1), 845.

Feng, L., Wang, Y., Hou, X., Qin, B., Kutser, T., Qu, F., et al. (2024). Harmful algal blooms in inland waters. *Nature Reviews Earth & Environment*, *5*(9), 631–644.

Flörl, L., Meyer, A., & Bokulich, N. A. (2025). Exploring sub-species variation in food microbiomes: A roadmap to reveal hidden diversity and functional potential. *Applied and Environmental Microbiology*, e00524–e00525.

Fotler, R. (2024). Toxicological profile of microcystins: Determination of the toxicokinetic and toxicodynamic mechanisms of microcystins in human cells.

Gao, G., Bai, D., Li, T., Li, J., Jia, Y., Li, J., et al. (2025). Understanding filamentous cyanobacteria and their adaptive niches in Lake Honghu, a shallow eutrophic lake. *Journal of Environmental Sciences, 152*, 219–234.

Gerhard, M., Koussoroplis, A. M., Raatz, M., Pansch, C., Fey, S. B., Vajedsamiei, J., et al. (2023). Environmental variability in aquatic ecosystems: Avenues for future multifactorial experiments. *Limnology and Oceanography Letters, 8*(2), 247–266.

Grasso, C. R., Pokrzywinski, K. L., Waechter, C., Rycroft, T., Zhang, Y., Aligata, A., et al. (2022). A review of cyanophage–host relationships: Highlighting cyanophages as a potential cyanobacteria control strategy. *Toxins, 14*(6), 385.

He, Y., Zhou, Y., Zhou, Z., He, J., Liu, Y., Xiao, Y., et al. (2023). Allelopathic effect of pyrogallic acid on cyanobacterium *Microcystis aeruginosa*: The regulatory role of nitric oxide and its significance for controlling harmful algal blooms (HABs). *Science of The Total Environment*, *858*, 159785.

Hwang, S. O., Cho, I. H., Kim, H. K., Hwang, E. A., Han, B. H., & Kim, B. H. (2024). Toward a brighter future: Enhanced sustainable methods for preventing algal blooms and improving water quality. *Hydrobiology, 3*(2), 100–118.

Ibrahim, H. A., Abdeltwab, A. M., Amer, M. S., Gad, A. A., & El Sayed, W. M. (2023). Geomicrobiology of African lakes. In *Lakes of Africa* (pp. 195–246). Elsevier.

Idowu, E. (2024). Microbial ecology: Understanding the dynamics of microbial communities and their impact on ecosystems.

Igwaran, A., Kayode, A. J., Moloantoa, K. M., Khetsha, Z. P., & Unuofin, J. O. (2024). Cyanobacteria harmful algae blooms: Causes, impacts, and risk management. *Water, Air, & Soil Pollution, 235*(1), 71.

Kang, M., Jeong, S., Ko, S. R., Kim, M. S., & Ahn, C. Y. (2024). Biotechnological approaches for suppressing microcystis blooms: Insights and challenges. *Applied Microbiology and Biotechnology, 108*(1), 466.

Krausfeldt, L. E., Shmakova, E., Lee, H. W., Mazzei, V., Loftin, K. A., Smith, R. P., et al. (2024). Microbial diversity, genomics, and phage–host interactions of cyanobacterial harmful algal blooms. *MSystems, 9*(7), e00709–e00723.

Kumar, A., Maniyar, C. B., Fiorentino, I. R., & Mishra, D. R. (2025). Remote sensing of cyanobacterial harmful algal blooms: Current trends and future directions. *Progress in Environmental Geography, 4*(1), 131–150.

Kumari, N., Mishra, S., Häder, D. P., & Sinha, R. P. (2022). Cyanobacterial blooms and their implications in the changing environment. *Advances in Environmental and Engineering Research, 3*(1), 1–41.

Kutlu, D., Kasalak, M. A., & Bahar, M. (2025). Assessing climate change impacts on outdoor recreation: Insights from visitor and business perspectives. *Sustainability, 17*(8), 3400.

Lakshmikandan, M., Li, M., & Pan, B. (2024). Cyanobacterial blooms in environmental water: Causes and solutions. *Current Pollution Reports, 10*(4), 606–627.

Lan, J., Liu, P., Hu, X., & Zhu, S. (2024). Harmful algal blooms in eutrophic marine environments: Causes, monitoring, and treatment. *Water, 16*(17), 2525.

Li, H., Jiang, M., Li, P., Xu, Z., Jiang, P., Chen, L., et al. (2024). Picocyanobacterial-bacterial interactions sustain cyanobacterial blooms in nutrient-limited aquatic environments. *Environmental Research, 260*, 119508.

Li, S., Yan, X., Chen, H., Jeppesen, E., Xiao, P., Jin, L., et al. (2025). Cyanobacterial blooms specifically alter the dispersal-mediated taxonomic and functional vertical similarity of microbial communities in a subtropical reservoir. *Water Research, 281*, 123574.

Li, F., Xu, M. B., Pan, L. H., Li, J., Lan, C. B., Li, Z., et al. (2025). Ammonifying and phosphorus-solubilizing function of *Aliikangiella maris* sp. nov. isolated from *Phaeocystis globosa* bloom and algal–bacterial interactions. *Frontiers in Microbiology, 16*, 1516993.

Li, S., Murava, R. T., Zhang, Q., Zhou, T., Omoregie, A. I., Rajasekar, A., & Ouahbi, T. (2025b). Linking antibiotic residues and antibiotic resistance genes to water quality parameters in urban reservoirs: A seasonal perspective. *Environments, 12*(3), 96.

Litchman, E., & de Tezanos Pinto, P. (2024). Ecology of algae and cyanobacteria (phytoplankton). In *Wetzel's limnology* (pp. 511–538). Academic Press.

Liu, L., Zeng, X., Zheng, J., Zou, Y., Qiu, S., & Dai, Y. (2022). AHL-mediated quorum sensing to regulate bacterial substance and energy metabolism: A review. *Microbiological Research, 262*, 127102.

Louer, E. M., Yi, G., Carmone, C., Robben, J., Stunnenberg, H. G., den Hollander, A. I., & Deen, P. M. (2020). Genes involved in energy metabolism are differentially expressed during the day–night cycle in murine retinal pigment epithelium. *Investigative Ophthalmology & Visual Science, 61*(5), 49–49.

Manganelli, M., Testai, E., Tazart, Z., Scardala, S., & Codd, G. A. (2023). Co-occurrence of taste and odor compounds and cyanotoxins in cyanobacterial blooms: Emerging risks to human health? *Microorganisms, 11*(4), 872.

Marečková, N. (2025). *The Diversity of Soil Cyanobacteria of the Dominican Republic* (Doctoral dissertation, Faculty of Science Department of Ecology and Environmental Sciences The Diversity of Soil Cyanobacteria of the Dominican Republic NatálieMarečková Bachelor's Thesis Submitted to the Department of Ecology and Environmental Sciences Faculty of Science, Palacký University Olomouc).

Massey, I. Y., Al Osman, M., & Yang, F. (2022). An overview on cyanobacterial blooms and toxins production: Their occurrence and influencing factors. *Toxin Reviews, 41*(1), 326–346.

Mishra, A., Rajput, S., Gupta, P. S., Goyal, V., Singh, S., Sharma, S., et al. (2021). Role of cyanobacteria in Rhizospheric nitrogen fixation. In *Soil nitrogen ecology* (pp. 497–519). Springer International Publishing.

Mohammed, V., &Arockiaraj, J. (2024). Unveiling the trifecta of cyanobacterial quorum sensing: LuxI, LuxR and LuxS as the intricate machinery for harmful algal bloom formation in freshwater ecosystems. *Science of the Total Environment*, 171644.

Mohanty, S., Paul, A., & Sahoo, P. K. (2025). Climate change: A strong driving force for cyanobacterial bloom. In *Handbook of aquatic microbiology* (pp. 28–39). CRC Press.

Munkes, B., Löptien, U., & Dietze, H. (2021). Cyanobacteria blooms in the Baltic Sea: A review of models and facts. *Biogeosciences, 18*(7), 2347–2378.

Paerl, H. W. (2023). Climate change, phytoplankton, and HABs. In *Climate change and estuaries* (pp. 315–334). CRC Press.

Parikh, H. S., Dave, G., Singh, P. K., & Tiwari, A. (2025). Exploring the significance of xenic and axenic cultures for diatoms: Methodological insights, scope, and biotechnological implications. *Systems Microbiology and Biomanufacturing*, 1–21.

Patrick, A. E., Florentine, S., Palmer, G. C., Mondon, J., & Mitchell, B. D. (2025). Nature of phytoplankton communities in the southern hemisphere shallow lakes, with an emphasis on the presence of submerged aquatic vegetation: A review. *Aquaculture, Fish and Fisheries, 5*(3), e70068.

Pei, Y., Xu, R., Hilt, S., & Chang, X. (2020). Effects of cyanobacterial secondary metabolites on phytoplankton community succession. *Co-Evolution of Secondary Metabolites*, 323–344.

Pereira, L. (2025). The unsung heroes of aquatic ecosystems: The vital roles of. *The Role of Plankton in Freshwater and Marine Ecology*, 3.

Philippot, L., Griffiths, B. S., & Langenheder, S. (2021). Microbial community resilience across ecosystems and multiple disturbances. *Microbiology and Molecular Biology Reviews, 85*(2), 10–1128.

Pimentel, J. D. S. M., Nunes, Á. C., Pereira, D. A., & Giani, A. (2025). Cyanobacteria gene expression in response to environmental stress and seasonal changes. *World Journal of Microbiology and Biotechnology, 41*(5), 1–14.

Raza, T., Qadir, M. F., Khan, K. S., Eash, N. S., Yousuf, M., Chatterjee, S., et al. (2023). Unraveling the potential of microbes in decomposition of organic matter and release of carbon in the ecosystem. *Journal of Environmental Management, 344*, 118529.

Ren, Z., Cai, T., Wan, Y., Zeng, Q., Li, C., Zhang, J., et al. (2023). Unintended consequences: Disrupting microbial communities of *Nilaparvata lugens* with non-target pesticides. *Pesticide Biochemistry and Physiology, 194*, 105522.

Ren, J., Ye, J., Cui, X., Zhang, X., Lang, C., Xie, W., et al. (2023). Bacterial community (free-living vs particle-attached) assembly driven by environmental factors and a more stable network in the pre-bloom period than post-bloom. *International Biodeterioration & Biodegradation, 180*, 105592.

Roemhild, R., Bollenbach, T., & Andersson, D. I. (2022). The physiology and genetics of bacterial responses to antibiotic combinations. *Nature Reviews Microbiology, 20*(8), 478–490.

Saud, S., Nawaz, T., Hassan, S., Ur Rahman, T., Rasheed, M. N., Hussain, S., & Fahad, S. (2024). Nitrogen-fixing cyanobacteria and soil enrichment for a greener future. In *Environment, climate, plant and vegetation growth* (pp. 391–424). Springer Nature Switzerland.

Schreckinger, J. (2023). *Short-and long-term succession in intermittent rivers and ephemeral streams: Sediment microbial communities structure and biogeochemical trajectories* (Doctoral dissertation, BTU Cottbus-Senftenberg).

Semple, A., & Pandhal, J. (2025). Engineering the phycosphere: Fundamental concepts and tools for the bottom-up design of microalgal-bacterial consortia. *Applied Phycology, 6*(1), 21–51.

Shafik, W. (2025). Ecological health impact in the marine world. In *Radiation status in the marine world* (pp. 307–335). Springer Nature Switzerland.

Shahni, Y. S., Singh, K. P., Pandey, S., Panigrahi, C. K., Das, N., Panotra, N., et al. (2025). Role of quorum sensing molecules in plant-microbe interaction for sustainable agriculture: A review. *Journal of Advances in Biology & Biotechnology, 28*(5), 889–903.

Simkin, A. J., Kapoor, L., Doss, C. G. P., Hofmann, T. A., Lawson, T., & Ramamoorthy, S. (2022). The role of photosynthesis related pigments in light harvesting, photoprotection and enhancement of photosynthetic yield in planta. *Photosynthesis Research, 152*(1), 23–42.

Singh, V. K., Jha, S., Rana, P., Mishra, S., Kumari, N., Singh, S. C., et al. (2023). Resilience and mitigation strategies of cyanobacteria under ultraviolet radiation stress. *International Journal of Molecular Sciences, 24*(15), 12381.

Singh, K., Chandra, R., & Purchase, D. (2022). Unraveling the secrets of rhizobacteria signaling in rhizosphere. *Rhizosphere, 21*, 100484.

Sukenik, A., & Kaplan, A. (2021). Cyanobacterial harmful algal blooms in aquatic ecosystems: A comprehensive outlook on current and emerging mitigation and control approaches. *Microorganisms, 9*(7), 1472.

Šulčius, S., Alzbutas, G., & Lukashevich, V. (2025). Cyanophage lysis of the cyanobacterium *Nodularia spumigena* affects the variability and fitness of the host-associated microbiome. *Environmental Microbiology, 27*(4), e70042.

Sun, W., Xing, J., Li, X., & Wu, S. (2025). Effects of planting density on water restoration performance of *Vallisneria spinulosa* Yan growth system constructed by enclosure. *Water, 17*(11), 1603.

Teikari, J. E., Russo, D. A., Heuser, M., Baumann, O., Zedler, J. A., Liaimer, A., & Dittmann, E. (2025). Competition and interdependence define interactions of *Nostoc* sp. and *Agrobacterium* sp. under inorganic carbon limitation. *NPJ Biofilms and Microbiomes, 11*(1), 42.

Thawabteh, A. M., Naseef, H. A., Karaman, D., Bufo, S. A., Scrano, L., & Karaman, R. (2023). Understanding the risks of diffusion of cyanobacteria toxins in rivers, lakes, and potable water. *Toxins, 15*(9), 582.

Tong, C. Y., Honda, K., & Derek, C. J. C. (2023). A review on microalgal-bacterial co-culture: The multifaceted role of beneficial bacteria towards enhancement of microalgal metabolite production. *Environmental Research, 228*, 115872.

Van Le, V., Ko, S. R., Kang, M., Shin, Y., Lim, B., Kang, Y. H., et al. (2023). Periphyton reduces cyanobacterial blooms by promoting potentially cyanobactericidal bacteria. *Journal of Applied Phycology, 35*(3), 1285–1299.

Villalobos, T., Suárez-Isla, B., & Garcia, C. (2025). Health and environmental impacts of cyanobacteria and cyanotoxins from freshwater to seawater. *Toxins, 17*(3), 126.

Wan, L., Cao, L., Song, C., Cao, X., & Zhou, Y. (2022). Regulation of the nutrient cycle pathway and the microbial loop structure by different types of dissolved organic matter decomposition in lakes. *Environmental Science & Technology, 57*(1), 297–309.

Wang, Z., Akbar, S., Sun, Y., Gu, L., Zhang, L., Lyu, K., et al. (2021). Cyanobacterial dominance and succession: Factors, mechanisms, predictions, and managements. *Journal of Environmental Management, 297*, 113281.

Wang, Y., Xu, C., Lin, Q., Xiao, W., Huang, B., Lu, W., et al. (2024). Modeling of algal blooms: Advances, applications and prospects. *Ocean & Coastal Management, 255*, 107250.

Wang, Y., Yu, H., Cao, X., Qi, W., Peng, J., Liu, H., & Qu, J. (2025). Coexistence of cyanobacteria and microplastics in eutrophic lake ecosystems: Interactions and combined effects. *Critical Reviews in Environmental Science and Technology*, 1–24.

Wang, K., Mou, X., Cao, H., Struewing, I., Allen, J., & Lu, J. (2021a). Co-occurring microorganisms regulate the succession of cyanobacterial harmful algal blooms. *Environmental Pollution, 288*, 117682.

Wang, L., Yi, Z., Zhang, P., Xiong, Z., Zhang, G., & Zhang, W. (2024a). Comprehensive strategies for microcystin degradation: A review of the physical, chemical, and biological methods and genetic engineering. *Journal of Environmental Management, 365*, 121707.

Wang, W., Li, M., Chen, P., Yuan, S., Wang, K., Wang, S., & Jiang, X. (2025a). Role of nitrogen cycling functional genes and their key influencing factors in eutrophic aquatic ecosystems. *Environmental Reviews, 33*, 1–10.

Wen, C., Zhang, X., Chen, W., Li, J., Ding, L., & Yang, S. (2025). The effect of temperature on carbon, nitrogen elements, and chlorophyll-a content in harmful algal blooms and their role in marine carbon and nitrogen cycles. *Marine Environmental Research*, 106978.

West, J. J., Järnberg, L., Berdalet, E., & Cusack, C. (2021). Understanding and managing harmful algal bloom risks in a changing climate: Lessons from the European CoCliME project. *Frontiers in Climate, 3*, 636723.

WIJESINGHE, W. A. A. N. (2024). The allelopathic effect of macrophyte extracts on *Microcystis aeruginosa* inhibition.

Xia, P., Peng, J., Gu, T., Yu, S., & Xu, Z. (2025). Lake-related ecosystem services facing social–ecological risks. *People and Nature, 7*(4), 734–751.

Xiong, J., Li, Z., &Fantke, P. (2025). Modeling human exposure to chemicals via different freshwater pathways. *Exposure and Health*, 1–19.

Xiong, W., Wang, S., Jin, Y., Wu, Z., Liu, D., & Su, H. (2023). Insights into nitrogen and phosphorus metabolic mechanisms of algal-bacterial aerobic granular sludge via metagenomics: Performance, microbial community and functional genes. *Bioresource Technology, 369*, 128442.

Xu, S., He, C., Kang, Z., Song, S., & Li, C. (2022). The ecological responses of bacterioplankton during a *Phaeocystis globosa* bloom in Beibu Gulf, China highlighted by integrated metagenomics and metatranscriptomics. *Marine Biology, 169*(3), 36.

Yan, Z., Jin, Y., Li, T., Zhang, X., Yang, Q., Ren, C., & Qiao, L. (2025). Monthly variation, environmental drivers, and ecological functions of marine bacterial community in a eutrophic coastal area of China. *Microorganisms, 13*(4), 837.

Yang, L., Cao, X. Y., Chen, X. Y., Deng, Q. H., Wan, L. L., Li, X. W., et al. (2021). Community composition and functional genes explain different ecological roles of heterotrophic bacteria attached to two bloom-forming cyanobacterial genera. *Science of the Total Environment, 758*, 143850.

Yang, Y., Chen, J., Chen, X., Jiang, Q., Liu, Y., & Xie, S. (2021b). Cyanobacterial bloom induces structural and functional succession of microbial communities in eutrophic lake sediments. *Environmental Pollution, 284*, 117157.

Ye, T., Zhao, Z., Bai, L., Song, N., & Jiang, H. (2020). Characteristics and bacterial community dynamics during extracellular polymeric substance (EPS) degradation of cyanobacterial blooms. *Science of the Total Environment, 748*, 142309.

Yin, H., Bao, Y., Huang, T., Zhang, Y., Sun, T., Tao, P., et al. (2025). Effects of cyanobacterial growth and decline on dissolved organic matter and endogenous nutrients release at the sediment–water interface. *Carbon Research, 4*(1), 1–13.

Yu, H. (2025). Climate change unveils hidden microbial dangers. *Environmental Science and Ecotechnology, 24*, 100544.

Zahir, M., Su, Y., Shahzad, M. I., Ayub, G., Rehman, S. U., & Ijaz, J. (2024). A review on monitoring, forecasting, and early warning of harmful algal bloom. *Aquaculture*, 741351.

Zeng, M. (2024). The mutual effect of nutrients on plant–herbivore interactions. *Plant Ecology, 225*(10), 1035–1045.

Zhang, X., Ye, X., Chen, L., Zhao, H., Shi, Q., Xiao, Y., et al. (2020). Functional role of bloom-forming cyanobacterium Planktothrix in ecologically shaping aquatic environments. *Science of the Total Environment, 710*, 136314.

Zhang, Z., Nair, S., Tang, L., Zhao, H., Hu, Z., Chen, M., et al. (2021). Long-term survival of Synechococcus and heterotrophic bacteria without external nutrient supply after changes in their relationship from antagonism to mutualism. *mBio, 12*(4), e0161421.

Zhang, H., Xie, Y., Zhang, R., Zhang, Z., Hu, X., Cheng, Y., et al. (2023). Discovery of a high-efficient algicidal bacterium against *Microcystis aeruginosa* based on examinations toward culture strains and natural bloom samples. *Toxins, 15*(3), 220.

Zhang, Y., Hou, J., Gu, Y., Zhu, X., Xia, J., Wu, J., et al. (2025). Spatiotemporal variation assessment and improved prediction of cyanobacteria blooms in lakes using improved machine learning model based on multivariate data. *Environmental Management*, 1–16.

Zhong, X., Duan, R., Hou, S., Chen, M., Tan, X., Hess, W. R., & Shi, T. (2025). Transcriptome remodeling drives acclimation to iron availability in the marine N2-fixing cyanobacterium *Trichodesmium erythraeum* IMS101. *mSystems*, e01499–24.

Zhou, J., Lao, Y. M., Song, J. T., Jin, H., Zhu, J. M., & Cai, Z. H. (2020). Temporal heterogeneity of microbial communities and metabolic activities during a natural algal bloom. *Water Research, 183*, 116020.

Zhu, C., Zhang, J., Wang, X., Yang, Y., Chen, N., Lu, Z., et al. (2021). Responses of cyanobacterial aggregate microbial communities to algal blooms. *Water Research, 196*, 117014.

Zoccarato, L., & Grossart, H. P. (2019). Relationship between lifestyle and structure of bacterial communities and their functionality in aquatic systems. *The Structure and Function of Aquatic Microbial Communities* (pp. 13–52). Springer International Publishing.

Chapter 13
Metagenomic Discoveries and the Expanding Horizons of Cyanobacterial Research

Taufiq Nawaz, Liping Gu, Shah Fahad, Shah Saud, and Ruanbao Zhou

Abstract Cyanobacteria, commonly known as blue-green algae, have played a fundamental role in shaping Earth's evolutionary history, particularly through their contribution to the oxygenation of the atmosphere. This chapter explores their ancient lineage, morphological diversity, and central role in enabling complex life via oxygenic photosynthesis. From their emergence approximately 3.0 billion years ago to their involvement in the Great Oxidation Event, cyanobacteria have been instrumental in the development of life on Earth. Recent metagenomic studies have uncovered previously uncharacterized lineages, such as *Melainabacteria* and *Sericytochromatia,* revealing novel metabolic capacities and evolutionary relationships. This chapter also highlights cyanobacteria's influence on global biogeochemical cycles and their adaptation to diverse ecological niches, including extreme environments. Molecular insights particularly from genomic analyses and horizontal gene transfer shed light on their evolutionary innovations and ecological versatility. Additionally, the environmental footprint of cyanobacteria, from harmful algal blooms to their potential role in astrobiology, is discussed. Overall, this chapter emphasizes the importance of continued research into cyanobacterial diversity and genomics to fully harness their biotechnological and industrial potential.

T. Nawaz (✉) · L. Gu · S. Fahad · R. Zhou
Department of Biology/Microbiology, South Dakota State University, Brookings, USA
e-mail: Taufiq.nawaz@jacks.sdstate.edu

S. Saud
College of Life Science, Linyi University, Linyi, Shandong, China

S. Fahad et al. (eds.), *Cyanobacterial Blooms: Ecology, Evolution and Biogeochemical Impacts*, https://doi.org/10.1007/978-3-032-06042-6_13

Graphical Abstract

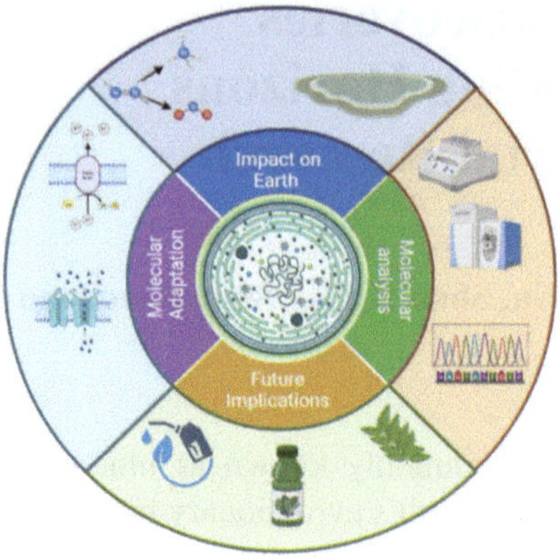

Keywords Cyanobacteria · Evolution · Genomics · Oxygenic photosynthesis · Biogeochemical cycles · Industrial applications

13.1 Introduction

Cyanobacteria, the primary oxygen producers, commonly known as blue-green algae, are prokaryotic organisms classified within the domain Bacteria. The chronological evidence from both 2.6-billion-year-old and 3.5-billion-year-old carbonate rocks underscores the central role of cyanobacteria in Earth's evolutionary history. The difference between these two ages lies in the progressive revelation of evidence over geological time. The 2.6 billion-year-old rocks provide concrete evidence of biomineralization associated with cyanobacteria during the Archean, marking a crucial phase in their evolutionary journey. Conversely, the 3.5 billion-year-old examples shed light on the ancient origins of cyanobacteria and provide essential insights into their origins and early development on Earth (Shestakov & Karbysheva, 2017). Taken together, these results highlight the importance of temporal evidence in elucidating the evolutionary history of cyanobacteria from their ancient beginnings to their central role in shaping Earth's biosphere. They are considered to be very important for their role as primary producers in both the present and past oceans and are thought to be the ancestors of eukaryotic creatures' chloroplasts. These bacteria are distinct from other prokaryotes due to their diverse morphologies and cellular development. They engage in oxygenic photosynthesis, which fosters the development of complex life forms and supports the oxygenation of land and aquatic ecosystems (Altermann & Kazmierczak, 2003; Layer, 2023). Cyanobacteria are ubiquitous Gram-negative eukaryotic bacteria known as Cyanophyta. There are many different morphological types of cyanobacteria, including unicellular, colonial, filament-forming, surface-anchored, and mat-making species. Their complex anatomical and physical nature enables them to survive in a wide variety of environments including terrestrial, freshwater, wastewater, and marine ones (El-Seedi et al., 2023).

Cyanobacteria have been essential in molding the evolution of our planet throughout Earth's history. These ancient organisms, which are thought to have existed as early as 3.0 billion years ago, set the stage for oxygenic photosynthesis during Archean Aeon. Cyanobacteria might have replaced other primary producers as the major ones by the late Archean epoch. After the Great Oxidation Event in Proterozoic Aeon, significant groups of cyanobacterial variety, including those leading to chloroplasts, appeared. The Neoproterozoic Oxidation Event, which occurred during the Neoproterozoic Era, greatly increased atmospheric oxygen levels and facilitated the creation of mammals and marine planktonic groups. Using metagenomic data, current research has discovered early divergent lineages of Cyanobacteria, including Melainabacteria and Sericytochromatia. Exploring the metabolic potential of these lineages requires their isolation and investigation (Cardona, 2017; Och & Shields-Zhou, 2012).

In the course of Earth's existence, cyanobacteria have played a vital role as primary producers and main source of atmospheric oxygen. Studying the evolution of cyanobacteria can provide insights into their origins and evolutionary milestones, shedding light on key events in Earth's environmental history. The diversity of cyanobacteria, including their cell types and differentiation, has been instrumental in categorizing both extant and extinct species, with implications for developmental and paleoenvironmental studies. Furthermore, cyanobacteria significantly influence the balances of oxygen, carbon, and global nitrogen. Their connection to the evolution of eukaryotes is evident through their believed ancestry to chloroplasts in plants and algae (Tomitani et al., 2006). A visual representation outlining the sections covered in this chapter is presented in Fig. 13.1.

13.2 Cyanobacteria: Ancient Microbial Pioneers

Cyanobacteria, among the oldest and most widespread microbes on Earth, provide valuable insights into the evolution of life and the planet's biogeochemical cycles (Sánchez-Baracaldo et al., 2022). With a diverse fossil record dating back over 3.5 billion years, these ancient organisms offer a glimpse into our planet's past. However, challenges arise in interpreting these fossils due to limited preservable features, changes in paleoecology, and potential morphological similarities. Despite these difficulties, several cyanobacteria fossils, like those of Eoentophysalis, Polybessurus, and Eohyella, have been clearly recognized. These fossils have different forms of division that make it possible to identify them as cyanobacteria with certainty (Demoulin et al., 2019). Cyanobacteria can be preserved in several ways, such as mineralized castings, organic films, or microscopic fossils inside rock sediments. Numerous elements, including the depositional environment, the presence of minerals, and

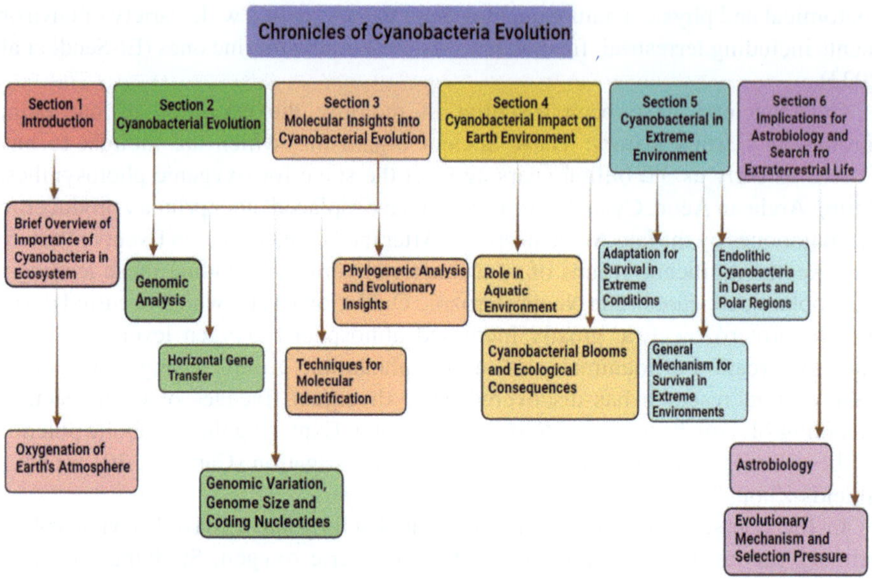

Fig. 13.1 The flow chart showing the sections of cyanobacterial evolution

the level of microbial degradation, affect the preservation of cyanobacteria fossils (Golubic & Seong-Joo, 1999; Sánchez-Baracaldo et al., 2022).

The discovery of cyanobacteria-induced biomineralization in 2.6 billion-year-old carbonate rocks provides evidence for the existence of the earliest cyanobacteria throughout the Archean. The difference between 2.6 and 3.5 billion years lies in the evolutionary timeline of cyanobacteria. The discovery of cyanobacteria-induced biomineralization in 2.6 billion-year-old carbonate rocks indicates the presence of early cyanobacteria during the Archean. This finding suggests that cyanobacteria were already thriving and involved in biomineralization processes around 2.6 billion years ago, underscoring their ancient origin and important role in Earth's history. Additionally, 2.7 billion years old Western Australian shales containing biomarkers like methyl-hopanes and steranes point to the presence of eukaryotes and oxygenic photosynthesis throughout the Neoarchean. Neoarchean rocks (2.7–2.5 Ga) were the source of the earliest reported cyanobacterial microfossils. These microscopic remains show that there was life on Earth before the Archean Aeon and cyanobacteria, which are photosynthetic microorganisms, may have existed then. However, the lack of well-preserved rocks and the change of fundamental structures by geological processes make it difficult to identify Archean microfossils. There are few fossils from the Archean period. Although they show early Earth life, Archean stromatolites cannot be considered fossils. Stromatolites are largely thought of as environmental indicators by a few researchers, while others see them as sedimentary or sedimentary structures (Benzerara et al., 2014; Kamennaya et al., 2012). The stratigraphic value given to stromatolites in the Precambrian is similarly influenced by these divergent viewpoints (Altermann & Kazmierczak, 2003; Brasier et al., 2006). The late Archean epoch, when oxygenic photosynthesis grew more complicated, is when crown group Cyanobacteria predominated as primary producers. The lineage that gave rise to chloroplasts and the majority of the major cyanobacterial groupings all emerged after the Great Oxidation Event (GOE). The final stages of the period brought the emergence of marine planktonic groupings, notably marine eukaryotic green algae (Sánchez-Baracaldo & Cardona, 2020).

It is a challenging and persistent field of study to date the genesis and diversification of cyanobacteria. The chronology of evolutionary events has been estimated using molecular clocks and fossil-based methods, although the accuracy of these estimates depends on different variables. Our understanding of the past history or evolution of these important microbes is set to improve in tune with the development of different analytical techniques and the integration of many data sources (Hipsley & Müller, 2014; Sánchez-Baracaldo et al., 2022).

13.2.1 Cyanobacteria's Role in the Oxygenation of Earth's Atmosphere

According to researchers, two significant oxygenation events occurred in Earth's history. The Great Oxidation Event (GOE), the first event, is regarded as the minimum age for the start of oxygenic photosynthesis. It took place between 2.32 and 2.4 billion years ago. As a result of this occurrence, the amount of oxygen in the atmosphere increased, and oxygen-dependent metabolic and biosynthetic pathways emerged (Lyons et al., 2014). The Neoproterozoic Oxidation Event (NOE), which occurred roughly 800–600 million years ago, markedly raised atmospheric oxygen levels. The NOE has been connected to the creation of marine planktonic groups, major glacial periods, the origin of animals, and perturbations in the carbon cycle. By the procedure of oxygen-dependent photosynthesis, cyanobacteria were a key component in the oxygenation of the Earth's atmosphere. The Great Oxidation Event (GOE) caused an increase in atmospheric oxygen levels owing partly to the oxygen generated by cyanobacteria in the course of photosynthesis. The development of life on Earth was significantly impacted by this event, which was the first huge increase in atmospheric oxygen (Planavsky et al., 2021; Sánchez-Baracaldo & Cardona, 2020).

It is believed that the development of oxygenic photosynthesis occurred at the same time as the emergence of Cyanobacteria, because oxygenic photosynthesis is only found in Cyanobacteria and other bacterial groups have developed as anoxygenic photosynthetic organisms (Hohmann-Marriott & Blankenship, 2011, 2012). The photosynthetic apparatus in cyanobacteria is made up of two essential parts: Photosystem I (PSI) and Photosystem II (PSII). The first stage of aerobic photosynthesis, featuring the oxidation of water and the release of oxygen, is carried out by PSII. The antenna subunits CP43 and CP47, as well as the core proteins D1 and D2, make up PSII. On the other side, PSI is involved in the conversion of NADP +

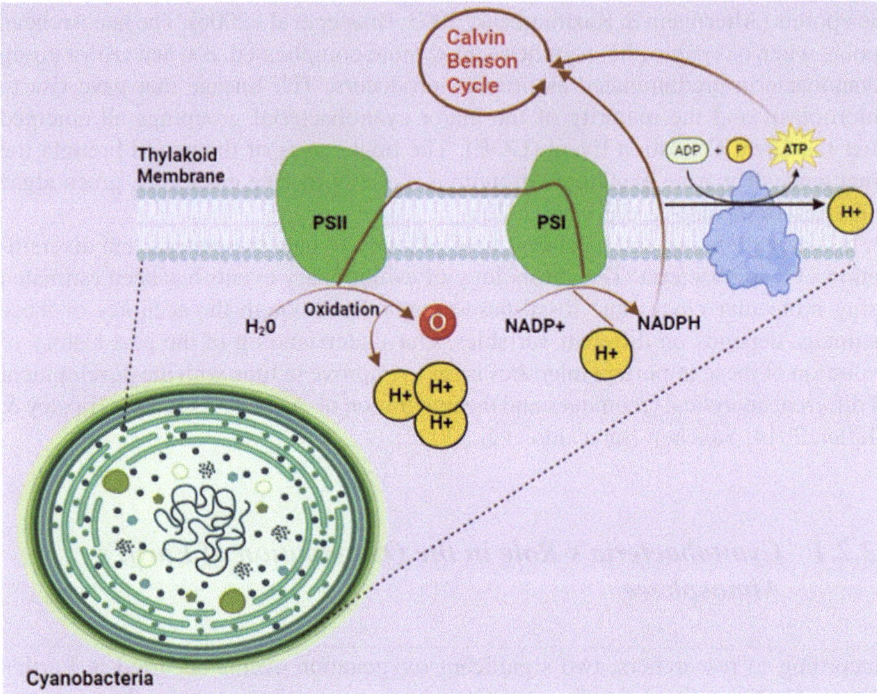

Fig. 13.2 Mechanism of aerobic photosynthesis in thylakoid membrane of cyanobacteria

to NADPH, which is the last stage of photosynthesis. Along with antenna subunits PshA and PshB, PSI also consists of the core proteins PsaA and PsaB. One distinguishing feature of the crown group Cyanobacteria is the heterodimeric structure of PSI (Cardona et al., 2019; Jensen et al., 2007). The aerobic photosynthesis in thylakoid membranes of cyanobacteria is shown (Fig. 13.2).

Theoretical models that elucidate the evolution of oxygenic photosynthesis in cyanobacteria greatly enrich our understanding of cyanobacteria evolution while providing invaluable guidance for research in diverse fields such as biotechnology, synthetic biology, ecology, and environmental sciences. To shed light on this evolutionary process, three basic theoretical models have been proposed: the fusion model (Olson & Blankenship, 2004), the selective loss model (Cardona et al., 2017), and the cyanobacterial export model (Mulkidjanian et al., 2006). These models not only deepen our understanding of how cyanobacteria evolved the capacity for oxygenic photosynthesis, but also serve as a fundamental framework for innovative research efforts aimed at harnessing the capabilities of cyanobacteria for diverse applications from biotechnology to environmental sustainability. According to the fusion model, two distinct nonphotosynthetic cyanobacterial ancestors provided the reaction centers I and II (RCI and RCII) in cyanobacteria via horizontal gene transfer (HGT). The selective loss paradigm postulates that, except for cyanobacteria, an original photosynthetic organism had both RCI and RCII, but one was lost in later lineages (Alexander & Svetlana, 2021; Soo et al., 2019). On the contrary, the cyanobacterial export theory proposes that photosynthesis started within the cyanobacterial lineage and that anoxygenic photosynthetic lineages acquired their reaction centers by HGT from progenitors of living cyanobacteria, or "procyanobacteria" (Drews, 2011; Mulkidjanian et al., 2006).

13.2.2 Diversification of Cyanobacteria

Cyanobacteria are a fascinating group of prokaryotic organisms that exhibit an impressive diversity of growth forms. Their growth strategies range from unicellular to multicellular and can include reversible or irreversible cell differentiation. These photosynthetic bacteria, commonly known as blue-green algae, can vary in size from tiny single cells to large filamentous thalli that can be several centimeters long. With their unique characteristics and tremendous potential for applications in various fields, cyanobacteria have captured the imagination of scientists and researchers worldwide (Pathak et al., 2018; Rastogi & Sinha, 2009). Cyanobacteria are found in almost every terrestrial and aquatic environment and have successfully adapted to survive in diverse habitats, from hot springs to polar regions (Gaysina et al., 2019). They have a long evolutionary history and have played a significant role in shaping the Earth's atmosphere and environment. Cyanobacteria have also been studied extensively for their potential applications in biotechnology, mariculture, agriculture, food and fuel, and biomedical. Their ability to fix atmospheric nitrogen and produce various secondary metabolites makes them a valuable resource for many industries (Schirrmeister et al., 2013). Cyanobacteria's importance in agriculture is shown in Fig. 13.3.

Cyanobacteria are classified based on their morphological attributes and developmental characteristics. The classification system recognizes five sections: Section 13.1 for unicellular cyanobacteria that divide by binary fission or budding, Section 13.2 for multiple fission resulting in baeocyte development, and Sections 13.3–13.5 for filamentous cyanobacteria. Section 13.3 includes filamentous non-heterocystous cyanobacteria that proliferate by trichome breakage. Sections 13.4 and 13.5 represent heterocystous cyanobacteria that can develop akinetes and heterocysts and reproduce mainly via hormogonia formation. These two sections further divide into Subsection A (Rippka et al. 1979), which includes filamentous cyanobacteria with heterocysts and akinetes, and Subsection B (Castenholz 2001), which includes filamentous cyanobacteria with heterocysts but without akinetes. Although this classification system has been useful, recent studies suggest that taxonomy based on morphologic traits alone may not fully reflect the evolutionary relationships among cyanobacteria. Therefore, new classification systems are being developed using molecular data (Pathak et al., 2022; Singh et al., 2013; Springstein et al., 2020). The conventional classification of cyanobacteria is shown (Fig. 13.4).

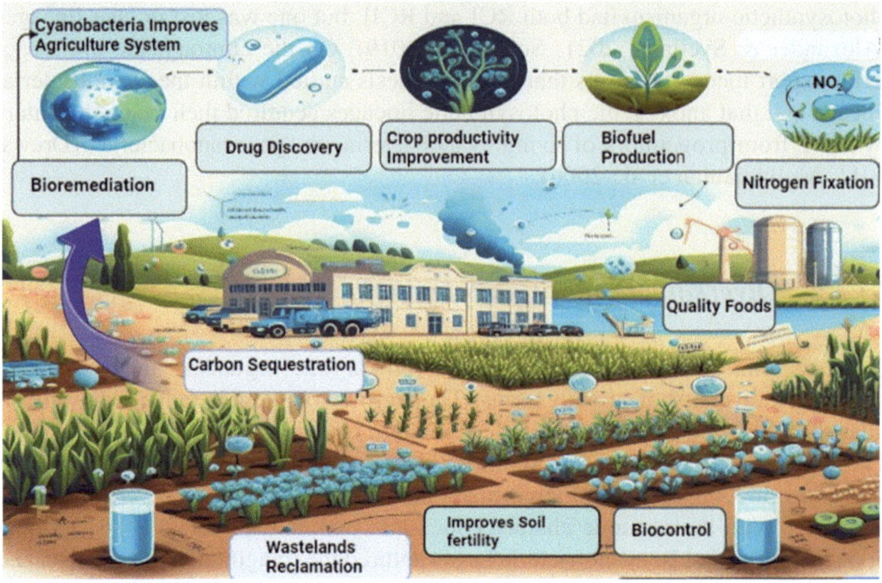

Fig. 13.3 Immense application of cyanobacteria in agriculture

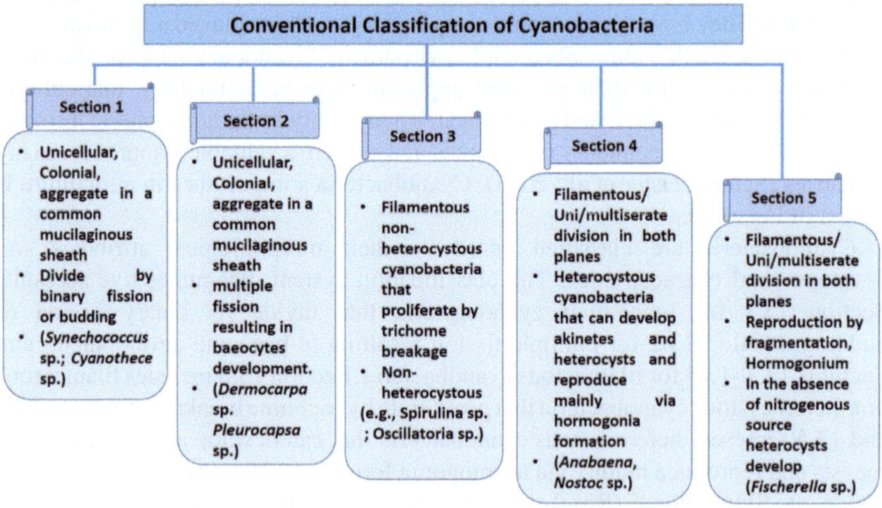

Fig. 13.4 Conventional classification of cyanobacteria

However, traditional taxonomy based on morphologic traits does not fully reflect the results of phylogenetic analyses in cyanobacteria. Additionally, 16S rRNA gene sequences lack the sensitivity for evolutionary changes that occur in ecological dynamics. A study was conducted to reconstruct phylogenetic trees based on 16S rRNA sequences and estimate the divergence times of cyanobacteria. This study also tested shifts in diversification rates, incorporating information on 281 species and 4194 strains (Hugenholtz et al., 2021). This event corresponded to a marked increase in the diversification rates of cyanobacteria. Two significant shifts in diversification rates were detected, one at the node where multicellularity evolved and the other at a later node. The "node" refers to a point on a phylogenetic tree where a lineage splits into two or more distinct lineages. These findings offer valuable insights into the temporal evolution of cyanobacteria and shed light on the early history of life (Schirrmeister et al., 2013).

Cyanobacteria have distinct morphotypes that have been present for over two billion years based on the fossil record. They are believed to have existed as early as 2.45–2.32 billion years ago and were responsible for the accumulation of atmospheric oxygen levels during the Great Oxidation Event. Cyanobacteria also show diversity at the genomic level, with significant variation in genome size, coding nucleotide proportion, G + C content, and the number of protein-coding genes. Two routes of genome development have been suggested based on multicopy gene abundance and different rates of genome size evolution (Demoulin et al., 2019; Larsson et al., 2011).

The first cyanobacteria are believed to have appeared around 2.7–2.6 billion years ago in warm shallow water bodies that formed small oxygen "oases" within the biosphere (Blank & Sanchez-Baracaldo, 2010). Microfossils similar to filamentous cyanobacteria have been found in Archean sediments (Buick, 2008). Previously, ancient carbonate stromatolites were considered as analogs of modern cyanobacterial mats, but later studies suggest that they were probably formed by filamentous anaerobic bacteria (Bosak et al., 2013). Oxygenic photosynthesis spread at least 1.05 billion years ago through the primary endosymbiosis of a cyanobacterium within a unicellular eukaryote (Demoulin et al., 2019). However, there are still questions about the origin of cyanobacteria, the origin of oxygenic photosynthesis, and the pattern and timing of cyanobacterial diversification (Fischer et al., 2016). The discrepancy between the cyanobacterial fossil record and older geochemical data suggestive of oxygenic photosynthesis is a crucial problem that needs careful consideration. Various types of evidence, such as cyanobacterial fossil stromatolites and microfossils, have been used to reconstruct the fossil record of cyanobacteria; however, they have their challenges and limitations. Addressing these challenges requires advances in analytical techniques and methods to identify cyanobacterial signatures with greater precision and reliability. Industrial research and development efforts aimed at improving the resolution of microfossil identification and understanding the unique signatures of cyanobacteria activity in sedimentary formations can improve the applicability of cyanobacteria in various industrial sectors while reducing potential risks associated with misinterpretations and unreliable assessments; geological and biological data are connected (Gupta et al., 2021; Pathak et al., 2022; Schopf, 2012). Identifying cyanobacteria activity and photosynthesis in stromatolites and microbially induced sedimentary structures presents challenges that have implications for various industrial applications. Cyanobacteria play a crucial role in the formation of microbial cells, particularly stromatolites, by influencing the shape of growing structures by trapping sediment particles and forming laminations. Filamentous cyanobacteria can bind sediment particles, while coccoid cyanobacteria contribute to lamination formation. In addition, phototrophic purple bacteria may have been important stromatolite formers before the advent of oxygenic photosynthesis (Petrescu & Ungureanu, 2022).

The formation of microbial carbonate by cyanobacteria has implications for the creation of artificial environments for the construction of modern microbialities, including stromatolites. Stromatolites are layered biochemical structures formed by the capture, binding and cementing of sediment particles by microorganisms such as cyanobacteria. Built over billions of years, these structures played a crucial role in transforming Earth into a habitable planet capable of supporting diverse life forms through photosynthesis (Gérard et al., 2018).

In industrial applications, the physicochemical processes produced by cyanobacteria in stromatolites could inspire the production of high-quality cement directly from sand. By mimicking these natural processes, humans could potentially develop innovative construction materials and learn valuable lessons for environmental sustainability (Gérard et al., 2018).

13.3 Molecular Insights into Cyanobacterial Evolution

13.3.1 Genomic Analysis of Cyanobacteria

In recent years, the pivotal role of DNA and RNA as carriers of genetic information has been increasingly acknowledged, serving as the cornerstone for a wide array of molecular methodologies employed in the analysis of microbial populations. These nucleic acids not only store genetic data but also play essential roles in gene expression, regulation, and cellular functions, making them indispensable tools for unraveling the complexities of microbial communities and their interactions within diverse ecosystems. These methods determine the number of gene variations present in any sample rather than directly characterizing each cell in an environmental sample. A molecular biology technique called DNA profiling is used to quickly anticipate the variety of the microbial population in any environment at any given time. Microbiologists have investigated a wide range of microbial systems using these fingerprinting approaches, and even the dynamics of their biodiversity or changes in community structure over time may be easily analyzed. Examples include marine, freshwater, soil, and human microbial communities. The development and sequencing of 16S rRNA gene libraries, microarrays, and metagenomics are just a few examples of the high-throughput sequencing techniques known as next-generation sequencing (NGS) that have largely replaced these older techniques (Kumari & Rai, 2020; Minchin & Lodge, 2019).

Kumari et al. (2009) categorized molecular-based analysis to study cyanobacterial diversity and evolutionary lineages into two groups. The first group includes PCR-dependent strategies, and the second group includes genome-based or PCR-independent strategies. All these molecular methods are employed for the investigation of cyanobacterial diversity. Multiple obstacles in culture and characterization while studying microbes could be avoided through these molecular approaches, which provided an enormous number of novel perspectives into the complex design of microbial ecosystems.

PCR-based DNA profiling is a powerful tool for studying microbial diversity in environmental samples, as it allows for the rapid and accurate identification of different microbial species present in a sample. Two PCR-based DNA profiling techniques are utilized for microbial diversity study: Terminal-restriction fragment length polymorphism (T-RFLP) and Automated ribosomal intergenic spacer analysis (ARISA) (Kumar et al., 2021; Siqueira et al., 2010). A fingerprinting method called T-RFLP is used to evaluate the variety of microbes. It is useful to estimate the microbial community structure and richness. Semiquantitative estimates of the relative significance of each discovered phylotype are provided by T-RFLP. It can be used to determine the species or genus that each electropherogram's fragment length corresponds to. Another fingerprinting method for determining microbial diversity is ARISA. The intergenic region between the 16S and 23S rRNA genes being amplified. Compared to T-RFLP, ARISA offers several advantages. Less common taxa within microbial communities can be more easily found by ARISA. It makes it possible to identify microbial taxa that comprise less than 5% of the entire amplified product, which T-RFLP might not be able to identify. Compared to T-RFLP, ARISA offers a greater diversity of microorganisms. Accordingly, ARISA is generally more efficient and reliable at identifying microbial diversity in aquatic environments (Danovaro et al., 2006). Using the RISA method, the abundance of cyanobacteria, particularly Scytonema and Leptolyngbya species has been reported by Srivastava et al. (2016). Prochlorococcus and Synechococcus, Picocyanobacteria, were examined by Lavin et al. (2008) using terminal restriction fragment length polymorphisms of 16S-23S rRNA sequences to determine their diversity.

The 16S rDNA library preparation exploits the variable region of 16S rDNA to construct amplicon primers. It is a two-step amplicon-PCR process followed by NGS. On comparison of the target sequences with those available in sequence databases, the information about the identity and relatedness of the new sequences is decoded. The 16S rRNA gene is used as the standard signature sequence for taxonomic characterization of microorganisms because it is ubiquitously present in the genome of every microbe except viruses and shows appropriate variations. Semenova et al. (2001) examined laboratory cultures of cyanobacteria through 16S rDNA cloning and sequencing. The cloning and sequencing technique has been used in various ecosystems for studying prokaryotic diversity (Zwart et al., 2005). Svenning et al. (2005) studied the diversity of symbiotic cyanobacterial strains within the genus Nostoc based on 16S rDNA sequence analysis.

Microarrays are an effective molecular approach for the high-throughput study of biological samples. Microarrays and DNA-DNA hybridization allowed for the identification and measurement of the diversity of bacterial species (Cho & Tiedje, 2001; Greene et al., 2003). With hundreds of very specific DNA sequences, microarrays with specific target genes have offered useful insights into the assessment of microbial diversity. A microarray is an arrangement of two dimensions on a rigid support that uses large-scale detection techniques to examine an extensive collection of biological samples. It is a remarkable PCR-independent strategy for studying cyanobacterial diversity and their interaction with surroundings (Singh et al., 2020).

The technology of metagenomics enables the investigation of the collective genomes of microbial populations found in environmental samples. Metagenomics provides significantly more details on the physiological functioning and diversification of microbial populations than methods based on a single gene, like 16S rRNA. It offers data on the relative abundance of various organisms and the community structure and is less biased than PCR. The polymorphism that exists in natural populations is captured by metagenomics, which makes sequence assembly more challenging but contains more data (Kumari & Rai, 2020). Metagenomics has immense potential to reveal the diversity and functional roles of cyanobacteria in various environments, including freshwater, marine, and terrestrial ecosystems. Metagenomics helps in studying cyanobacteria diversity by allowing researchers to analyze the genetic material of entire microbial communities, rather than just individual organisms. This approach enables the identification of a wide range of cyanobacterial taxa, including those that are difficult to culture or have not yet been described. Metagenomic data can also provide insights into the functional roles of cyanobacteria in various environments, such as their contributions to primary production, nutrient cycling, and carbon fixation. Additionally, metagenomics can reveal the interactions between cyanobacteria and other microorganisms, including symbiotic and pathogenic relationships. Overall, metagenomic approaches have proven to be a powerful tool for studying the diversity and ecology of cyanobacteria in natural and artificial environments (Alvarenga et al., 2017). The 16S rRNA gene is essential for determining phylogenetic relationships, identifying cyanobacteria, and assessing molecular diversity. The 16S rRNA gene possesses the properties of a suitable phylogenetic marker gene, including universal distribution in prokaryotes, functional consistency, both variable and conserved sections, big size, and thus relatively high information content. In order to assess molecular diversity and classify cyanobacteria phylogenetically, the 16S rRNA sequence analysis is currently the most promising method, because it does not depend on the culture conditions that affect growth (Johnson et al., 2019; Shariatmadari et al., 2017). Cyanobacterial isolates were characterized microscopically, sequenced for the 16S rRNA gene and associated ITS region, and phylogenetically analyzed (Akagha et al., 2023). The commonly used methods to identify diversity among cyanobacteria are shown (Table 13.1).

Table 13.1 Techniques to identify diversity of cyanobacteria

Method of identification		Species	Source
PCR-based DNA profiling techniques	Terminal-restriction fragment length polymorphism (T-RFLP)	Prochlorococcus and Synechococcus, picocyanobacteria	Lavin et al., (2008)
	Automated ribosomal intergenic spacer analysis (ARISA)	*Scytonema* and *Leptolyngbya* species	Srivastava et al., (2016)
	16 s rDNA	Symbiotic cyanobacterial strains within the genus *Nostoc*	Svenning et al., (2005)
	Denaturing gradient gel electrophoresis (DGGE profile)	*Microcystis* species	Kardinaal et al., (2007)
	Single-strand conformational polymorphism (SSCP)	*Xenococcus* and *Chroococcidiopsis*	Lee et al., (1996)
	Randomly amplified polymorphic DNA (RAPD)	*Phormidium retzii*	Eskew et al., (1993)
		Chroococcales	Mikolajczak et al., (1999)
Nucleic acid reassociation and hybridization analysis	Phenotyping	Nodularia strains	Lehtimäki et al., (2000)
	DNA-DNA hybridization	Bloom-forming oscillatorioid cyanobacteria	Suda et al., (2002)
	G + C contents	*Anabaena*	Baudouin-Cornu et al., (2004)
DNA microarray		*Prochlorococcus*	Muyzer and Smalla (1998)

13.4 Horizontal Gene Transfer and Its Impact on Cyanobacterial Evolution

Horizontal gene transfer (HGT), an important evolutionary mechanism observed in prokaryotes, is the transmission of genetic material across phylogenetically distant species. In recent years, the availability of complete genomes has facilitated the comprehensive analysis of HGT and highlighted its emerging role in the adaptation and evolution of eukaryotes (Tiwari & Bae, 2020).

The process of Horizontal Gene Transfer (HGT) plays a crucial role in evolution, involving the acquisition of foreign genes by any genome through transformation, transduction, or conjugation 1. In freshwater cyanobacteria, two primary modes of gene transfer have been identified: natural transformation such as *Synechococcus* spp., *Synechocystis, Nostoc muscorum,* and *Anabaena* spp. and *conjugation* such as *Anabaena* strains. Cyanobacteria are of particular interest in HGT studies as they are the ancestors of the endosymbiotic event that led to the evolution of plastid organelles (Yerrapragada et al., 2009). Horizontal gene transfer (HGT) is an essential mechanism for cyanobacteria to acquire genes that provide them with selective advantages in their habitats. HGT has a significant impact on habitat adaptation, and it is crucial to understand its role in the evolution of cyanobacteria (Chen et al., 2021b).

The detailed evolutionary history of symbiotic cyanobacteria, particularly those of the order Nostocales, offers potential applications in industrial contexts, particularly with regard to understanding the mechanisms of horizontal gene transfer (HGT) and their effects. Research into the evolution of cyanobacteria through the fossil record provides insight into the early development of Earth and life. Stromatolites, often associated with cyanobacteria activity, provide evidence of oxygenic photosynthesis, but may also have been produced by non-cyanobacterial lineages or in association with other microorganisms such as methanotrophs (Demoulin et al., 2019).

Cyanobacteria mats, dominated by cyanobacterial species, are widespread in shallow marine environments and play a crucial role in creating oxygenated environments through photosynthetic activity (Stal, 2012). As architects of sedimentary structures, cyanobacteria influence sediment production through carbonate precipitation driven by photosynthesis. Their growth, movement and behavioral responses control depositional processes and shape the resulting sedimentary structures. In addition, cyanobacteria are involved in the biogeochemical cycling of elements and can directly or indirectly lead to the destruction or alteration of sediments (Golubic et al., 2000).

Understanding the evolutionary history of Cyanobacteria, including mechanisms like HGT, can provide valuable insights for industrial applications. By studying how these ancient organisms have evolved and interacted with their environment over time, researchers can potentially apply this knowledge to various industrial contexts where understanding genetic transfer mechanisms and environmental interactions is crucial.

The work examined the fine-scale evolutionary history of symbiotic Cyanobacteria, specifically those belonging to the order Nostocales. Because horizontal gene transfers (HGT) are common in the evolution of bacteria, the study used a unique synteny-based method to find genes less likely to have experienced HGT by focusing on Collinear Orthologous Regions (COR). After sixteen references Nostocales genomes were examined, and 99 genes comprising 28 COR were found (Cornet et al., 2021).

13.5 Cyanobacteria's Impact on Earth's Environment

Cyanobacteria represent the most abundant, diverse, and widely distributed group of photosynthetic organisms. They pioneered the process of splitting water molecules and releasing oxygen into the atmosphere via oxygenic photosynthesis, playing a pivotal role in shaping various biogeochemical processes. Recent research has highlighted their significance in not only oxygen production but also in nitrogen fixation, carbon sequestration, and even potential applications in bioremediation and biofuel production. Their versatile metabolic capabilities and ecological impact continue to intrigue scientists exploring their multifaceted roles in global ecosystems (Sukenik & Kaplan, 2021).

In aquatic environments, cyanobacteria play a crucial role as primary producers in the phytoplankton community. They form a significant part of the phytoplankton biomass and contribute to the production of organic compounds that support the food web. Cyanobacteria can also form biofilms and mats in benthic habitats, further contributing to primary production in aquatic ecosystems. Cyanobacteria are capable of nitrogen fixation, converting atmospheric nitrogen into a form that can be used by other organisms. This ability to fix nitrogen allows cyanobacteria to contribute to nutrient cycling in aquatic ecosystems, providing a vital source of nitrogen for other organisms (Khalifa et al., 2021; Stal, 2015). Mechanism of nitrogen fixation carried out by cyanobacteria (Fig. 13.5). Important nitrogen fixing cyanobacterial genera are mentioned (Table 13.2).

13.5.1 Cyanobacterial Blooms and Ecological Consequences

The ancient cyanobacteria, as the earliest oxygenic photoautotrophs on Earth, have profoundly shaped the planet's ecosystem. Their extensive evolutionary timeline has enabled them to adapt to various environmental shifts, including changes in climate, geochemistry, and human-induced alterations to aquatic ecosystems. Recent studies have delved into the resilience of cyanobacteria in the face of anthropogenic pressures such as salinization, water withdrawals, diversions, and nutrient overenrichment (eutrophication), shedding light on their adaptive strategies and ecological responses to contemporary challenges. Global warming is largely responsible for the growth and persistence of many cyanobacterial species, which show their best growth rates and bloom potentials at relatively high-water temperatures. Because they outcompete beneficial phytoplankton, reduce oxygen during bloom senescence, and produce a range of poisonous secondary metabolites, bloom-forming cyanobacterial species may be detrimental to the environment, organisms, and human health (Paerl & Otten, 2013).

Moreover, cyanobacteria blooms, such as members of the genus *Aphanizomenon, Cylindrospermopsis, Dolichospermum, Microcystis, Nodularia, Planktothrix* and *Trichodesmium* (Huisman et al., 2018), can increase water turbidity and raise

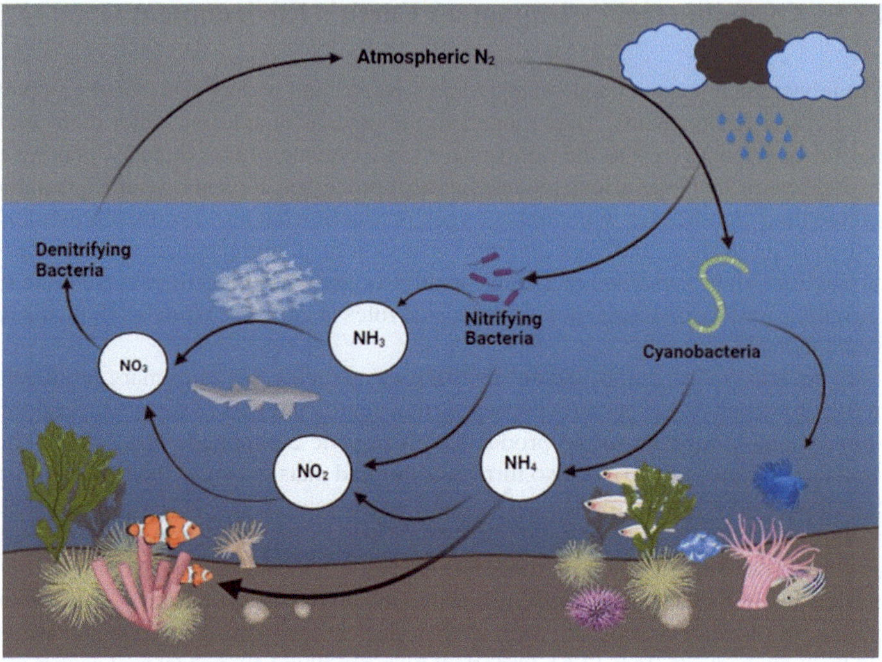

Fig. 13.5 Nitrogen fixation mechanism by cyanobacteria

pH levels, while also producing taste and odor compounds, ultimately reducing ecosystem services such as recreational value and drinking water potability (Jüttner & Watson, 2007). These challenges persist during bloom senescence, causing intense microbial degradation of algal biomass and leading to anoxia and aquatic animal mortality (Rabalais et al., 2010). Cyanobacteria are recognized as the most harmful bloom-forming taxa for aquatic ecosystems and trophic structures. While they can increase the resource use efficiency of phytoplankton, they block energy transfer from primary producers to zooplankton (Filstrup et al., 2014). The grazing of zooplankton on phytoplankton is usually disrupted during cyanobacteria blooms due to cyanotoxin production, decreased nutritional quality, and the formation of large filaments and colonies (Moustaka-Gouni & Sommer, 2020). This shift in energy transfer is supported by stable isotope analyses, demonstrating that zooplankton shifts its diet to assimilate carbon from small particles during cyanobacterial blooms, particularly from microbial sources (Berry et al., 2017; de Kluijver et al., 2012).

Cyanobacteria blooms lead to substantial modifications in plankton structure and composition, especially during bloom senescence, as bacteria and ciliates proliferate. These changes also impact the composition of zooplankton, favoring small forms specialized in consuming these alternative food sources. While the effects of Cyanobacteria blooms on phytoplankton and zooplankton structure and ecosystem functioning are well understood, the effects of other bloom-forming taxa on aquatic food webs remain largely unexplored (Amorim & do Nascimento Moura, 2021).

Table 13.2 Important nitrogen-fixing cyanobacteria

Forms of cyanobacteria	Cyanobacterial members
Unicellular	*Aphanothece*, *Chroococcidiopsis*, *Dermocapsa*, *Synechococcous*, *Gleocapsa* (*Gloethece*)*, *Myxosarcina*, *Pleurocapsa**, *Xenococcus*
Filamentous heterocystous	*Anabaena**, *Anabaenopsis*, *Aulosira*, *Calothrix**, *Camptylonema*, *cholorogloea*, *Cholorogloeopsis*, *Cylindrospermum*, *Fischerella**, *Gloeotrichia*, *Hapalosiphon*, *Masstigocaldus*, *Nodularia*, *Nostoc**, *Nostochopsis*, *Rivularia*, *Scytonema**, *Scytonematopsis*, *Stigonema*, *Tolypothrix*, *Westiella*, *Wesiellopsis*
Filamentous non-heterocystous	*Lynbya*, *Microcoleus chthonoplastes*, *Myxosarcina*, *Oscillartoria*, *Plectonema boryanum*, *Pseudanabaena*, *Schizothrix*, *Trichodesmium*

*Some of the strains of these genera live symbiotically with other plants (Sinha & Hader, 1996)

The study examined the origins and effects of phytoplankton dominance in 50 waterbodies in the area, paying particular attention to the abundance of Chlorophyta and Cyanobacteria. The findings revealed that 32–52% of communities had a predominance of phytoplankton, with the main contributors being Chlorophyta (35.5–40.6%) and Cyanobacteria (30.3–36.5%). Waterbody features and water column parameters were discovered as important predictors of dominance through the use of multinomial logistic regression. The effects of dominance on the biological characteristics and operations of phytoplankton communities differed according to the dominant organisms. Particularly, cyanobacteria dominance was linked to more significant changes in co-occurrence networks, suggesting a more severe disturbance in the composition and efficiency of phytoplankton populations. It may be inferred from this those cyanobacterial blooms, which are probably impacted by environmental conditions, have ecological ramifications that go beyond variations in biomass and richness. These include a major modification of the network structure of phytoplankton communities. The study highlights how crucial it is to comprehend the intricate relationships that exist among dominance patterns, environmental causes, and the ecological effects of phytoplankton populations in urban waterbodies (Escalas et al., 2019; Tanvir et al., 2021). The dynamics of bloom forming cyanobacteria is shown in Fig. 13.6.

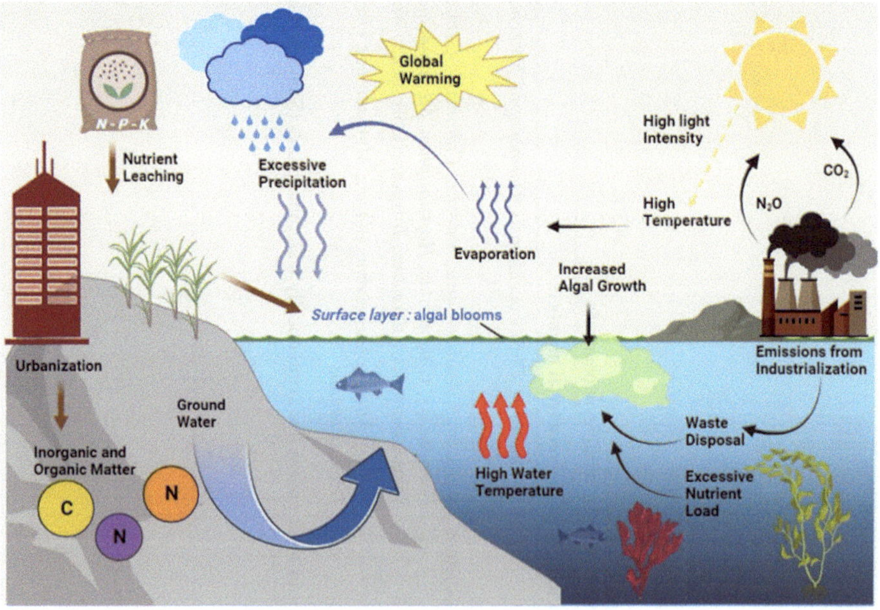

Fig. 13.6 Dynamics of cyanobacterial blooms in warm oceanic environment

13.6 Cyanobacteria in Extreme Environments

13.6.1 Adaptations for Survival in Extreme Conditions

Cyanobacteria, known for their remarkable diversity as photosynthetic prokaryotes, have been instrumental in catalyzing the transition of Earth's environmental conditions from anaerobic to aerobic. Recent research has highlighted their adaptive capabilities in thriving across a spectrum of extreme environments, showcasing their resilience to varying stressors and their potential applications in bioremediation, biofuel production, and even space exploration studies. The cyanobacteria are typically the first oxygen-evolving photoautotrophic colonizers of primary deglaciated environments, kicking off the first succession process. They also helped to stabilize the soil by forming crusts on the top. Cyanobacteria are ubiquitous and can be found in aquatic and terrestrial ecosystems, including hot springs, deserts, and Polar Regions. Although these places are well known to have harsh climatic conditions, cyanobacteria showed their ability to cope up with the stresses of low temperatures and constant irradiance of high and low intensity of light, ultraviolet radiation (U.V.R.), freezing, and desiccation (Tashyreva & Elster, 2016; Zakhia et al., 2008).'

Cyanobacteria are known for their strong adaptability under different environmental conditions, such as exposure to Gamma or UV radiation, heat, salinity, desiccation, and heavy metals. However, some of the cyanobacterial strains are resistant to these stress conditions. These environmental conditions affect the morphology (cell wall alteration), biochemicals (production of screening pigments), or physiology (state transitions), or there may be some genetic changes that occur as a result of acclimatization under the environmental extremes (Šabacká & Elster, 2006; Singh, 2018).

13.6.2 General Mechanisms Adapted by Cyanobacteria to Survive Under Extreme Habitat

Abiotic stresses, including salt, heavy metals, light, drought, and temperature extremes, have a significant impact on the physiology and metabolic behavior of cyanobacteria (Pandey et al., 2022). However, these stressors trigger a range of stimuli and defense systems that allow cyanobacteria to adapt and cope up with these conditions. During stress, the osmotic pressure changes and metabolic behavior undergoes changes, such as the enhanced synthesis of polyunsaturated fatty acids and increased concentration of photosystem units in the thylakoid membrane (Pandey et al., 2022; Singh et al., 2022). Short-term reactions take less than a minute, while long-term reactions can last for hours to several days, involving de novo protein formation. Examples of long-term responses include changes in photosystem stoichiometry and alterations in pigment quantity (Fujita et al., 1994).

Exposure to UV radiation can have a significant impact on the physical structure and biological processes of cyanobacteria. To cope up with the stress of UV-B radiation and photooxidative damage caused by reactive oxygen species (ROS), cyanobacteria synthesize metabolites like mycosporine-like amino acids (M.A.A.s) and scytonemin. These metabolites protect the cyanobacteria from ROS and help repair DNA damage caused by UV radiation (Ehling-Schulz & Scherer, 1999).

Temperature changes can have a significant impact on various biological processes, including the physiology and metabolism of cyanobacteria (Rezayian et al., 2019). The cells' components and metabolism can be affected by the duration and intensity of temperature changes. One of the most immediate effects of temperature-induced changes is the alteration of membrane fluidity. In response to temperature stress, the cells' membrane composition may change such as modifications in membrane fluidity, the presence of heat shock proteins (HSPs), fatty acid saturation, and protein integrity (Yadav et al., 2022).

The physiological and metabolic processes of cyanobacteria, as well as their growth, are affected by pH. pH variations have an impact on metabolic processes and cellular physiology, including the transport of substances across cytoplasmic membranes and the bioavailability of nutrients. It is well recognized that the enzymes oxalate decarboxylase and carbonic anhydrase are essential for maintaining and regulating pH. A drastic increase in the ATP-substrate binding proteins suggests that the phosphate absorption pathway is extremely susceptible to increase in the medium pH (Sanz-Luque et al., 2020; Yadav et al., 2022).

13.7 Endolithic Cyanobacteria in Deserts and Polar Regions

Cyanobacteria and algae are the primary producers of endolithic communities and, as such, are essential to these ecosystems (Gelsinger et al., 2020; Qi et al., 2021; Wierzchos et al., 2015). Cyanobacteria are oxygenic photosynthesizers that convert photosynthetically visible light (400–700 nm) into chemical energy using complex molecular machinery, including light-harvesting and energy-transducing complexes, pigment biosynthesis, photo-sensory proteins, and associated response regulators (Bhaduri, 2019). Endolithic cyanobacteria from arid deserts are primarily members of the orders Chroococcales (Chroococcidiopsis and Gloeocapsa), Nostocales, and Oscillatoriales (Meslier & DiRuggiero, 2019; Murray et al., 2021, 2022).

Friedmann and Ocampo (1976) reported the first occurrence of polar crypto endolithic microorganisms in the McMurdo Dry Valleys of Antarctica living as vertically stratified communities beneath the surfaces of sandstone rocks. They have since received much attention to understand community composition, survival strategies, physiological requirements, associated microbe-mineral interactions, as well as their potential as analogs for life on Mars.

The orders Chroococcales (including Chroococcidiopsis and Gloeocapsa), Nostocales, and Oscillatoriales are the main groups of endolithic cyanobacteria found in arid deserts (Hu et al., 2012; Murray et al., 2022; Omelon, 2008). In the endolithic habitat, phototrophs must be able to react to the characteristics of light, since the rock substrate places severe physical restrictions on light transmission. Cyanobacteria have been found to employ a range of light-harvesting techniques to achieve low quantum flux densities and FRL, including the production of substituted forms of chlorophyll a and the spectral tuning of light-harvesting antennae (Croce & Van Amerongen, 2014; Wolf, 2019).

Cyanobacteria exposed to far-red light have been identified to modify the core components of their photosynthesis machinery and produce both Chl f and Chl d, which are red-shifted chlorophylls that allow photons in the FRL range to be harvested. This process is known as the far-red-light photoacclimation response (Wolf, 2019). Different low-light habitats, such as dense microbial mats, lakes, caverns, soil, stromatolites, beach rock biofilms, and numerous niches in subtropical forest ecosystems, have been found to harbor chlorophyll-producing cyanobacteria (Antonaru et al., 2020; Neilan et al., 2002; Ohkubo & Miyashita, 2017). Employing the phycobilisome linker gene as a marker of FRL-photosynthesis systems, it was recently shown that Chl f-synthesizing cyanobacteria are widely distributed (Antonaru et al., 2020). The FaRLiP response eventually reduces photosynthetic efficiency, yet it allows cyanobacteria to survive in a variety of low-light environments (Hopton, 2022) (Table 13.3).

13.7.1 Implications in Astrobiology and the Search for Extraterrestrial Life

Astrobiology represents a cutting-edge interdisciplinary field at the intersection of astronomy and biology, with a focus on exploring the possibility of life beyond Earth. The exploration mission has identified nearly 200 planets and satellites within our solar system, along with over 5000 exoplanets orbiting distant stars. This vast celestial diversity has fueled scientific interest in understanding planetary environments that could potentially support life. Despite these discoveries, Earth remains the sole known living planetary body, serving as a guiding reference in the quest for extraterrestrial life (Schultz et al., 2023). Extremophiles, microorganisms capable of thriving in extreme conditions, play a pivotal role in advancing our understanding of potential life beyond Earth. Among them, polyextremophiles organisms adapted to multiple extremes emerge as a crucial avenue of research for astrobiology and space exploration (Curtis-Harper, 2017). These resilient microbes serve as valuable models, offering insights into the search for indigenous extraterrestrial life and the detection of life-produced metabolites outside our planet (Rekadwad et al., 2023).

The primary goal of astrobiology is to search for life on planetary bodies beyond Earth. Planets such as Mars, Venus, and the moons Europa, Enceladus, and Titan

Table 13.3 Different mechanisms employed by cyanobacteria to survive under extreme conditions

	Stress	Mechanism	References
1	UV light	Synthesize metabolites, which offer immense protection from ROS and DNA damage	Ehling-Schulz and Scherer (1999)
2	High temperature	Cyanobacteria change their protein integrity fatty acid saturation and HSPs	Rezayian et al., (2019), Yadav et al., (2022)
3	Low temperature	Antifreezing proteins Structural modifications	Cockell et al., (1999), Yadav et al., (2022)
4	pH variations	Cyanobacteria use enzymes such as oxalate decarboxylase and carbonic anhydrase	Sanz-Luque et al., (2020), Yadav et al., (2022)
5	Osmotic stress and salinity	Intracellular trafficking, controlled sodium absorption, and active sodium efflux	Kanesaki et al., (2002), Molitor et al., (1986), Singh (1950), Srivastava et al., (2005)
6	Desiccation	Carbon dioxide fixation	Palmer and Friedmann (1990)
7	Metals	Changes in membrane structure, activation or deactivation of transfer pumps, metabolic transformation, and confinement	Hantke (2005)

provide unique opportunities to investigate the chemical evolution and origin of life. Their biochemistry features, resembling those supporting life on Earth, make them prime targets for extensive research (Board et al., 2018). To discern the potential for extraterrestrial life, defining the boundary conditions conducive to life is essential. Outer space presents harsh and inhospitable environmental factors, including high radiation doses, extreme temperatures, varied gravity, pressure, pH, salinity, energy source variations, and nutrient scarcity (DasSarma et al., 2020). However, the adaptability of extremophilic microbes on Earth, thriving in diverse and extreme habitats, suggests the possibility of survival in space's harsh conditions. Therefore, understanding the molecular mechanisms and unique physiological characteristics of extremophiles becomes paramount in defining the limits of life on Earth and identifying conditions favorable to life on other planetary bodies (Merino et al., 2019; Orosz et al., 2019). Microbial life is constrained by environmental parameters such as temperature, pH, salinity, water availability, pressure, radiation, and nutrient levels. Extremophiles, specifically those capable of thriving under multiple extreme stresses simultaneously (poly-extremophiles), demonstrate optimal growth in these challenging habitats (Capece et al., 2013).

Most extremophiles are associated with the Archaea and Bacteria domains, with only a few belonging to Eukarya. Their adaptability and resilience make isolated extremophiles promising candidates for astrobiology studies. Earth's extremophile habitats, resembling planetary bodies in outer space, provide critical insights into adaptation mechanisms and help identify biosignatures applicable in habitable zones beyond Earth (Lammer et al., 2009). Exploring modern extremophiles on Earth is crucial for understanding adaptation mechanisms and identifying novel biosignatures applicable in habitable zones beyond Earth (Taubner et al., 2020). Additionally, studying extreme microbiomes analogous to Earth can offer insights into potential extraterrestrial colonizers and how these microbes might contribute to the transformation of planets facing constant extreme conditions (Coleine et al., 2022; Núñez-López & Moskal, 2019). In summary, (poly) extremophiles are invaluable tools for predicting the boundaries of living organisms and deciphering the mechanisms and strategies behind survival in extreme environments.

13.7.2 Evolutionary Mechanisms and Selection Pressures

In addition, cyanobacteria exhibit remarkable versatility, thriving in diverse environmental conditions, which positions them as ideal candidates for colonizing terrestrial ecosystems. Recent studies have delved into the mechanisms underlying their adaptability and ecological significance in shaping terrestrial habitats, shedding light on their pivotal role in ecosystem dynamics and biogeochemical cycles. During colonization, they offer a range of benefits, such as protecting the soil from erosion, improving its texture, promoting carbon and nitrogen fixation, and increasing soil organic matter. Moreover, they collaborate with other microorganisms like bacteria, algae, archaea, and fungi to promote the growth of lichens, mosses, and vascular plants. In natural environments, filamentous cyanobacteria initiate colonization at the top millimeters of the soil and form BCSs, attracting other species for second-line colonization (Ferrenberg et al., 2017; Stal, 2007).

The only predominant and most extensively dispersed photoautotrophs that support the structural stability, moisture retention, and fertility of neighboring microorganisms are cyanobacteria found in microbial communities. They are significant in balancing ecosystem functions (Dabravolski & Isayenkov, 2022).

A study involved a comparative genomic analysis of 650 cyanobacterial genomes to explore the genetic basis of cyanobacterial habitat adaptation in marine, freshwater, and terrestrial environments. It is found that genomic size variations among cyanobacterial strains are strongly associated with their adaptation to specific habitats, which suggests that the expansion of gene families is a common strategy used by terrestrial cyanobacteria to cope with fluctuating environments, while the genomes of many marine strains have undergone contraction to adapt to nutrient-poor conditions. Furthermore, researchers identified hundreds of genes that are strongly associated with specific habitats, including genes involved in light sensing and absorption, chemotaxis, nutrient transporters, and responses to osmotic stress. These genes are crucial for the survival and adaptation of organisms in specific habitats, which they adapted by horizontal gene transfer, and such genetic exchanges are more frequent in terrestrial cyanobacteria (Chen et al., 2021a).

A recent study delved into the genes responsible for stress response in microbial mat communities dominated by cyanobacteria in Arctic and Antarctic areas including sigma B, EPS, cold shock proteins, and membrane modifications; low temperature may damage protein' structure, while nucleic acid content becomes more stable, making it difficult to support cellular processes like replication, transcription, and translation (Varin et al., 2012). To survive, cyanobacteria produce specialized proteins such as cold shock proteins, antifreeze protein, and glycine betaine (D'Amico et al., 2006; Dasauni et al., 2022).

They also increase the production of unsaturated fatty acids to maintain membrane fluidity, which is crucial for nutrient transport (Los et al., 2013). Cyanobacteria have adapted to freezing and desiccation by entering a dormant state with low metabolic activity. In addition, they produce exopolymer matrixes that provide support for biological and biogeochemical interactions. These matrixes consist of exopolysaccharides (EPS) that play a structural role and are responsible for creating and maintaining the microenvironment necessary for the growth and metabolism of cyanobacterial consortia. It has been discovered that EPS provides cryoprotection and desiccation protection for a diverse range of microorganisms (Dasauni et al., 2021; Makhalanyane et al., 2015; Rana & Upadhyay, 2020).

Pyschrophiles have the unique ability to not only survive but thrive in extremely low temperatures. To adapt to freeze–thaw conditions, they produce polyunsaturated fatty acids with a short chain length to incorporate into their membrane. Cyanobacteria, on the other hand, produce compatible solutes like trehalose, which helps to reduce the freezing point of intracellular solutes and prevent cell desiccation. Additionally, they secrete extracellular polymeric substances, which in turn help to minimize ice nucleation around the cells. This allows cyanobacterial consortiums to survive long-term seasonal dormancy phases in frozen environments. Upon thawing, these freeze-dried cyanobacterial mats can resume photosynthesis and experience high metabolic activity. Interestingly, the desiccation-resistant genus *Chroococcidiopsis* has also shown resistance against ionizing radiation, most likely due to an efficient DNA repair mechanism (Genuário et al., 2019; Klatt et al., 2020).

Photosynthetic organisms are sensitive to reactive oxygen species and oxygen, which are aggravated by light. Phototrophs have developed detoxification mechanisms and alternative oxygen-dependent enzymes to synthesize chlorophyll. Studies reveal that photosynthetic reaction centers in Cyanobacteria evolved to increase electron transfer efficiency and avoid photodamage caused by chlorophyll radicals. H2S-driven anoxygenic photosynthesis emerged in Cyanobacteria before oxygenic photosynthesis and played a role in fine-tuning the oxygen-evolving complex and repairing oxidative damage. Non-cyanobacterial anoxygenic phototrophs lack elaborate mechanisms to cope with oxygen, suggesting that they have occupied similar niche spaces throughout Earth's history (Hamilton, 2019; Olson & Straub, 2016).

13.8 Future Prospects and Implications

Cyanobacteria are a versatile source of macronutrients and micronutrients used as human food, protein supplements, and in agriculture, pollution control, bioenergy, and pharmaceuticals. They are easy to cultivate, can be manipulated genetically, and can be found in geothermally heated regions of the Earth. They are considered biotechnological treasures for high-temperature processes and for CO_2 sequestration, biofuel production, and bioremediation (Abbew et al., 2022; Mazard et al., 2016; Żymańczyk-Duda et al., 2022).

Cyanobacteria have unique capabilities to absorb and fix atmospheric nitrogen and CO_2, allowing them to thrive in harsh environmental conditions such as unfertile soils and saline waters. Due to these abilities, they are excellent candidates for the production of biodegradable plastics and biofuels (Malyan et al., 2020; Srivastava et al., 2021). Cyanobacteria have been studied as a potential source of biofuels due to their ability to produce triacylglycerol (TAG) and diacylglycerol (DAG), which can be used as biodiesel precursors (Savakis & Hellingwerf, 2015). Cyanobacteria species such as *Synechocystis, Spirulina, Anabaena,* and *N. muscorum* can function as bio-factories to produce biofuels. They provide a promising and renewable source of energy for the energy sector (Kaur & Reddersen, 2022). Researchers have successfully produced biofuels such as 2-methyl-1-butanol, isobutanol, 2,3-butanediol ethanol, isobutanol, and ethylene from engineered cyanobacteria (*Synechocystis* sp.) (Nozzi et al., 2013). Cyanobacteria have been considered a promising source for biofuel and bioplastic production due to their ability to produce diacylglycerol (DAG) and triacylglycerol (TAG), which can be used as precursors for biodiesel. Furthermore, the availability of genetic sequence information of these microbes makes them even more efficient for generating renewable biofuels (Agarwal et al., 2022).

Extensive research is necessary to improve the usefulness of cyanobacteria for human and animal exploitation. It is essential to profile cyanobacterial secondary metabolites, which are commonly used in different countries for nutraceutical and pharmaceutical applications, including antimicrobial and antitumor activities. Cyanobacterial enzymatic systems are currently used as biocatalysts to achieve various synthetic goals. This field is still not fully explored, and it offers biological tools with different features compared to light-independent organisms such as fungi and heterotrophic bacteria. Further research is needed to effectively exploit cyanobacteria using green technological advancements and gain efficient insights into their applications on larger scales. Future investigation is also promising to explore more beneficial roles of cyanobacteria in biotechnology (Hamida et al., 2020; Żymańczyk-Duda et al., 2022).

13.9 Conclusion

Cyanobacteria, among the most ancient life forms on Earth, have played a pivotal role in shaping the planet's biosphere most notably by contributing to the oxygenation of Earth's atmosphere. Their evolutionary development has provided key insights into the history of life, including the origin of chloroplasts in plants and algae through endosymbiosis. Advanced genomic techniques, both PCR-dependent and PCR-independent techniques, have been instrumental in elucidating the complex molecular structure of cyanobacteria, including insights from fossil records. As photosynthetic prokaryotes, cyanobacteria exhibit remarkable ecological and genetic diversity, with thousands of species identified and many more yet to be discovered. However, fully understanding their diversity remains a challenge due to difficulties in culturing many strains and ongoing debates over species classification. To date, only a fraction of cyanobacterial species have been sequenced. Recent advances in metagenomics, single-cell genomics, and bioinformatics offer powerful tools to overcome these limitations, enabling deeper insights into their evolutionary trajectories and ecological roles. Continued exploration of cyanobacterial diversity and genomics holds tremendous promise for both fundamental science and practical applications. Engineered cyanobacterial strains are being investigated for sustainable biofuel production, bioplastics, carbon capture, nitrogen fixation in agriculture, and environmental remediation. Understanding their metabolic potential and ecological adaptability not only informs early Earth biology but also paves the way for innovative solutions to pressing global challenges in energy, food security, and environmental sustainability.

References

Abbew, A.-W., Amadu, A. A., Qiu, S., Champagne, P., Adebayo, I., Anifowose, P. O., & Ge, S. (2022). Understanding the influence of free nitrous acid on microalgal-bacterial consortium in wastewater treatment: a critical review. *Bioresource Technology*, 127916.

Agarwal, P., Soni, R., Kaur, P., Madan, A., Mishra, R., Pandey, J., Singh, S., & Singh, G. (2022). Cyanobacteria as a promising alternative for sustainable environment: Synthesis of biofuel and biodegradable plastics. *Frontiers in Microbiology, 13*, Article 939347.

Akagha, M. U., Pietrasiak, N., Bustos, D. F., Vondrášková, A., Lamb, S. C., & Johansen, J. R. (2023). Albertania and Egbenema gen. nov. from Nigeria and the United States, expanding biodiversity in the Oculatellaceae (cyanobacteria). *Journal of Phycology*.

Alexander, P., & Svetlana, A. (2021). New life for old discovery: Amazing story about how bacterial predation on Chlorella resolved a paradox of dark cyanobacteria and gave the key to early history of oxygenic photosynthesis and aerobic respiration. *Protistology, 15*(3), 107–126.

Altermann, W., & Kazmierczak, J. (2003). Archean microfossils: A reappraisal of early life on Earth. *Research in Microbiology, 154*(9), 611–617.

Alvarenga, D. O., Fiore, M. F., & Varani, A. M. (2017). A metagenomic approach to cyanobacterial genomics. *Frontiers in Microbiology, 8*, 809.

Amorim, C. A., & do Nascimento Moura, A. (2021). Ecological impacts of freshwater algal blooms on water quality, plankton biodiversity, structure, and ecosystem functioning. *Science of the Total Environment, 758*, 143605.

Antonaru, L. A., Cardona, T., Larkum, A. W., & Nürnberg, D. J. (2020). Global distribution of a chlorophyll f cyanobacterial marker. *The ISME Journal, 14*(9), 2275–2287.

Baudouin-Cornu, P., Schuerer, K., Marlière, P., & Thomas, D. (2004). Intimate evolution of proteins: Proteome atomic content correlates with genome base composition. *Journal of Biological Chemistry, 279*(7), 5421–5428.

Benzerara, K., Skouri-Panet, F., Li, J., Férard, C., Gugger, M., Laurent, T., & Moreira, D. (2014). Intracellular Ca-carbonate biomineralization is widespread in cyanobacteria. *Proceedings of the National Academy of Sciences, 111*(30), 10933–10938.

Berry, M. A., Davis, T. W., Cory, R. M., Duhaime, M. B., Johengen, T. H., Kling, G. W., Marino, J. A., Den Uyl, P. A., Gossiaux, D., & Dick, G. J. (2017). Cyanobacterial harmful algal blooms are a biological disturbance to Western Lake Erie bacterial communities. *Environmental Microbiology, 19*(3), 1149–1162.

Bhaduri, S. (2019). *Structure-Function of Membrane Protein Complexes Involved in Oxygenic Photosynthesis* Purdue University].

Blank, C., & Sanchez-Baracaldo, P. (2010). Timing of morphological and ecological innovations in the cyanobacteria–A key to understanding the rise in atmospheric oxygen. *Geobiology, 8*(1), 1–23.

Bosak, T., Knoll, A. H., & Petroff, A. P. (2013). The meaning of stromatolites. *Annual Review of Earth and Planetary Sciences, 41*, 21–44.

Brasier, M., McLoughlin, N., Green, O., & Wacey, D. (2006). A fresh look at the fossil evidence for early Archaean cellular life. *Philosophical Transactions of the Royal Society B: Biological Sciences, 361*(1470), 887–902.

Buick, R. (2008). When did oxygenic photosynthesis evolve? *Philosophical Transactions of the Royal Society B: Biological Sciences, 363*(1504), 2731–2743.

Capece, M. C., Clark, E., Saleh, J. K., Halford, D., Heinl, N., Hoskins, S., & Rothschild, L. J. (2013). Polyextremophiles and the constraints for terrestrial habitability. In *Polyextremophiles: Life under multiple forms of stress* (pp. 3–59).

Cardona, T. (2017). Photosystem II is a chimera of reaction centers. *Journal of Molecular Evolution, 84*(2–3), 149–151.

Cardona, T., Sánchez-Baracaldo, P., Rutherford, A. W., & Larkum, A. W. (2019). Early Archean origin of photosystem II. *Geobiology, 17*(2), 127–150.

Cardona, W., Kunz, M., & Pettorino, V. (2017). Determining H0 with Bayesian hyper-parameters. *Journal of Cosmology and Astroparticle Physics, 2017*(03), 056.

Chen, M.-Y., Teng, W.-K., Zhao, L., Hu, C.-X., Zhou, Y.-K., Han, B.-P., Song, L.-R., & Shu, W.-S. (2021). Comparative genomics reveals insights into cyanobacterial evolution and habitat adaptation. *The ISME Journal, 15*(1), 211–227.

Chen, R., Huangfu, L., Lu, Y., Fang, H., Xu, Y., Li, P., Zhou, Y., Xu, C., Huang, J., & Yang, Z. (2021b). Adaptive innovation of green plants by horizontal gene transfer. *Biotechnology Advances, 46*, Article 107671.

Cho, J.-C., & Tiedje, J. M. (2001). Bacterial species determination from DNA-DNA hybridization by using genome fragments and DNA microarrays. *Applied and Environmental Microbiology, 67*(8), 3677–3682.

Cockell, C. S., Stokes, M. D., & Korsmeyer, K. E. (1999). Overwintering strategies of Antarctic organisms. *Environmental Reviews, 8*(1), 1–19.

Coleine, C., Stajich, J. E., & Selbmann, L. (2022). Fungi are key players in extreme ecosystems. In *Trends in ecology & evolution.*

Cornet, L., Magain, N., Baurain, D., & Lutzoni, F. (2021). Exploring syntenic conservation across genomes for phylogenetic studies of organisms subjected to horizontal gene transfers: A case study with Cyanobacteria and cyanolichens. *Molecular Phylogenetics and Evolution, 162*, Article 107100.

Croce, R., & Van Amerongen, H. (2014). Natural strategies for photosynthetic light harvesting. *Nature Chemical Biology, 10*(7), 492–501.

Curtis-Harper, E. (2017). *Potential microbial processes in an ancient martian environment: An investigation into bio-signature production and community ecology*. Open University (United Kingdom).

Dabravolski, S. A., & Isayenkov, S. V. (2022). Metabolites facilitating adaptation of desert cyanobacteria to extremely arid environments. *Plants, 11*(23), 3225.

D'Amico, S., Collins, T., Marx, J. C., Feller, G., Gerday, C., & Gerday, C. (2006). Psychrophilic microorganisms: Challenges for life. *EMBO Reports, 7*(4), 385–389.

Danovaro, R., Luna, G., Dell'Anno, A., & Pietrangeli, B. (2006). Comparison of two fingerprinting techniques, terminal restriction fragment length polymorphism and automated ribosomal intergenic spacer analysis, for determination of bacterial diversity in aquatic environments. *Applied and Environmental Microbiology, 72*(9), 5982–5989.

Dasauni, K., Bisht, D., & Nailwal, T. K. (2021). Rhizosphere modelling and nanotechnology: New outlooks in sustainable agriculture. In *Microbiological activity for soil and plant health management* (pp. 563–581).

Dasauni, K., Divya, & Nailwal, T. K. (2022). Cyanobacteria in cold ecosystem: Tolerance and adaptation. In *Survival strategies in cold-adapted microorganisms* (pp. 1–29).

DasSarma, P., Antunes, A., Simões, M. F., & DasSarma, S. (2020). Earth's stratosphere and microbial life. *Current Issues in Molecular Biology, 38*(1), 197–244.

de Kluijver, A., Yu, J., Houtekamer, M., Middelburg, J. J., & Liu, Z. (2012). Cyanobacteria as a carbon source for zooplankton in eutrophic Lake Taihu, China, measured by 13C labeling and fatty acid biomarkers. *Limnology and Oceanography, 57*(4), 1245–1254.

Demoulin, C. F., Lara, Y. J., Cornet, L., François, C., Baurain, D., Wilmotte, A., & Javaux, E. J. (2019). Cyanobacteria evolution: Insight from the fossil record. *Free Radical Biology and Medicine, 140*, 206–223.

Drews, G. (2011). The evolution of cyanobacteria and photosynthesis. In *Bioenergetic processes of cyanobacteria: From evolutionary singularity to ecological diversity* (pp. 265–284).

Ehling-Schulz, M., & Scherer, S. (1999). UV protection in cyanobacteria. *European Journal of Phycology, 34*(4), 329–338.

El-Seedi, H. R., El-Mallah, M. F., Yosri, N., Alajlani, M., Zhao, C., Mehmood, M. A., Du, M., Ullah, H., Daglia, M., & Guo, Z. (2023). Review of marine Cyanobacteria and the aspects related to their roles: Chemical, biological properties, nitrogen fixation and climate change. *Marine Drugs, 21*(8), 439.

Escalas, A., Hale, L., Voordeckers, J. W., Yang, Y., Firestone, M. K., Alvarez-Cohen, L., & Zhou, J. (2019). Microbial functional diversity: From concepts to applications. *Ecology and Evolution, 9*(20), 12000–12016.

Eskew, D., Caetano-Anolles, G., Bassam, B., & Gresshoff, P. (1993). DNA amplification fingerprinting of the Azolla-Anabaena symbiosis. *Plant Molecular Biology, 21*, 363–373.

Ferrenberg, S., Tucker, C. L., & Reed, S. C. (2017). Biological soil crusts: Diminutive communities of potential global importance. *Frontiers in Ecology and the Environment, 15*(3), 160–167.

Filstrup, C. T., Hillebrand, H., Heathcote, A. J., Harpole, W. S., & Downing, J. A. (2014). Cyanobacteria dominance influences resource use efficiency and community turnover in phytoplankton and zooplankton communities. *Ecology Letters, 17*(4), 464–474.

Fischer, W. W., Hemp, J., & Johnson, J. E. (2016). Evolution of oxygenic photosynthesis. *Annual Review of Earth and Planetary Sciences, 44*, 647–683.

Fujita, Y., Murakami, A., Aizawa, K., & Ohki, K. (1994). Short-term and long-term adaptation of the photosynthetic apparatus: Homeostatic properties of thylakoids. In *The molecular biology of cyanobacteria* (pp. 677–692).

Gaysina, L. A., Saraf, A., & Singh, P. (2019). Cyanobacteria in diverse habitats. In *Cyanobacteria* (pp. 1–28). Elsevier.

Gelsinger, D. R., Uritskiy, G., Reddy, R., Munn, A., Farney, K., & DiRuggiero, J. (2020). Regulatory noncoding small RNAs are diverse and abundant in an extremophilic microbial community. *Msystems, 5*(1), https://doi.org/10.1128/msystems.00584-00519

Genuário, D. B., Vaz, M. G., Santos, S. N., Kavamura, V. N., & Melo, I. S. (2019). Cyanobacteria From Brazilian extreme environments: toward functional exploitation. In *Microbial diversity in the genomic era* (pp. 265–284). Elsevier.

Gérard, E., De Goeyse, S., Hugoni, M., Agogué, H., Richard, L., Milesi, V., Guyot, F., Lecourt, L., Borensztajn, S., Joseph, M. B., Leclerc, T., Sarazin, G., Jézéquel, D., Leboulanger, C., & Ader, M. (2018). Key role of Alphaproteobacteria and cyanobacteria in the formation of stromatolites of Lake Dziani Dzaha (Mayotte, Western Indian Ocean). *Frontiers in Microbiology, 22*(9), 796. https://doi.org/10.3389/fmicb.2018.00796. PMID: 29872424; PMCID: PMC5972316.

Golubic, S., & Seong-Joo, L. (1999). Early cyanobacterial fossil record: Preservation, palaeoenvironments and identification. *European Journal of Phycology, 34*(4), 339–348.

Golubic, S., Seong-Joo, L., Browne, K.M. (2000). Cyanobacteria: Architects of sedimentary structures. In R. E. Riding, & S. M. Awramik (eds.), *Microbial sediments*. Springer. https://doi.org/10.1007/978-3-662-04036-2_8

Greene, E., Hubert, C., Nemati, M., Jenneman, G., & Voordouw, G. (2003). Nitrite reductase activity of sulphate-reducing bacteria prevents their inhibition by nitrate-reducing, sulphide-oxidizing bacteria. *Environmental Microbiology, 5*(7), 607–617.

Gupta, C., Rabani, M. S., Gupta, M. K., Habib, A., Pathak, A., Tripathi, S., & Singh, R. (2021). Cyanobacteria as sustainable microbiome for agricultural industries. In *Microbiota and biofertilizers, Vol 2: Ecofriendly tools for reclamation of degraded soil environs* (pp. 267–291).

Friedmann, E. I., & Ocampo, R. (1976). Endolithic blue-green algae in the dry valleys: primary producers in the Antarctic desert ecosystem. *Science, 193*(4259), 1247–1249.

Hamida, R. S., Ali, M. A., Redhwan, A., & Bin-Meferij, M. M. (2020). Cyanobacteria–a promising platform in green nanotechnology: A review on nanoparticles fabrication and their prospective applications. *International Journal of Nanomedicine*, 6033–6066.

Hamilton, T. L. (2019). The trouble with oxygen: The ecophysiology of extant phototrophs and implications for the evolution of oxygenic photosynthesis. *Free Radical Biology and Medicine, 140*, 233–249.

Hantke, K. (2005). Bacterial zinc uptake and regulators. *Current Opinion in Microbiology, 8*(2), 196–202.

Hipsley, C. A., & Müller, J. (2014). Beyond fossil calibrations: Realities of molecular clock practices in evolutionary biology. *Frontiers in Genetics, 5*, 138.

Hohmann-Marriott, M. F., & Blankenship, R. E. (2011). Evolution of photosynthesis. *Annual Review of Plant Biology, 62*, 515–548.

Hohmann-Marriott, M. F., & Blankenship, R. E. (2012). The photosynthetic world. In *Photosynthesis: Plastid biology, energy conversion and carbon assimilation* (pp. 3–32).

Hopton, C. M. (2022). *Genomic analysis of astrobiology-relevant adaptations to low light in far-red light utilising cyanobacteria* [University of Bristol].

Hu, C., Gao, K., & Whitton, B. A. (2012). Semi-arid regions and deserts. In *Ecology of cyanobacteria II: Their diversity in space and time* (pp. 345–369). Springer.

Hugenholtz, P., Chuvochina, M., Oren, A., Parks, D. H., & Soo, R. M. (2021). Prokaryotic taxonomy and nomenclature in the age of big sequence data. *The ISME Journal, 15*(7), 1879–1892.

Huisman, J., Codd, G. A., Paerl, H. W., Ibelings, B. W., Verspagen, J. M., & Visser, P. M. (2018). Cyanobacterial blooms. *Nature Reviews Microbiology, 16*(8), 471–483.

Jensen, P. E., Bassi, R., Boekema, E. J., Dekker, J. P., Jansson, S., Leister, D., Robinson, C., & Scheller, H. V. (2007). Structure, function and regulation of plant photosystem I. *Biochimica et Biophysica Acta (BBA)-Bioenergetics, 1767*(5), 335–352.

Johnson, J. S., Spakowicz, D. J., Hong, B.-Y., Petersen, L. M., Demkowicz, P., Chen, L., Leopold, S. R., Hanson, B. M., Agresta, H. O., & Gerstein, M. (2019). Evaluation of 16S rRNA gene sequencing for species and strain-level microbiome analysis. *Nature Communications, 10*(1), 5029.

Jüttner, F., & Watson, S. B. (2007). Biochemical and ecological control of geosmin and 2-methylisoborneol in source waters. *Applied and Environmental Microbiology, 73*(14), 4395–4406.

Kanesaki, Y., Suzuki, I., Allakhverdiev, S. I., Mikami, K., & Murata, N. (2002). Salt stress and hyperosmotic stress regulate the expression of different sets of genes in Synechocystis sp. PCC 6803. *Biochemical and Biophysical Research Communications, 290*(1), 339–348.

Kardinaal, W. E. A., Janse, I., Kamst-van Agterveld, M., Meima, M., Snoek, J., Mur, L. R., Huisman, J., Zwart, G., & Visser, P. M. (2007). Microcystis genotype succession in relation to microcystin concentrations in freshwater lakes. *Aquatic Microbial Ecology, 48*(1), 1–12.

Kamennaya, N. A., Ajo-Franklin, C. M., Northen, T., & Jansson, C. (2012). Cyanobacteria as biocatalysts for carbonate mineralization. *Minerals, 2*(4), 338–364.

Kaur, S., & Reddersen, B. (2022). Algae based solutions for polluted environments to restore ecosphere equilibrium. *International Journal of Environmental Pollution and Remediation, 9*, 09–18.

Khalifa, S. A., Shedid, E. S., Saied, E. M., Jassbi, A. R., Jamebozorgi, F. H., Rateb, M. E., Du, M., Abdel-Daim, M. M., Kai, G.-Y., & Al-Hammady, M. A. (2021). Cyanobacteria—From the oceans to the potential biotechnological and biomedical applications. *Marine Drugs, 19*(5), 241.

Klatt, J. M., Gomez-Saez, G. V., Meyer, S., Ristova, P. P., Yilmaz, P., Granitsiotis, M. S., Macalady, J. L., Lavik, G., Polerecky, L., & Bühring, S. I. (2020). Versatile cyanobacteria control the timing and extent of sulfide production in a Proterozoic analog microbial mat. *The ISME Journal, 14*(12), 3024–3037.

Kumar, P., Singh, A., Sachan, A., & Sachan, S. G. (2021). Molecular tools for microbial diversity analysis. In *Wastewater treatment* (pp. 55–65). Elsevier.

Kumari, N., & Rai, L. C. (2020). Cyanobacterial diversity: molecular insights under multifarious environmental conditions. In *Advances in cyanobacterial biology* (pp. 17–33). Elsevier.

Kumari, N., Srivastava, A., Bhargava, P., & Rai, L. L. (2009). Molecular approaches towards assessment of cyanobacterial biodiversity. *African Journal of Biotechnology, 8*(18).

Lammer, H., Bredehöft, J., Coustenis, A., Khodachenko, M., Kaltenegger, L., Grasset, O., Prieur, D., Raulin, F., Ehrenfreund, P., & Yamauchi, M. (2009). What makes a planet habitable? *The Astronomy and Astrophysics Review, 17*, 181–249.

Larsson, J., Nylander, J. A., & Bergman, B. (2011). Genome fluctuations in cyanobacteria reflect evolutionary, developmental and adaptive traits. *BMC Evolutionary Biology, 11*, 1–21.

Lavin, P., Gómez, P., Gonzále, B., & Ulloa, O. (2008). Diversidad de las picocianobacterias marinas Prochlorococcus y Synechococcus por medio de polimorfismos de longitud de fragmentos de restricción terminal en secuencias del espaciador transcrito interno del ARNr 16S–23S. *Revista Chilena De Historia Natural, 81*(4), 515–531.

Layer, A. (2023). *Phycobilisome abundance regulator psor and its conservation and function within cyanobacteria* [Michigan State University].

Lee, D.-H., Zo, Y.-G., & Kim, S.-J. (1996). Nonradioactive method to study genetic profiles of natural bacterial communities by PCR-single-strand-conformation polymorphism. *Applied and Environmental Microbiology, 62*(9), 3112–3120.

Lehtimäki, J., Lyra, C., Suomalainen, S., Sundman, P., Rouhiainen, L., Paulin, L., Salkinoja-Salonen, M., & Sivonen, K. (2000). Characterization of Nodularia strains, cyanobacteria from brackish waters, by genotypic and phenotypic methods. *International Journal of Systematic and Evolutionary Microbiology, 50*(3), 1043–1053.

Los, D. A., Mironov, K. S., & Allakhverdiev, S. I. (2013). Regulatory role of membrane fluidity in gene expression and physiological functions. *Photosynthesis Research, 116*, 489–509.

Lyons, T. W., Reinhard, C. T., & Planavsky, N. J. (2014). The rise of oxygen in Earth's early ocean and atmosphere. *Nature, 506*(7488), 307–315.

Makhalanyane, T. P., Valverde, A., Velázquez, D., Gunnigle, E., Van Goethem, M. W., Quesada, A., & Cowan, D. A. (2015). Ecology and biogeochemistry of cyanobacteria in soils, permafrost, aquatic and cryptic polar habitats. *Biodiversity and Conservation, 24*, 819–840.

Malyan, S. K., Singh, S., Bachheti, A., Chahar, M., Sah, M. K., Kumar, A., Yadav, A. N., & Kumar, S. S. (2020). Cyanobacteria: A perspective paradigm for agriculture and environment. In *New and future developments in microbial biotechnology and bioengineering* (pp. 215–224). Elsevier.

Mazard, S., Penesyan, A., Ostrowski, M., Paulsen, I. T., & Egan, S. (2016). Tiny microbes with a big impact: The role of cyanobacteria and their metabolites in shaping our future. *Marine Drugs, 14*(5), 97.

Merino, N., Aronson, H. S., Bojanova, D. P., Feyhl-Buska, J., Wong, M. L., Zhang, S., & Giovannelli, D. (2019). Living at the extremes: Extremophiles and the limits of life in a planetary context. *Frontiers in Microbiology, 10*, 780.

Meslier, V., & DiRuggiero, J. (2019). Endolithic microbial communities as model systems for ecology and astrobiology. In *Model ecosystems in extreme environments* (pp. 145–168). Elsevier.

Mikolajczak, A., Söller, R., Blohm, D., & Fischer, U. (1999). Molecular biological characterization of unicellular marine Cyanobacteria from the Baltic Sea. *Bulletin de l'Institut Océanographique (Monaco)*, 95–99.

Minchin, S., & Lodge, J. (2019). Understanding biochemistry: Structure and function of nucleic acids. *Essays in Biochemistry, 63*(4), 433–456.

Molitor, V., Erber, W., & Peschek, G. A. (1986). Increased levels of cytochrome oxidase and sodium-proton antiporter in the plasma membrane of Anacystis nidulans after growth in sodium-enriched media. *FEBS Letters, 204*(2), 251–256.

Moustaka-Gouni, M., & Sommer, U. (2020). Effects of harmful blooms of large-sized and colonial cyanobacteria on aquatic food webs. *Water, 12*(6), 1587.

Mulkidjanian, A. Y., Koonin, E. V., Makarova, K. S., Mekhedov, S. L., Sorokin, A., Wolf, Y. I., Dufresne, A., Partensky, F., Burd, H., & Kaznadzey, D. (2006). The cyanobacterial genome core and the origin of photosynthesis. *Proceedings of the National Academy of Sciences, 103*(35), 13126–13131.

Murray, B., Dailey, M., Ertekin, E., & DiRuggiero, J. (2021). Draft metagenomes of endolithic cyanobacteria and cohabitants from hyper-arid deserts. *Microbiology Resource Announcements, 10*(30), https://doi.org/10.1128/mra.00206-00221

Murray, B., Ertekin, E., Dailey, M., Soulier, N. T., Shen, G., Bryant, D. A., Perez-Fernandez, C., & DiRuggiero, J. (2022). Adaptation of cyanobacteria to the endolithic light spectrum in hyper-arid deserts. *Microorganisms, 10*(6), 1198.

Muyzer, G., & Smalla, K. (1998). Application of denaturing gradient gel electrophoresis (DGGE) and temperature gradient gel electrophoresis (TGGE) in microbial ecology. *Antonie Van Leeuwenhoek, 73*, 127–141.

Neilan, B. A., Burns, B. P., Relman, D. A., & Lowe, D. R. (2002). Molecular identification of cyanobacteria associated with stromatolites from distinct geographical locations. *Astrobiology, 2*(3), 271–280.

Nozzi, N. E., Oliver, J. W., & Atsumi, S. (2013). Cyanobacteria as a platform for biofuel production. *Frontiers in Bioengineering and Biotechnology, 1*, 7.

Núñez-López, V., & Moskal, E. (2019). Potential of CO2-EOR for near-term decarbonization. *Frontiers in Climate, 1*, 5.

Och, L. M., & Shields-Zhou, G. A. (2012). The Neoproterozoic oxygenation event: Environmental perturbations and biogeochemical cycling. *Earth-Science Reviews, 110*(1–4), 26–57.

Ohkubo, S., & Miyashita, H. (2017). A niche for cyanobacteria producing chlorophyll f within a microbial mat. *The ISME Journal, 11*(10), 2368–2378.

Olson, J. M., & Blankenship, R. E. (2004). Thinking about the evolution of photosynthesis. *Photosynthesis Research, 80*, 373–386.

Olson, K. R., & Straub, K. D. (2016). The role of hydrogen sulfide in evolution and the evolution of hydrogen sulfide in metabolism and signaling. *Physiology, 31*(1), 60–72.

Omelon, C. R. (2008). Endolithic microbial communities in polar desert habitats. *Geomicrobiology Journal, 25*(7–8), 404–414.

Orosz, J. A., Welsh, W. F., Haghighipour, N., Quarles, B., Short, D. R., Mills, S. M., Satyal, S., Torres, G., Agol, E., & Fabrycky, D. C. (2019). Discovery of a third transiting planet in the Kepler-47 circumbinary system. *The Astronomical Journal, 157*(5), 174.

Paerl, H. W., & Otten, T. G. (2013). Harmful cyanobacterial blooms: Causes, consequences, and controls. *Microbial Ecology, 65*, 995–1010.

Palmer, R. J., & Friedmann, E. I. (1990). Water relations and photosynthesis in the cryptoendolithic microbial habitat of hot and cold deserts. *Microbial Ecology, 19*, 111–118.

Pandey, A., Singh, G., Pandey, N., Patel, A., Tiwari, S., & Prasad, S. M. (2022). Impacts of environmental stress on physiology and biochemistry of cyanobacteria. In *Ecophysiology and biochemistry of cyanobacteria* (pp. 65–89). Springer.

Pathak, J., Rajneesh, Maurya, P. K., Singh, S. P., Haeder, D.-P., & Sinha, R. P. (2018). Cyanobacterial farming for environment friendly sustainable agriculture practices: Innovations and perspectives. *Frontiers in Environmental Science, 6*, 7.

Pathak, J., Singh, P. R., Sinha, R. P., & Rastogi, R. P. (2022). Evolution and distribution of cyanobacteria. In *Ecophysiology and biochemistry of cyanobacteria* (pp. 1–30). Springer.

Petrescu, F. I. T., & Ungureanu, L. M. (2022). About the cyanobacteria and stromatolites. *OnLine Journal of Biological Sciences, 22*(1), 87. https://doi.org/10.3844/ojbsci.2022.87.111

Planavsky, N. J., Crowe, S. A., Fakhraee, M., Beaty, B., Reinhard, C. T., Mills, B. J., Holstege, C., & Konhauser, K. O. (2021). Evolution of the structure and impact of Earth's biosphere. *Nature Reviews Earth & Environment, 2*(2), 123–139.

Qi, J., Ma, B., Miao, S., Liu, R., Hu, C., & Qu, J. (2021). Pre-oxidation enhanced cyanobacteria removal in drinking water treatment: A review. *Journal of Environmental Sciences, 110*, 160–168.

Rabalais, N., Diaz, R. J., Levin, L., Turner, R. E., Gilbert, D., & Zhang, J. (2010). Dynamics and distribution of natural and human-caused hypoxia. *Biogeosciences, 7*(2), 585–619.

Rana, S., & Upadhyay, L. S. B. (2020). Microbial exopolysaccharides: Synthesis pathways, types and their commercial applications. *International Journal of Biological Macromolecules, 157*, 577–583.

Rastogi, R. P., & Sinha, R. P. (2009). Biotechnological and industrial significance of cyanobacterial secondary metabolites. *Biotechnology Advances, 27*(4), 521–539.

Rekadwad, B. N., Li, W.-J., Gonzalez, J. M., Punchappady Devasya, R., Ananthapadmanabha Bhagwath, A., Urana, R., & Parwez, K. (2023). Extremophiles: The species that evolve and survive under hostile conditions. *3 Biotech, 13*(9), 316.

Rezayian, M., Niknam, V., & Ebrahimzadeh, H. (2019). Stress response in cyanobacteria. *Iranian Journal of Plant Physiology, 9*(3), 2773–2787.

Šabacká, M., & Elster, J. (2006). Response of cyanobacteria and algae from Antarctic wetland habitats to freezing and desiccation stress. *Polar Biology, 30*, 31–37.

Sánchez-Baracaldo, P., Bianchini, G., Wilson, J. D., & Knoll, A. H. (2022). Cyanobacteria and biogeochemical cycles through Earth history. *Trends in Microbiology, 30*(2), 143–157.

Sánchez-Baracaldo, P., & Cardona, T. (2020). On the origin of oxygenic photosynthesis and Cyanobacteria. *New Phytologist, 225*(4), 1440–1446.

Sanz-Luque, E., Bhaya, D., & Grossman, A. R. (2020). Polyphosphate: A multifunctional metabolite in cyanobacteria and algae. *Frontiers in Plant Science, 11*, 938.

Savakis, P., & Hellingwerf, K. J. (2015). Engineering cyanobacteria for direct biofuel production from CO_2. *Current Opinion in Biotechnology, 33*, 8–14.

Schirrmeister, B. E., de Vos, J. M., Antonelli, A., & Bagheri, H. C. (2013). Evolution of multicellularity coincided with increased diversification of cyanobacteria and the Great Oxidation Event. *Proceedings of the National Academy of Sciences, 110*(5), 1791–1796.

Schopf, J. W. (2012). The fossil record of cyanobacteria. In *Ecology of cyanobacteria II: Their diversity in space and time* (pp. 15–36). Springer.

Schultz, J., dos Santos, A., Patel, N., & Rosado, A. S. (2023). Life on the edge: Bioprospecting extremophiles for astrobiology. *Journal of the Indian Institute of Science*, 1–17.

Semenova, E., Kuznedelov, K., & Grachev, M. (2001). Nucleotide sequences of fragments of 16S rRNA of the Baikal natural populations and laboratory cultures of cyanobacteria. *Molecular Biology, 35*, 405–410.

Shariatmadari, Z., Moharrek, F., Riahi, H., Heidari, F., & Aslan, E. (2017). Efficiency of partial 16S rRNA gene sequencing as molecular marker for phylogenetic study of cyanobacteria, with emphasis on some complex taxa. *Acta Biologica Szegediensis, 61*(1), 59–68.

Shestakov, S. V., & Karbysheva, E. A. (2017). The origin and evolution of cyanobacteria. *Biology Bulletin Reviews, 7*(4), 259–272.

Singh, A., Pal, D. B., Mohammad, A., Alhazmi, A., Haque, S., Yoon, T., Srivastava, N., & Gupta, V. K. (2022). Biological remediation technologies for dyes and heavy metals in wastewater treatment: New insight. *Bioresource Technology, 343*, Article 126154.

Singh, H. (2018). Desiccation and radiation stress tolerance in cyanobacteria. *Journal of Basic Microbiology, 58*(10), 813–826.

Singh, P. K., Kumar, A., Singh, V. K., & Shrivistava, A. K. (2020). *Advances in cyanobacterial biology*. Academic Press.

Singh, P., Singh, S. S., Elster, J., & Mishra, A. K. (2013). Molecular phylogeny, population genetics, and evolution of heterocystous cyanobacteria using nif H gene sequences. *Protoplasma, 250*, 751–764.

Singh, R. N. (1950). Reclamation of 'Usar' lands in India through blue-green algae. *Nature, 165*(4191), 325–326.

Sinha, R. P., & Hader, D.-P. (1996). Invited review photobiology and ecophysiology of rice field cyanobacteria. *Photochemistry and Photobiology, 64*(6), 887–896.

Siqueira, J. F., Sakamoto, M., & Rosado, A. S. (2010). Microbial community profiling using terminal restriction fragment length polymorphism (T-RFLP) and denaturing gradient gel electrophoresis (DGGE). In *Oral biology: Molecular techniques and applications* (pp. 71–85).

Soo, R. M., Hemp, J., & Hugenholtz, P. (2019). Evolution of photosynthesis and aerobic respiration in the cyanobacteria. *Free Radical Biology and Medicine, 140*, 200–205.

Space Studies Board & National Academies of Sciences, Medicine, Division on Engineering, Physical Science. (2018). An astrobiology strategy for the search for life in the universe.

Springstein, B. L., Nürnberg, D. J., Weiss, G. L., Pilhofer, M., & Stucken, K. (2020). Structural determinants and their role in cyanobacterial morphogenesis. *Life, 10*(12), 355.

Srivastava, A. K., Bhargava, P., & Rai, L. C. (2005). Salinity and copper-induced oxidative damage and changes in the antioxidative defence systems of Anabaena doliolum. *World Journal of Microbiology and Biotechnology, 21*, 1291–1298.

Srivastava, M., Kaushik, M. S., & Mishra, A. K. (2016). Linking the physicochemical properties with the abundance and diversity of rhizospheric bacterial population inhabiting paddy soil based on a concerted multivariate analysis of PCR-DGGE and RISA. *Geomicrobiology Journal, 33*(10), 894–905.

Srivastava, R., Kanda, T., Yadav, S., Mishra, R., & Atri, N. (2021). Cyanobacteria in rhizosphere: Dynamics, diversity, and symbiosis. In *Plant, soil and microbes in tropical ecosystems* (pp. 51–69).

Stal, L. J. (2007). Cyanobacteria: diversity and versatility, clues to life in extreme environments. In *Algae and cyanobacteria in extreme environments* (pp. 659–680). Springer.

Stal, L. J. (2012). Cyanobacterial mats and stromatolites. In B. Whitton (eds.), *Ecology of cyanobacteria II*. Springer. https://doi.org/10.1007/978-94-007-3855-3_4

Stal, L. J. (2015). Nitrogen fixation in cyanobacteria. *eLS*, 1–9.

Suda, S., Watanabe, M. M., Otsuka, S., Mahakahant, A., Yongmanitchai, W., Nopartnaraporn, N., Liu, Y., & Day, J. G. (2002). Taxonomic revision of water-bloom-forming species of oscillatorioid cyanobacteria. *International Journal of Systematic and Evolutionary Microbiology, 52*(5), 1577–1595.

Sukenik, A., & Kaplan, A. (2021). Cyanobacterial harmful algal blooms in aquatic ecosystems: A comprehensive outlook on current and emerging mitigation and control approaches. *Microorganisms, 9*(7), 1472.

Svenning, M. M., Eriksson, T., & Rasmussen, U. (2005). Phylogeny of symbiotic cyanobacteria within the genus Nostoc based on 16S rDNA sequence analyses. *Archives of Microbiology, 183*, 19–26.

Tanvir, R. U., Hu, Z., Zhang, Y., & Lu, J. (2021). Cyanobacterial community succession and associated cyanotoxin production in hypereutrophic and eutrophic freshwaters. *Environmental Pollution, 290*, Article 118056.

Tashyreva, D., & Elster, J. (2016). Annual cycles of two cyanobacterial mat communities in hydro-terrestrial habitats of the High Arctic. *Microbial Ecology, 71*, 887–900.

Taubner, R.-S., Olsson-Francis, K., Vance, S. D., Ramkissoon, N. K., Postberg, F., de Vera, J.-P., Antunes, A., Camprubi Casas, E., Sekine, Y., & Noack, L. (2020). Experimental and simulation efforts in the astrobiological exploration of exooceans. *Space Science Reviews, 216*, 1–41.

Tiwari, P., & Bae, H. (2020). Horizontal gene transfer and endophytes: An implication for the acquisition of novel traits. *Plants, 9*(3), 305.

Tomitani, A., Knoll, A. H., Cavanaugh, C. M., & Ohno, T. (2006). The evolutionary diversification of cyanobacteria: Molecular–phylogenetic and paleontological perspectives. *Proceedings of the National Academy of Sciences, 103*(14), 5442–5447.

Varin, T., Lovejoy, C., Jungblut, A. D., Vincent, W. F., & Corbeil, J. (2012). Metagenomic analysis of stress genes in microbial mat communities from Antarctica and the High Arctic. *Applied and Environmental Microbiology, 78*(2), 549–559.

Wierzchos, J., DiRuggiero, J., Vítek, P., Artieda, O., Souza-Egipsy, V., Škaloud, P., Tisza, M., Davila, A. F., Vílchez, C., & Garbayo, I. (2015). Adaptation strategies of endolithic chlorophototrophs to survive the hyperarid and extreme solar radiation environment of the Atacama Desert. *Frontiers in Microbiology, 6*, 934.

Wolf, B. M. (2019). *Diverse far-red light utilization strategies in cyanobacteria and algae.* Washington University.

Yadav, M. R., Choudhary, M., Singh, J., Lal, M. K., Jha, P. K., Udawat, P., Gupta, N. K., Rajput, V. D., Garg, N. K., & Maheshwari, C. (2022). Impacts, tolerance, adaptation, and mitigation of heat stress on wheat under changing climates. *International Journal of Molecular Sciences, 23*(5), 2838.

Yerrapragada, S., Siefert, J. L., & Fox, G. E. (2009). Horizontal gene transfer in cyanobacterial signature genes. In *Horizontal gene transfer: Genomes in flux* (pp. 339–366).

Zakhia, F., Jungblut, A.-D., Taton, A., Vincent, W. F., & Wilmotte, A. (2008). Cyanobacteria in cold ecosystems. In *Psychrophiles: From biodiversity to biotechnology* (pp. 121–135).

Zwart, G., Kamst-van Agterveld, M. P., Van Der Werff-Staverman, I., Hagen, F., Hoogveld, H. L., & Gons, H. J. (2005). Molecular characterization of cyanobacterial diversity in a shallow eutrophic lake. *Environmental Microbiology, 7*(3), 365–377.

Żymańczyk-Duda, E., Samson, S. O., Brzezińska-Rodak, M., & Klimek-Ochab, M. (2022). Versatile applications of cyanobacteria in biotechnology. *Microorganisms, 10*(12), 2318.

Chapter 14
Photosynthetic Adaptations and Light Harvesting in Bloom-Forming Cyanobacteria

Taufiq Nawaz, Sirmast Faiz, Shah Fahad, Nitish Joshi, Shah Saud, Junaid Yousaf, Muhammad Aqil, Muhammad Adnan, Muhammad Amjad, and Imran Khan

Abstract Cyanobacteria are among the oldest extant oxygenic phototrophs and remain integral to global biogeochemical cycles, particularly carbon and nitrogen fluxes in aquatic ecosystems. A subset of these organisms, the bloom-forming cyanobacteria, have adapted unique photosynthetic strategies that enable ecological dominance in nutrient-rich and light-variable environments. This chapter presents a comprehensive examination of the structural, physiological, and molecular mechanisms that facilitate efficient light capture and energy conversion in these taxa. Key adaptations include the deployment of specialized pigment–protein complexes,

T. Nawaz (✉) · S. Fahad
Department of Biology and Microbiology, South Dakota State University, Brookings, SD, USA
e-mail: Taufiq.nawaz@jacks.sdstate.edu

S. Faiz
National Institute of Virology, Dr. Panjwani Center for Molecular Medicine and Drug Research, International Center of Chemical and Biological Sciences, University of Karachi, Karachi, Pakistan

N. Joshi
Department of Agronomy, Horticulture and Plant Science, South Dakota State University, Brookings, SD, USA

S. Saud
College of Life Science, Linyi University, Linyi, Shandong, China

J. Yousaf
Department of Food Science and Engineering, Beijing Technology and Business University, Beijing, China

M. Aqil
College of Food Science and Technology, Yunnan Agricultural University, Kunming, China

M. Adnan · M. Amjad
Department of Agriculture, University of Swabi, Ambar, Pakistan

I. Khan
Center for Life Science Ventures, Cornell University, Ithaca, NY, USA

S. Fahad et al. (eds.), *Cyanobacterial Blooms: Ecology, Evolution and Biogeochemical Impacts*, https://doi.org/10.1007/978-3-032-06042-6_14

such as *phycobilisomes*, and dynamic regulatory processes like *state transitions* and *chromatic acclimation* (e.g., CA4, CA6 systems), which allow modulation of light energy distribution between *Photosystem I* and *Photosystem II*. Technological advances particularly in *time-resolved fluorescence spectroscopy*, *fluorescence lifetime imaging microscopy* (*FLIM*), and *Förster resonance energy transfer (FRET)* have enabled unprecedented insights into real-time energy transfer and protein–protein interactions within cyanobacterial photosystems. Concurrently, the advent of *CRISPR-Cas* and synthetic biology has opened new frontiers in dissecting and engineering photosynthetic pathways for enhanced CO_2 fixation and biofuel synthesis. The ecological relevance of these adaptations is further contextualized through their roles in harmful algal bloom (HAB) dynamics, competition for light and nutrients, and climate resilience. Bloom-forming genera such as *Microcystis*, *Dolichospermum*, and *Planktothrix* are discussed in terms of their photophysiological plasticity and environmental triggers. We also explore emerging applications, including the development of biohybrid systems, *biophotovoltaics*, and carbon-negative biofactories utilizing engineered cyanobacterial strains. The chapter concludes by identifying critical gaps in our understanding of regulatory networks governing photosynthetic responses under multifactorial stressors and proposes integrative approaches combining *omics*, remote sensing, and ecological modeling. Collectively, these insights underscore the centrality of cyanobacterial photosynthesis not only in ecosystem functioning but also in biotechnological solutions to energy, climate, and sustainability challenges.

Keywords *Cyanobacteria* · Photosynthesis · Light harvesting · *Phycobilisome* · *Chromatic acclimation* · Harmful algal blooms · Synthetic biology · *CRISPR-Cas* · Climate resilience · *Photosystem II*

14.1 Introduction

14.1.1 Overview of Cyanobacteria

Cyanobacteria are unicellular prokaryotic organisms with single cells, and blue-green algae is another name for cyanobacteria because they possess both blue and green photosynthetic pigments (Jassim et al., 2023). They are not true algae but belong to the domain Bacteria, phylum Cyanobacteria. Cyanobacteria are represented by their ability to carry out oxygenic photosynthesis, which is a feature they have in common with higher plants and algae (Kupriyanova et al., 2023). Morphologically, cyanobacteria show vast diversity in their forms from unicellular to colonial forms and from multicellular filamentous organisms to specialized cells such as heterocysts for fixation of nitrogen and akinetes for enduring unfavorable conditions.

According to Nath (2025), the role of cyanobacteria in evolution cannot be emphasized excessively because cyanobacteria are thought to be among the oldest photosynthetic organisms to inhabit Earth, with evidence of existence through fossils for more than 2.5 billion years. Together with this early ability, they were instrumental in creating the Earth's early anoxic atmosphere into an oxygenated world, a process referred to as the Great Oxidation Event. Therefore, cyanobacteria are not only ecological innovators but also foundational figures in the evolutionary path of aerobic life forms (Pathak et al., 2022). In modern ecosystems, cyanobacteria remain highly important. They are among the major primary producers in aquatic systems, providing the foundation of food webs and fostering biodiversity. In poor nutrient conditions, some cyanobacteria can fix nitrogen from the atmosphere, thus improving soil fertility and aquatic productivity (Ramakrishnan et al., 2023). This greatly reflects their high adaptability and ecological resilience, which is the ability to evolve in very different ecological places from freshwater lakes to the sea and to deserts to polar conditions.

14.1.2 Importance of Photosynthesis in Cyanobacteria

Photosynthesis is the foundation of cyanobacterial ecological role and metabolic activity. Cyanobacteria undergo oxygenic photosynthesis with water as the electron donor and the release of molecular oxygen as a byproduct (Upadhyay et al., 2021). This is achieved through a complicated photosynthetic machinery that contains two photosystems, such as Photosystem I (PSI) and Photosystem II (PSII), which are located within internal thylakoid membranes (Xuan & Li, 2021). These photosystems are embedded with chlorophyll a and accessory pigments such as phycobiliproteins, which extend the scope of light absorption and maximize energy trapping under diverse environmental conditions.

Cyanobacterial photosynthetic machinery maintains a deep connection to its chloroplast analog in eukaryotic plants and algae (Stadnichuk & Kusnetsov, 2021). This indicates the prevalence of the well-held endosymbiotic hypothesis whereby chloroplasts are assumed to have arisen as a product of being overcome by a non-photosynthetic eukaryotic cell ancestor, cyanobacterium. For this reason, cyanobacterial photosynthesis not only has important implications regarding primary productivity and geochemical cycling but also illuminates the origin of photosynthesis on Earth. Furthermore, cyanobacteria have an essential position in the Earth's carbon cycle. They photosynthesize to fix significant amounts of atmospheric carbon dioxide (CO_2), thus sequestering carbon and playing a role in climate control. Their efficiency of photosynthesis, tolerance of varying light intensities, and ability to withstand stressors of the environment make them important actors in aquatic productivity and carbon flux dynamics.

14.1.3 Focus on Bloom-Forming Cyanobacteria

Among the taxonomically diverse cyanobacteria, there is a subset that is well known for producing dense, observable combinations referred to as blooms (Salmaso et al., 2024). These bloom-producing cyanobacteria involve genera such as *Microcystis, Dolichospermum* (previous name *Anabaena*), *Aphanizomenon*, and *Planktothrix*. These blooms tend to be present in nutrient-enriched (eutrophic) water bodies and are controlled by a variety of factors such as temperature, light intensity, and levels of nutrients (most notably nitrogen and phosphorus). Bloom-inducing cyanobacteria have various unique features facilitating their dominance during favorable conditions (Wang et al., 2021). These include buoyancy-controlling gas vesicles, enabling the cells to maintain their position within the water column for maximum illumination; high-efficiency nutrient acquisition; and capabilities to outgrow other phytoplankton when exposed to limited light or extreme temperatures. In addition, most bloom-forming species produce secondary metabolites referred to as cyanotoxins, including microcystins and anatoxins, which are toxic to aquatic life, terrestrial animals, and human beings.

The ecological role of cyanobacterial blooms is considerable and frequently negative. Harmful algal blooms (HABs), especially, may cause hypoxia (oxygen depletion), fish kills, and disturbance of aquatic ecosystems (Geletu, 2023). From a socio-economic viewpoint, HABs impact water quality, fisheries, recreation, and public health and entail significant economic losses and management problems. Due to the accelerating frequency and magnitude of blooms, a consequence of anthropogenic nutrient loading and climate change, the importance of learning about the biology and ecology of bloom-forming cyanobacteria has heightened (Fig. 14.1).

14.1.4 Relevance of Studying Photosynthetic Adaptations in Bloom-Forming Species

It is crucial to understand the photosynthetic mechanisms of bloom-causing cyanobacteria to estimate bloom behavior and plan for mitigation strategies. These bacteria usually have high-light-harvesting ability and photoprotective devices to enable them to maximize photosynthesis under fluctuating and, at times, extreme conditions, such as irradiance or deficiency in nutrients (Sreeharsha & Mohan, 2024). For example, certain species can regulate the makeup of their light-harvesting complexes or the architecture of their thylakoid membranes upon experiencing environmental stress, hence sustaining photosynthetic performance and cellular homeostasis (Pivato, 2023). In addition, the plasticity of photosynthesis in bloom-forming cyanobacteria makes them ecologically competitive. Fine-tuning of photosynthetic processes allows them to survive and grow in environments that are stressful to other phototrophs. Studying such adaptations not only provides insight into cyanobacterial

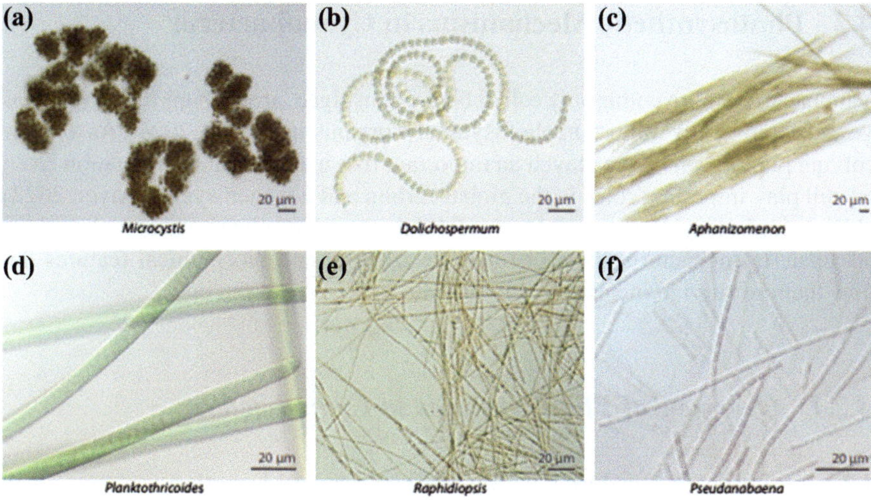

Fig. 14.1 Major bloom-forming cyanobacterial genera, such as *Microcystis* (**a**), *Dolichospermum* (**b**), *Aphanizomenon* (**c**), *Planktothricoides* (**d**), *Raphidiopsis* (**e**), and *Pseudanabaena* (**f**). Scale bar: 20 µm. *Source* Song et al. (2023)

physiology but also has applications in biotechnology, including biofuel production, carbon capture, and the use of bioindicators for ecosystem monitoring.

14.1.5 Key Goals and Objectives of This Chapter

This chapter examines the photosynthetic strategies of bloom-dominating cyanobacteria and focuses on their ecological significance and evolutionary implications. Precisely, the objectives are as follows:

- To explain the structural and functional features of the cyanobacterial photosynthetic equipment, emphasizing dissimilarities and similarities with other photosynthetic organisms
- To explore the physiological features that contribute to the capability of bloom-dominating cyanobacteria to bloom in eutrophic and dynamic aquatic habitats, especially light usage and resistance to stress
- To examine the effect of environmental parameters such as light intensity, availability of nutrients, and temperature on cyanobacteria's photosynthesis and bloom dynamics
- To expound on the overall ecological and biogeochemical implications of cyanobacterial blooms, such as participation in global carbon cycling and ecosystem health and management implications.

14.2 Photosynthetic Mechanisms in Cyanobacteria

Cyanobacteria, more commonly called blue-green algae, are perhaps the most primitive and ecologically important photosynthetic organisms (Büdel, 2024). As oxygen-evolving phototrophs, they played an important role in the Great Oxygenation Event and still play important roles in the global carbon and nitrogen cycles (Aiyer, 2022). While their photosynthetic machinery is fundamentally similar to that of higher plants and algae, cyanobacteria show distinctive structural and biochemical features that allow them to survive in varied environments, such as extreme ones.

14.2.1 Overview of Photosynthesis in Cyanobacteria

14.2.1.1 Photosystems I and II: Basic Structure and Function

Two membrane-embedded integral protein complexes, Photosystem I (PSI) and Photosystem II (PSII), drive cyanobacterial photosynthesis within the internal thylakoid membranes (Rolo et al., 2024). The photosystems operate synergistically in light-dependent reactions. PSII starts the process off by absorbing a photon and tie together the energy to remove electrons from water molecules, generating protons and eliminating molecular oxygen as a waste product (Shevela et al., 2021). The energized electrons are then carried through an electron transport chain, which consists of plastoquinone, cytochrome b6f complex, and plastocyanin. PSI accepts these electrons and re-energizes them with the help of some more photons to reduce $NADP^+$ to NADPH (Viola et al., 2022). This electron movement, along with the translocation of protons across the thylakoid membrane, results in ATP synthesis by chemiosmosis.

14.2.1.2 Light Reactions and Carbon Fixation

The Calvin–Benson cycle, which is fueled by the light reaction products ATP and NADPH, fixes atmospheric carbon dioxide into organic molecules (Distefano, 2022). It starts with the carboxylation of ribulose-1,5-bisphosphate catalyzed by RuBisCO (Meloni et al., 2023). Cyanobacteria typically boost carbon fixation efficiency through carboxysomes, such as protein microcompartments that bring CO_2 close to RuBisCO, reducing the enzyme's oxygenase activity. These are especially beneficial under low CO_2 or during photorespiration conditions.

14.2.2 Unique Features of Cyanobacterial Photosynthesis

14.2.2.1 Phycobilisomes and Chlorophylls

One characteristic feature of cyanobacteria is that they employ phycobilisomes, highly arranged, water-soluble light-harvesting antenna complexes that relay energy to PSI and PSII (Elias, 2024). These complexes are attached to the surface of thylakoid membranes and hugely expand the variety of wavelengths of light available for photosynthesis. Cyanobacteria mainly make use of chlorophyll a; however, chlorophyll d or f is synthesized by some cyanobacteria that absorb far-red light (Gisriel, 2024). Such adaptation enables photosynthetic function during low or filtered light conditions, for example, under dense biofilms or shade aquatic environments.

The Role of Carotenoids in Light Harvesting

Carotenoids, including β-carotene and zeaxanthin, serve dual functions in cyanobacteria (Huang et al., 2025). In addition to assisting in light capture, they protect the photosynthetic machinery from photooxidative destruction by reducing single oxygen and eliminating excess energy as heat (Sachdev et al., 2023). These pigments are central to the stability of photosystems and impart resistance to high-light stress and thus are critical for photosynthetic efficacy and cellular lifespan.

14.2.3 Photosynthetic Pigments in Cyanobacteria

In cyanobacteria, photosynthesis occurs through the chlorophyll a pigment that mainly absorbs blue and red wavelengths of light (Mendili & Khadhri, 2025). Cyanobacteria change from plants along green algae by possessing accessory phycobiliproteins that serve as additional pigment components (Chini Zittelli et al., 2023). The three accessory pigments found in cyanobacteria include phycocyanin, which absorbs orange and red light, phycoerythrin, which absorbs green light, and allophycocyanin, which absorbs red light (Glibert, 2024). The absorption spectrum of phycocyanin matches orange and red wavelengths at 620–630 nm, but phycoerythrin directly absorbs green wavelengths from 495 to 570 nm (Thevarajah et al., 2024). Allophycocyanin shows its light absorption peak at wavelengths ranging from 650 to 655 nm within the red spectrum (Merican, 2025). The pigments enable cyanobacteria to obtain lighting beyond normal ranges when they live in aquatic habitats that face changes in light quality based on depth and water turbidity.

Through diverse pigmentation, cyanobacteria gain access to ecological spaces that other photosynthetic organisms cannot enter (Novoveská et al., 2023). Cyanobacteria with sufficient phycoerythrin gain dominance in deep or turbid waters because this pigment enables better light absorption of green wavelengths compared to red or blue wavelengths (Dufour, 2023). Cyanobacteria, which live in surface and high-light environments, contain higher amounts of phycocyanin and chlorophyll a for effective light absorption. Through a mechanism known as complementary chromatic adaptation (CCA), organisms have the power to modify their pigment mix adaptively (Borella et al., 2023). Cyanobacteria regulate their pigment expression through CCA to adapt their spectral light responses for maximizing photosynthetic efficiency under different environmental conditions.

14.2.4 Pigment–Protein Complexes and Light-Harvesting Mechanisms

Large pigment–protein complexes enable effective light harvesting through the proper integration of phycobiliproteins (Dodson et al., 2023). Phycobilisomes are surface antenna complexes composed of ordered arrays of pigments that bind to thylakoid membranes. Phycobilisomes incorporate allophycocyanin into their core structure alongside phycocyanin rods and phycoerythrin rods, depending on species (Stadnichuk & Kusnetsov, 2023). Linker proteins act as stabilizing elements that maintain the functional integrity of these complexes while facilitating directional energy transfer. Energy from light reaches phycoerythrin, which transfers its energy to phycocyanin and then to allophycocyanin before reaching chlorophyll a in Photosystem II (PSII) (Jha et al., 2024). The Förster resonance energy transfer (FRET)-based energy transfer cascade operates to minimize energy waste and produce efficient quantum transfer in these systems.

Cyanobacteria depend on phycobilisomes to collect and transfer light energy effectively when the lighting quality remains poor (Contreras-Ropero et al., 2024). The structured elements of these complexes show excellent adaptability through which they adjust their composition as well as their dimensions based on detected environmental signals. The ability of cyanobacteria to modify phycobilisomes functions as an essential survival mechanism in environments that present decreased light availability or complex light spectra, such as underwater shaded zones.

14.2.5 Light-Harvesting Complexes and Photosystems

The prokaryotic organism, cyanobacteria, utilizes Photosystem I and Photosystem II to carry out oxygen-generating oxygenic photosynthesis through which they cleave water molecules using solar power (Lee & Kim, 2024). The electron transport chain operates through the shared function of these photosystems to produce ATP and NADPH. The slight energy primarily transfers through phycobilisomes to reach PSII, as this complex activates the water-splitting reaction and releases oxygen products (Sreeharsha & Venkata Mohan, 2024). During different light and metabolic conditions, cyanobacteria redirect their energy supply through a process called state transitions. The cellular need determines the shift in energy distribution between PSII and PSI as phycobilisomes direct most energy to PSII in State 1, yet move energy to PSI in State 2 (van Stokkum et al., 2023). Energy flow regulation serves two crucial purposes by facilitating appropriate photosynthetic efficiency while avoiding redox imbalances during electron transport. The elasticity in the distribution of excitation energy reflects cyanobacteria's very high physiological flexibility while acting as a crucial survival mechanism within different and unstable ecological niches.

14.2.6 Photosynthetic Adaptations to Environmental Conditions

One of the distinctive features of cyanobacterial photosynthesis is its capacity to adapt to changing light intensities and spectral properties (Simkin et al., 2022). Chromatic acclimation (CA), a primary adaptive mechanism, has been found to comprise several types (CA1 to CA7) (Montgomery, 2022). These mechanisms include changes in phycobilisome composition and structure as a response to ambient light color, thus increasing the efficiency of light absorption under fluctuating environmental conditions. For instance, when under green light, the synthesis of phycoerythrin is increased, whereas in red light, phycocyanin synthesis is enhanced.

Apart from spectral adaptation, cyanobacteria are also photoacclimated through the control of phycobilisome size. During high light, phycobilisome size is minimized to prevent photodamage, whereas in low light, large phycobilisomes are produced to enhance light-harvesting capability (Li et al., 2023). Furthermore, photoprotection mechanisms such as non-photochemical quenching (NPQ) are induced under conditions of excess light (Zuo, 2025). These processes release excess excitation energy as heat, thus preventing the photosystems from oxidative damage due to the overgeneration of reactive oxygen species (ROS).

Other cyanobacteria also have alternative chlorophylls, such as chlorophyll d and chlorophyll f, which are capable of absorbing far-red light outside the range of chlorophyll a (Bataeva et al., 2024). These pigments are generally present in cyanobacteria that live in extremely low-light conditions such as in the shade aquatic environments. The availability of these pigments even further extends the usable light range, demonstrating the incredible flexibility of cyanobacterial photosynthesis.

14.2.6.1 Responses to Environmental Stress and Photosynthetic Resilience

Although cyanobacteria are excellent photosynthesizers when conditions are best, they are also impressively resilient when environmental stress is encountered. In environments including high salinity, temperature extremes, ultraviolet light, and nutrient scarcity, cyanobacteria implement a range of protective mechanisms to maintain photosynthetic activity (Yadav et al., 2022). One of these mechanisms is the utilization of alternative electron pathways, including cyclic electron flow around PSI or flavodiiron protein involvement, which assist in dissolving excess electrons and avoiding over-reduction of the photosynthetic electron transport chain when the Calvin cycle finds the middle ground. Induced by stress, heat shock proteins (HSPs) and orange carotenoid proteins (OCPs) are induced under stress conditions and are involved in stabilizing the photosynthetic apparatus and minimizing oxidative damage (Mehdizadeh Allaf & Peerhossaini, 2022). Some filamentous cyanobacteria also produce akinetes, thick-walled, resting cells that can survive long-term desiccation, cold, or nutrient deprivation. These survival cells allow cyanobacteria to survive extreme environmental conditions and recolonize when conditions return to normal.

14.3 Light Harvesting and the Role of Phycobilisomes in Cyanobacteria

Phycobilisomes (PBS) are specialized light-harvesting antenna complexes that are crucial to the photosynthetic efficiency of cyanobacteria (Bryant & Gisriel, 2024). In contrast, plants and algae that use chlorophyll-based light-harvesting complexes embedded within their thylakoid membranes, cyanobacteria utilize PBS to capture a broader light spectrum. These supramolecular structures allow efficient absorption and transfer of light energy to the photosynthetic reaction centers, particularly in low-light or spectrally shifted aquatic environments (Frehan et al., 2023). This essay examines the structure and function of PBS, their energy transfer mechanisms, the diversity of their composition in bloom-forming cyanobacteria, and how environmental and physiological factors modulate their function and expression (Fig. 14.2).

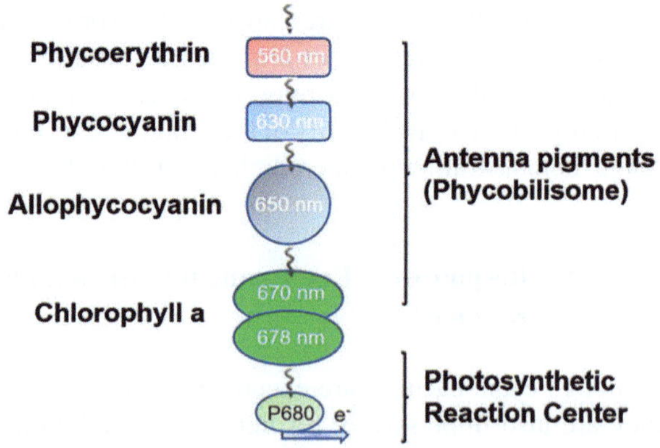

Fig. 14.2 The peripheral rods and central core of PBSs pass on the light energy to the reaction center. The solar energy absorbed by the antenna pigments covalently attached to PBPs is channeled to the photosynthetic reaction center. *Source* Li et al., (2023)

14.3.1 Structure and Function of Phycobilisomes (PBS)

14.3.1.1 General Structure of Phycobilisomes

Large multi-protein complexes known as phycobilisomes anchor to the thylakoid membrane surface of cyanobacteria. The primary components of phycobilisomes consist of the three main phycobiliproteins: phycocyanin (PC), phycoerythrin (PE), and allophycocyanin (APC) (Sui, 2021). Multiple pigment–protein complexes organize within a series of concentric structures where central rods extend outward from a base core. The rods consist of PC and PE components, yet the core contains APC elements, which form a direct connection to the photosystems. Phycobiliproteins attach bilins (linear tetrapyrroles) chromophores to become the foundation for their distinctive colors and extensive light wavelength capture abilities (Taniguchi & Lindsey, 2023). The absorption spectrum of photosynthesis becomes wider because of the phycobiliproteins, while aquatic ecology benefits from this capability due to water dilution variations across depths.

14.3.1.2 Interaction Between PBS and Reaction Centers in Cyanobacteria

Phycobilisomes operate as functional components that connect to Photosystem II (PSII) and, importantly, link to Photosystem I (PSI) (Ejaz et al., 2023). The PBS core maintains its position adjacent to PSII reaction centers, which allows swift energy movement between the components. Energy from light passes through peripheral rods (mainly PE and PC) before reaching the APC core and subsequently reaches chlorophyll a in PSII. The special state transitions cause PBS to transmit energy from PSI, which helps maintain an equal balance of ATP and NADPH production.

14.3.2 Energy Transfer Mechanism in PBS

14.3.2.1 Role of Energy Transfer from Light-Harvesting Pigments to Reaction Centers

Energy transfer in PBS utilizes Förster resonance energy transfer (FRET) as its fundamental mechanism. The quantum mechanical transfer of energy occurs between pigments using this mechanism without requiring photon emission (Bassi & Dall'Osto, 2021). Successive energy absorption by PE in outer rods leads to energy transfer through PC before arriving at APC in the core structure. Chlorophyll a in photosystem reaction centers receives the final energy transfer that originates from the APC (Akhtar et al., 2022). The system operates with exceptional efficiency because it generates minimal energy losses when transferring power. The structured energy transfer route keeps energy received from a broad surface area precisely focused at the photochemical reaction site. The special construct offers valuable advantages for habitats where light is limited or spectral resources are restricted.

14.3.3 Variations in Phycobilisome Composition in Bloom-Forming Cyanobacteria

14.3.3.1 Through Adaptation Processes

Microcystis, *Anabaena*, and *Nodularia* display diverse PBS structures that help them capture light efficiently in their particular ecological settings (Deepika et al., 2022). The light spectrum of *Synechococcus* and *Prochlorococcus* determines the exact composition of their PBS complexes. Surface water cyanobacteria develop PE control because they absorb blue-green light, whereas deeper water strains favor PC and APC to capture the dominant red-shifted wavelengths (Lovindeer, 2021). Cyanobacteria modify PBS rod dimensions together with their chemical composition during this adaptation process. The scientific name for such organisms includes longer photopigment rods containing additional PE subunits to better capture the restricted available light when they reside in deep or turbid habitats. Cyanobacteria species achieve better light-harvesting ability through their superior light-harvesting capacity while establishing dominant positions within eutrophic and stratified water ecosystems.

14.3.3.2 The Role of PBS in Enhancing Photosynthetic Efficiency

The presence of PBS creates an advantageous situation for cyanobacteria in photosynthetic competitions. The wide spectral range coverage enables cyanobacteria to keep photosynthesizing in situations where other organisms experience light restriction or photoinhibition (Simkin et al., 2022). The improved light-harvesting capability of bloom-forming cyanobacteria supports their speedy ecosystem dominance, which occasionally manifests as harmful algal blooms (HABs). PBS complexes operate at high densities which enables their mobile nature to change their positions according to shifts in light conditions (Krawczyk et al., 2021). The photosynthetic systems uphold their maximal efficiency levels across different environmental conditions because of this arrangement.

14.3.4 Environmental Modulation of PBS Composition

14.3.4.1 The Effect of Nutrient Availability (Nitrogen, Phosphorus) on PBS

The availability of nutrients functions as a crucial modulator in the biological process that produces PBS. The presence of nitrogen as an essential phycobiliproteins ingredient means that reduced nitrogen levels result in PBS degradation through chlorotic processes (Yoshihara & Kobayashi, 2022). Moonlighting phycobiliproteins function as nitrogen recyclers, by which cyanobacteria break down these proteins to produce yellowish pigmented cells. Phosphorus directly affects PBS expression because its influence extends to total cellular metabolic processes (Kritmetapak & Kumar, 2021). Insufficient phosphorus causes reduced ATP synthesis, which affects the transcription of PBS-related genes and results in improper functioning of light-harvesting antennae. When phosphorus levels in their environment are abundant, some cyanobacteria create polyphosphate granules inside their cells as they take up surplus mineral matter. The ability to store phosphorus internally as polyphosphate during nutrient-rich periods enables cyanobacteria to maintain PBS functionality during phosphorus-shortage periods which provides them better survival in variable ecosystems.

14.3.4.2 Response to Light Intensity and Other Abiotic Factors (Temperature, pH)

The composition of phosphorus-binding proteins and their structural arrangement respond directly to light intensity levels as part of the photoacclimation function (Devkota and Durnford, (2025). Cyanobacteria make their pigment-binding antenna system bigger as well as develop additional PE-rich structures to optimally absorb scarce light under dim conditions. When light illumination remains intense, cyanobacteria minimize PBS size to reduce the risk of overstimulation and phototoxicity. Temperature and pH also influence PBS function. When temperature increases, it destabilizes the interactions between pigments and proteins, thus reducing energy transfer efficiency (Nowruzi et al., 2022). Changes in pH levels affect protonation states of PBS amino acid residues, resulting in format changes or decreased functional capability.

14.3.5 *Functional Redundancy and Adaptation to Changing Light Conditions*

14.3.5.1 Functional Flexibility in Light Harvesting and Adaptation to Fluctuating Light Levels

Cyanobacteria develop comprehensive functional diversification and adaptable light-harvesting capabilities in their system. The photosynthetic organism demonstrates this functionality through its ability to shift excitation energy between PSII and PSI based on the electron transport chain redox state (Gu, 2023). Light-dependent transitions balance the ATP and NADPH molecular production across different environmental conditions with changing metabolic rates.

Cyanobacteria possess additional photosynthetic pigments along with chlorophyll d and chlorophyll f that allow far-red light absorption (Gisriel et al., 2022). The pigments in these bacteria enable far-red light photoacclimation, which enables photosynthetic activity in environments where standard photosynthetic pigments prove ineffective. The ability of cyanobacteria to modify PBS structure, combined with orientation adjustment in shallow ponds or rocky shore zones, enables them to sustain optimal photosynthetic activity (Kumar et al., 2024). Survival and competitive advantages in unstable ecological settings heavily depend on this plasticity mechanism.

14.4 Photosynthetic Adaptations in Response to Light Conditions

Photoautotrophic organisms known as cyanobacteria depend significantly on light to power their process of photosynthesis. Aquatic ecosystems experience broad variations in light exposure because of differences in water depth and turbidity levels, together with daily patterns and yearly changes (Zanghi & Ioannou, 2025). Cyanobacteria developed complex photosynthetic adaptations that maximize their light-harvesting potential, together with the conversion of energy and photoprotection systems (Li et al., 2023). Largely because of their adaptive capabilities, cyanobacteria survive wide-ranging light conditions and accumulate population levels that affect ecological systems. The study examines cyanobacterial adaptations to light availability, including pigment changes for bloom development and space-dependent behaviors, along with environmental effects on bloom systems caused by changing light conditions.

14.4.1 Adaptation to High-Light Conditions

Cyanobacteria must overcome conflicting demands when operating within bright light conditions. Cyanobacteria need light for photosynthesis, up till now high-light levels activate photoinhibition and form reactive oxygen species (ROS) that harm their cells (Grettenberger et al., 2024). Various photoprotection mechanisms operate in cyanobacteria to defend their sensitive photosynthetic machinery, especially Photosystem II, from oxidative stress damage. The core photoprotection approach induces the activation of non-photochemical quenching (NPQ) (Lu et al., 2022). Heating the excess absorbed light energy functions through the NPQ mechanism to avoid damaging photosynthetic reaction centers. Under intense light conditions, cyanobacteria use two protective strategies by reducing both phycobilisome number and size to control the transfer of energy to PSII. The photosynthetic output can be optimized under varying light conditions because of this structural adjustment which uses the state transition mechanism to move excitation energy between PSI and PSII. The protective function of carotenoids and other proteins plays an essential role in these structures. The carotenoid compounds zeaxanthin, echinenone, and myxoxanthophyll operate as antioxidant molecules that protect against ROS and deexcite triplet chlorophyll states (Polyakov et al., 2023). The photoprotection of cyanobacteria depends heavily on proteins especially the orange carotenoid protein (OCP). The blue light activates OCP, which allows it to bind phycobilisomes which result in thermal deactivation for energy dissipation. As an exceptional regulatory system, OCP demonstrates how photosensitive proteins create superior survival outcomes during light-related stress situations.

14.4.2 Adaptation to Low-Light Conditions

Cyanobacteria adapt their light-harvesting ability upward during reduced light conditions that occur in turbid or stratified waters to maintain their photosynthetic processes (Wu, 2024). Phycobilisomes experience substantial growth as one of the main adaptations when cyanobacteria encounter these conditions. The environment of low-light intensity induces cyanobacteria to build larger PBS systems by adding additional phycobiliproteins, including phycoerythrin (Li et al., 2023). The cellular structure expansion increases the overall light photon-capturing ability of cells by enhancing their absorption cross-section. Adjustments in cell pigmentation patterns enhance the ability to capture light photons. The light spectrum affects the pigment properties of cyanobacteria as they adjust their phycobiliprotein concentration ratios according to light wavelength (Tayebati et al., 2024). The light color affects which strains increase PE synthesis because various cyanobacterial strains increase production of PC and APC specifically under red-light conditions. The modifications lead to improved energy harvesting capacity during particular light conditions, while these adjustments are controlled by light-responsive proteins RcaE and CcaS/R systems at the

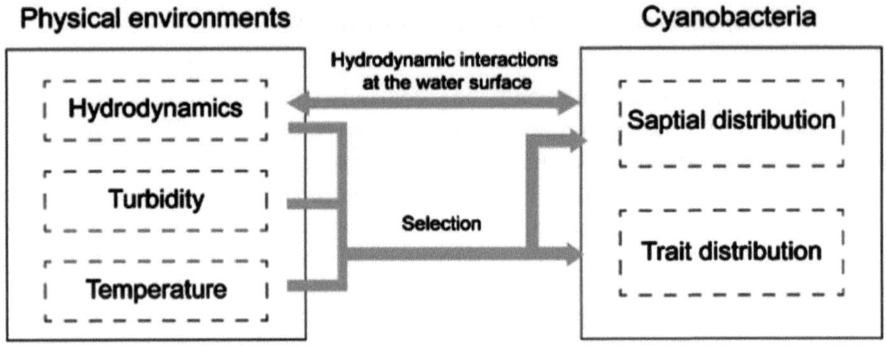

Fig. 14.3 Schematic diagram showing the interactions of cyanobacteria with physical environments, as shown in this figure. *Source* Wu (2024)

transcriptional level (La Rocca et al., 2024). Cyanobacteria make additional chlorophyll variants through their modification processes. Some cyanobacteria possess the genetic ability to produce chlorophyll d and chlorophyll f to capture far-red light that exceeds traditional chlorophyll a wavelengths. Photoreceptors using these variants enable photosynthetic processes to function in deeply shaded areas such as beneath algal blooms, where light penetration remains minimal (Fig. 14.3).

14.4.3 Shifts in Pigment Composition with Bloom Formation

Cyanobacterial blooms display marked alterations in pigment content because these changes react directly to the light environment, along with water conditions (Chatterjee & More, 2023). Self-shading emerges as a dominant factor when cyanobacterial blooms become denser and spread throughout the water column (Wu, 2024). During algal bloom growth, the top layers experience maximum light exposure while lower layers experience progressively reduced light availability. Cyanobacteria adapt their pigment distribution to preserve photosynthesis within every water layer. Light availability creates fundamental changes in bloom developmental processes. When cyanobacteria proliferation starts but light levels remain high, cyanobacteria commonly reduce PBS expression to protect themselves from light-induced damage. When the cellular density reaches high levels, light availability becomes restricted, so cells respond through enlarged PBS content plus reorganized phycobiliprotein distribution (Bryant & Gisriel, 2024). Bloom cells maintain efficient photosynthetic capabilities within darkened regions of the bloom environment thanks to these adaptations. The specific locations of photosynthetic pigments form a vertical structure that fosters biomass growth under particular lighting conditions (Grasso et al., 2024). Surface cells in denser blooms produce additional carotenoid protection, yet deeper cells express high PE levels together with chlorophyll variants to maximize light absorption potential. Bloom persistence and collective growth are helped by spatial pigment distribution patterns that promote survival under both stratified conditions and nutrient-rich environments.

14.4.4 Vertical and Horizontal Migration in Cyanobacterial Blooms

Cyanobacteria adapt their light exposure through both biochemical and structural modifications, with strategic behaviors on top (Gupta et al., 2023). Vertical migration enables cells to ascend in the water column to position themselves perfectly for maximum light exposure. The presence of gas vesicles in numerous cyanobacteria cells enables buoyancy through protein-bound structures. The size of internal gas in these vesicles will decide whether cyanobacteria cells will float to the water surface or sink from the surface (Greening et al., 2025). Vertical positioning serves as a critical adaptation in daily patterns of light and dark conditions. The life cycle of cyanobacteria involves upward migration to the water surface for light acquisition, while descent happens at night or when light intensity becomes excessive to prevent photoinhibition. Many cyanobacteria accomplish this synchronization by coordinating their movements and creating periodic vertical shifts, which optimize photosynthesis while reducing harm. Horizontal migration affects cyanobacteria populations most prominently within areas with shallow waters or flowing systems because transport occurs naturally without active movement (Nawaz et al., 2024a, 2024b, 2024c). Some cyanobacterial species demonstrate phototactic behaviors that enable them to navigate toward or further from light sources. The ability of cyanobacteria to change their behavior helps form blooms in favorable light and nutrient areas while potentially shaping the growth or decline patterns of bloom regions during periods.

14.4.5 Ecophysiological Impacts of Light Availability

Light presence impacts cyanobacteria in various ways throughout their life cycles, from cellular function to community actions in aquatic systems. Light functions as the main controller for cyanobacterial metabolic functions by overseeing the operations of photosynthesis as well as nitrogen fixation and carbon assimilation, while regulating secondary metabolite synthesis (Lucius & Hagemann, 2024). Light conditions exercise significant control over cyanobacterial bloom frequency patterns within freshwater aquatic environments. When surface light intensifies in stable stratified clear water environments, *Microcystis aeruginosa* floats to the top since it monopolizes the sunlit photic zone, thus outcompeting other phytoplankton species. Turbid waters, along with overcast conditions, promote dominance of cyanobacteria that express high levels of PBS while using alternative chlorophylls for survival. Seasonal differences in light availability also influence the formation and sequence of blooms.

In temperate climates, blooms of cyanobacteria tend to occur at maximum intensity in late summer, following peak solar irradiance and stratification of the water column. Low-light conditions and mixing in winter or early spring inhibit the development of blooms. Climate change, however, affects temperature and precipitation patterns and is causing the timing and light intensity of availability to change, resulting in modification of bloom phenology and geography.

14.5 Environmental and Ecological Factors Influencing Bloom Formation

Cyanobacterial blooms are an urgent ecological concern in freshwater and marine ecosystems, fueled by a multifaceted interaction of environmental and ecological drivers (Mohammed & Arockiaraj, 2024). Among the most important contributors are light intensity, temperature, nutrient supply, and the adaptive potential of cyanobacteria. These drivers shape bloom initiation, size, and duration, and their impacts are being amplified by climate change.

14.5.1 Impact of Light and Temperature on Bloom Formation

14.5.1.1 How Light and Temperature Interact to Trigger Cyanobacterial Blooms

Cyanobacteria are photosynthetic bacteria that rely largely on light and temperature for growth and propagation. Synergy between these two factors has the potential to create a suitable environment for bloom development. Light supplies the energy for photosynthesis, while temperature influences metabolic activities and enzymatic reactions. Warmer temperatures, especially those greater than 20 °C, stimulate the growth of most cyanobacterial species, particularly bloom formers such as *Microcystis* and *Anabaena* (Lakshmikandan et al., 2024). At the same time, higher light availability in the surface water column, usually induced by stratification during summer, promotes photosynthesis. Combined, higher temperature and light form a feedback loop that encourages rapid biomass growth and scum formation on the surface.

14.5.1.2 The Role of Light Intensity and Temperature in Bloom Size and Frequency

Light intensity has a direct effect on the photosynthetic rate and hence biomass formation. Low levels of light favor faster growth, as long as all other conditions, such as

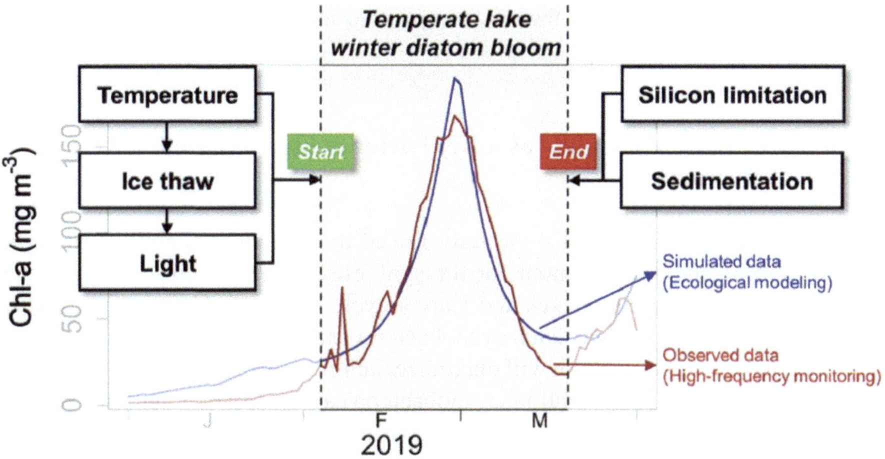

Fig. 14.4 Light intensity and temperature in bloom size and frequency. *Source* Kong et al. (2021)

nutrients, are also appropriate. Nonetheless, too much light induces photoinhibition, even though numerous cyanobacteria have established protection mechanisms, such as phycobilisomes and photoprotective pigments, to counteract this effect. Temperature not only affects bloom frequency, increasing blooms to be more frequent in summer months, but also impacts the dominance of certain cyanobacterial taxa (Reinl et al., 2023). For example, *M. aeruginosa* prefers warm, stagnant water and tends to dominate other phytoplankton under such conditions. Seasonal and geographic differences in light and temperature regimes therefore have a significant impact on determining the timing and size of blooms (Fig. 14.4).

14.5.2 Influence of Nutrient Availability

14.5.2.1 The Role of Nitrogen and Phosphorus in Bloom-Forming Species

Nutrient enrichment, specifically with nitrogen (N) and phosphorus (P), is among the most powerful forces behind cyanobacterial blooms. Nutrients are necessary for cellular processes, including nucleic acid synthesis, ATP production, and chlorophyll development (Ali et al., 2023). Agricultural runoff, wastewater effluent, and urban stormwater are typical sources of nutrient loading in aquatic ecosystems. Phosphorus is frequently the limiting nutrient in aquatic ecosystems, but nitrogen can be limiting in the marine system. Cyanobacteria can survive under low-nitrogen conditions through nitrogen fixation, a feature that few other groups of phytoplankton have. Heterocysts enable genera such as *Anabaena* and *Aphanizomenon* to fix atmospheric

N_2 into usable forms, giving them a competitive advantage under nitrogen-poor waters.

14.5.2.2 Nutrient Availability as a Key Factor in Photosynthetic Efficiency

Photosynthetic efficiency is also strongly linked to nutrient availability. Nitrogen and phosphorus are both essential for the synthesis of the proteins and pigments of the light-harvesting complexes and Calvin cycle enzymes (Croce et al., 2024). Under conditions of high nutrients, cyanobacteria can sustain high chlorophyll and accessory pigment levels, which will maximize their photosynthetic rates and growth capacities. But in nutrient limitation, cyanobacteria can modify their internal nutrient quotas and reorganize their cellular equipment to maximize photosynthesis. Such physiological plasticity allows them to flourish under a broad spectrum of nutrient regimes and is one of the reasons they are ecologically successful.

14.5.3 Interactions Between Light Harvesting and Ecological Success

14.5.3.1 How Photosynthetic Adaptations Help Cyanobacteria Dominate Specific Environments

Cyanobacteria have several adaptations that improve their light-harvesting capacity and enable them to thrive in a variety of aquatic environments (Kumar et al., 2024). These are the presence of phycobiliproteins enables them to capture a wider range of light wavelengths than other phytoplankton. This is particularly beneficial in turbid or shaded waters where the quality and intensity of light change. In addition to this, the buoyancy of cyanobacteria is controlled by gas vesicles to place them ideally in the water column for best light capture (Igwaran et al., 2024). This vertical migration tactic enables them to escape from light saturation closer to the water surface in intense sunlight and move to nutrient-poor layers further down when required.

14.5.3.2 Competitive Advantage in Nutrient-Poor or Nutrient-Rich Waters

In nutrient-deficient waters, the capacity to fix nitrogen provides cyanobacteria with a clear competitive advantage over non-fixing phytoplankton (Opdahl, 2023). Their low growth rates under these conditions are compensated by their long-term survival and plasticity. Conversely, in nutrient-rich waters, cyanobacteria can quickly grow, outcompeting other species through better nutrient acquisition and effective energy conversion (Paquette, 2025). Additionally, certain cyanobacteria synthesize secondary metabolites, including allelopathic compounds and toxins that suppress the growth of rival organisms and prevent grazing by zooplankton. Such chemical defense further strengthens their competitive status in various aquatic ecosystems.

14.5.4 Climate Change and Bloom Dynamics

14.5.4.1 Effects of Altered Light and Temperature Regimes Due to Climate Change on Bloom Formation

Climate change is expected to enhance cyanobacterial bloom formation through various mechanisms. Global warming will probably increase the thermal growing season of cyanobacteria, especially in temperate zones (Kumar et al., 2024). Higher surface water temperatures enhance water column stratification, limiting mixing and forcing cyanobacteria to concentrate more in the photic zone. Changes in precipitation patterns, such as more intense and frequent storms, can cause more nutrient excess, which will further contribute to blooms (Byl et al., 2025). Temporarily, deficiencies can decrease freshwater inflow, raising water residence times and creating conditions that favor bloom resolution. Light regimes are also impacted as ice cover decreases and cloud patterns change, differing underwater light availability (Schlegel et al., 2023). These changes will produce more auspicious conditions for cyanobacterial dominance, particularly in higher latitude areas where blooms were uncommon before.

14.5.4.2 Long-Term Ecological Impacts of Blooms on Aquatic Ecosystems

The long-term implications of repeated cyanobacterial blooms are significant as well as they can induce hypoxia or anoxia of bottom water through organic decomposition, leading to fish kills and benthic ecosystem disruption (Dai et al., 2023). Blooms disrupt food web structure by lowering diversity and replacing edible algae with toxic or inedible species. Toxin production by specific cyanobacteria harms human and animal health, impacting drinking water sources, recreational waters, and food webs in aquatic systems (Mishra, 2023). Long-term exposure to bloom-related toxins such as microcystins and anatoxins has been associated with liver toxicity and neurotoxicity. In addition, bloom events are associated with important economic losses from water treatment, fisheries, and tourism. As climate change becomes more intense, cyanobacterial bloom management will need to incorporate cohesive approaches to address nutrient inputs, land use, and ecosystem resilience.

14.6 Technological Advances and Future Directions

Cyanobacteria, the venerable photosynthetic microbes, are integral players in global carbon cycling and aquatic ecosystem processes (Dwivedi & Ahmad, 2023). Their ability to photosynthesize more efficiently with their capacity for adaptation to an array of habitats has rendered them not only a point of ecological interest as causative agents of harmful algal blooms but also a viable bet for biotechnological advances. Technological breakthroughs in spectroscopy, genetic engineering, and synthetic biology have greatly increased our knowledge of cyanobacterial photosynthesis and introduced new prospects for applied research (Nawaz et al., 2024a, 2024b, 2024c). This chapter discusses the technologies to investigate light harvesting in cyanobacteria, advancements in synthetic biology, and the potential for cyanobacteria to solve future environmental and energy challenges.

14.6.1 Technologies for Studying Light Harvesting in Cyanobacteria

14.6.1.1 Recent Advancements in Spectroscopy and Fluorescence Techniques

The investigation of cyanobacterial light harvesting has been transformed by the creation of powerful spectroscopic and fluorescence-based techniques (Khayatian et al., 2023). Methods such as time-resolved fluorescence spectroscopy, chlorophyll fluorescence analysis (PAM fluorometry), and single-molecule fluorescence microscopy allow scientists to study the energy capture and transfer dynamics in photosynthetic complexes at an exceptional level. Spectroscopy, especially emission and absorption spectroscopy, aids in the identification of pigments such as phycobiliproteins and chlorophylls used in light harvesting. Time-resolved fluorescence spectroscopy can uncover energy transfer kinetics among these pigments and reaction centers, which allows mapping of the efficiency and responsiveness of the photosynthetic process against different light conditions (Keil et al., 2025). Fluorescence lifetime imaging microscopy (FLIM) and Förster resonance energy transfer (FRET) have further enhanced the study by enabling real-time imaging of photosynthetic protein interactions within living cells. These methods are crucial for determining how cyanobacteria adapt to varying light intensity and quality, shedding insight into their ecological success and bloom formation patterns.

14.6.1.2 Use of Genetic Engineering to Study Photosynthetic Proteins

Genetic modification has also turned out to be a convenient way of probing the function and control of photosynthetic proteins within cyanobacteria (Cheng et al.,

2023). By making use of CRISPR-Cas systems as well as other gene editing systems, researchers have been able to manipulate genes for proteins involved in the photosynthetic machinery, such as the Photosystem II D1 protein or light-harvesting antenna proteins. Mutational analyses and knockout or overexpression experiments allow scientists to specify the roles of particular proteins in light harvesting, photoprotection, and carbon fixation. Additionally, the insertion of fluorescent markers or reporter genes joined to photosynthetic genes allows gene expression and protein targeting to be traced in real time under different environmental conditions (Li et al., 2025). These genetic approaches not only illuminate cyanobacterial biology's fundamentals but also provide the basis for synthetic biology approaches to promote photosynthetic efficiency for industrial purposes.

14.6.2 Synthetic Biology and Bioengineering of Cyanobacterial Photosynthesis

14.6.2.1 Potential for Enhancing Cyanobacterial Productivity for Biofuel Production

Cyanobacteria are promising organisms for biofuel production because they have rapid growth, can utilize sunlight and CO_2 as inputs, and have relatively uncomplicated genomes (Arshad et al., 2025). Synthetic biology has enabled their metabolic pathways to be reprogrammed to maximize the yield of biofuels and valuable bioproducts such as ethanol, butanol, hydrogen, and isoprenoids (Singh et al., 2024). Through combining metabolic pathway engineering and synthetic gene circuits, scientists can divert carbon fluxes away from biomass generation toward desired products. For example, heterologous expression of pathways in plants or other microorganisms can improve the synthesis of lipids or hydrocarbons in cyanobacteria to mirror the composition of standard fossil fuels (Simkovsky & Carcellar, 2023). Another important strategy is to enhance the efficiency of the photosynthetic light reactions. This can be made possible by redesigning light-harvesting antenna complexes so that they are more effective in capturing and converting solar energy or by incorporating more efficient forms of the RuBisCO enzyme, which is responsible for CO_2 fixation in the Calvin cycle.

14.6.2.2 Engineering Cyanobacteria for Improved Light Harvesting in Artificial Systems

With increasing demand for sustainable energy and resource-saving agriculture, there is a strong urge to construct artificial systems to imitate or integrate biological photosynthesis. Engineered cyanobacteria are under investigation as critical components of biohybrid devices such as bio-photovoltaics, artificial leaves, and microbial fuel cells

(Dwivedi & Dwivedi, 2023). Through adjustment of the light-absorbing abilities of cyanobacteria through protein and genetic engineering, researchers can develop optimized strains for individual light conditions, maximizing solar energy capture and conversion. The engineered strains may be paired with semiconductors or photonic materials to produce hybrid devices that convert sunlight into electrical or chemical energy. Furthermore, coating engineered cyanobacteria within clear hydrogels or nanomaterials makes it possible to produce photosynthetically active panels and surfaces that can develop self-sufficient bioenergy systems for off-grid or remote operations.

14.6.3 Future Perspectives

14.6.3.1 The Role of Cyanobacterial Photosynthetic Adaptations in Climate Change Mitigation

Cyanobacteria have an important part to play in international climate change mitigation strategies. Their carbon fixation and oxygen generation capabilities make them natural partners for lowering atmospheric CO_2 levels (Correa et al., 2023). With improved cultivation and utilization of engineered cyanobacteria within carbon capture systems, it is possible to use them for large-scale industrial sequestration of CO_2. Cyanobacterial consortia may also be incorporated in engineered ecosystems such as algal turf scrubbers, photobioreactors, or aquaponic systems to scavenge excess nutrients and CO_2 and yield biomass for biofuel or feedstock (Wicker, 2023). They have high photosynthetic efficiency, particularly in conditions of high light and temperature, which is optimal for operating in the climate change-hit regions. Additionally, as warming and ocean acidification modify marine ecosystems, knowledge and harnessing cyanobacterial resistance and photosynthetic responses may assist in formulating conservation plans for maintaining marine diversity and enhancing ecosystem recovery.

14.6.3.2 Potential for Biotechnological Applications Based on Cyanobacterial Bloom Dynamics

Whereas blooms are usually considered environmental hazards, their dynamics and biomass hold promising biotechnological utilization (Abate et al., 2024). High cell densities inherent to blooms can be exploited for harvesting biomass in huge quantities that, in turn, can be manufactured into biofertilizers, animal feedstock, or bioplastics. Understanding of bloom-forming species' triggers and life cycles would enable controlled growth of particular strains for application in biotechnology (Anabtawi et al., 2024). For instance, strains that can yield high-value chemicals such as phycocyanin, antioxidants, or bioactive molecules might be selectively favored under specially designed environmental conditions. Kundu et al. (2023) revealed that the processing of bloom biomass to biochar or incorporating it into anaerobic digestion facilities for biogas production. Such methods not only provide clean energy but also present a sustainable approach toward bloom mitigation as well as transforming ecological threats into valuable resources.

14.7 Conclusion

14.7.1 Summary of Key Findings

This research has examined the impressive photosynthetic innovations of bloom-creating cyanobacteria, especially those that allow them to survive and thrive in various aquatic ecosystems. Most important among these innovations is the capacity of cyanobacteria to maximize light gathering by adopting different mechanisms such as the evolution of specialized light-gathering complexes (LHCs), successful utilization of phycobilisomes, and the control of pigment organization to accommodate different light intensities and spectra. Furthermore, the work has emphasized the functions of state changes and photoacclimation tactics that enable cyanobacteria to dynamically reallocate excitation energy between photosystems based on fluctuating environmental conditions.

Cyanobacteria have developed special adaptations, such as chromatic acclimation, which allows them to adjust their phycobiliprotein composition according to the spectral quality of the surrounding light. This ability is particularly important in turbid or stratified water bodies where light quality and availability change with depth and season. In addition, some bloom-forming species contain high-light-inducible proteins (HLIPs) and other protective mechanisms that reduce photooxidative stress, allowing for photosynthetic efficiency even under high irradiance. Concurrently, physiological characteristics such as gas vesicle regulation of buoyancy and effective nutrient acquisition systems add to the competitive edge of these organisms. Cumulatively, these photosynthetic and ecological adaptations underpin the success of cyanobacteria in creating lasting and often toxic blooms in aquatic and marine ecosystems.

14.7.2 Ecological Significance

Understanding bloom-forming cyanobacterial photosynthetic processes and light-harvesting strategies is necessary for various ecological and environmental considerations (Saleemet al., 2024). In the first instance, cyanobacterial blooms may exert significant influences on water quality, aquatic species richness, and ecosystem processes. By dominating light and nutrients, cyanobacterial blooms exclude other phytoplankton species from co-occurrence, thereby creating lower species richness and novel food web processes. Therefore, the spread of toxin-forming cyanobacteria such as *Microcystis*, *Dolichospermum*, and *Planktothrix* constitutes a direct hazard to animal and human health (Faulkner et al., 2025). Cyanobacteria, from a biogeochemical point of view, are of central importance in global primary production and nutrient cycling, especially nitrogen and carbon. Their photosynthesis is a key contributor to oxygen evolution in aquatic ecosystems and impacts carbon sequestration. The diazotrophic potential of some bloom-forming genera also enables atmospheric nitrogen fixation, bringing bioavailable nitrogen into ecosystems that would otherwise be nitrogen limited. While desirable at low levels, high levels of nitrogen fixation may enhance eutrophication and encourage bloom persistence.

The photosynthetic capability and competence of cyanobacteria to cope with diverse climatic and environmental conditions are in close relation to their success. Consequently, the increasing dominance of noxious blooms over the last few decades is more and more attributed to anthropogenically driven stressors such as nutrient loading, global warming, and changes in hydrological regimes. Warmer global temperatures and longer durations of water column stratification benefit cyanobacterial growth as they are well adapted to withstand high temperatures and low-light intensities (Litchman & de Tezanos Pinto, 2024). Hence, knowledge of the basic processes of cyanobacterial photosynthesis is important not just for ecological studies but also for guiding water resource management and mitigation measures.

14.7.3 Future Research Needs

Although significant advances have been made in understanding the molecular and physiological underpinnings of cyanobacterial photosynthesis, a few knowledge gaps still exist. One of the areas for future research is the specific regulatory networks that modulate photosynthetic acclimation and pigment expression in bloom-causing cyanobacteria (Yu et al., 2024). Although research has described chromatic acclimation systems such as CA4 and CA6, the genetic regulation and ecological relevance of these systems in a wide range of environments remain poorly resolved.

Another important research area is the relationship between environmental stressors such as nutrient availability, temperature, and oxidative stress and photosynthetic adaptation. An examination of how cyanobacteria combine information from more than one stress pathway to control photosynthetic function could shed light on their bloom under projected climates.

In addition, omics technologies, especially transcriptomics, proteomics, and metabolomics, hold promising directions for deciphering cyanobacteria's intricate responses to light and nutrient conditions (Zammit et al., 2023). These high-throughput methods can be combined with ecological modeling and remote sensing data to predict bloom occurrence and persistence with higher precision. There is also an urgent requirement to investigate the role of microbial interactions, including symbiotic relationships with heterotrophic bacteria, in facilitating or inhibiting cyanobacterial photosynthesis and bloom formation. Such interactions can influence nutrient dynamics, toxin production, and even photosynthetic efficiency in natural assemblages. Lastly, applied research to develop sustainable methods for preventing cyanobacterial blooms should take into consideration the lessons of photosynthetic biology (Nawaz et al., 2024a, 2024b, 2024c). For example, manipulation of light, e.g., through artificial mixing or shading technology, can interfere with cyanobacterial light acquisition tactics and decrease bloom formation. Such as insights into the factors that induce or suppress chromatic acclimation could result in ecosystem-level treatments that promote non-bloom-forming phytoplankton.

14.7.4 Concluding Remarks

The investigation of bloom-forming cyanobacterial adaptations to photosynthesis is not abstractly interesting for its benefit; it has very real significance to the solution of some most pressing ecological and environmental dilemmas. Freshwater resource sustainability, aquatic ecosystem health, and human population safety are all facing increased risk around the globe as harmful cyanobacterial blooms accelerate. With ever-rising global temperatures and widespread nutrient pollution, cyanobacteria's adaptive capability will become more significant. Through investigation of light-harvesting mechanisms, pigment control, and photosynthetic acclimation, scientists can gain deeper insight into competitive processes that fuel cyanobacterial success. It is essential knowledge to guide predictive models, early warning syste Canuti and Penna ms, and evidence-based management practices intended to control or reduce bloom occurrence. Moreover, research on cyanobacterial photosynthesis contributes to more general understanding of microbial ecology, evolutionary biology, and climate resilience. Cyanobacteria are among the most ancient and ecologically important photoautotrophs on the planet, their fossil history extending back more than 2.5 billion years. Their ability to survive environmental change opens a window onto evolutionary innovations that have constructed life on this planet—and potentially continue to do so in the Anthropocene. Expanding our knowledge of photosynthetic adaptations in bloom-forming cyanobacteria is an essential step toward effectively managing aquatic ecosystems in a changing world. It requires an integrated, multidisciplinary approach uniting molecular biology, ecology, environmental science, and public health. Only through these collaborative efforts can we potentially predict and respond to the complex issues generated by cyanobacterial blooms in the coming decades.

References

Abate, R., Oon, Y. L., Oon, Y. S., Bi, Y., Mi, W., Song, G., & Gao, Y. (2024). Diverse interactions between bacteria and microalgae: A review for enhancing harmful algal bloom mitigation and biomass processing efficiency. *Heliyon*. https://doi.org/10.1016/j.heliyon.2024.e36503

Aiyer, K. (2022). The great oxidation event: How cyanobacteria changed life. American Society for Microbiology. https://asm.org/Articles/2022/February/The-Great-Oxidation-Event-How-Cyanobacteria-Change. Accessed on 24–04–2025.

Akhtar, P., Biswas, A., Balog-Vig, F., Domonkos, I., Kovács, L., & Lambrev, P. H. (2022). Trimeric photosystem I facilitates energy transfer from phycobilisomes in Synechocystis sp. PCC 6803. *Plant Physiology, 189*(2), 827–838. https://doi.org/10.1093/plphys/kiac130

Ali, S., Riaz, A., Mamtaz, S., & Haider, H. (2023). Nutrients and crop production. *Current Research in Agriculture and Farming, 4*(2), 1–15.

Anabtawi, H. M., Lee, W. H., Al-Anazi, A., Mohamed, M. M., & Aly Hassan, A. (2024). Advancements in biological strategies for controlling harmful algal blooms (HABs). *Water, 16*(2), 224. https://doi.org/10.3390/w16020224

Arshad, S., Qadir, M. L., Hussain, N., Ali, Q., Han, S., & Ali, D. (2025). Advances in CRISPR/Cas9 technology: Shaping the future of photosynthetic microorganisms for biofuel production. *Functional Plant Biology, 52*(2). https://doi.org/10.1071/FP24255

Bassi, R., & Dall'Osto, L. (2021). Dissipation of light energy absorbed in excess: The molecular mechanisms. *Annual Review of Plant Biology, 72*(1), 47–76. https://doi.org/10.1146/annurev-arplant-071720-015522

Bataeva, Y. V., Sinetova, M. A., Kurashov, E. A., Krylova, J. V., Kolombet, L. V., & Grigoryan, L. N. (2024). Characterization of biological activity and evaluation of exogenous metabolites of cyanobacteria 'Anabaena'sp. IPPAS B-2020. *Microbiology, 93*(5), 537–550.

Borella, L., Marchese, D., Trivellin, N., & Sforza, E. (2023). Complementary chromatic adaptation as a strategy to increase energy conversion efficiency of microalgae-cyanobacteria consortia in continuous LED photobioreactors. *Energy Conversion and Management, 294*, Article 117549.

Bryant, D. A., & Gisriel, C. J. (2024). The structural basis for light harvesting in organisms producing phycobiliproteins. *The Plant Cell, 36*(10), 4036–4064. https://doi.org/10.1093/plcell/koae126

Büdel, B. (2024). Cyanobacteria/blue-green algae. In *Biology of algae, lichens and bryophytes* (pp. 25–99). Springer Berlin Heidelberg.

Byl, T. D., Moore, D. M., Cunningham, C., & Young, D. E. (2025). Science needs for determining the effects of climate change on harmful algal blooms in the southeastern United States (No. 2025–1004). US Geological Survey. https://doi.org/10.3133/ofr20251004

Canuti, E., & Penna, A. (2024). Dynamics of phytoplankton communities in the Baltic Sea: Insights from a multi-dimensional analysis of pigment and spectral data—part I, pigment dataset. *Frontiers in Marine Science, 11*, 1425347.

Chatterjee, S., & More, M. (2023). Cyanobacterial harmful algal bloom toxin microcystin and increased vibrio occurrence as climate-change-induced biological co-stressors: Exposure and disease outcomes via their interaction with gut–liver–brain axis. *Toxins, 15*(4), 289. https://doi.org/10.3390/toxins15040289

Cheng, J., Zhang, K., & Hou, Y. (2023). The current situations and limitations of genetic engineering in cyanobacteria: A mini review. *Molecular Biology Reports, 50*(6), 5481–5487. https://doi.org/10.1007/s11033-023-08456-8

Chini Zittelli, G., Lauceri, R., Faraloni, C., Silva Benavides, A. M., & Torzillo, G. (2023). Valuable pigments from microalgae: Phycobiliproteins, primary carotenoids, and fucoxanthin. *Photochemical & Photobiological Sciences, 22*(8), 1733–1789.

Contreras-Ropero, J. E., Barajas-Solano, A. F., García-Martínez, J. B., Barajas-Ferreira, C., & Zuorro, A. (2024). Optimization of phycobiliprotein biosynthesis in thermotolerant cyanobacteria through light parameter adjustment. *Results in Engineering, 23*, Article 102644. https://doi.org/10.1016/j.rineng.2024.102644

Correa, S. S., Schultz, J., Lauersen, K. J., & Rosado, A. S. (2023). Natural carbon fixation and advances in synthetic engineering for redesigning and creating new fixation pathways. *Journal of Advanced Research, 47*, 75–92. https://doi.org/10.1016/j.jare.2022.07.011

Croce, R., Carmo-Silva, E., Cho, Y. B., Ermakova, M., Harbinson, J., Lawson, T., Zhu, X. G., et al. (2024). Perspectives on improving photosynthesis to increase crop yield. *The Plant Cell, 36*(10), 3944–3973. https://doi.org/10.1093/plcell/koae132

Dai, M., Zhao, Y., Chai, F., Chen, M., Chen, N., Chen, Y., Zhang, Z., et al. (2023). Persistent eutrophication and hypoxia in the coastal ocean. *Cambridge Prisms: Coastal Futures, 1*, e19.xcvgh.

Deepika, C., Wolf, J., Roles, J., Ross, I., & Hankamer, B. (2022). Sustainable production of pigments from cyanobacteria. In *Cyanobacteria in biotechnology: Applications and quantitative perspectives* (pp. 171–251). Springer International Publishing. https://doi.org/10.1007/10_2022_211

Devkota, S., & Durnford, D. G. (2025). Photoacclimation strategies of Chlamydomonas reinhardtii in response to high-light stress in stationary phase. *Journal of Photochemistry and Photobiology B: Biology, 262*, Article 113082.

Distefano, L. (2022). *Regulation of starch metabolism in guard cells of Arabidopsis thaliana* [Doctoral dissertation, ETH Zurich].

Dodson, E. J., Ma, J., Suissa Szlejf, M., Maroudas-Sklare, N., Paltiel, Y., Adir, N., Sun, S., Sui, S. F., & Keren, N. (2023). The structural basis for light acclimation in phycobilisome light harvesting systems systems in Porphyridium purpureum. *Communications Biology, 6*(1), 1210. https://doi.org/10.1038/s42003-023-05586-4

Dufour, L. (2023). *Phytoplankton and water color: multiscale analysis of the adaptive advantage conferred by chromatic acclimation in marine cyanobacteria* [Doctoral dissertation, Sorbonne Université].

Dwivedi, N., & Dwivedi, S. (2023). Bio-inspired nanomaterials: applications in artificial photosynthesis and energy generation. In *Bionanotechnology towards green energy* (pp. 33–58). CRC Press.

Dwivedi, S., & Ahmad, I. Z. (2023). Cyanobacteria in ocean. In *Current status of marine water microbiology* (pp. 47–66). https://doi.org/10.1007/978-981-99-5022-5_4

Elias, E. G. M. (2024). Coloring outside the lines: Harvesting far-red photons for photosynthesis. https://doi.org/10.5463/thesis.753

Faulkner, S., Sweetman, C., Hutson, J., Soole, K., Hobson, P., & Fallowfield, H. (2025). Uptake of the cyanobacterial toxin microcystin by crop plants irrigated with contaminated wastewater: A review. *Reviews in Environmental Science and Bio/technology*. https://doi.org/10.1007/s11157-024-09716-0

Ejaz, A., Sutter, M., Lechno-Yossef, S., Kerfeld, C. A., & Squires, A. H. (2025). Phycobilisome core architecture influences photoprotective quenching by the orange carotenoid protein. *Proceedings of the National Academy of Sciences, 122*(41), e2420355122.

Frehan, S. K., Dsouza, L., Li, X., Eríc, V., Jansen, T. L., Mul, G., Holzwarth, A. R., Buda, F., Sevink, G. A., de Groot, H.J., & Huijser, A. (2023). Photon energy-dependent ultrafast exciton transfer in chlorosomes of Chlorobium tepidum and the role of supramolecular dynamics. *The Journal of Physical Chemistry B, 127*(35), 7581–7589.

Greening, C., & Lithgow, T. (2020). Formation and function of bacterial organelles. *Nature Reviews Microbiology, 18*(12), 677-689.

Geletu, T. T. (2023). Lake eutrophication: Control of phytoplankton overgrowth and invasive aquatic weeds. *Lakes & Reservoirs: Research & Management, 28*(1), Article e12425. https://doi.org/10.1111/lre.12425

Gisriel, C. J. (2024). Recent structural discoveries of photosystems I and II acclimated to absorb far-red light. *Biochimica et Biophysica Acta (BBA)-Bioenergetics, 1865*(3), 149032. https://doi.org/10.1016/j.bbabio.2024.149032

Gisriel, C. J., Shen, G., Ho, M. Y., Kurashov, V., Flesher, D. A., Wang, J., Bryant, D. A., et al. (2022). Structure of a monomeric photosystem II core complex from a cyanobacterium acclimated to far-red light reveals the functions of chlorophylls d and f. *Journal of Biological Chemistry, 298*(1).

Glibert, P. M. (2024). Capturing light and the diversity of pigments. In *Phytoplankton whispering: An introduction to the physiology and ecology of microalgae* (pp. 85–103).

Grasso, N., Fasciolo, B., Awouda, A. M. M., & Bruno, G. (2024). A smart aeroponic chamber: Structure and architecture for an efficient production and resource management. In *Hydroponics:*

The future of sustainable farming (pp. 353–380). Springer US. https://doi.org/10.1007/978-1-0716-3993-1_18

Grettenberger, C. L., Abou-Shanab, R., & Hamilton, T. L. (2024). Limiting factors in the operation of photosystems I and II in cyanobacteria. *Microbial Biotechnology, 17*(8), Article e14519. https://doi.org/10.1111/1751-7915.14519

Gu, L. (2023). Optimizing the electron transport chain to sustainably improve photosynthesis. *Plant Physiology, 193*(4), 2398–2412. https://doi.org/10.1093/plphys/kiad490

Gupta, A., Pandey, P., Gupta, R., Tiwari, S., & Singh, S. P. (2023). Responding to light signals: A comprehensive update on photomorphogenesis in cyanobacteria. *Physiology and Molecular Biology of Plants, 29*(12), 1915–1930. https://doi.org/10.1007/s12298-023-01386-6

Huang, J. J., Xu, W., Lin, S., & Cheung, P. C. K. (2025). The bioactivities and biotechnological production approaches of carotenoids derived from microalgae and cyanobacteria. *Critical Reviews in Biotechnology, 45*(2), 276–304. https://doi.org/10.1080/07388551.2024.2359966

Igwaran, A., Kayode, A. J., Moloantoa, K. M., Khetsha, Z. P., & Unuofin, J. O. (2024). Cyanobacteria harmful algae blooms: Causes, impacts, and risk management. *Water, Air, & Soil Pollution, 235*(1), 71. https://doi.org/10.1007/s11270-023-06782-y

Jassim, Y. A., Awadh, E. F. A., & Al-Amery, S. M. H. (2023). A review of general properties of blue-green algae (Cyanobacteria). *Biomedicine and Chemical Sciences, 2*(2), 143–148. https://doi.org/10.48112/bcs.v2i2.397

Jha, S., Singh, V. K., Singh, A. P., Gupta, A., Rana, P., & Sinha, R. P. (2024). The radiant world of cyanobacterial phycobiliproteins: Examining their structure, functions, and biomedical potentials. *Targets, 2*(1), 32–51. https://doi.org/10.3390/targets2010002

Keil, E., Kumar, A., Bäuml, L., Reiter, S., Thyrhaug, E., Moser, S., Duffy, C. D., de Vivie-Riedle, R., & Hauer, J. (2025). Reassessing the role and lifetime of Q x in the energy transfer dynamics of chlorophyll a. *Chemical Science, 16*(4), 1684–1695. https://doi.org/10.1039/D4SC06441K

Khayatian, S., Bavali, A., Moradi, S., Farhadi, M., & Jameie, S. B. (2023). Fluorescence inner filters of Arthrospira platensis: Novel perspective for precise fluorescence-based sensors. *Spectrochimica Acta Part a: Molecular and Biomolecular Spectroscopy, 284*, Article 121791. https://doi.org/10.1016/j.saa.2022.121791

Kong, X., Seewald, M., Dadi, T., Friese, K., Mi, C., Boehrer, B., Schultze, M., Rinke, K., & Shatwell, T. (2021). Unravelling winter diatom blooms in temperate lakes using high frequency data and ecological modeling. *Water Research, 190*, 116681.

Krawczyk, K. M., Sarracini, A., Green, P. B., Hasham, M., Tang, K., Pare-Labrosse, O., Voznyy, O., Wilson, M. W., & Miller, R. D. (2021). Anisotropic, nonthermal lattice disordering observed in photoexcited PbS quantum dots. *The Journal of Physical Chemistry C, 125*(40), 22120–22132.

Kritmetapak, K., & Kumar, R. (2021). Phosphate as a signaling molecule. *Calcified Tissue International, 108*(1), 16–31. https://doi.org/10.1007/s00223-019-00636-8

Kumar, V., Kedia, V., Jaiswal, S., & Kharwar, S. (2024). Underwater photosynthesis in cyanobacteria: Challenges and adaptations. In *Stress biology in photosynthetic organisms: Molecular insights and cellular responses* (pp. 249–266). Springer Nature Singapore. https://doi.org/10.1007/978-981-97-1883-2_11

Kundu, R., Kunnoth, B., Pilli, S., Polisetty, V. R., & Tyagi, R. D. (2023). Biochar symbiosis in anaerobic digestion to enhance biogas production: A comprehensive review. *Journal of Environmental Management, 344*, Article 118743. https://doi.org/10.1016/j.jenvman.2023.118743

Kupriyanova, E. V., Pronina, N. A., & Los, D. A. (2023). Adapting from low to high: An update to CO_2-concentrating mechanisms of cyanobacteria and microalgae. *Plants, 12*(7), 1569. https://doi.org/10.3390/plants12071569

La Rocca, N., Moro, I., Battistuzzi, M., & Rascio, N. (2024). Cyanobacteria 16 responses to different and microalgae light regimes and CO_2 availability. In *Handbook of photosynthesis*, 296.

Lakshmikandan, M., Li, M., & Pan, B. (2024). Cyanobacterial blooms in environmental water: Causes and solutions. *Current Pollution Reports, 10*(4), 606–627.

Lee, K. P., & Kim, C. (2024). Photosynthetic ROS and retrograde signaling pathways. *New Phytologist, 244*(4), 1183–1198. https://doi.org/10.1111/nph.20134

Li, X., Hou, W., Lei, J., Chen, H., & Wang, Q. (2023). The unique light-harvesting system of the algal phycobilisome: Structure, assembly components, and functions. *International Journal of Molecular Sciences, 24*(11), 9733. https://doi.org/10.3390/ijms24119733

Li, Y., Deng, L., Walker, E. J. L., Karas, B. J., & Mock, T. (2025). Genetic engineering in diatoms: Advances and prospects. *The Plant Journal, 121*(6), Article e70102. https://doi.org/10.1111/tpj.70102

Litchman, E., & de Tezanos Pinto, P. (2024). Ecology of algae and Cyanobacteria (phytoplankton). In *Wetzel's limnology* (pp. 511–538). Academic Press. https://doi.org/10.1016/B978-0-12-822701-5.00018-5

Lovindeer, R. A. E. (2021). *Ecological significance of cyanobacteria pigment color acclimation in the ocean.* University of California.

Lu, D., Zhang, Y., Zhang, A., & Lu, C. (2022). Non-photochemical quenching: From light perception to photoprotective gene expression. *International Journal of Molecular Sciences, 23*(2), 687. https://doi.org/10.3390/ijms23020687

Lucius, S., & Hagemann, M. (2024). The primary carbon metabolism in cyanobacteria and its regulation. *Frontiers in Plant Science, 15*, 1417680. https://doi.org/10.3389/fpls.2024.1417680

Mehdizadeh Allaf, M., & Peerhossaini, H. (2022). Cyanobacteria: Model microorganisms and beyond. *Microorganisms, 10*(4), 696. https://doi.org/10.3390/microorganisms10040696

Meloni, M., Gurrieri, L., Fermani, S., Velie, L., Sparla, F., Crozet, P., Zaffagnini, M., et al. (2023). Ribulose-1, 5-bisphosphate regeneration in the Calvin-Benson-Bassham cycle: Focus on the last three enzymatic steps that allow the formation of Rubisco substrate. *Frontiers in Plant Science, 14*, 1130430.

Mendili, M., & Khadhri, A. (2025). Chlorophylls: The verdant world of photosynthetic pigments. In *Microbial colorants: Chemistry, biosynthesis and applications* (pp. 223–239). https://doi.org/10.1002/9781394287888.ch9

Merican, F. (2025). Lecturer | PhD | University of Science Malaysia, George Town | USM | School of Biological Sciences | Research profile. https://www.researchgate.net/profile/Faradina-Merican. Accessed on 24–04–2025

Mishra, R. K. (2023). The effect of eutrophication on drinking water. *British Journal of Multidisciplinary and Advanced Studies, 4*(1), 7–20. https://doi.org/10.37745/bjmas.2022.0096

Mohammed, V., & Arockiaraj, J. (2024). Unveiling the trifecta of cyanobacterial quorum sensing: LuxI, LuxR and LuxS as the intricate machinery for harmful algal bloom formation in freshwater ecosystems. *Science of the Total Environment*, 171644.

Montgomery, B. L. (2022). Reflections on cyanobacterial chromatic acclimation: Exploring the molecular bases of organismal acclimation and motivation for rethinking the promotion of equity in stem. *Microbiology and Molecular Biology Reviews, 86*(3), e00106-e121.

Nath, B. B. (2025). *Vanishing life on earth.* Blue Rose Publishers. https://books.google.com/books?id=4ddSEQAAQBAJ&lpg=PA1&ots=0TSnssLd4M&dq=The%20role%20of%20cyanobacteria%20in%20evolution%20cannot%20be%20emphasised%20excessively%20because%20cyanobacteria%20are%20thought%20to%20be%20among%20the%20oldest%20photosynthetic%20organisms%20to%20inhabit%20Earth%2C%20with%20evidence%20of%20existence%20through%20fossils%20for%20more%20than%202.5%20billion%20years&lr&pg=PA1#v=onepage&q&f=false

Nawaz, T., Gu, L., Fahad, S., Saud, S., Bleakley, B., & Zhou, R. (2024a). Exploring sustainable agriculture with nitrogen-fixing cyanobacteria and nanotechnology. *Molecules, 29*(11), 2534. https://doi.org/10.3390/molecules29112534

Nawaz, T., Gu, L., Fahad, S., Saud, S., Harrison, M. T., & Zhou, R. (2024b). Sustainable protein production through genetic engineering of cyanobacteria and use of atmospheric N_2 gas. *Food and Energy Security, 13*(2), Article e536. https://doi.org/10.1002/fes3.536

Nawaz, T., Gu, L., Hu, Z., Fahad, S., Saud, S., & Zhou, R. (2024c). Advancements in synthetic biology for enhancing cyanobacterial capabilities in sustainable plastic production: A green horizon perspective. *Fuels, 5*(3), 394–438.

Novoveská, L., Nielsen, S. L., Eroldoğan, O. T., Haznedaroglu, B. Z., Rinkevich, B., Fazi, S., Robbens, J., Vasquez, M., & Einarsson, H. (2023). Overview and challenges of large-scale cultivation of photosynthetic microalgae and cyanobacteria. *Marine Drugs, 21*(8), 445. https://doi.org/10.3390/md21080445

Nowruzi, B., Konur, O., & Anvar, S. A. A. (2022). The stability of the phycobiliproteins in the adverse environmental conditions relevant to the food storage. *Food and Bioprocess Technology, 15*(12), 2646–2663. https://doi.org/10.1007/s11947-022-02855-8

Opdahl, L. J. (2023). Lactate-bound metals: A sustainable alternative to inorganic micronutrient fertilizers. Washington State University.

Paquette, A. (2025). Growth, turnover, and recycling of cyanobacteria from soda lakes.

Pathak, J., Singh, P. R., Sinha, R. P., & Rastogi, R. P. (2022). Ecophysiology and biochemistry of cyanobacteria.

Pivato, M. (2023). The molecular basis of Chlamydomonas reinhardtii responses to the environment: The role of intracellular Ca^{2+} signalling and minor antenna proteins. https://hdl.handle.net/11562/1103546

Polyakov, N. E., Focsan, A. L., Gao, Y., & Kispert, L. D. (2023). The endless world of carotenoids—Structural, chemical and biological aspects of some rare carotenoids. *International Journal of Molecular Sciences, 24*(12), 9885. https://doi.org/10.3390/ijms24129885

Ramakrishnan, B., Maddela, N. R., Venkateswarlu, K., & Megharaj, M. (2023). Potential of microalgae and cyanobacteria to improve soil health and agricultural productivity: A critical view. *Environmental Science: Advances, 2*(4), 586–611. https://doi.org/10.1039/D2VA00158F

Reinl, K. L., Harris, T. D., North, R. L., Almela, P., Berger, S. A., Bizic, M., Yokota, K., et al. (2023). Blooms also like it cold. *Limnology and Oceanography Letters, 8*(4), 546–564. https://doi.org/10.1002/lol2.10316

Rolo, D., Schöttler, M. A., Sandoval-Ibáñez, O., & Bock, R. (2024). Structure, function, and assembly of PSI in thylakoid membranes of vascular plants. *The Plant Cell, 36*(10), 4080–4108. https://doi.org/10.1093/plcell/koae169

Saleem, A., Anwar, S., Saud, S., Kamal, T., Fahad, S., & Nawaz, T. (2025). Cyanobacteria diversity and ecological roles: insights into cyanobacterial adaptations and environmental implications. *Journal of Umm Al-Qura University for Applied Sciences*, 1–19.

Sachdev, S., Ansari, S. A., & Ansari, M. I. (2023). Photosynthetic apparatus: major Site of oxidative damage. In *Reactive oxygen species in plants: The right balance* (pp. 75–92). Springer Nature Singapore.

Salmaso, N., Cerasino, L., Pindo, M., & Boscaini, A. (2024). Taxonomic and functional metagenomic assessment of a Dolichospermum bloom in a large and deep lake south of the Alps. *FEMS Microbiology Ecology, 100*(10), fiae117. https://doi.org/10.1093/femsec/fiae117

Schlegel, R. W., Singh, R. K., Gentili, B., Bélanger, S., Castro de la Guardia, L., Krause-Jensen, D., Miller, C. A., Sejr, M., & Gattuso, J. P. (2023). Underwater light environment in Arctic fjords. *Earth System Science Data Discussions*, 2023, 1–27. https://doi.org/10.5194/essd-16-2773-2024

Shevela, D., Kern, J. F., Govindjee, G., Whitmarsh, J., & Messinger, J. (2021). Photosystem II. *eLS, 2*(7), 1–16. https://doi.org/10.1002/9780470015902.a0029372

Simkin, A. J., Kapoor, L., Doss, C. G. P., Hofmann, T. A., Lawson, T., & Ramamoorthy, S. (2022). The role of photosynthesis related pigments in light harvesting, photoprotection and enhancement of photosynthetic yield in planta. *Photosynthesis Research, 152*(1), 23–42. https://doi.org/10.1007/s11120-021-00892-6

Simkovsky, R., & Carcellar, F. (2023). Why algae? In *Rethinking polyester polyurethanes* (pp. 29–65). Elsevier. https://doi.org/10.1016/B978-0-323-99982-3.00002-X

Singh, V. K., Jha, S., Rana, P., Soni, R., Lalnunpuii, R., Singh, P. K., Singh, R. P., & Singh, G. (2024). Cyanobacteria as a biocatalyst for sustainable production of biofuels and chemicals. *Energies, 17*(2), 408. https://doi.org/10.3390/en17020408

Sreeharsha, R. V., & Mohan, S. V. (2024). Microbial photosynthesis. Springer Singapore.

Sreeharsha, R. V., & Venkata Mohan, S. (2024). Structural and functional dynamics of microbial photosystem complexes. In *Microbial photosynthesis: From basic biology to artificial cell factories and industrial applications* (pp. 23–41). Springer Nature Singapore.

Stadnichuk, I. N., & Kusnetsov, V. V. (2021). Endosymbiotic origin of chloroplasts in plant cells' evolution. *Russian Journal of Plant Physiology, 68*, 1–16.

Stadnichuk, I. N., & Kusnetsov, V. V. (2023). Phycobilisomes and phycobiliproteins in the pigment apparatus of oxygenic photosynthetics: From cyanobacteria to tertiary endosymbiosis. *International Journal of Molecular Sciences, 24*(3), 2290. https://doi.org/10.3390/ijms24032290

Song, L., Jia, Y., Qin, B., Li, R., Carmichael, W. W., Gan, N., & Sukenik, A. (2023). Harmful cyanobacterial blooms: biological traits, mechanisms, risks, and control strategies. *Annual Review of Environment and Resources, 48*(1), 123–147.

Sui, S. F. (2021). Structure of phycobilisomes. *Annual Review of Biophysics, 50*(1), 53–72. https://doi.org/10.1146/annurev-biophys-062920-063657

Taniguchi, M., & Lindsey, J. S. (2023). Absorption and fluorescence spectra of open-chain tetrapyrrole pigments–bilirubins, biliverdins, phycobilins, and synthetic analogues. *Journal of Photochemistry and Photobiology C: Photochemistry Reviews, 55*, Article 100585. https://doi.org/10.1016/j.jphotochemrev.2023.100585

Tayebati, H., Pajoum Shariati, F., Soltani, N., & Sepasi Tehrani, H. (2024). Effect of various light spectra on amino acids and pigment production of Arthrospira platensis using flat-plate photobioreactor. *Preparative Biochemistry & Biotechnology, 54*(8), 1028–1039.

Thevarajah, B., Piyatilleke, S., Nimarshana, P. H. V., Koushalya, S., Malik, A., & Ariyadasa, T. U. (2024). Exploring effective light spectral conversion techniques for enhanced production of Spirulina-derived blue pigment protein, c-phycocyanin. *Bioresource Technology*, 130612. https://doi.org/10.1016/j.biortech.2024.130612

Upadhyay, D., Shukla, K., Mishra, A., Jindal, T., Sharma, S., & Shukla, S. (2021). Molecular aspects and oxygen relations of nitrogen fixation in cyanobacteria. *Soil nitrogen ecology* (pp. 521–568). Springer International Publishing.

van Stokkum, I. H., Akhtar, P., Biswas, A., & Lambrev, P. H. (2023). Energy transfer from phycobilisomes to photosystem I at 77 K. *Frontiers in Plant Science, 14*, 1293813.

Viola, S., Roseby, W., Santabarbara, S., Nürnberg, D., Assunção, R., Dau, H., Sellés, J., Boussac, A., Fantuzzi, A., & Rutherford, A. W. (2022). Impact of energy limitations on function and resilience in long-wavelength Photosystem II. *Elife, 11*, e79890.

Wang, Z., Akbar, S., Sun, Y., Gu, L., Zhang, L., Lyu, K., Huang, Y., & Yang, Z. (2021). Cyanobacterial dominance and succession: Factors, mechanisms, predictions, and managements. *Journal of Environmental Management, 297*, Article 113281. https://doi.org/10.1016/j.jenvman.2021.113281

Wicker, R. J. (2023). Bioprospecting for nordic photosynthetic consortia for wastewater treatment, carbon capture, and value creation. https://urn.fi/URN:ISBN:978-952-412-037-1

Wu, H. (2024). *Interactions of cyanobacterial blooms with physical processes in lakes* [Doctoral dissertation, Rheinland-Pfälzische Technische Universität Kaiserslautern-Landau].

Xuan, M., & Li, J. (2021). Photosystem II-based biomimetic assembly for enhanced photosynthesis. *National Science Review, 8*(8), nwab051. https://doi.org/10.1093/nsr/nwab051

Yadav, P., Singh, R. P., Rana, S., Joshi, D., Kumar, D., Bhardwaj, N., Gupta, R. K., & Kumar, A. (2022). Mechanisms of stress tolerance in cyanobacteria under extreme conditions. *Stresses, 2*(4), 531–549. https://doi.org/10.3390/stresses2040036

Yoshihara, A., & Kobayashi, K. (2022). Photosynthesis and cell growth trigger degradation of Phycobilisomes during nitrogen limitation. *Plant and Cell Physiology, 63*(2), 189–199. https://doi.org/10.1093/pcp/pcab159

Yu, J., Xu, W., Wang, J., Gao, Q., Xiu, L., Yan, Q., & Huang, L. (2024). Contact-mediated algicidal mechanism of Vibrio coralliirubri ACE001 against the harmful alga Karenia mikimotoi. *Iscience, 27*(11).

Zammit, G., Zammit, M. G., & Buttigieg, K. G. (2023). Emerging technologies for the discovery of novel diversity in cyanobacteria and algae and the elucidation of their valuable metabolites. *Diversity, 15*(11), 1142.

Zanghi, C., & Ioannou, C. C. (2025). The impact of increasing turbidity on the predator–prey interactions of freshwater fishes. *Freshwater Biology, 70*(1), Article e14354. https://doi.org/10.1111/fwb.14354

Zuo, G. (2025). Non-photochemical quenching (NPQ) in photoprotection: Insights into NPQ levels required to avoid photoinactivation and photoinhibition. *New Phytologist*. https://doi.org/10.1111/nph.70121

Chapter 15
Evolutionary Strategies for Bloom Formation and Persistence

Zohaib Younas, Ilyas Ahmad, Faiz Ullah, and Zia-ur-Rehman Mashwani

Abstract Harmful algal blooms (HABs) refer to uncontrolled and rapid growths of algae or cyanobacteria in aquatic and marine ecosystems, frequently settling at or near the water surface. These microorganisms are incredibly resilient and tolerant, enabling them to grow under diverse environmental conditions. Their capacity to maximize physiological as well as ecological performance as a response to abiotic conditions is evidence of their competitive supremacy in aquatic environments. Recent years have seen a growth in the prevalence and intensity of HABs, leading scientists and policymakers to advance monitoring and mitigation measures. Remote sensing devices like satellite-based systems (e.g., NASA's MODIS) allow for large-scale observation of bloom dynamics through chlorophyll concentration and thermal pattern detection. At the same time, in situ sensors give continuous information on water quality parameters such as temperature, nutrient levels, and toxin concentrations, which are synthesized into early warning systems using predictive models. Preventative action aims at minimizing anthropogenic sources of eutrophication by limiting agricultural runoff, enhancing wastewater treatment, and recreating natural filtration mechanisms such as wetlands. We covered in this chapter emerging control methods such as ultrasonic disruption and clay flocculation are designed to prevent the formation of blooms prior to the development of ecological and public health effects. Knowledge of the adaptive mechanisms of bloom-forming species is essential to further improve predictive models and implement more successful management strategies. Since HABs continue to present challenges to aquatic biodiversity, water resources, and human health, interdisciplinary studies and advanced technologies must continue being necessary in order to curb their spread in a changing climate. Future research must focus on sustainable means that tackle both the ecological causes of HABs and their socioeconomic impacts.

Keywords Harmful algal blooms (HABs) · Anthropogenic activities · Climate change · Monitoring strategies

---

Z. Younas · I. Ahmad · F. Ullah · Z.-R. Mashwani (✉)
Department of Botany, PMAS Arid Agriculture University, Rawalpindi, Pakistan
e-mail: mashwani@uaar.edu

S. Fahad et al. (eds.), *Cyanobacterial Blooms: Ecology, Evolution and Biogeochemical Impacts*, https://doi.org/10.1007/978-3-032-06042-6_15

15.1 Introduction

A bloom can be described as a sudden and spectacular increase in the abundance of microscopic aquatic organisms primarily phytoplankton like algae and cyanobacteria in estuarine, freshwater, or marine ecosystems (Glibert, 2016). The growth and sporadic dominance of specific poisonous or dangerous algal genera are features associated with HABs incidents (Anderson, 2009). These events are commonly detected by color changes in water, which may turn green, brown, or red due to excessive growth of chlorophyll a, and are typically realized through abrupt changes in the chemical and ecological balance of the surrounding water (Sarkar & Sarkar, 2018). The huge economic growth of coastal regions in the last few decades has led to the discharges of massive amount of domestic sewage, industrial wastewater, and agricultural wastewater into the ocean (Shen et al., 2024; Zhang et al., 2022). Eutrophication pollution induces bloom of phytoplankton, creating worldwide ecological complications that may profoundly change ambient biogeochemical conditions (Andersson et al., 2010). Various investigations with molecular approaches concluded that following algal blooms, the structure and response to biological and environmental factors of bacterioplankton communities changed (Choi et al., 2018). However, few investigations are performed on the process governing the formation of bacterioplankton communities and interspecific relationships during algal blooms (Shao et al., 2020).

Under normal circumstances, phytoplankton are present in regulated and balanced amounts in aquatic ecosystems and act as the basis of food webs. Some algal species causing harmful algal blooms (HABs) secrete toxic substances that interfere with biological processes throughout entire ecosystems, from plankton to humans (Sellner et al., 2003). These toxins destroy coastal resources, such as submerged aquatic vegetation and benthic communities (epifauna and infauna). These most serious impacts are mass mortality in fish, birds, and marine mammals and severe impact in humans including respiratory distress, gastrointestinal diseases, neurological effects, skin lesions, and irritation. The acute effects render HABs not only ecologically devastating but also a public health issue of major concerns (Sellner et al., 2003). Previous literature established that a survey made during (1987–1992) approximated HAB-related damage at about $49 million per year to the US economy. In addition to explicit monetary loss, long-term toxicity from repeated algal blooms has resulted in year-round closure of potentially productive wild shell fisheries on area such as Alaska and Georges Bank (Hoagland et al., 2002).

These closures are due to two key factors: chronic toxin exposure which makes shellfish inedible and inefficacy of purging stored toxins from tainted shellfish, even which blooms are absent (Hoagland et al., 2002). Fluctuations in environmental condition, such as nutrient availability (i.e., nitrogen and phosphorus), temperature, high illumination levels, and low turbulence of water prevail, allow phytoplankton to grow rapidly and form blooms (Halsey & Jones, 2015). While some blooms are natural part of seasonal cycle, human-induced climate change due to crop runoff, sewage, and overall climatic shifts has accelerated the frequency, severity, and area covered by these events over the last several years (Zhang et al., 2022). Particularly, algal or cyanobacterial blooms have both the beneficial and adverse functions, which identify the ecological complexity of these organisms in freshwater habitats (Havens, 2008).

15.2 Ecological Significance and Consequences of Algal Bloom Formation

Phytoplankton blooms of non-harmful and harmful types are of great ecological and environmental significance. Non-harmful algal blooms (nHABs) are a constituent of primary production and may serve as zooplankton and filter feeder diet. Harmful algal blooms (HABs) are sudden, uncontrolled algal or cyanobacterial blooms in freshwater or aquatic environments that accumulate on or beneath the water's surface. These blooms produce toxins, deplete oxygen, discolor water, and present serious hazards to humans, wildlife, ecosystems, and local economies. Their occurrence results from intricate dynamics of nutrient loading (particularly nitrogen and phosphorus), conductive hydrological conditions, and climatic factors and becomes more intense and more frequent under changing environmental conditions (Brooks et al., 2016).

Phytoplankton produce nearly 50% of the world's net primary production and contain less than 1% of the planet's photosynthetically active biomass (Malone et al., 2015). Blooms provide a source of food for zooplankton, fish, and benthic filter feeders. They are important to carbon fixation and oxygenation and serve as the basis of the aquatic food web. In the majority of eutrophic environments, phytoplankton blooms enhance biological productivity and structure trophic dynamics. In seasonally fluctuating systems like temperate lakes or upwelling regions, spring blooms signal the beginning of the growing season and trigger higher trophic interactions (Peeters et al., 2007). Blooms also significantly affect biogeochemical cycling of the elements carbon (C), nitrogen (N), phosphorus (P), and silica (Si). Marine high-density blooms such as diatom blooms are components of the biological carbon pump that transports organic carbon from surface waters to the deep sea (Mallimadugula and Hameed, 2023). Blooms of cyanobacteria, such as those made up of *Anabaena* and *Nodularia*, assimilate atmospheric nitrogen (N_2) into forms usable by living things, altering nitrogen dynamics in impoverished-nutrient waters (Wang et al., 2023).

However, harmful algal blooms have a number of negative ecological impacts. *Microcystis*, *Alexandrium*, and *Karenia brevis* synthesize neurotoxin, hepatotoxins, or paralytic shellfish toxins that effect aquatic organisms and humans (James et al., 2010). HABs can kill other phytoplankton species by allelopathy, shading, or nutrient competition, thereby taking on monoculture-type dominance and reducing species diversity (Worm et al., 2006). Recurrent blooms are a cause and an effect of eutrophication, reducing water clarity, increasing turbidity, and altering pH and temperature regimes. Toxins like microcystins contaminate municipal water supplies and pose a considerable threat to human health, while HABs upset tourism, fisheries, and aquaculture, costing an estimated billion annually worldwide (Vantarakis, 2021). Ciguatera poisoning is a human disease caused by poison from minute algae known as *Gambierdiscus* and *Fukuyoa.* These algae inhibit surfaces such as seaweed, corals, and rocks in coral reefs. Fish become poisoned after consuming these algae, and the poison accumulates in their bodies which led to cause of human ill on consumption of these fish (Anderson et al., 2021). In addition, more global warming, changes in participation, and increased carbon dioxide levels support growth and expansion of blooms (Fig. 15.1).

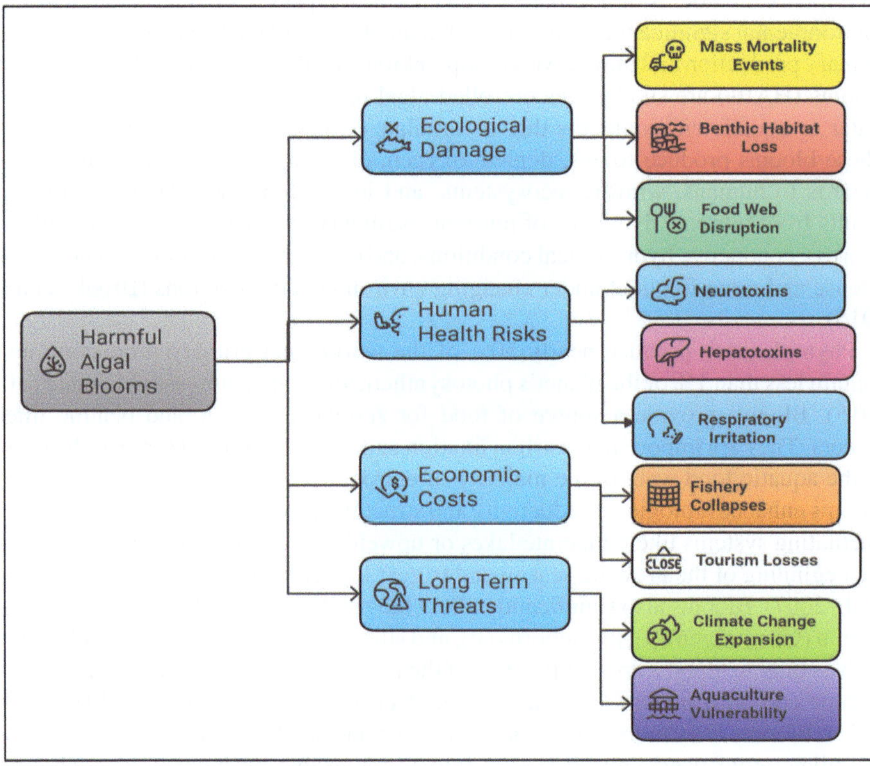

Fig. 15.1 Consequences of harmful algal blooms (HABs) formation

15.3 Anthropogenic Load on Harmful Algal Bloom (HAB) Formation

Harmful algal blooms (HABs) are more frequently taking place in freshwater ecosystems, estuaries, and marine coast around the globe owing to excessive nutrient pollution. Both developed and developing countries across all continents, including Japan and Israel in Asia and countries in Europe, Africa, Australia, and the America, are being affected by this expanding environmental disaster. The spread of these poisonous phytoplankton blooms has become a worldwide water quality issue, posing an international threat to ecosystem, economies, and public health. Excessive algal blooms of toxic cyanobacterial *Microcystis aeruginosa* are found to be abundant in Brazil highly eutrophic coastal Bay. Likewise, its occurrence has escalated in headwaters of Chesapeake Bay, posing public issues (Ferrao-Filho et al., 2002). In contrast, along the Louisiana coast, the diatom Pseudo-nitzschia responsible for producing the neurotoxin domoic acid causing amnesic shellfish poisoning expanded greatly over the last 50 years. This expansion is highly associated with high nitrate load from the Mississippi River, and it underscores the significance of nutrient pollution in influencing HAB processes (Parsons & Dortch, 2002). The Japanese Ministry of Environment had to endorse concerted efforts for bloom management and water quality preservation because of massive blooms in Lake Kasumigaura. The Kasumigaura bloom has substantially plugged up fisheries and water purifying machinery, impacting socioeconomic and local activities (Nwankwegu et al., 2019). Most importantly, everyone agrees that HABs are complex events that are generally initiated by several simultaneous environmental conditions rather than one single environmental factor (Heisler et al., 2008). For example, the primary drivers of HAB events in the Beibu Gulf, Yangtze River Estuary, Atlantic Coast of Europe, and Tampa Bay, Florida were the applications of agricultural fertilizers, industrialization, and anthropogenic activity-related nutrient runoff (Desmit et al., 2018; Xu et al., 2019).

Among the most urgent world pollution in coastal water bodies is nutrient contamination, fueled by anthropogenic forces like population expansion, agricultural development, and energy production and use (Howarth et al., 2002). Aquatic environments experience extensive degradation from other human activities such as urbanization, industrialization, global change, and sustainable utilization of aquatic resources. The increase in nitrogen and phosphorus, released to coastal water, driven by industrial growth, agricultural intensification, and increasing population along the coast, has also increased these effects (Ferreira et al., 2011). It has been published that the most severe economic losses in South America were due to tuna mortalities linked to *Tripos furca* and *Chattonella* in the Mexican Pacific and salmon mortalities attributed to *Pseudochattonella verruculosa* and *Alexandrium catenella* in Chile (Sunesen et al., 2021). A study reported by Hallegraeff et al. (2021) revealed that the fish-killing algal blooms due to species such as *Chattonella*, *Karlodinium*, *Margalefidinium*, and *Prorocentrum cordatum* have become increasingly important issues in the recent times in Malaysia and the Philippines. At the same time, excess marine death caused by anoxia, initiated by the collapse of massive dinoflagellate blooms, has been correlated with South African high-biomass blooms of *Gonyaulax*, *Lingulodinium*, *Prorocentrum*, *Protoceratium*, and *Tripos* (Pitcher & Louw, 2021).

15.4 Historical Overview of Algal Bloom-Forming Species

Harmful algal blooms (HABs) are now recognized as a serious global issue. They affect aquatic life, human health, and water resources. This is a global concern that has prevailed from hundreds of years ago and has not yet been solved, possessing a fascinating history. With the passage of time, the understanding of algal blooms has changed dramatically; from ancient observations to modern genetic research, the journey has been remarkable. Aside from dinoflagellates, other toxin-forming groups are cyanobacteria, raphidophytes, diatoms, and some others with smaller numbers of toxic members (e.g., prymnesiophytes) reported by Carmichael (2001). The principal groups of harmful algal bloom (HAB) toxins, sorted by their related syndromes, include paralytic shellfish poisoning (PSP), neurotoxic shellfish poisoning (NSP), amnesic shellfish poisoning (ASP), ciguatera fish poisoning, and diarrheal shellfish poisoning (DSP) (Lin & Carpenter, 1996). Those toxins contain neurotoxins, carcinogens, and other toxic substances. Their impacts result from various chemical mechanisms (e.g., free radical formation) and are injurious to living organisms—humans included—when exposure happens directly or indirectly through ingestion of contaminated filter-feeding shellfish or planktivorous fish. Long before microscopes existed, humans observed strange events in water bodies. People living near lakes, rivers, and oceans noticed sudden changes. Sometimes, the water turned red, green, or brown. At other times, the air above the water smelled bad. In many cases, large numbers of fish died without warning. Ancient people could not explain these events scientifically. They had no knowledge of microorganisms. So, they often blamed these strange happenings on supernatural causes. In Greece, Rome, China, and other regions, people thought these were signs of divine anger. Myths and legends described the sea turning into blood or being cursed (Horner, 2018).

By the 1900s, communities and scientists could see the detrimental effects caused by the prevalence of algal blooms. Red tides and great fish kills were occurring more frequently by unusual discoloration of water at coastlines regions. All of these events cause intensifying situation, with one occurred in 1840 when Florida experienced a red tide. Later, it was confirmed that causative agents are *K. brevis* which is usually known for the release of brevetoxins. These toxins poison marine life and make people ill, particularly by targeting the lungs. Another significant incident was documented in 1927 at California, where some people were found ill after consuming shellfish. Scientists learned that the culprit was Pseudo-nitzschia, which is a toxic diatom that yields domoic acid, a neurotoxin (Molin et al., 1992). After World War II, there was a surge in industrialization and agricultural growth, which resulted in more runoff of fertilizers and waste products into lakes and rivers. These contaminants, high in nitrogen (N) and phosphorus (P), caused extensive eutrophication—a process that a nutrient "banquet" for algae, which promotes out-of-control growth. Dense layers of algae soon blanketed water surfaces, with toxic cyanobacteria such as *M. aeruginosa* being a significant concern. This genus secretes microcystins, highly toxic compounds that were shown to cause liver damage. One of the most infamous instances was Lake Erie in North America, which was badly polluted through the 1960s and 1970s and suffered from repeated algal blooms (see Fig. 15.2). This chapter further discovered that a comparable outbreak occurred in Europe and Asia, where members of the genus *Anabaena, Nodularia,* and *Aphanizomenon* were frequent offenders, damaging aquatic habitats, drinking water sources, and public health. At this point, algal blooms were no longer a localized phenomenon but an environmental crisis worldwide, reported by Finni et al. (2001).

In the late twentieth century, researchers have directed attention to laboratory experiments to demystify bloom-forming algae. By growing algae under lab conditions, scientists have developed key information on their growth habits, reproductive life cycles, and toxin manufacturing cycles. The improvement in analytical tools like high-performance liquid chromatography (LC–MS) to separate chemical compounds and enzyme-linked immunosorbent assay (ELISA) to identify trace toxins such as domoic acid, brevetoxins, and microcystins. Governments responded by initiating large-scale monitoring programs, public health advisories, and short time closure of fisheries to alleviate risks (Brush, 2009). Furthermore, scientists used DNA sequencing, PCR, and metagenomics to analyze algae at the genetic level and found "cryptic species" which were indistinguishable under microscopes but genetically different, having different toxicity. The major breakthrough was finding the mcy gene cluster in *Microcystis* responsible for microcystin production. This made it possible to detect the toxic blooms early when the toxins did not even occur

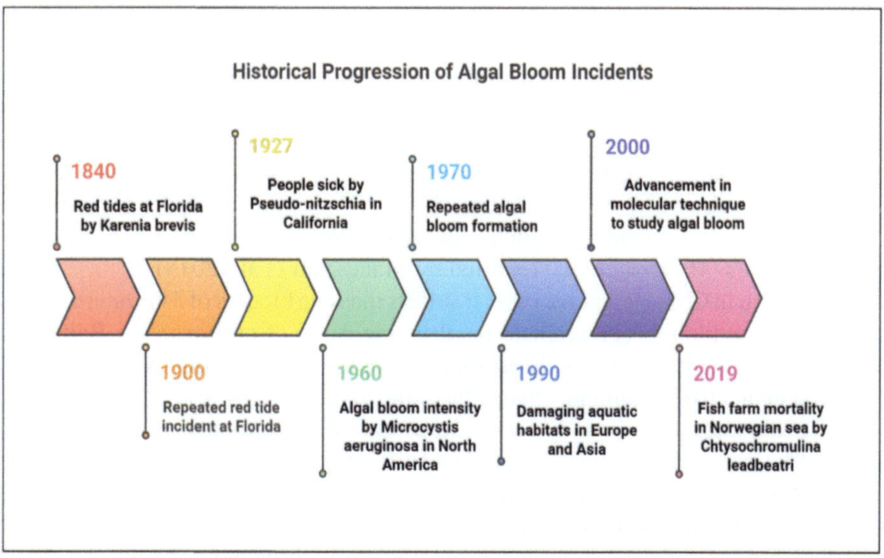

Fig. 15.2 Historical progression of algal bloom formation all over the world

yet (Kim et al., 2014). Karlson et al. (2021) reported that *Chrysochromulina leadbeateri* was the primary cause of fish farm mortality in the Norwegian Sea, Eastern North Sea, and Kattegat-Skagerrak in 1991 and 2019 as shown in Fig. 15.2.

15.5 Harmful Algal Blooms in a Changing Climate: Surveillance and Intervention Strategies

Particularly, adaptation in the case of water quality of resources means anticipating the negative impacts of climate change and responding suitably to mitigate the damage (Johnson et al., 2018). CyanoHABs can create or capitalize on opportunities. The first IPCC report to confer a "high degree of confidence" link between CyanoHABs and climate change was the special study on the Ocean and Cryosphere in a Changing Climate (SROCC), which was endorsed by the UN's Intergovernmental Panel on Climate Change in September 2019. Accordingly, it is essential that adaptation measures to maintain and improve water quality should be based on an understanding of the relationship between climate change and CyanoHABs (Barruffa et al., 2021). The chapter covers the four-stage methods for prioritizing climate change adaptation options with a focus on CyanoHABs: (1) adaptation to climate change; (2) assessing vulnerability; (3) targeting adaptation practices; and (4) formulating adaptation practices.

The first stage focused on climate change affects the frequency and global occurrence of toxic blooms in the aquatic ecosystems (Prakash et al., 2024). Severe weather is also expected to intensify under climate change, which will enhance transient hydrodynamic fluctuations that are conductive to the growth of toxic cyanobacteria (CyanoHABs) reported by Marrone et al. (2023). CyanoHABs have adapted to unfavorable weather conditions which thrive in environments with low nutrient availability and high carbon dioxide levels. In such circumstances, higher-rank consumer animals that utilized the affected waters for drinking and recreation, including fish, zooplankton, and mammals, can be intoxicated by the secondary metabolites derived from CyanoHABs (Chorus & Bartram, 1999). The dominant species flourish in shallow lakes belong to order Oscillatoriales (*Planktothrix agardhii*), Chroococcales, and Nostocales (*M. aeruginosa*) reported by Mantzouki et al. (2018).

Lake Taihu in China is a grave case. It suffers massive blooms of *Microcystis* every summer. These blooms contaminate the drinking water of millions. The Baltic Sea too suffers blooms due to *Nodularia spumigena*, a nitrogen-fixing cyanobacterium (Johnson et al., 2018). Climate change has exacerbated things. Warm water aids the growth of algae at a faster rate. Excessive rains wash more nutrients into water bodies. Drought lowers the movement of water, and it is easier for algae to thrive. Excessive carbon dioxide will also enhance algal growth. Increasing temperatures decrease water layers' resistance to vertical displacement by phytoplankton. In the presence of turbulence, cyanobacteria are dispersed uniformly throughout the water column (Barruffa et al., 2021). With further warming and reduced wind-driven mixing, the water column stabilizes and allows buoyant cyanobacteria to ascend and collect at the surface in dense blooms. Some harmful cyanobacterial taxa contain defensive pigments that enable them to withstand the intense illumination of surface waters (Paerl et al., 1983). The surface blooms also shade sunlight from entering deeper layers of water, restricting the development of non-buoyant eukaryotic phytoplankton competitors (Huisman et al., 2018). Some cyanobacteria also flourish under stratified conditions, facilitating the occurrence of harmful algal blooms (HABs) indirectly. In temperate lakes, stratification usually starts in spring, but rising temperatures can cause the process to initiate earlier and last longer into the fall season.

Across lengthy timescales of evolution, some algal species, particularly those recognized for their ability to create dense blooms, have adapted remarkably well to a variety of nonliving environmental factors. These environmental factors, termed abiotic factors, have a basic function of influencing algal survival, dominance, and distribution patterns throughout freshwater and marine ecosystems. By making these evolutionary adaptations, algae have become extremely successful at colonizing environments that have vastly different physical characteristics. One of the most crucial abiotic factors affecting algal existence is sunlight. Algae are photosynthetic organisms, and they depend significantly on light to generate energy. In order to cope with changing light regimes, particularly within deep or turbid bodies of water, most algae have a capacity to alter their internal composition of photosynthetic pigments. This is beneficial for them to absorb light of various wavelengths more efficiently (Kaur et al., 2022). Therefore, they are able to maintain effective photosynthesis even at decreased light intensities brought about by water depth or suspended material. Aside from light regulation, most algae can also shift their position within the water column by modulating their buoyancy devices. Some of them, especially cyanobacteria, have gas vesicles—small internal bodies that enable them to rise towards the top or descend further, depending on the availability of light and the density of the water (Rani et al., 2021).

With this action, they are able to stay close to light-rich areas, particularly when surface light is at its best. Temperature fluctuations also affect algal behavior. Most of the bloom producers are well adapted to tolerate vast thermal fluctuations. They keep growing and propagating during hot seasons, heatwaves, and even in instances of violent seasonal change. Actually, global warming has made it possible for some algae to extend their original range to areas formerly too cold, showing a clear connection between algal geographic extent and climate change (Mayer, 2020). Another critical physical variable is water flow or turbulence. Various algae react differently to hydrodynamic conditions. Some prefer tranquil, quiet waters with stable sedimentation and penetration of light. Others prefer regions of moderate to high water movement with the benefits of vertical mixing that ensures nutrients from deeper levels of water are carried to the surface (Wiśniewska et al., 2021). Salinity, or the amount of salt in water, is also key. Some bloom-forming genera like *Prymnesium* and *Alexandrium* have high tolerance levels for brackish or saline conditions. These algae employ specialized osmoregulatory and metabolic mechanisms to regulate ionic balance within cells (Rybak & Gąbka, 2018). This enables them to thrive and survive in estuaries and coastal regions where freshwater and seawater interact—conditions that would be fatal to most other organisms. Altogether, these adaptive characteristics illustrate how algae have optimized their physiological and ecological responses to fit a diverse range of abiotic environments. This ecological adaptability provides them with an advantage within aquatic ecosystems in which other organisms cannot thrive.

15.5.1 Modern Monitoring and Prevention

In recent years, scientists and policymakers have implemented various strategies to monitor and mitigate harmful algal blooms (HABs). Satellites like NASA's MODIS track blooms from space by analyzing chlorophyll concentrations and water temperature fluctuations, enabling predictions of bloom formation and spread. On the ground, real-time sensors deployed in lakes and coastal waters continuously monitor key parameters such as temperature, light exposure, nutrient levels, and toxin concentrations. This data feeds into advanced computer models that improve bloom forecasting and early warning systems.

To prevent HABs, several measures are being adopted:

- **Reducing fertilizer runoff** by implementing better agricultural practices
- **Upgrading wastewater treatment** to minimize nutrient discharge
- **Restoring wetlands** to naturally filter excess nutrients from runoff
- **Applying ultrasonic treatment** to disrupt algal cells
- **Using clay dispersion** to weigh down algae and prevent surface scums.

15.6 Life Cycle and Survival Strategies of Bloom-Forming Algal Species

The life history of bloom-forming algae reveals a fine balance between the environment and the survival mechanism of the algae itself. The organisms are found to be very resilient and can adapt extremely well to varying conditions. They not only thrive well in favorable environments but even survive during adverse conditions. Their life history patterns may be explained by the r/K selection theory, a phenomenon that categorizes species according to their reproductive activity and ecological strategy (Reznick et al., 2002). Some bloom-forming algae are r-strategists. They are species such as *M. aeruginosa* and *Dolichospermum* (Mariani et al., 2015). They thrive in unstable systems that are momentarily enriched with nutrients. They reproduce rapidly. They possess brief life cycles. When light, temperature, and resources such as nitrogen and phosphorus are plentiful, their populations can skyrocket. These algae overwhelm aquatic systems temporarily before vanishing when the environment shifts (Caswell, 1982). In comparison, K-strategist species are slower growing.

Some examples of such species include certain diatoms and dinoflagellates such as *Alexandrium* and *Ceratium* (Papanikolopoulou et al., 2018). Such species are well suited to stable conditions with lower nutrients. Even though they do not grow quicker, they make the best use of limited resources. These species can live longer and form blooms that recur yearly. These blooms are more stable and longer lasting than r-strategist blooms (Caswell, 1982).

With changes in seasons, these algae go through varying biological phases. Under unfavorable conditions—like low light, low temperature, or nutrient scarcity—they enter a resting or quiescent phase. This allows them to survive until favorable conditions arise. Freshwater cyanobacteria such as *Anabaena* undergo the formation of akinetes, thick-walled cells. These are cold-resistant, resistant to desiccation, and nutrient-poor. Likewise, marine dinoflagellates and diatoms form resting cysts or spores. These deposit on the lake or sea bed and can survive for months or even years (Arora & Sahoo, 2015). During this dormant phase, the algae are inactive, for the most part. Yet, they react to messages from their environment. During spring in temperate climates, the days get longer, the water warms, and there is more sunlight. These conditions trigger the resting cells to become active. When they revive, they move into the water column and resume growth. This is the beginning of the subsequent bloom cycle (Anderson et al., 2003). When conditions are suitable, the active cells replicate. The bloom starts once the algae develop rapidly, particularly if the surface water is warm and tranquil. Certain algae, such as cyanobacteria, employ gas vesicles within their cells to buoyancy. These assist them in drifting to the water surface where there is maximum light availability. Under quiet water conditions, they may create thick surface mats or scums covering extensive areas (Yongsheng et al., 2013).

Some species dominate during this bloom period. These species are referred to as bloomers. They tend to outcompete others either by growing more quickly or by being toxic. Other species, referred to as performers, do not dominate yet still add something to the system. They help in photosynthesis and maintain the ecosystem running. The duration and composition of nutrients also have a significant contribution to bloom development. Nutrients such as nitrate, ammonia, and phosphorus are critical for growth. Availability is based on rainfall, fertilizer runoff, and industrial waste. The nitrogen to phosphorus ratio (N:P) also determines what type of algae tends to grow better. Even trace elements such as iron have an impact on bloom development. Blooms are being made worse by climate change. There is an increase in temperature, increased CO_2 levels in the atmosphere, altering rainfall, and droughts, all of which make bloom-forming algae more favorable. Warm, slow-flowing, or still bodies of water are particularly vulnerable. Bloom-forming algae have a selective advantage over other organisms under these circumstances (Asadian et al., 2018). Blooms eventually begin to subside. This occurs when nutrients are depleted, light is reduced, or too much waste products are produced by the algae. This is when the algae resume their resting phase. Some drift downwards once more and develop cysts or spores. The cyclical pattern of the seasons ceases and is ready to repeat itself when the time is right. All this from rest to blooming and then back again is strongly orchestrated. It illustrates just how sophisticated even rudimentary organisms can be. Bloom-forming algae are not haphazard in their action. They react to light, temperature, nutrient cues, and seasonal fluctuations in a very intelligent manner (Asadian et al., 2018). In summary, the life strategy of bloom-forming algae speaks volumes about their success in their environment. They are able to maximize favorable conditions. They are also well equipped to survive unfavorable conditions. Their flexibility, plasticity, and resilience have enabled them to colonize lakes, rivers, and oceans globally. Even with human disturbance and global warming, these algae persist.

15.6.1 Nutrient Absorption and Utilization Efficiency

Algal species responsible for creating heavy blooms have developed highly effective nutrient uptake mechanisms. These species are capable of taking up essential nutrients like nitrogen and phosphorus even if they become available at extremely low levels in the water. They become highly favored due to their capacity to grow in nutrient-deprived conditions, particularly in ecosystems that become frequently subjected to variations in nutrient levels. Certain bloom-forming algae display a physiological trait called luxury consumption. With this tactic, they absorb nutrients in amounts greater than their current metabolic requirements. Rather than utilizing all the absorbed nutrients immediately, they sequester the excess inside their cells in a reserved form for future use. This proves especially advantageous at times when external sources of nutrients are minimal or for some reason unavailable. By such a mechanism, they survive and grow while other organisms are left to face the limitation (Anabtawi et al., 2024). In addition, some algal species exhibit a mixotrophic life. Mixotrophy is a nutritional mode in which algae undergo photosynthesis such as normal autotrophs and also take in dissolved organic particles or even tiny organisms such as bacteria. This dual strategy enables them to produce energy from light when it is possible and acquire organic nutrients when photosynthetic efficiency is minimal. Such metabolic flexibility confers stress resilience and survival in varied ecological conditions (Anabtawi et al., 2024).

15.6.2 Regulation of Buoyancy in Water

Another major characteristic noted in bloom-forming algae is their capacity for buoyancy regulation. Such an adaptation permits them to float or sink in the water column as per environmental influences and metabolic requirements. Cyanobacteria, particularly *Microcystis* species, are known to be well adapted for this purpose. They have specialized intracellular structures known as gas vesicles that consist of protein and contain gas. These vesicles lower the cells' total density so they can float towards the water surface (Lakshmikandan et al., 2024). This vertical travel is critical for maximizing access to light and nutrients. During the day, algae travel upwards to the surface to absorb sunlight, which is crucial in photosynthesis. They can fall to deeper water layers during nighttime or periods of nutrient scarcity at the surface. This up-and-down traveling increases their potential to get all the resources they need for further growth. Aside from gas vesicles, most algal cells also store lipids as reserve energy. Lipids are less dense than water; thus, their presence contributes to buoyancy. These lipid bodies play two roles. They not only aid in flotation but also function as energy reserves that can be used by the algae when photosynthesis is reduced or demand for growth increases. All these mechanisms confer a highly adaptive advantage within stratified aquatic systems (Lakshmikandan et al., 2024).

15.6.3 Structural and Morphological Adaptations

Algae that often create blooms are also marked by specific structural characteristics that contribute to their survival, defense, and ecological success. One of these characteristics is the formation of colonies, in which individual cells aggregate to create filaments or clusters. *Anabaena* and *Aphanizomenon* are good examples, which aggregate into long filamentous forms. Colonies are more resistant to predators since they are too large to be easily engulfed by most grazers (Yan et al., 2021). Other organisms form defensive features such as spines, reinforced cell walls, or layers of gelatinous mucilage. These morphological defenses shield them against zooplankton predators and will also slow down the rate of sinking of cells so that they can stay in the productive upper layers of water. The mucilage can also serve as a defense against damaging environmental agents such as UV radiation or chemical poisons (McCourt et al., 2023). Another interesting adjustment is plasticity of cell size, which refers to the ability of some algal species to change their size according to environmental factors. Large-sized algal cells have a competitive advantage since they can override light and space so that it becomes challenging for small cells to share those resources. In contrast, small species gain from high reproduction rates and are capable of rapid adaptation to changes in their environment. This size variation enables algae to fill a broad variety of ecological niches and enables various species to compete under varying conditions (Yan et al., 2021).

15.6.4 Horizontal Gene Transfer and Coordinated Genetic Clusters

Algal species that are recurrently engaged in large-scale blooms exhibit enormous flexibility in their genetic structure. One of the most fascinating attributes among these species is that they can execute horizontal gene transfer (HGT)—a process of gene exchange by which they can capture completely new genes from other microorganisms present in their habitat, such as bacteria or other microbial species. This transfer of genetic information allows algae to acquire entirely new metabolic characteristics that were not passed on via normal evolutionary mechanisms. As an instance, they can acquire the ability to trap and utilize nitrogen from inorganic resources, take up phosphorus more effectively, or even produce toxic substances that function as biological poisons (Qiu et al., 2013). Interestingly, the acquired genes in this case are not dispersed randomly across their genome. Rather, they are often grouped in gene clusters, which are densely packed arrays of genes that function together to carry out sophisticated biochemical activities. In organisms like *M. aeruginosa*, a notorious bloom former, the mcy gene cluster has been directly correlated with microcystin synthesis, a potent toxin (Hunsperger et al., 2015). This group consists of a sequence of enzyme genes that cooperate to build the molecular shape of the toxin. Additionally, such clusters of genes are programmed to be highly responsive to shifts in environmental conditions. They function as one integrated apparatus. Likewise, in organisms like *K. brevis* that are dinoflagellates, genes involved in brevetoxin synthesis are arranged in the same modular structure. Such genes remain inactive except on the receipt of particular environmental stimuli like seasonal oscillations or chemical cues. This kind of structural organization in the genome armors algae with means to change rapidly into aggressive or toxic metabolic modes whenever necessary.

15.7 Regulation of Gene Expression Under Stressful Environmental Conditions

Aside from the mere presence of genes, what really makes an organism adaptable is the manner, timing, and degree by which these genes are expressed. Algae that cause blooms exhibit high sensitivity towards changes in their immediate environment. When they experience situations like the depletion of nutrients or strong sunlight, the algae trigger quick and specific alterations in their gene expression levels. For example, during laboratory tests to mimic nutrient deficiency, often termed as nutrient stress assays, algae have been found to turn on particular genes when essential elements are lacking. When phosphorus becomes scarce, they turn on genes such as phoA and pstS. These genes code for proteins to take up phosphate more efficiently. Conversely, when there is not enough nitrogen, genes like narB and nirA are activated (Bacova et al., 2020). These facilitate the conversion of nitrate into forms utilizable by metabolic pathways. All of these genetic responses are stringently controlled by environmental signals that trigger change. Under oxidative stress, which is caused by too much light when algae are exposed or overactive photosynthesis, toxic by-products called reactive oxygen species (ROS) start to build up inside cells. These molecules, left unchecked, have the potential to injure vital biomolecules such as DNA, proteins, and cell membranes. To counteract this, algae turn on a set of defense genes, including superoxide dismutase, catalase, and peroxiredoxin (Liu & Wei, 2024). These enzymes assist in counteracting the toxic activities of ROS so that the algae can still grow even under extreme stress, as is often experienced during peak bloom periods.

15.8 Epigenetic Responses and Molecular Flexibility

Alongside gene acquisition and regulation, the advances of the last few decades have revealed a more subtle mechanism of control known as epigenetic regulation. This area studies how the activity of genes can be changed without affecting the actual DNA sequence. In processes such as DNA methylation, histone adjustment, and noncoding RNA interference, algae are allowed to selectively silence or activate specific genetic processes to respond to exterior pressures. When algae are exposed to long-term environmental stress like chronic nutrient starvation, shifts in salinity, or fluctuating temperature conditions, they go through epigenetic modifications that effectively "reprogram" their cell behavior. Some genes are shut down permanently, while others are transcribed more actively (Ferrari et al., 2023). Methylation of DNA serves as a type of molecular memory that allows the cell to "remember" previous stress events and respond more effectively if those conditions recur in the future. At the same time, bloom-forming algae demonstrate exceptional molecular plasticity. This is understood as their ability to quickly modulate internal metabolic and structural activities to accommodate changing environmental factors. This organism can switch from autotrophic and mixotrophic modes of nutrition such that they use sunlight via photosynthesis or break down organic substances when light or nutrients are in short supply (Zhao et al., 2024). They can also shift the way they store energy from carbohydrate storage to lipid accumulation and produce stress-resistance proteins within a very short time. All of this is controlled by internal regulatory circuits, which include feedback mechanisms that are constantly checking the environment and adjusting gene expression accordingly. This dynamic molecular plasticity allows algae not just to cope short term but also to survive and dominate other species over long ecological cycles. The occurrence and recurrence of algal blooms depend not only on external factors like nutrient levels or temperatures but also reflect their intrinsic molecular strategies. Algae utilize horizontal gene transfer to absorb foreign genetic elements and incorporate novel survival mechanisms, which are organizationally advanced in the form of adaptive gene clusters that allow them to respond rapidly and effectively to environmental cues. In addition, their gene expression is highly regulated to respond to particular stress situations like nutrient deficiencies and oxidative stress. On top of this, epigenetic regulation adds a different dimension of long-term flexibility, where algae have the capability to have memory of environmental stressors. Their molecular plasticity enables them to constantly adjust their metabolism, form, and protective mechanisms to survive in changing environments. Collectively, this chapter concluded these internal strategies render bloom-forming algae not only survivors but also superior competitors in aquatic systems. Knowledge on these advanced mechanisms is crucial to forecasting bloom occurrences and developing more efficient management and prevention strategies in the future.

15.9 Chemical Defense Mechanisms and Allelopathic Interactions

Apart from physical adaptations, algae also involve extremely specialized biochemical warfare tactics to outcompete other organisms present in their habitat. Among the most important biochemical interactions found in algae is allelopathy, a process by which organisms secrete and release chemical compounds into their environment to inhibit competitors (Allen et al., 2018). These bioactive molecules, referred to as allelochemicals, are emitted by some algal species as a mode of survival strategy. The presence of such chemicals in the water can suppress the growth of other algae, aquatic vegetation, cyanobacteria, or even microscopic aquatic organisms like zooplankton, which are phytoplankton consumers (Leflaive & Ten-hage, 2007). Through suppression of the photosynthetic capacity or nutrient uptake potential of surrounding organisms, allelochemicals enable the producing algae to become ecologically dominant. In more severe instances, some toxic algal species, such as *K. brevis* and *Prymnesium parvum*, are even able to produce highly toxic substances—for instance, brevetoxins and prymnesins. These may lead to high mortality among aquatic fauna, even whole fish populations. Their influence is not just focused on animal life; they can also have long-term implications on aquatic food webs and overall ecosystem health. These poisonous compounds play a twofold function: they protect the algae from predators and inhibit competing species (Jonsson et al., 2009). Surprisingly, the release of allelopathic compounds is frequently dependent and responsive to environmental stress. If nutrient levels decline or population densities rise, the synthesis of such protective compounds can be very greatly augmented (Jiang et al., 2019). Such a response indicates that algae have the capability to dynamically control their chemical release in relation to competitive stress, emphasizing the evolutionary wisdom inherent in their survival tactics. Long-term effects of allelopathic interactions are significant. By modifying the aquatic food web structure, these compounds have the ability to decrease biodiversity and increase the likelihood of repeated blooms in ecosystems (Bacellar Mendes & Vermelho, 2013). Thus, chemical defense not only helps some particular algae but also enhances the ecological dominance of these organisms.

15.9.1 Biofilm Formation and Surface Attachment Mechanisms

In addition to floating freely in water, most algae have the special ability to attach themselves to surfaces and form biofilm complex microbial communities that develop on submerged surfaces (Sun et al., 2023). This form of existence has a number of survival benefits and is another vital ecological approach employed by bloom-forming species. A biofilm is essentially a tight aggregate of algae within a matrix self-secreted extracellularly, made mostly of adhesive polysaccharides and other organic substances. This slime coating shields the cells from extreme environmental conditions, predators, and even chemicals. In addition, it traps the nutrients, forming a localized microenvironment where the algae are able to survive even when the external water is nutrient limited (Schnurr & Allen, 2015). These biofilms can adhere to a broad range of natural and man-made substrata—like rocks, underwater vegetation, soil grains, plastic, and even metal or concrete surfaces. Shallow shoreline areas (littoral zones) or benthic environments at the bottoms of lakes and oceans are especially conducive to biofilm growth (Gu et al., 2022). Here, water flow slows down, and light remains available so that attached algae can survive where free-swimming algae would perish. Certain cyanobacteria and green algae form mat-like groups that have the capability to make a major impact on their microhabitat. They modify vital environmental factors like pH, concentration of oxygen, and localized nutrient cycling. Due to this, they can inhibit the growth of other microbial populations, thus claiming space and resources for themselves (Zheng et al., 2017).

Moreover, these biofilms tend to be long-term reservoirs of life. In unfavorable times, the cells embedded within them can be shielded within the biofilm matrix. When conditions become favorable once more, such cells may again be activated and produce new blooms. This process has usually been referred to as a "seed bank" function, highlighting their contribution to bloom re-establishment (Brush et al., 2020). In short, the capacity of algae to develop biofilms and become attached to many surfaces greatly increases their survival potential. It enables them to move into particular niches, survive during stress in the environment, and come back with a vengeance when times become favorable again. This tactic, along with their other traits, makes bloom-forming algae highly specialized for long-term ecological success.

15.10 Conclusion

In recent years, scientists and policymakers have implemented various strategies to monitor and mitigate harmful algal blooms (HABs) in freshwater ecosystems, estuaries, and marine coasts around the globe. These strategies are based on the fact that some algal species causing HABs secrete toxic substances that interfere with biological processes throughout entire ecosystems from plankton to humans. The occurrence and recurrence of HAB depend not only on external factors like nutrient levels or temperatures but also reflect their intrinsic molecular strategies. These internal strategies render bloom-forming algae not only survivors but also superior competitors in aquatic systems. Knowledge on these advanced mechanisms is crucial to forecasting bloom occurrences and developing more efficient management and prevention strategies in the future.

References

Allen, J. L., Ten-Hage, L., & Leflaive, J. (2018). Regulation of fatty acid production and release in benthic algae: Could parallel allelopathy be explained with plant defence theories? *Microbial Ecology, 75*(3), 609–621.

Anabtawi, H. M., Lee, W. H., Al-Anazi, A., Mohamed, M. M., & Aly Hassan, A. (2024). Advancements in biological strategies for controlling harmful algal blooms (HABs). *Water, 16*(2), 224.

Anderson, D. M. (2009). Approaches to monitoring, control and management of harmful algal blooms (HABs). *Ocean & Coastal Management, 52*(7), 342–347.

Anderson, D. M., Fensin, E., Gobler, C. J., Hoeglund, A. E., Hubbard, K. A., Kulis, D. M., Trainer, V. L., et al. (2021). Marine harmful algal blooms (HABs) in the United States: History, current status and future trends. *Harmful Algae, 102*, Article 101975.

Anderson, D. M., Fukuyo, Y., & Matsuoka, K. (2003). Cyst methodologies. In *Manual on harmful marine microalgae, Monographs on oceanographic methodology, 11* (pp. 165–190).

Andersson, A. F., Riemann, L., & Bertilsson, S. (2010). Pyrosequencing reveals contrasting seasonal dynamics of taxa within Baltic Sea bacterioplankton communities. *The ISME Journal, 4*(2), 171–181.

Arora, M., & Sahoo, D. (2015). Growth forms and life histories in green algae. *The algae world* (pp. 121–175). Springer, Netherlands.

Asadian, M., Fakheri, B. A., Mahdinezhad, N., Gharanjik, S., Beardal, J., & Talebi, A. F. (2018). Algal communities: an answer to global climate change. *CLEAN–Soil, Air, Water*, *46*(10), 1800032.

Bacellar Mendes, L. B., & Vermelho, A. B. (2013). Allelopathy as a potential strategy to improve microalgae cultivation. *Biotechnology for Biofuels, 6*(1), 152.

Bacova, R., Kolackova, M., Klejdus, B., Adam, V., & Huska, D. (2020). Epigenetic mechanisms leading to genetic flexibility during abiotic stress responses in microalgae: A review. *Algal Research, 50*, Article 101999.

Barruffa, A. S., Sposito, V., & Faggian, R. (2021). Climate change and cyanobacteria harmful algae blooms: Adaptation practices for developing countries. *Marine and Freshwater Research, 72*(12), 1722–1734.

Brooks, B. W., Lazorchak, J. M., Howard, M. D., Johnson, M. V. V., Morton, S. L., Perkins, D. A., Reavie, E. D., Scott, G. I., Smith, S. A., & Steevens, J. A. (2016). Are harmful algal blooms becoming the greatest inland water quality threat to public health and aquatic ecosystems? *Environmental Toxicology and Chemistry, 35*(1), 6–13.

Brush, G. S. (2009). Historical land use, nitrogen, and coastal eutrophication: A paleoecological perspective. *Estuaries and Coasts, 32*(1), 18–28.

Brush, M. J., Giani, M., Totti, C., Testa, J. M., Faganeli, J., Ogrinc, N., Kemp, W. M., & Umani, S. F. (2020). Eutrophication, harmful algae, oxygen depletion, and acidification. In *Coastal ecosystems in transition: A comparative analysis of the northern Adriatic and Chesapeake bay* (pp. 75–104).

Carmichael, W. W. (2001). Health effects of toxin-producing cyanobacteria: "The CyanoHABs." *Human and Ecological Risk Assessment: An International Journal, 7*(5), 1393–1407.

Caswell, H. (1982). Life history theory and the equilibrium status of populations. *The American Naturalist, 120*(3), 317–339.

Choi, D. H., An, S. M., Yang, E. C., Lee, H., Shim, J., Jeong, J., & Noh, J. H. (2018). Daily variation in the prokaryotic community during a spring bloom in shelf waters of the East China Sea. *FEMS Microbiology Ecology, 94*(9), fiy134.

Chorus, I., & Bartram, J. (1999). Toxic cyanobacteria in water. E & FN Spon. World Health Organization.

Desmit, X., Thieu, V., Billen, G., Campuzano, F., Dulière, V., Garnier, J., Lacroix, G., et al. (2018). Reducing marine eutrophication may require a paradigmatic change. *Science of the Total Environment, 635*, 1444–1466.

Ferrao-Filho, A. S., Domingos, P., & Azevedo, S. M. (2002). Influences of a Microcystis aeruginosa Kützing bloom on zooplankton populations in Jacarepaguá Lagoon (Rio de Janeiro, Brazil). *Limnologica, 32*(4), 295–308.

Ferrari, M., Muto, A., Bruno, L., & Cozza, R. (2023). DNA methylation in algae and its impact on abiotic stress responses. *Plants, 12*(2), 241.

Ferreira, J. G., Andersen, J. H., Borja, A., Bricker, S. B., Camp, J., Da Silva, M. C., Claussen, U., et al. (2011). Overview of eutrophication indicators to assess environmental status within the European Marine Strategy Framework Directive. *Estuarine, Coastal and Shelf Science, 93*(2), 117–131.

Finni, T., Laurila, S., & Laakkonen, S. (2001). The history of eutrophication in the sea area of Helsinki in the 20th century. *AMBIO: A Journal of the Human Environment, 30*(4), 264–271.

Glibert, P. M. (2016). Algal blooms. In *Encyclopedia of estuaries* (pp. 7–16). Springer.

Gu, P., Wu, H., Zhang, Z., Li, Q., Zhang, W., Zheng, Z., Yang, K., Miao, H., & Xu, J. (2022). Biological effects of harvesting harmful algal blooms on submerged macrophytes and leaf biofilms: A mesocosm experiment. *Journal of Cleaner Production, 361*, 132256.

Hallegraeff, G., Enevoldsen, H., & Zingone, A. (2021). Global harmful algal bloom status reporting. *Harmful Algae, 102*, Article 101992.

Halsey, K. H., & Jones, B. M. (2015). Phytoplankton strategies for photosynthetic energy allocation. *Annual Review of Marine Science, 7*(1), 265–297.

Havens, K. E. (2008). Cyanobacteria blooms: effects on aquatic ecosystems. In *Cyanobacterial harmful algal blooms: state of the science and research needs* (pp. 733–747).

Heisler, J., Glibert, P. M., Burkholder, J. M., Anderson, D. M., Cochlan, W., Dennison, W. C., Suddleson, M., et al. (2008). Eutrophication and harmful algal blooms: A scientific consensus. *Harmful Algae, 8*(1), 3–13.

Hoagland, P. A. D. M., Anderson, D. M., Kaoru, Y., & White, A. W. (2002). The economic effects of harmful algal blooms in the United States: Estimates, assessment issues, and information needs. *Estuaries, 25*(4), 819–837.

Horner, R. (2018). History of ice algal investigations. In *Sea ice biota* (pp. 1–19). CRC Press.

Howarth, R. W., Sharpley, A., & Walker, D. (2002). Sources of nutrient pollution to coastal waters in the United States: Implications for achieving coastal water quality goals. *Estuaries, 25*(4), 656–676.

Huisman, J., Codd, G. A., Paerl, H. W., Ibelings, B. W., Verspagen, J. M., & Visser, P. M. (2018). Cyanobacterial blooms. *Nature Reviews Microbiology, 16*(8), 471–483.

Hunsperger, H. M., Randhawa, T., & Cattolico, R. A. (2015). Extensive horizontal gene transfer, duplication, and loss of chlorophyll synthesis genes in the algae. *BMC Evolutionary Biology, 15*(1), 16.

James, K. J., Carey, B., O'halloran, J., & Škrabáková, Z. (2010). Shellfish toxicity: Human health implications of marine algal toxins. *Epidemiology & Infection, 138*(7), 927–940.

Jiang, M., Zhou, Y., Wang, N., Xu, L., Zheng, Z., & Zhang, J. (2019). Allelopathic effects of harmful algal extracts and exudates on biofilms on leaves of Vallisneria natans. *Science of the Total Environment, 655*, 823–830.

Johnson, B. E., Noble, P. J., Heyvaert, A. C., Chandra, S., & Karlin, R. (2018). Anthropogenic and climatic influences on the diatom flora within the Fallen Leaf Lake watershed, Lake Tahoe Basin, California over the last millennium. *Journal of Paleolimnology, 59*(2), 159–173.

Jonsson, P. R., Pavia, H., & Toth, G. (2009). Formation of harmful algal blooms cannot be explained by allelopathic interactions. *Proceedings of the National Academy of Sciences, 106*(27), 11177–11182.

Karlson, B., Andersen, P., Arneborg, L., Cembella, A., Eikrem, W., John, U., Suikkanen, S., et al. (2021). Harmful algal blooms and their effects in coastal seas of Northern Europe. *Harmful Algae, 102*, Article 101989.

Kaur, M., Saini, K. C., Ojah, H., Sahoo, R., Gupta, K., Kumar, A., & Bast, F. (2022). Abiotic stress in algae: Response, signaling and transgenic approaches. *Journal of Applied Phycology, 34*(4), 1843–1869.

Kim, K. M., Park, J. H., Bhattacharya, D., & Yoon, H. S. (2014). Applications of next-generation sequencing to unravelling the evolutionary history of algae. *International Journal of Systematic and Evolutionary Microbiology*, *64*(Pt_2), 333–345.

Lakshmikandan, M., Li, M., & Pan, B. (2024). Cyanobacterial blooms in environmental water: Causes and solutions. *Current Pollution Reports, 10*(4), 606–627.

Leflaive, J. P., & Ten-Hage, L. O. Ï. C. (2007). Algal and cyanobacterial secondary metabolites in freshwaters: a comparison of allelopathic compounds and toxins. *Freshwater Biology*, *52*(2), 199–214.

Lin, S., & Carpenter, E. J. (1996). An empirical protocol for whole-cell immunofluorescence of marine phytoplankton. *Journal of Phycology, 32*(6), 1083–1094.

Liu, D., & Wei, L. (2024). Epigenetic regulation in response to CO_2 fluctuation in marine microalga Nannochloropsis oceanica. *Microbial Ecology, 87*(1), 4.

Malone, T., Azzaro, M., Bode, A., Brown, E., Duce, R., Kamykowski, D., Eghtesadi, P., et al. (2015). Primary Production, Cycling of Nutrients, Surface Layer and Plankton. *Centro Oceanográfico de A Coruña.*

Mariani, M. A., Padedda, B. M., Kaštovský, J., Buscarinu, P., Sechi, N., Virdis, T., & Lugliè, A. (2015). Effects of trophic status on microcystin production and the dominance of cyanobacteria in the phytoplankton assemblage of Mediterranean reservoirs. *Scientific Reports, 5*(1), 17964.

Mantzouki, E., Lürling, M., Fastner, J., de Senerpont Domis, L., Wilk-Woźniak, E., Koreivienė, J., & Warming, T. P. (2018). Temperature effects explain continental scale distribution of cyanobacterial toxins. *Toxins, 10*(4), 156.

Mallimadugula, B. V., & Hameed, A. (2023). Diatom algae for carbon sequestration in oceans. In *Insights into the World of Diatoms: From Essentials to Applications* (pp. 113–119). Singapore: Springer Nature Singapore.

Marrone, B. L., Banerjee, S., Talapatra, A., Gonzalez-Esquer, C. R., & Pilania, G. (2023). Toward a predictive understanding of cyanobacterial harmful algal blooms through AI integration of physical, chemical, and biological data. *ACS Es&t Water, 4*(3), 844–858.

Mayer, T. (2020). Interactions of fish, algae, and abiotic factors in a shallow, tropical pond. *Hydrobiologia, 847*(20), 4145–4160.

McCourt, R. M., Lewis, L. A., Strother, P. K., Delwiche, C. F., Wickett, N. J., de Vries, J., & Bowman, J. L. (2023). Green land: Multiple perspectives on green algal evolution and the earliest land plants. *American Journal of Botany, 110*(5), Article e16175.

Molin, D., Guidoboni, E., & Lodovisi, A. (1992). Mucilage and the phenomena of algae in the history of the Adriatic: periodization and the anthropic context (17th–20th centuries). In *Marine coastal eutrophication* (pp. 511–524). Elsevier.

Nwankwegu, A. S., Li, Y., Huang, Y., Wei, J., Norgbey, E., Sarpong, L., Lai, Q., & Wang, K. (2019). Harmful algal blooms under changing climate and constantly increasing anthropogenic actions: the review of management implications. *3 Biotech, 9*(12), 449.

Paerl, H. W., Tucker, J., & Bland, P. T. (1983). Carotenoid enhancement and its role in maintaining blue-green algal (Microcystis aeruginosa) surface blooms 1. *Limnology and Oceanography, 28*(5), 847–857.

Papanikolopoulou, L. A., Smeti, E., Roelke, D. L., Dimitrakopoulos, P. G., Kokkoris, G. D., Danielidis, D. B., & Spatharis, S. (2018). Interplay between r-and K-strategists leads to phytoplankton underyielding under pulsed resource supply. *Oecologia, 186*(3), 755–764.

Parsons, M. L., & Dortch, Q. (2002). Sedimentological evidence of an increase in Pseudo-nitzschia (Bacillariophyceae) abundance in response to coastal eutrophication. *Limnology and Oceanography, 47*(2), 551–558.

Peeters, F., Straile, D., Lorke, A., & Livingstone, D. M. (2007). Earlier onset of the spring phytoplankton bloom in lakes of the temperate zone in a warmer climate. *Global Change Biology, 13*(9), 1898–1909.

Pitcher, G. C., & Louw, D. C. (2021). Harmful algal blooms of the Benguela eastern boundary upwelling system. *Harmful Algae, 102*, Article 101898.

Prakash, R., Lee, J., Kim, S. H., Lee, H. Y., & Lee, J. (2024). Insight into the interaction between ionic nanobubbles and cyanobacteria: Elucidating the role of surfactant and zeta potential in the inactivation of harmful cyanobacteria. *Chemical Engineering Science, 287*, Article 119699.

Qiu, H., Yoon, H. S., & Bhattacharya, D. (2013). Algal endosymbionts as vectors of horizontal gene transfer in photosynthetic eukaryotes. *Frontiers in Plant Science, 4*, 366.

Rani, S., Gunjyal, N., Ojha, C. S. P., & Singh, R. P. (2021). Review of challenges for algae-based wastewater treatment: Strain selection, wastewater characteristics, abiotic, and biotic factors. *Journal of Hazardous, Toxic, and Radioactive Waste, 25*(2), 03120004.

Reznick, D., Bryant, M. J., & Bashey, F. (2002). R-and K-selection revisited: The role of population regulation in life-history evolution. *Ecology, 83*(6), 1509–1520.

Rybak, A. S., & Gąbka, M. (2018). The influence of abiotic factors on the bloom-forming alga Ulva flexuosa (Ulvaceae, Chlorophyta): Possibilities for the control of the green tides in freshwater ecosystems. *Journal of Applied Phycology, 30*(2), 1405–1416.

Sarkar, S. K., & Sarkar, S. K. (2018). Algal blooms: Potential drivers, occurrences and impact. In *Marine algal bloom: characteristics, causes and climate change impacts* (pp. 53–109).

Schnurr, P. J., & Allen, D. G. (2015). Factors affecting algae biofilm growth and lipid production: A review. *Renewable and Sustainable Energy Reviews, 52*, 418–429.

Sellner, K. G., Doucette, G. J., & Kirkpatrick, G. J. (2003). Harmful algal blooms: Causes, impacts and detection. *Journal of Industrial Microbiology and Biotechnology, 30*(7), 383–406.

Shao, Q., Lin, Z., Zhou, C., & Yan, X. (2020). Bacterioplankton assembly and interspecies interactions follow trajectories of Gymnodinium– diatom bloom. *Marine Environmental Research, 160*, Article 105010.

Shen, A., Liu, H., Zhu, Y., & Zeng, J. (2024). Long-term response of interspecific competition among three typical bloom-forming species to changes in phosphorus and temperature. *Marine Environmental Research, 196*, Article 106421.

Sun, K. M., Wang, J., Ju, Q., Zhao, Y., Kong, X., Yuan, C., & Tian, Y. (2023). The mitigating effects of diatom-bacteria biofilm on coastal harmful algal blooms: A lab-based study concerning species-specific competition and biofilm formation. *Journal of Environmental Management, 335*, Article 117544.

Sunesen, I., Méndez, S. M., Mancera-Pineda, J. E., Bottein, M. Y. D., & Enevoldsen, H. (2021). The Latin America and Caribbean HAB status report based on OBIS and HAEDAT maps and databases. *Harmful Algae, 102*, Article 101920.

Vantarakis, A. (2021). Eutrophication and public health. *Chemical lake restoration: Technologies, innovations and economic perspectives* (pp. 23–47). Springer International Publishing.

Wang, X., Yin, Z., Chen, J., & Liu, J. (2023). Phytoplankton carbon utilization strategies and effects on carbon fixation. *Water, 15*(11), 2137.

Wiśniewska, K., Śliwińska-Wilczewska, S., Lewandowska, A., & Konik, M. (2021). The effect of abiotic factors on abundance and photosynthetic performance of airborne cyanobacteria and microalgae isolated from the southern Baltic Sea region. *Cells, 10*(1), 103.

Worm, B., Barbier, E. B., Beaumont, N., Emmett Duffy, J., Folke, C., Halpern, B. S., Jackson, J. B. C., et al. (2006). Impacts of biodiversity loss on ocean ecosystem services. *Science, 314*(5800), 787–790.

Xu, Y., Zhang, T., & Zhou, J. (2019). Historical occurrence of algal blooms in the northern Beibu Gulf of China and implications for future trends. *Frontiers in Microbiology, 10*, 451.

Yan, P., Guo, J. S., Zhang, P., Xiao, Y., Li, Z., Zhang, S. Q., Zhang, Y. X., & He, S. X. (2021). The role of morphological changes in algae adaptation to nutrient stress at the single-cell level. *Science of the Total Environment, 754*, Article 142076.

Yongsheng, Z., Binghui, Z., Xia, J., & Hao, Z. (2013). Role of gas vesicles and intra-colony spaces during the process of algal bloom formation. *Water Environment Research, 85*(6), 539–548.

Zhang, P., Peng, C., Zhang, J., Zhang, J., Chen, J., & Zhao, H. (2022). Long-term harmful algal blooms and nutrients patterns affected by climate change and anthropogenic pressures in the Zhanjiang bay, China. *Frontiers in Marine Science, 9*, Article 849819.

Zhao, Z., Chen, G., & Zhang, L. (2024). Epigenetic regulation in algae: Implications for growth, development, and stress response. *International Journal of Aquaculture, 14*.

Zheng, Y., Huang, Y., Liao, Q., Fu, Q., Xia, A., & Zhu, X. (2017). Impact of the accumulation and adhesion of released oxygen during Scenedesmus obliquus photosynthesis on biofilm formation and growth. *Bioresource Technology, 244*, 198–205.

Chapter 16
Future Directions in Cyanobacterial Bloom Research and Management Strategies

Waseem Ahmed Khattak, Muhammad Anas, Erdogan E. Hakki, Mehmet Ali Avci, Javed Iqbal, Banzeer Ahsan Abbasi, Shazia Rehman, and Shah Fahad

Abstract Cyanobacterial blooms represent a growing global challenge, driven by climate change, nutrient enrichment, and hydrological modifications. This chapter explores future directions in research and management strategies to address these complex blooms. Key knowledge gaps include strain-specific toxin production, bloom dynamics in tropical and polar ecosystems, and climate–bloom interactions. Advanced monitoring technologies such as hyperspectral remote sensing, biosensors, artificial intelligence (AI)-driven modeling, and genomic tools are highlighted as critical for early detection and proactive management. Sustainable solutions like nature-based interventions, circular nutrient management, and biotechnological innovations

W. A. Khattak
Department of Plant Sciences, Faculty of Biological Sciences, Quaid-I-Azam University, Islamabad, Pakistan
e-mail: waseemkhattack@bs.qau.edu.pk

M. Anas · E. E. Hakki
Department of Soil Science and Plant Nutrition, Selçuk University, Konya, Türkiye
e-mail: muhammed.anas@selcuk.edu.tr

E. E. Hakki
e-mail: eehakki@selcuk.edu.tr

M. A. Avci
Department of Field Crops, Faculty of Agriculture, Selcuk University, Konya, Türkiye
e-mail: mavci@se7lcuk.edu.tr

J. Iqbal (✉)
Department of Botany, Bacha Khan University, Charsadda, Khyber Pakhtunkhwa, Pakistan
e-mail: javed89qau@gmail.com

B. A. Abbasi · S. Rehman
Department of Botany, Rawalpindi Women University, Rawalpindi, Pakistan
e-mail: shazia.rehman@f.rwu.edu.pk

S. Fahad
Department of Agronomy, Abdul Wali Khan University Mardan, Mardan, Khyber Pakhtunkhwa, Pakistan

S. Fahad et al. (eds.), *Cyanobacterial Blooms: Ecology, Evolution and Biogeochemical Impacts*, https://doi.org/10.1007/978-3-032-06042-6_16

are proposed to replace reactive measures. The chapter emphasizes the need for integrated governance, policy alignment, and community engagement through citizen science and risk communication. Global cooperation is essential to harmonize standards, share knowledge, and support vulnerable regions. By combining interdisciplinary research, technological advancements, and inclusive governance, a paradigm shift toward resilient and equitable bloom mitigation can be achieved.

Keywords Cyanobacterial blooms · Climate change · Monitoring technologies · Sustainable management · Global cooperation

16.1 Introduction

The problem of cyanobacterial blooms has become one of the most recurrent and complicated water quality burden of the twenty-first century. These are of serious concern to the aquatic ecosystems, human health, and water security around the world and they have been occurring more frequently, spreading to new geographies, and becoming more severe due to the interaction of climate change, nutrient enrichment, and hydrological modifications. Although significant strides have been implemented in terms of gaining insights and developing short-term mitigation measures, the current ones are usually reactive, inconsistent, and geographically targeted (Mishra et al., 2020). Multifactorial nature of cyanobacterial blooms requires an outlook research and management agenda that is interdisciplinary, anticipatory, and flexible. Such efforts should progressively reconsider the usual limnological measures, and future attempts should incorporate the use of complex techniques that can be used to assess the water bodies remotely, and some of these tools include things like genomics, remote sensing, predictive modeling, and using the data to make decisions as well. Also, there is a need to shift the focus to the social, economic, and policy aspects of bloom mitigation to add more inclusive and sustainable measures (Barruffa et al., 2021a). This chapter sums up the major potential research and management approaches to cyanobacterial blooms. It emphasizes the necessity to seal the existing gaps in knowledge, implement next-generation technologies and reconsider governance patterns, and integrate bloom risk into wider climate change and sustainability strategies. Integrating innovation, including encouraging cross-sectoral interactions, the scientific and policy community will find the way to more successful and sustainable solutions to the increasingly serious problem of harmful cyanobacterial blooms (Carson et al., 2024).

16.2 Knowledge Gaps in Cyanobacterial Bloom Research

Even with massive progress in the science of cyanobacterial blooms, there are numerous unresolved questions in science. The lack of such knowledge does not allow the creation of solid, predictive, and sustainable mitigation frameworks. To perfect the management strategies in responding to the changing environmental scenario, it is crucial to uncover these gaps and determine the methods of addressing them to build ecological resilience (Davoudi et al., 2012).

16.2.1 Strain-Specific Toxin Production and Genetic Variability

A significant gap encloses the lack of knowledge of strain-specific variations in toxin manufacturing. Although some genera such as *Microcystis*, *Anabaena* (*Dolichospermum*), and *Planktothrix* are notorious for being toxic producers, they do not all strains of a particular species produce toxins. The genetic and environmental events leading to the reprogramming of cells in toxin production are not well described; thus, it is hard to model bloom toxicity. Genomic and transcriptomic research are also required to unravel the regulation of cyanotoxin production along the multiple environmental gradients (D'Agostino et al., 2016).

16.2.2 Bloom Dynamics in Tropical and Polar Ecosystems

Majority of the research of blooms has concentrated on the temperate lakes and reservoirs. On the contrary the tropics and the poles have not received much attention in cyanobacterial research. The tropical systems commonly have the entire year growth of algal blooms due to the warmth and abundance of nutrients, but the species structure and toxin spectrum and causes have not been characterized. On the same lines, areas that were thought to be free of blooms as they were too cold, polar ecosystems are now being affected by seasonal blooms since they are warming up, yet little is known about how they work. Increasing the monitoring systems and wild research in such underdeveloped regions is important (Danielsen et al., 2005).

16.2.3 Uncertainty in Climate–Bloom Interactions

The interactions among the variables of climate (e.g., temperature, precipitation, and extreme events) with the bloom development remain not clear yet. Most existing models assume linear relationships, although cyanobacteria might react to changing climatic conditions nonlinearly or at certain thresholds. To give one example, short-term heatwaves, the concentration of nutrients caused by drought conditions, and alteration of water column stratification can dramatically alter the course of a bloom almost overnight. To capture these dynamics, a long-term, multifactorial field experiment, as well as ecosystem-level study, is needed (Caldararu et al., 2023).

16.2.4 Lack of Interdisciplinary and Systems-Level Studies

The issue of cyanobacterial bloom has historically remained partitioned in the ecological or the limnological field. But drivers and factors behind the blooms overlap with hydrology, climatology, microbiology, agricultural, economical, and people health aspects. The researchers of the future have to take systems-level approach by involving ecological, social, and technological aspects in the research. The multi-sector solutions must be synthesized by interdisciplinary teamwork capable of developing functioning solutions in the real world about trade-offs and feedback (Srikrishnan et al., 2022).

16.3 Advancing Monitoring Technologies

The potential of managing cyanobacterial bloom greatly relies on how revolutionary the progress of environmental monitoring becomes (Glibert et al., 2018). The conventional methods of water sampling, pigment analysis, microscopic, and so on are quite reliable but are restricted with spatial and time aspects. They are labor intensive and time-consuming and cannot be used to ensure round the clock coverage of large bodies of water or distant water bodies. With complexities associated with the state of climatic conditions and the rate of bloom expansion in recent times, there is a dire need of high-resolution real-time and scalable monitoring systems, capable of not only recognizing the occurrence of the blooms but also its intensity at an early stage, prior to its elevation to a crisis level. New technologies such as hyperspectral imaging and UAV-based monitoring as well as biosensors, artificial intelligence, and molecular-level diagnostics are now providing suitable and potent alternatives (Ali et al., 2024). Such tools also help to increase detection accuracy as well as models of dynamic decision-making framework running early warning and responses to management. The inclusions of these next-generation technologies into routine surveillance will play important roles in making the transition of bloom management skills based on proactive prevention out of the prevailing reactionary mitigation model (Fig. 16.1) (Tahmasebi, 2024).

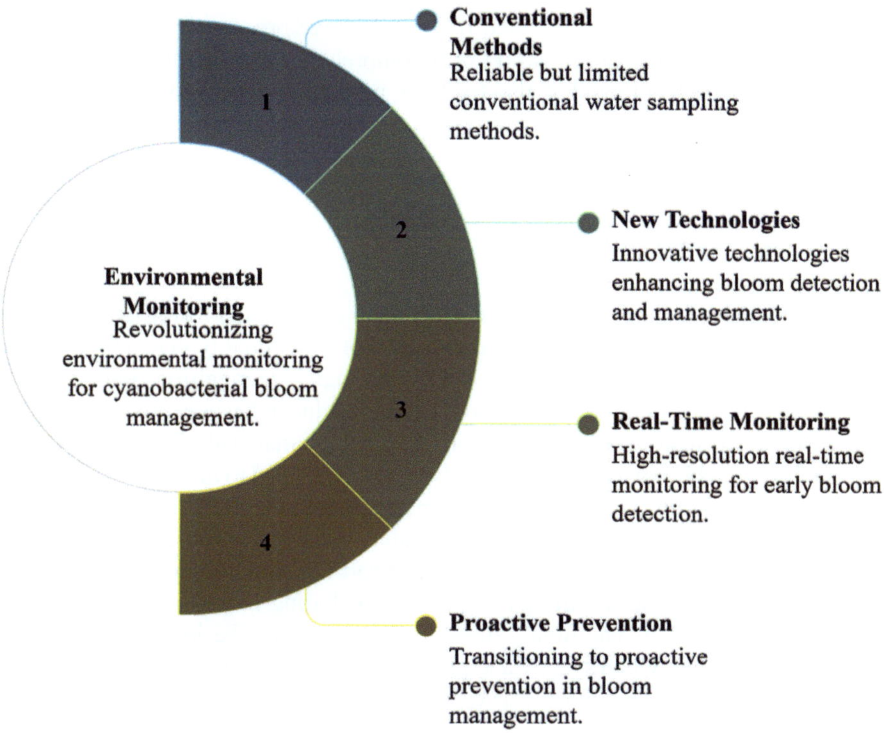

Fig. 16.1 Advancing cyanobacterial bloom management

16.3.1 Next-Generation Remote Sensing

Remote sensing which was once an adjunct to other ways of observing things has become one of the main elements of the current environmental observation (Chuvieco, 2020). Widespread sensors such as MODIS and Landsat are traditional multispectral sensors whose measurements have been of great help because it has been able to identify surface-level algal bloom using broad chlorophyll proxies. These sensors also provide difficulty in separating cyanobacteria and other algal groups particularly in optically complicated inland waters. Since the advent of hyperspectral remote sensing, much greater level of spectral resolution is now possible, to the extent of separating individual pigments including phycocyanin and phycoerythrin which are cyanobacteria specific. This can provide better bloom detection, especially coupled with thermal and meteorological data to decide of the drivers of the bloom. Besides, aerial platforms using drones and compact hyperspectral cameras with unprecedented flexibility capable of carrying out high-resolution monitoring of small, spatially varied systems like ponds, irrigation canals, and urban reservoirs are available. These airborne systems can be used on a regular basis, cheaply, and based on local incidents, addressing serious gaps caused by satellite-based solutions. When combined with in situ calibration and robotized image processing, remote sensing will become an irreplaceable method of a comprehensive surveillance of blooms, not only helping detect them but also tracking, modeling in the passing of time, and across a wide range of freshwater ecosystems (Döll et al., 2016).

16.3.2 Biosensors and Nanobiosensors

Biosensors are the technology of the future in water quality monitoring as they would allow determining the most important parameters of water quality in real time and on the spot. They consist of such devices as using biological recognition elements, e.g., antibodies, enzymes, or nucleic acids, bound to a transducer that converts biochemical interaction to an electrical signal or an optical signal (Kumar et al., 2022). Biosensors in cyanobacterial bloom monitoring have been also designed to measure cyanotoxins including microcystins, cylindrospermopsin, and anatoxins in the nanomolar range levels much less than human and ecological safety levels. Nanobiosensors and nanomaterials, e.g., gold nanoparticle, graphene, or carbon nanotube, which make them more sensitive, are able to amplify the signal, including more recent innovations (Kulkarni et al., 2022). The highly developed sensors may be incorporated into a portable use through which speedy and precise readouts can be measured in the field without any laboratory facilities. It is especially useful in a remote or poor resource environment where it is too costly to have enables a conventional analytical tool. Moreover, the biosensors may be used in fixed locations to continuously monitor and send out automatic warnings to the managers of possible bloom activities way before they are visible. Although they have problems with both standardization and durability, they are being further refined both in their robustness and their ability to multiplex (detect multiple toxins or species on the same run) and expense. Being more available and developed, these tools can transform the bloom risk assessment and can serve as an instrument to advocate community-level water safety efforts (Sullivan et al., 2019).

16.3.3 Artificial Intelligence and Machine Learning Integration

Introduction of artificial intelligence (AI) and machine learning (ML) elements in the monitoring of bloom is a paradigm change in the study and prediction of environmental data (Sheik et al., 2024). Contrary to the statistical models of the past, ML algorithms do not use predefined relationships and assumptions, but they can analyze the data presented within complex, high-dimensional systems to draw a conclusion and make predictions without the systematic knowledge of a system. In the study of cyanobacterial bloods, AI is used to observe time series collected by satellite, weather stations, water quality in situ monitors, and/or hydrological sensors to identify anomalies, characterize events based on occurring blooms, and forecast occurrence, intensity, and length of validity. The nature of bloom formation is dynamic and can be represented by these models with nonlinear interactions of variables (temperature, light, nutrient levels, wind patterns, and more). Additionally, AI applications may be constantly upgraded with fresh information, and they will be willing to change depending on the local situation and get better as time progresses (Bughin et al., 2017; Javaid et al., 2022). Real-time processing of massive data is now possible in the cloud-based platforms and through edge computing, offering the opportunity of allowing water managers to receive early warnings, risk maps interaction, and scenario modeling. Despite these issues that are yet to be solved, including the shortage of data, the interpretation of models, and combining the newly found algorithm with systems of decision-making, AI can provide an effective prism through which one can predict and react to the blooms in circulating climatic and ecological conditions (Wang et al., 2024).

16.3.4 Molecular and Genomic Tools

Molecular and genomic tools are turning over a new leaf in the world of knowledge of cyanobacterial communities and their capability of blooms. Many times, even identification using traditional microscopy mechanism might be restricted due to the ambiguity in taxonomy as well as fail to distinguish between toxigenic and non-toxigenic strains (Delgado et al., 2021). Publication of genomic sequences and newer molecular methods including quantitative polymerase chain reaction (qPCR), digital droplet PCR (ddPCR), and metagenomics have been developed to give specific, sensitive, high-throughput detection of Pacific Northwest cyanobacterial DNA and toxin genes. These equipment can alert before the appearance of blooms, so there is the ability to take mitigation measures. One method in particular, environmental DNA (eDNA) sampling, is proving to be a less-destructive method of describing whole mixtures of microbial life not normally measurable by traditional means, including rare or novel cyanobacterial groups that would otherwise be outside of detection. Moreover, transcriptomic and proteomic methods are the use of providing information about the pattern of gene expression, metabolic activity, and toxin production as environmental differences change (Tomanek, 2011). This type of functional information is vital toward a differentiation between the simple presence of cyanobacteria and the active phases of the blooms. Molecular diagnostics are promising to be incorporated into regular monitoring systems that will enhance our ecological knowledge and advance risk evaluation and target interventions. With further advances in sequencing technologies to become more accessible and user friendly, molecular tools will become much more central to the next generation of bloom forecasting and adaptive management systems (Satam et al., 2023).

16.4 Predictive Modeling Innovations

Overshooting cyanobacterial blooms is still one of the most crucial issues in freshwater ecosystems prediction. Although monitoring technologies present data in real time or retrospective, predictive models can provide, uniquely, the opportunity to simulate bloom development in a variety of environments conditions and management decisions, which allows the stakeholders time to prepare and be ready to respond. The existing models can, however, be in most cases limited in their scale, too simplistic to be able to cover the complexity of climate-driven changes, nutrient inputs at the catchment level, and indeed the socio-economic aspects. Modeling a cyanobacterial bloom should involve building dynamic, integrated, and highly datasets that combine environmental reality with computational fastness and saliency to policy makers (Wang et al., 2024). Higher resolution (in space and time) coupled physical biology models are under development representing the various interactions present between hydrodynamics, nutrient cycling, phytoplankton physiology, and toxin production. CE-QUAL-W2, Delft3D, and AQUATOX, which mimic lake-specific processes, by considering variables such as water temperature profiles, light attenuation, mixing regimes, and sediment water nutrient exchange, are becoming more popular. These models are useful in the determination of the crucial thresholds and the effects of variability of climate, change in land use, and management interventions. To capture the changes due to climate change better in the future, it will be necessary to represent more of the extreme events including heatwaves, droughts, and flash floods, as the changes associated with these disturbances will be more important and frequent in the bloom dynamics (Behrenfeld et al., 2013).

There is also one of the most promising ways, the use of big data and long-term ecological dataset in predictive modeling (LaDeau et al., 2017). The 30 years of satellite imagery, weather data, and in situ monitoring data can now be used to construct powerful machine learning models able to capture nonlinear, lagged, and threshold-based connections. Convolutional neural networks (CNNs) and recurrent neural networks (RNNs), deep learning frameworks, are currently being investigated because of their ability to learn time series and spatial data, which increases the precision of bloom forecasts. Such tools can combine data of varying origins, such as nutrient loading and river inflow rates, land cover and land cover change of the watershed, and even public health reports, to give probabilistic prediction, risk maps, and scenario comparisons based on specific lakes or areas (Ren et al., 2019). The other frontier is the absorption of socio-economic as well as institutional variables in predictive frameworks. These incorporate modeling of the impact of various policy decisions (e.g., agricultural runoff control and wastewater treatment improvements), land use changes, and economic actions (e.g., fertilizer prices or irrigation patterns) on bloom drivers. These models can be of value in decision-making as they assess the costs and viability and long-term results of alternative management solutions in various climate scenarios. Examples of these include agent-based models which can simulate the actions of actors such as farmers, water managers, and policymakers in different environmental and regulatory conditions; such models can be used to make more realistic quantifications of bloom risk and governance outcomes. Finally, scenario analysis is an ever-critical element to predictive modeling, particularly that of adaptive water management. Several possible future scenarios under various greenhouse gas emission trajectories (e.g., RCP 2.6 vs. RCP 8.5), water use demands, or the policies on restoration may be simulated by the models. It enables the stakeholders to assess, examine trade-offs, and develop adaptive capacity in response to uncertainty. The next step is to connect predictive models to an interactive interface and a real-time data feed, letting water managers, researchers, and policymakers interact with forecasts and assess their response directly, in real time and as they should respond.

16.5 Sustainable and Innovative Management Approaches

Conventional methods of controlling these cyanobacterial blooms seem to be failing to adequately manage this problem, especially with the two forces of climate change and anthropogenic nutrient enrichment on the rise, and the latter most efforts may only be treating the symptoms but not the causes. There has to be a system-wide sustainable, ecologically savvy and cross-sectorial long-term solution. This part identifies new strategies (ecological and technological) that present scalable, resilient, and environmentally sustainable solutions toward the management of cyanobacterial blooms in the future (Barruffa et al., 2021a).

16.5.1 Nature-Based Solutions for Watershed and In-Lake Management

Nature-based solutions (NbS) are becoming popular as a form of sustainable measures on controlling nutrient inflows and restoring the balance of the ecosystem. Nitrogen and phosphorus runoff at agricultural landscapes could be reduced through the structural use of constructed wetlands, vegetated buffer strips, as well as riparian forest areas that could allow the reduction of the runoff by large percentages. The systems act as natural filters to better take up nutrients that support vegetation and microbial communities, as well as delivering biodiversity and climate co-benefits that include habitat-enhancing functions, carbon sequestration, and erosion (Ballesteros Possu, 2015). Ecological management options of lakes including biomanipulation include changes in food web structure to subdue the phytoplankton. This may involve the more abundant growth of zooplankton grazer species (e.g., *Daphnia*) or suppression of planktivorous fish, otherwise suppressive of these grazers. Such interventions can structurally change the trophic balance and bring down the biomass of cyanobacteria when strategic (timing, place, and action). Nevertheless, the effectiveness of NbS is also associated with the long-term decrease of nutrient loading on a watershed level and the adaptive management founded on the specifics of the local environment (Robotham et al., 2023).

16.5.2 Emerging Biotechnological and Synthetic Biology Tools

The new developments in the field of biotechnology are providing new ways of specific control of cyanobacteria. They are researching the development of the engineered microbial consortia that would compete with/inhibit harmful cyanobacteria by nutrient competition, toxin degradation, or quorum sensing interference. Virus A potential alternative to drugs that has particular promise is phage therapy, the use of viruses that infect cyanobacteria selectively. Phage therapy is being investigated as a species-specific method of bloom suppression, but its scalability and environmental effects are under investigation (Grasso et al., 2022). Potentially, in the future, the genetic disruption of toxin biosynthesis pathways in cyanobacteria will be possible by using synthetic biology tools like the CRISPR-Cas systems. Such interventions might help in deterring toxic blooms without touching non-toxic cyanobacteria, at the same time that they are still theoretical. Deployment of such technologies will be vital in terms of ethical and ecological safety checks and controls and the regulating forces along with the population acceptance (Lescrauwaet et al., 2022).

Simultaneously, there is a development of advanced materials to absorb excessive phosphorus, clean up water column turbidity, or serve as slow-release cyanobacterial blast inhibitors, including modified clays, biochar composites, and biodegradable nanomaterials. These materials provide lake-specific solutions (which can be customized) and can be used to support larger nutrient management strategies (Tammeorg et al., 2024).

16.5.3 Circular Nutrient Management and Precision Agriculture

The control of the long-term blooms requires attention to be given to the source of nutrient enrichment. Such precision agriculture-related technologies, such as GPS-assisted application of fertilizers, mapping of soil nutrients, and automated irrigation systems, are useful in streamlining the efficiency of nutrient use, which will lead to the reduction of runoffs into nearby water bodies. Fertilizers that are not applied freely are also used, which are released depending on the demand of crop; this helps in reducing the loss that is due to leaching and volatilization (Pan et al., 2016). Together with this, the circular nutrient management approaches seek to complete the nutrient loops by recovering them and reusing them. Recycling of phosphorus can be done through incorporating technology to recycle the waste in wastewater to enhance agricultural reconstruction through phosphorus recycling technologies like bioelectrochemical nutrient recovery and struvite precipitation. Composting of manure, biogas digesters, and constructed wetlands as effluent polishing are being incorporated more frequently into sources of nutrient-smart farming systems, which relate to both water quality conservation and sustainable production of food.

16.5.4 Policy Innovation and Integrated Governance

The correct management of bloom entails both administrative structures and technical devices. Policy should be turned around to more integrated plan management, based on watersheds, that knits together the activities of agriculture, industry and municipalities, and conservation organizations. The legal tools to impose nutrient limits, such as the total maximum daily loads (TMDLs), have to be reinforced and accompanied by the open monitoring systems (Harding, 2015). Economic instruments such as nutrient trading programs or pollutant taxes are based on market mechanisms and provide economic incentives to comply together with encouraging innovation. No less important are institutional arrangements to promote coordination between jurisdictions and across levels of administration (e.g., river basin organizations or inter-agency task forces) (Priscoli, 2009). Most importantly, governance should be encompassing, which means that the rights, knowledge, and demands of the marginalized people should also be taken into consideration as they are frequently and disproportionally impacted by water insecurity caused by bloom. The basis of realizing legitimacy and equity, as well as sustainable accomplishment, is through the application of public involvement, stakeholder engagement, and capacity building (Fig. 16.2).

16.6 Climate Adaptation and Resilience Planning

The problem of cyanobacterial blooms is directly linked to the overall trends of climate changes, as it is connected to temperature rise, precipitation variability, the increase of the number of extremely oriented weather conditions, and the factors of long-term droughts (Paerl et al., 2019). Subsequently, the management of cyanobacterial bloom cannot be isolated from climate adaptation programming. Preparation of multi-year resilience has to bring into national adaptation planning, local water management, and infrastructure upstream, the risk of bloom. The present section describes some main lines along which the mitigation of cyanobacterial blooms can be incorporated into the strategies of climate-resilient planning (Nawaz et al., 2024).

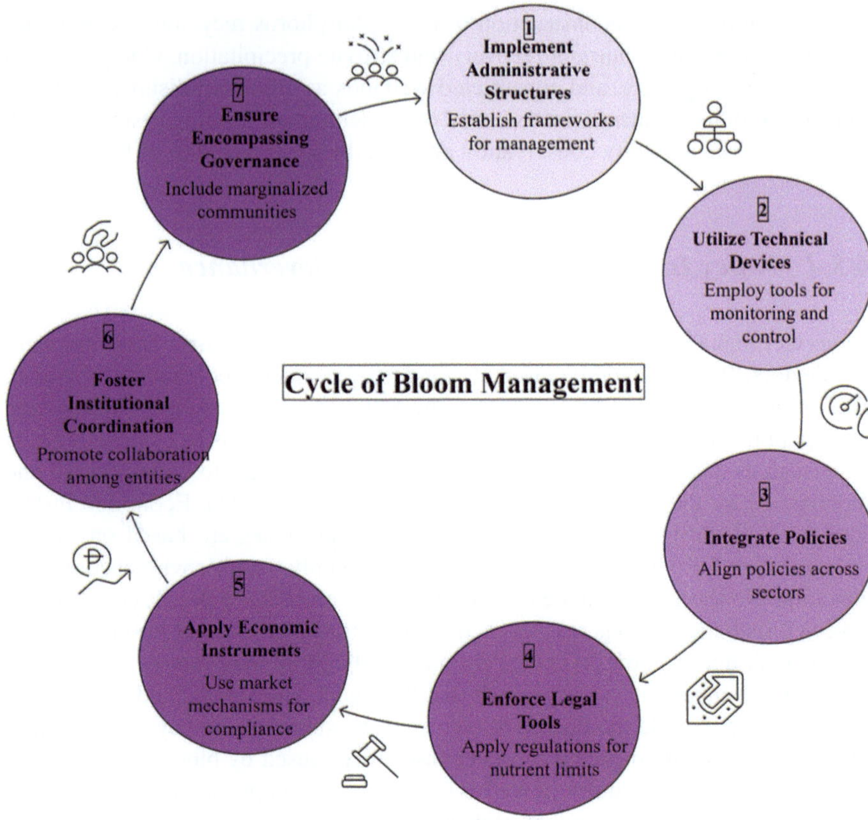

Fig. 16.2 Framework for effective cyanobacterial bloom management, highlighting key components such as governance inclusivity, policy integration, technical monitoring tools, institutional collaboration, and economic incentives to ensure sustainable and coordinated mitigation strategies

16.6.1 Integrating Bloom Risk into Climate Adaptation Policies

Whereas it is evident that climate change is interconnected with the dynamics of bloom, majority of national and regional climate change adaptation strategies have not specifically included the risks of harmful algal bloom (HAB). This oversight may result into reactive crisis management instead of proactive resilience building. The context of the identified adaptation strategies needs to take into consideration the nature of blooms as the climate-related hazard and prioritize their consideration in the policies of water security, general health, and diversity preservation. The scope of vulnerability assessments ought to be increased to target high-hazard water bodies, infrastructure at-risk (e.g., drinking water reservoirs), and populations at heightened risk, especially in areas with limited or decentralized governance structures, or with limited monitoring capacity. The connection of bloom risk to climatic effects, be that a heatwave that encourages the onset of bloom or a flood event that causes nutrient runoff mobilization, allows the government to develop more extensive hazard mitigation strategies. Eutrophication control and restoration of wetlands or developing an early warning system could also be put into the nationally determined contributions (NDCs) of the Paris Agreement as their co-benefit to combat climate change and improve the state of water quality (Dovie, 2019).

16.6.2 Designing Climate-Resilient Water Infrastructure

Water infrastructure planning and management should be transformed to handle the two stresses of the rise in cyanobacterial blooms and hydrological fluctuations. Dams, reservoirs, canals, and water treatment facilities ought to be assessed on their susceptibility to effect of blooms which incorporates adsorption of contaminants to treatment intakes, adsorption of the toxins to treatment filters, and additional treatment expenses in order to eliminate the toxins. Some of the climate-resilient design considerations are versatile water intakes that are able to bypass bloom-laden surface layers, increased real-time monitoring of key locations, and the use of adaptive treatment technologies like advanced oxidation processes or activated carbon filtration.

The change in seasonality and extreme inflows which may influence residence time, stratification patterns, and nutrient loading are also to be considered by storage infrastructure. Hard infrastructure solutions can be supplemented by green infrastructure at the upstreaming retention basins and repaired floodplains that serve as a buffer against surges of nutrients during storm events. All in all, the infrastructure planning should be anticipatory and dynamic and should be directed by parameters of modeling scenarios and with the help of climate and bloom forecasts (Asadabadi & Miller-Hooks, 2017).

16.6.3 Cross-Sectoral Collaboration for Resilience Building

Cyanobacterial blooms are not solely a water management issue; they intersect with agriculture, public health, biodiversity conservation, and disaster risk reduction. Building resilience to blooms requires coordinated action across these sectors. For example, agriculture departments must align fertilizer use regulations with watershed protection goals, while public health agencies should establish surveillance systems for toxin outbreaks and coordinate water utilities on response protocols. Similarly, biodiversity agencies can promote wetland conservation and native vegetation buffers that reduce nutrient runoff and improve ecological stability (Mander et al., 2017). Integrated climate adaptation platforms—such as national climate councils, interministerial task forces, or transboundary river basin commissions—can serve as coordination mechanisms to align bloom-related interventions with other climate resilience strategies. Shared databases, joint funding schemes, and harmonized monitoring indicators can facilitate synergies and reduce duplication of efforts. Multisectoral collaboration is especially vital in countries facing resource constraints, where maximizing co-benefits and minimizing trade-offs are critical (Table 16.1) (Kim, 2024).

16.7 Community Engagement and Citizen Science

Long-term solutions to the growing challenge of cyanobacterial blooms will not succeed without the active involvement of local communities. While much of the research and technological development in bloom management has occurred within scientific and governmental domains, community-based approaches are increasingly recognized as indispensable for early detection, risk communication, policy support, and long-term stewardship of water resources. Citizen participation enhances monitoring capacity, fosters public trust, and builds societal resilience by translating scientific knowledge into meaningful action at the local level (Sánchez-Hernández, 2024).

16.7.1 Empowering Citizens in Monitoring and Reporting

Citizen science initiatives are transforming the way environmental data is collected, interpreted, and acted upon. In the context of cyanobacterial blooms, trained volunteers, recreational water users, farmers, and lakeshore residents can play a pivotal role in identifying bloom events, reporting changes in water quality, and collecting water samples. Many citizen science programs now use standardized protocols and mobile applications that allow for real-time geo-referenced data submission, photos, and qualitative descriptions of bloom severity (Huang et al., 2024).

This crowdsourced information complements formal monitoring by covering a broader spatial range and enabling rapid local response, particularly in underserved or remote regions where regulatory monitoring is infrequent. When combined with remote sensing and automated systems, citizen observations help verify satellite detections and fill temporal gaps in official datasets. Importantly, involving citizens in the scientific process enhances environmental literacy and encourages long-term behavioral change, including reduced fertilizer use, septic system maintenance, and support for conservation initiatives (Thompson et al., 2024).

Table 16.1 Key strategies and interventions for integrating cyanobacterial bloom management into climate adaptation and resilience planning frameworks

No.	Focus area	Strategy/action	Target level	Expected outcome	References
1	National climate policy alignment	Include HABs in National Adaptation Plans (NAPs) and NDCs	National government	Systematic recognition of bloom risks in climate agendas	UN-Water. (2019). Climate change and water
2	Climate–bloom linkage	Explicitly link bloom emergence to climate hazards (e.g., heatwaves and floods)	Policy and planning agencies	Integrated hazard mitigation strategies	Poulain et al. (2018)
3	Vulnerability mapping	Identify bloom-prone water bodies, high-risk infrastructure, and communities	Environmental and health agencies	Prioritized intervention zones	Barange et al. (2018). Climate impacts on fisheries
4	Early warning integration	Develop climate-smart early warning systems for bloom prediction	Meteorological and water agencies	Improved preparedness and response	West et al. (2021)
5	Eutrophication control targets	Set reduction goals for nutrient inputs under climate commitments	National climate/ environment authorities	Reduced bloom drivers and better water quality	Engler (2024)
6	Wetland restoration	Use wetlands as climate–bloom buffers	Ministries of environment/ agriculture	Enhanced nutrient retention and flood absorption	Barruffa et al. (2021b)
7	Flexible water intakes	Design infrastructure to access bloom-free water layers	Water utilities, engineers	Minimized cyanotoxin intake	Campbell et al. (2018)
8	Real-time water quality monitoring	Install continuous sensors in key water systems	Infrastructure operators	Early detection and process optimization	Pham et al. (2021)
9	Adaptive treatment technologies	Advanced treatment: ozonation, membranes, carbon filters	Water authorities, municipalities	Safe drinking water during blooms	WHO (2023a, 2023b, 2023c)
10	Upstream retention basins	Control stormwater nutrient surges via wetlands	Urban planning, hydrology units	Runoff-driven bloom prevention	Cooley et al. (2019)

(continued)

Table 16.1 (continued)

No.	Focus area	Strategy/action	Target level	Expected outcome	References
11	Scenario-based infrastructure	Use models to guide infrastructure location/design	Civil engineers, climate experts	Long-term resilience and efficiency	Thomas et al. (2020)
12	Floodplain restoration	Reconnect rivers to floodplains for nutrient trapping	River basin authorities	Natural nutrient regulation	Stephens (2022)
13	Multisectoral coordination	Form water–climate–agriculture–health platforms	Governmental task forces	Coherent cross-sector implementation	March et al. (2024)
14	Fertilizer regulation alignment	Align agri-nutrient policies with water protection	Agriculture ministries	Reduced nutrient runoff	UN ESCAP (2020)
15	Public health surveillance	Monitor toxin outbreaks and communicate risks	Health ministries, WHO offices	Health protection for exposed populations	WHO (2023a, 2023b, 2023c)
16	Ecosystem-based adaptation	Use nature-based nutrient buffers	Conservation organizations	Cost-effective, eco-friendly adaptation	Tirado and Raj (2023)
17	Biodiversity integration	Protect aquatic biodiversity and riparian buffers	Forestry and wildlife departments	Habitat and bloom mitigation co-benefits	Dabbadie et al. (2019)
18	Climate risk mapping platforms	Develop dashboards for climate–bloom hazard mapping	National disaster agencies	Data-informed decision making	AFMESI (2022)
19	Joint funding schemes	Combine water and climate finance	Finance ministries, donors	Efficient, sustainable funding streams	West et al. (2021)
20	Harmonized indicators and metrics	Shared resilience metrics and evaluation methods	Monitoring and evaluation units	Comparable, traceable outcomes	Barange et al. (2018)

16.7.2 Developing User-Friendly Tools and Platforms

To facilitate effective community participation, accessible and intuitive digital platforms must be developed. Mobile apps such as "BloomWatch," "CyanoTracker," and other locally adapted systems allow users to submit bloom sightings, access real-time risk maps, and receive alerts or safety advisories. These platforms often include built-in educational content on bloom causes, health risks, and safe water practices (King, 2021). In addition, web-based dashboards that integrate citizen reports with scientific monitoring, satellite data, and forecast models provide an interactive and transparent means of engaging the public. Some programs also integrate gamification elements to encourage consistent participation and data sharing. Open data access is essential for ensuring transparency, empowering local decision-makers, and fostering collaborative problem-solving between citizens, researchers, and policymakers (Mahmoudi et al., 2024).

16.7.3 Building Capacity and Enhancing Risk Communication

Beyond data collection, community engagement must focus on strengthening local capacity for bloom response and risk mitigation. This includes educational campaigns targeting schools, community organizations, and farmers on the ecological impacts of blooms and the role of human activities in nutrient pollution. Workshops, training sessions, and on-site demonstrations can help build local skills in water sampling, use of monitoring kits, and safe response practices during bloom outbreaks (Stauffer et al., 2019). Risk communication is particularly important in high-risk or marginalized communities that rely directly on surface waters for drinking, irrigation, or recreation. Communication strategies must be culturally appropriate, linguistically accessible, and tailored to different audience segments—including children, elders, tourists, and health professionals. Community radio, social media, and local leaders play critical roles in disseminating alerts and preventive messages, especially during bloom emergencies (Panagiotou & Nikezis, 2024).

16.7.4 Creating Long-Term Stewardship Models

Ultimately, community engagement must evolve beyond short-term involvement toward long-term water stewardship. Co-management models, where communities participate in decision-making, planning, and enforcement alongside governmental agencies, foster ownership and accountability. Establishing lake or watershed councils, involving indigenous and local knowledge systems, and integrating traditional ecological practices can all contribute to sustainable governance structures. Funding mechanisms—such as microgrants, volunteer stipends, or support for local NGOs—can sustain participation and incentivize innovation. Legal recognition of citizen-generated data and the inclusion of community perspectives in policy formulation ensure that bloom management strategies reflect ground-level realities and priorities (Synnes & Vartdal, n.d.).

16.8 Global Cooperation and Policy Harmonization

Cyanobacterial blooms are not confined by national borders—they affect transboundary rivers, international lakes, and shared aquifers. In a globally interconnected climate system, atmospheric and hydrological processes, agricultural trade, and cross-border water usage all contribute to the spread and intensification of blooms. This reality necessitates a coordinated international response grounded in policy harmonization, cooperative research, data sharing, and joint management of freshwater systems. Global cooperation is essential not only for scientific advancement but also for addressing inequities in monitoring capacity, access to clean water, and institutional readiness across countries and regions (Evaristo et al., 2023).

16.8.1 Strengthening Transboundary Water Governance

Shared water bodies such as Lake Victoria, the Mekong River, and the Great Lakes of North America highlight the importance of collaborative governance. In many cases, upstream nutrient pollution drives downstream bloom outbreaks, leading to ecological damage, public health concerns, and diplomatic tensions. Effective transboundary management requires institutional frameworks that clearly define roles, responsibilities, enforcement mechanisms, and equitable benefit sharing (Morgera, 2016). River basin organizations (RBOs) such as the International Commission for the Protection of the Danube River (ICPDR) or the Mekong River Commission offer models of cooperative governance. These bodies facilitate integrated water resources management (IWRM), support joint monitoring programs, and provide forums for negotiation and conflict resolution. Future efforts should aim to embed cyanobacterial bloom mitigation explicitly into the mandates of such organizations and ensure that upstream–downstream dynamics are addressed in nutrient reduction strategies (Barruffa et al., 2021a).

16.8.2 Harmonizing Water Quality Standards and Reporting Protocols

Significant disparities exist in how countries define, monitor, and respond to harmful algal blooms. Some nations have detailed guidelines for microcystin levels in drinking and recreational water, while others lack any formal standards. This inconsistency hampers data comparison, risk communication, and coordinated response across borders (Kalogiannidis, 2024). Global standardization—led by organizations such as the World Health Organization (WHO), the United Nations Environment Programme (UNEP), and the International Organization for Standardization (ISO)—is urgently needed. Clear benchmarks for cyanotoxin thresholds, standardized monitoring methodologies, and shared classification systems for bloom severity will facilitate harmonized reporting and improve early warning systems. Additionally, developing global repositories of bloom data can support comparative research, inform regional risk assessments, and drive coordinated international action (Anderson et al., 2019).

16.8.3 Expanding International Research and Knowledge Exchange

International collaboration in research is vital to overcome knowledge gaps, pool technical expertise, and ensure that innovations are globally accessible. Networks such as the Global Lake Ecological Observatory Network (GLEON) and the Global Water Partnership (GWP) facilitate cross-border research, data harmonization, and training programs. These platforms should be expanded to include dedicated working groups on cyanobacterial bloom modeling, toxin monitoring, remote sensing applications, and citizen science integration. Exchange programs, fellowships, and joint research grants can enhance capacity building, especially for early-career scientists and institutions in the Global South, where bloom risks are high, but resources are limited. Equitable partnerships that respect local knowledge and promote co-development of tools and policies are essential to ensuring mutual benefit (Terrado et al., 2023).

16.8.4 Financing and Technical Support for Low-Income Countries

Addressing cyanobacterial blooms requires significant investment in monitoring infrastructure, laboratory capacity, treatment technologies, and institutional development. Low- and middle-income countries often face the greatest burden of bloom-related impacts but lack the financial and technical means to respond effectively.

Climate finance mechanisms—such as the Green Climate Fund (GCF), Global Environment Facility (GEF), or development bank-supported water programs—should prioritize HAB risk mitigation and provide targeted support for vulnerable regions. Blended finance approaches that leverage public and private sector contributions, alongside technical assistance, can accelerate technology transfer and enable scalable interventions. Supporting community-based monitoring, open-source tools, and decentralized management systems ensures that investments reach the grassroots level and build long-term resilience (Table 16.2) (Kangana et al., 2024).

16.9 Conclusion

Cyanobacterial blooms have evolved from a localized water quality issue into a pressing global environmental challenge. Their rising frequency, geographic spread, and intensity—exacerbated by climate change, nutrient loading, and hydrological alterations—underscore the urgency of adopting comprehensive, forward-looking strategies. This chapter has outlined the future directions needed to address the complex, multifactorial nature of cyanobacterial blooms through innovations in science, governance, technology, and community action. Closing critical knowledge gaps remains a top priority. Future research must advance our understanding of species-specific toxin production, bloom dynamics in underexplored ecosystems, and the interaction between climatic extremes and bloom proliferation. Investments in molecular diagnostics, high-frequency monitoring, and next-generation remote sensing are essential to support real-time detection and early intervention. Equally vital are predictive models that integrate ecological, hydrological, and socio-economic variables—enabling anticipatory planning and adaptive water resource management. Management strategies must shift from reactive control to long-term resilience building. Nature-based solutions, circular nutrient management, biotechnological innovations, and inclusive governance structures offer promising avenues for mitigating bloom risk sustainably. These approaches should be supported by policy reforms, institutional coordination, and regulatory instruments that enforce nutrient limits while incentivizing innovation and cooperation. Moreover, the role of communities cannot be overstated. Citizen science, public education, and local stewardship models not only enhance monitoring capacity but also strengthen public ownership of water protection. Fostering participatory governance ensures that bloom mitigation efforts are context specific, culturally appropriate, and socially equitable. Finally, international cooperation and policy harmonization will be indispensable as bloom risks extend across borders and disproportionately affect low-resource regions. Collaborative research, global data infrastructure, and equitable funding mechanisms must underpin a unified response to this planetary issue. In essence, managing

Table 16.2 Key priorities, strategies, and actors involved in global cooperation and policy harmonization for cyanobacterial bloom mitigation

No.	Focus area	Strategy/action	Key actors/organizations	Intended outcome	References
1	Transboundary water governance	Establish integrated basin-wide management plans	River Basin Organizations, national governments	Coordinated upstream–downstream nutrient control	Shrivastava et al. (2025)
2	Role clarity and enforcement	Define roles, responsibilities, enforcement mechanisms	ICPDR, MRC, local agencies	Effective governance and conflict resolution	Shrivastava et al. (2025)
3	Integrated water resources management	Mainstream bloom mitigation into IWRM plans	UN-Water, GWP, RBOs	Holistic water resource planning	UN-Water (2018)
4	Monitoring collaboration	Implement joint monitoring programs across borders	Neighboring states, regional consortia	Shared data and early warning	Cooley et al. (2019)
5	Bloom severity classifications	Adopt standard severity levels for bloom alerts	WHO, UNEP, ISO	Harmonized communication and response	WHO (2023a, 2023b, 2023c
6	Cyanotoxin thresholds	Agree on safe toxin levels in water	WHO, national health bodies	Improved public health protection	WHO (2011)
7	Monitoring methods	Standardize sampling and analysis methods	ISO, UNEP, water agencies	Comparable and reliable data	UNEP (2021)
8	Global bloom databases	Create open-access HAB event repositories	UNEP, GLEON	Global risk analysis, hotspot tracking	GLEON (2020)
9	Cross-border risk communication	Develop protocols for joint health advisories	WHO regional offices, health ministries	Timely alerts for at-risk populations	WHO WHO (2023a, 2023b, 2023c)
10	Capacity disparities	Identify low-capacity areas for targeted support	UNEP, World Bank	Equitable management support	World Bank (2020)
11	Technical training	Training in remote sensing and molecular tools	UNESCO-IHP, regional institutes	Strengthened monitoring and diagnostic capacity	UNESCO (2019)
12	Knowledge sharing platforms	Expand networks like GLEON to focus on HABs	GLEON, GWP, universities	Improved interdisciplinary collaboration	GLEON (2020)

(continued)

Table 16.2 (continued)

No.	Focus area	Strategy/action	Key actors/organizations	Intended outcome	References
13	Cross-disciplinary fellowships	Support Global South scientist exchange programs	TWAS, GEF, national science councils	Research equity and capacity building	TWAS (2022)
14	Co-development of tools	Joint development of bloom detection software, kits	NGOs, universities, regional labs	Locally adapted technology	Barruffa et al. (2021b)
15	Funding for infrastructure	Prioritize bloom-related projects in climate finance	GCF, GEF, dev banks	Improved monitoring and treatment	GEF (2021)
16	Blended finance models	Use mixed public–private funding for HAB interventions	IFIs, private sector, climate funds	Scalable and sustained impact	OECD (2023)
17	Technology transfer	Support deployment of sensors and remote systems	International donors, NGOs	Equitable access to innovation	UNEP (2021)
18	Open-source tools	Promote open-source HAB monitoring solutions	Research institutes, developers	Democratized monitoring access	Tirado and Raj (2023)
19	Support for citizen science	Scale up community-based water quality monitoring	NGOs, UNESCO, ministries	Community empowerment and local data	UNESCO (2020)
20	Inclusion in adaptation finance	Integrate HABs into climate finance and policy instruments	UNFCCC, national authorities	Climate-aligned HAB responses	UNFCCC (2021)
21	Shared response protocols	Develop harmonized emergency HAB protocols	WHO, health alliances	Coordinated health protection	WHO (2023a, 2023b, 2023c)
22	Legal harmonization	Align environmental laws across regions	EU, ASEAN, regional blocs	Consistent regulation	EU Commission (2020)
23	Regional policy dialogues	Create platforms for regional policy exchange	UNEP regional offices, river basin orgs	Innovation and mutual learning	UNEP (2022)
24	Language/accessibility barriers	Translate HAB protocols into diverse languages	UN bodies, NGOs	Inclusive participation	FAO (2020)

(continued)

Table 16.2 (continued)

No.	Focus area	Strategy/action	Key actors/organizations	Intended outcome	References
25	Gender-sensitive approaches	Promote women's leadership in bloom management	UN Women, ministries	Gender-equitable water governance	UN Women (2021)
26	Indigenous knowledge integration	Integrate traditional ecological knowledge	Indigenous councils, heritage bodies	Context-aware and culturally respectful adaptation	FAO (2021)
27	Monitoring trade-related nutrient flows	Analyze fertilizer and food trade nutrient impacts	WTO, FAO, UNEP	Global accountability in nutrient flows	WTO (2022)
28	Emergency funding for outbreaks	Establish contingency funds for bloom crises	UN OCHA, GEF-SGP	Swift intervention capability	UN OCHA (2023)
29	Evaluation frameworks	Track global progress on HAB risk and policy harmonization	UNEP, OECD, legal experts	Transparent global monitoring	OECD (2023)
30	Integration into SDG tracking	Include bloom metrics in SDG 6 and SDG 13 systems	UN Stats Division, national SDG units	Global policy visibility	UN Stats Division (2022)

cyanobacterial blooms in the future will require nothing less than a paradigm shift—one that embraces interdisciplinary knowledge, cross-sector collaboration, climate-smart planning, and empowered communities. With coordinated global action and sustained commitment, it is possible to restore the health of freshwater ecosystems and protect the essential services they provide for all life.

References

AFMESI. (2022). *Integration of ocean-based adaptation and mitigation actions into regional and national climate policies in Africa.* https://afmesi.org/wp-content/uploads/2023/05/Ocean-Evidence-Base.pdf

Ali, F., Razzaq, A., Tariq, W., Hameed, A., Rehman, A., Razzaq, K., Sarfraz, S., Rajput, N. A., Zaki, H. E., & Shahid, M. S. (2024). Spectral intelligence: AI-driven hyperspectral imaging for agricultural and ecosystem applications. *Agronomy, 14*(10), 2260.

Anderson, C. R., Berdalet, E., Kudela, R. M., Cusack, C. K., Silke, J., O'Rourke, E., Dugan, D., McCammon, M., Newton, J. A., & Moore, S. K. (2019). Scaling up from regional case studies to a global harmful algal bloom observing system. *Frontiers in Marine Science, 6*, 250.

Asadabadi, A., & Miller-Hooks, E. (2017). Optimal transportation and shoreline infrastructure investment planning under a stochastic climate future. *Transportation Research Part b: Methodological, 100*, 156–174.

Ballesteros Possu, W. (2015). *Carbon storage potential of windbreaks on agricultural lands of the continental United States.*

Barange, M., Bahri, T., Beveridge, M. C. M., Cochrane, K. L., Funge-Smith, S., & Poulain, F. (2018). *Impacts of climate change on fisheries and aquaculture: Synthesis of current knowledge, adaptation and mitigation options.* FAO Technical Paper No. 627. https://www.researchgate.net/publication/325871167

Barruffa, A. S., Sposito, V., & Faggian, R. (2021a). Climate change and cyanobacteria harmful algae blooms: Adaptation practices for developing countries. *Marine and Freshwater Research, 72*(12), 1722–1734.

Barruffa, A. S., Sposito, V., & Faggian, R. (2021b). Climate change and cyanobacteria harmful algae blooms: Adaptation practices for developing countries. *Marine and Freshwater Research, 72*(9), 1195–1209. https://doi.org/10.1071/MF21081

Behrenfeld, M. J., Doney, S. C., Lima, I., Boss, E. S., & Siegel, D. A. (2013). Annual cycles of ecological disturbance and recovery underlying the subarctic Atlantic spring plankton bloom. *Global Biogeochemical Cycles, 27*(2), 526–540.

Bughin, J., Hazan, E., Sree Ramaswamy, P., DC, W., & Chu, M. (2017). *Artificial intelligence the next digital frontier.*

Caldararu, S., Rolo, V., Stocker, B. D., Gimeno, T. E., & Nair, R. (2023). Ideas and perspectives: Beyond model evaluation–combining experiments and models to advance terrestrial ecosystem science. *Biogeosciences Discussions, 2023*, 1–22.

Campbell, B. M., Hansen, J., Rioux, J., Stirling, C. M., & Twomlow, S. (2018). Urgent action to combat climate change and its impacts (SDG 13): Transforming agriculture and food systems. *Current Opinion in Environmental Sustainability, 34*, 13–20. https://doi.org/10.1016/j.cosust.2018.06.005

Carson, S., Sægrov, S., Sempere, R., & Kujala, P. (2024). *Closing the blue loops.*

Chuvieco, E. (2020). *Fundamentals of satellite remote sensing: An environmental approach.* CRC Press.

Cooley, S. R., Bello, B., Bodansky, D., & Mansell, A. (2019). Overlooked ocean strategies to address climate change. *Global Environmental Change, 59*, Article 101984. https://doi.org/10.1016/j.gloenvcha.2019.101984

D'Agostino, P. M., Woodhouse, J. N., Makower, A. K., Yeung, A. C., Ongley, S. E., Micallef, M. L., Moffitt, M. C., & Neilan, B. A. (2016). Advances in genomics, transcriptomics and proteomics of toxin-producing cyanobacteria. *Environmental Microbiology Reports, 8*(1), 3–13.

Dabbadie, L., Aguilar-Manjarrez, J., & Soto, D. (2019). Effects of climate change on aquaculture: Drivers, impacts and policies. In *Impacts of climate change on fisheries and aquaculture: Synthesis of current knowledge* (FAO). https://books.google.com/books?id=1jl8DwAAQBAJ

Danielsen, F., Burgess, N. D., & Balmford, A. (2005). Monitoring matters: Examining the potential of locally-based approaches. *Biodiversity & Conservation, 14*, 2507–2542.

Davoudi, S., Shaw, K., Haider, L. J., Quinlan, A. E., Peterson, G. D., Wilkinson, C., Fünfgeld, H., McEvoy, D., Porter, L., & Davoudi, S. (2012). Resilience: A bridging concept or a dead end? "Reframing" resilience: Challenges for planning theory and practice interacting traps: Resilience assessment of a pasture management system in Northern Afghanistan urban resilience: What does it mean in planning practice? Resilience as a useful concept for climate change adaptation? The politics of resilience for planning: A cautionary note: Edited by Simin Davoudi and Libby Porter. *Planning Theory & Practice, 13*(2), 299–333.

Delgado, J., Bermúdez, E., Asensio, M. A., & Núñez, F. (2021). Molecular identification and subtyping of toxigenic and pathogenic penicillium and talaromyces. In *Molecular food microbiology* (pp. 289–307). CRC Press.

Döll, P., Douville, H., Güntner, A., Müller Schmied, H., & Wada, Y. (2016). Modelling freshwater resources at the global scale: Challenges and prospects. *Surveys in Geophysics, 37*, 195–221.

Dovie, D. B. K. (2019). Case for equity between Paris Climate agreement's Co-benefits and adaptation. *Science of the Total Environment, 656*, 732–739.

Engler, C. (2024). Fit for purpose? Evaluating climate change adaptation laws and policies for marine aquaculture in Chile. *Frontiers in Marine Science, 11*, 1386545. https://doi.org/10.3389/fmars.2024.1386545

EU Commission. (2020). *Environmental compliance assurance and EU water law*. https://environment.ec.europa.eu

Evaristo, J., Jameel, Y., Tortajada, C., Wang, R. Y., Horne, J., Neukrug, H., David, C. P., Fasnacht, A. M., Ziegler, A. D., & Biswas, A. (2023). Water woes: The institutional challenges in achieving SDG 6. *Sustainable Earth Reviews, 6*(1), 13.

FAO. (2020). *Gender and climate change in agriculture and water*. https://www.fao.org

FAO. (2021). *Indigenous peoples and climate change*. https://www.fao.org/indigenous-peoples

GEF. (2021). *Strengthening global capacity to control harmful algal blooms*. https://www.thegef.org

GLEON. (2020). *Global lake ecological observatory network HABs working group*. https://gleon.org

Glibert, P. M., Pitcher, G. C., Bernard, S., & Li, M. (2018). Advancements and continuing challenges of emerging technologies and tools for detecting harmful algal blooms, their antecedent conditions and toxins, and applications in predictive models. In *Global ecology and oceanography of harmful algal blooms* (pp. 339–357).

Grasso, C. R., Pokrzywinski, K. L., Waechter, C., Rycroft, T., Zhang, Y., Aligata, A., Kramer, M., & Lamsal, A. (2022). A review of cyanophage–host relationships: Highlighting cyanophages as a potential cyanobacteria control strategy. *Toxins, 14*(6), 385.

Harding, W. R. (2015). *A feasibility evaluation of the total maximum daily (pollutant) load (TMDL) approach for managing eutrophication in South African Dams: Report to the water research commission*. Water Research Commission Pretoria, South Africa.

Huang, X., Wang, S., Yang, D., Hu, T., Chen, M., Zhang, M., Zhang, G., Biljecki, F., Lu, T., & Zou, L. (2024). Crowdsourcing geospatial data for earth and human observations: A review. *Journal of Remote Sensing, 4*, 0105.

Javaid, M., Haleem, A., Singh, R. P., & Suman, R. (2022). Artificial intelligence applications for industry 4.0: A literature-based study. *Journal of Industrial Integration and Management, 7*(01), 83–111.

Kalogiannidis, S. (2024). Analyzing the challenges of cross-border disaster response and management: A European perspective. *Journal of Risk Analysis and Crisis Response, 14*(2).

Kangana, N., Kankanamge, N., De Silva, C., Goonetilleke, A., Mahamood, R., & Ranasinghe, D. (2024). Bridging community engagement and technological innovation for creating smart and resilient cities: A systematic literature review. *Smart Cities, 7*(6), 3823–3852.

Kim, K.-G. (2024). Co-benefits of combining adaptation and mitigation frameworks: A cross-cutting approach. In *Cross-cultural perspectives on climate change adaptation: Adapting to flood risk* (pp. 389–400). Springer.

King, J. H. (2021). *Health communication blindspot: A case study of harmful algal blooms in the South (HABITS).*

Kulkarni, M. B., Ayachit, N. H., & Aminabhavi, T. M. (2022). Recent advancements in nanobiosensors: Current trends, challenges, applications, and future scope. *Biosensors, 12*(10), 892.

Kumar, D., Prasad, M., & Mohan, H. (2022). Biological recognition elements. In *Electrochemical sensors* (pp. 213–239). Elsevier.

LaDeau, S., Han, B., Rosi-Marshall, E., & Weathers, K. (2017). The next decade of big data in ecosystem science. *Ecosystems, 20*, 274–283.

Lescrauwaet, L., Wagner, H., Yoon, C., & Shukla, S. (2022). Adaptive legal frameworks and economic dynamics in emerging tech-nologies: Navigating the intersection for responsible innovation. *Law and Economics, 16*(3), 202–220.

Mahmoudi, M., Eskandari, S., & Esmaili Givi, M. R. (2024). Challenges and complexities in leveraging data for evidence-based policy making a scoping review. طلاعات ا تبریمدو پردازش ژوهشنامه پ, *39*(انگلیسی نامه ویژه 2), 27–52.

Mander, Ü., Tournebize, J., Sauvage, S., & Sánchez-Perez, J. M. (2017). Wetlands and buffer zones in watershed management. *Ecological Engineering, 103*, 289–295.

March, A., Woolley, M., & Failler, P. (2024). Integration of climate change mitigation and adaptation in Blue Economy planning in Africa. *Mitigation and Adaptation Strategies for Global Change.* https://doi.org/10.1007/s11027-024-10133-5

Mishra, S., Anderson, K., Miller, B., Boyer, K., & Warren, A. (2020). Microgrid resilience: A holistic approach for assessing threats, identifying vulnerabilities, and designing corresponding mitigation strategies. *Applied Energy, 264*, Article 114726.

Morgera, E. (2016). The need for an international legal concept of fair and equitable benefit sharing. *European Journal of International Law, 27*(2), 353–383.

Nawaz, T., Fahad, S., Saud, S., Hassan, S., & Gu, L. (2024). Cyanobacterial solutions for climate-resilient agriculture and global food security. In *Environment, climate, plant and vegetation growth* (pp. 1–39). Springer.

OECD. (2023). *Blended finance for water-related ecosystems.* https://www.oecd.org/water

Paerl, H. W., Havens, K. E., Hall, N. S., Otten, T. G., Zhu, M., Xu, H., Zhu, G., & Qin, B. (2019). Mitigating a global expansion of toxic cyanobacterial blooms: Confounding effects and challenges posed by climate change. *Marine and Freshwater Research, 71*(5), 579–592.

Pan, B., Lam, S. K., Mosier, A., Luo, Y., & Chen, D. (2016). Ammonia volatilization from synthetic fertilizers and its mitigation strategies: A global synthesis. *Agriculture, Ecosystems & Environment, 232*, 283–289.

Panagiotou, N., & Nikezis, I. (2024). *D 7.3: Assessment of the role of community preparedness and engagement in risk and crisis communication.*

Pham, T. T. T., Friðriksdóttir, R., Weber, C. T., & Viðarsson, J. R. (2021). Guidelines for co-creating climate adaptation plans for fisheries and aquaculture. *Climatic Change, 166*, 47. https://doi.org/10.1007/s10584-021-03041-z

Poulain, F., Himes-Cornell, A., & Shelton, C. (2018). *Methods and tools for climate change adaptation in fisheries and aquaculture.* FAO Fisheries and Aquaculture Technical Paper No. 627. https://www.academia.edu/download/57265580/I9705EN.pdf#page=559

Priscoli, J. D. (2009). River basin organizations. In *Managing and transforming water conflicts* (pp. 135–165).

Ren, Y., Lü, Y., Comber, A., Fu, B., Harris, P., & Wu, L. (2019). Spatially explicit simulation of land use/land cover changes: Current coverage and future prospects. *Earth-Science Reviews, 190*, 398–415.

Robotham, J., Old, G., Rameshwaran, P., Sear, D., Trill, E., Bishop, J., Gasca-Tucker, D., Old, J., & McKnight, D. (2023). Nature-based solutions enhance sediment and nutrient storage in an agricultural lowland catchment. *Earth Surface Processes and Landforms, 48*(2), 243–258.

Sánchez-Hernández, M. I. (2024). Strengthening resilience: Social responsibility and citizen participation in local governance. *Administrative Sciences, 14*(10), 260.

Satam, H., Joshi, K., Mangrolia, U., Waghoo, S., Zaidi, G., Rawool, S., Thakare, R. P., Banday, S., Mishra, A. K., & Das, G. (2023). Next-generation sequencing technology: Current trends and advancements. *Biology, 12*(7), 997.

Sheik, A. G., Kumar, A., Patnaik, R., Kumari, S., & Bux, F. (2024). Machine learning-based design and monitoring of algae blooms: Recent trends and future perspectives–A short review. *Critical Reviews in Environmental Science and Technology, 54*(7), 509–532.

Shrivastava, S., Khan, T. K. H., & Sunanda, M. (2025). Climate change and its impact on water quality in major river basins worldwide. *International Journal of Environmental Studies*. http://theaspd.com/index.php/ijes/article/view/1365

Srikrishnan, V., Lafferty, D. C., Wong, T. E., Lamontagne, J. R., Quinn, J. D., Sharma, S., Molla, N. J., Herman, J. D., Sriver, R. L., & Morris, J. F. (2022). Uncertainty analysis in multi-sector systems: Considerations for risk analysis, projection, and planning for complex systems. *Earth's Future, 10*(8), e2021EF002644.

Stauffer, B. A., Bowers, H. A., Buckley, E., Davis, T. W., Johengen, T. H., Kudela, R., McManus, M. A., Purcell, H., Smith, G. J., & Vander Woude, A. (2019). Considerations in harmful algal bloom research and monitoring: Perspectives from a consensus-building workshop and technology testing. *Frontiers in Marine Science, 6*, 399.

Stephens, T. (2022). Climate change adaptation in marine and coastal areas and international law. In *Research handbook on climate change adaptation law* (pp. 225–244). Edward Elgar. https://www.elgaronline.com/view/book/9781800371491/book-part-9781800371491-15.xml

Sullivan, J., Croisant, S., Howarth, M., Subra, W., Orr, M., & Elferink, C. (2019). Implications of the GC-HARMS Fishermen's citizen science network: Issues raised, lessons learned, and next steps for the network and citizen science. *NEW SOLUTIONS: A Journal of Environmental and Occupational Health Policy, 28*(4), 570–598.

Synnes, A., & Vartdal, A. (n.d.). *Exploring participation by citizens, public organisations and businesses in an open data initiative: A Scandinavian case study of ocean data.*

Tahmasebi, M. (2024). Beyond defense: Proactive approaches to disaster recovery and threat intelligence in modern enterprises. *Journal of Information Security, 15*(2), 106–133.

Tammeorg, O., Chorus, I., Spears, B., Nõges, P., Nürnberg, G. K., Tammeorg, P., Søndergaard, M., Jeppesen, E., Paerl, H., & Huser, B. (2024). Sustainable lake restoration: From challenges to solutions. *Wiley Interdisciplinary Reviews: Water, 11*(2), Article e1689.

Terrado, M., Bojovic, D., Octenjak, S., Christel, I., & Clair, A. L. S. (2023). Good practice for knowledge co-development through climate related case studies. *Climate Risk Management, 40*, Article 100513.

Thomas, A., Baptiste, A., Martyr-Koller, R. C., Pringle, P., & Rhiney, K. (2020). Climate change and small island developing states. *Annual Review of Environment and Resources, 45*, 1–27. https://doi.org/10.1146/annurev-environ-012320-083355

Thompson, O. P., Breitenstein, R. S., Barnes, B. R., & Glory, R. (2024). *Promoting behavioral change and public awareness for sustainable clean air*. Springer.

Tirado, M. C., & Raj, S. A. (2023). Climate change and emerging risks to food safety: Building climate resilience. *Food Safety Magazine*. https://www.food-safety.com/articles/9099-climate-change-and-emerging-risks-to-food-safety-building-climate-resilience

Tomanek, L. (2011). Environmental proteomics: Changes in the proteome of marine organisms in response to environmental stress, pollutants, infection, symbiosis, and development. *Annual Review of Marine Science, 3*(1), 373–399.

TWAS. (2022). *South–South fellowships for environmental science*. https://twas.org

UN OCHA. (2023). *Rapid response humanitarian fund guidelines*. https://www.unocha.org

UN Stats Division. (2022). *SDG indicators metadata repository*. https://unstats.un.org

UN Women. (2021). *Making climate action inclusive: Gender and water*. https://www.unwomen.org

UNEP. (2021). *Assessment of freshwater HAB risks in developing countries*. https://www.unep.org

UNEP. (2022). *Regional dialogue on water governance*. https://www.unep.org

UNESCO. (2019). *IHP training manual on freshwater monitoring*. https://www.unesco.org

UNESCO. (2020). *Citizen science for water monitoring: Guide for practitioners*. https://www.unesco.org

UNFCCC. (2021). *NAP technical guidelines*. https://unfccc.int

United Nations ESCAP. (2020). *Water and climate change*. UN Regional Report. https://repository.unescap.org/handle/20.500.12870/7402

UN-Water. (2018). *Policy brief on integrated water resources management*. https://www.unwater.org

UN-Water. (2019). *Climate change and water: UN-Water policy brief*. UN-Water. https://www.unwater.org/sites/default/files/app/uploads/2019/07/UN-Water_Water-and-Climate-Policy-Brief_unedited.pdf

Wang, Y., Xu, C., Lin, Q., Xiao, W., Huang, B., Lu, W., Chen, N., & Chen, J. (2024). Modeling of algal blooms: Advances, applications and prospects. *Ocean & Coastal Management, 255*, Article 107250.

West, J. J., Järnberg, L., Berdalet, E., & Cusack, C. (2021). Understanding and managing harmful algal bloom risks in a changing climate: Lessons from the European CoCliME project. *Frontiers in Climate, 3*, Article 636723. https://doi.org/10.3389/fclim.2021.636723

World Bank. (2020). *Water security diagnostic for transboundary river basins*. https://www.worldbank.org

World Health Organization. (2011). *Guidelines for drinking-water quality* (4th ed.). https://www.who.int

World Health Organization. (2023a). *Climate change and health risk Communication*. https://www.who.int

World Health Organization. (2023b). *Addressing climate change: Supplement to the WHO water, sanitation and hygiene strategy 2018–2025*. WHO. https://books.google.com/books?hl=en&id=rRDAEAAAQBAJ

World Health Organization. (2023c). *Quantifying and valuing health outcomes of climate adaptation interventions*. WHO. https://books.google.com/books?id=vnsOEQAAQBAJ

WTO. (2022). *Trade and the environment: Fertilizer and nutrient flow*. https://www.wto.org

Chapter 17
Role of Cyanobacteria in Bioremediation and Environmental Restoration

Muhammad Anas, Waseem Ahmed Khattak, Mehmet Ali Avci, Erdogan E. Hakki, and Shah Fahad

Abstract Cyanobacteria, ancient and photosynthetically active microorganisms, are gaining attention as powerful agents in environmental bioremediation due to their diverse metabolic pathways and ecological adaptability. This chapter provides a comprehensive overview of the mechanisms, applications, and future prospects of cyanobacteria in remediating polluted environments. It explores how cyanobacteria contribute to nutrient removal, heavy metal sequestration, organic pollutant degradation, and carbon capture all essential processes in restoring the health and functionality of aquatic and terrestrial ecosystems. The chapter also highlights their role in environmental restoration efforts, including wastewater treatment, soil rehabilitation, recovery of eutrophic water bodies, and oil spill remediation. Innovative approaches such as genetic engineering, immobilization techniques, and integrated bioremediation systems are examined to showcase how technological advancements are improving the efficiency and scalability of cyanobacterial applications. Furthermore, future research directions emphasize the need for climate-resilient strains, large-scale deployment strategies, and supportive policy frameworks to overcome existing challenges like toxin production and environmental variability. The chapter concludes that

M. Anas (✉) · E. E. Hakki
Department of Soil Science and Plant Nutrition, Selçuk University, Konya, Türkiye
e-mail: muhammed.anas@selcuk.edu.tr

E. E. Hakki
e-mail: eehakki@selcuk.edu.tr

W. A. Khattak
Department of Plant Sciences, Faculty of Biological Sciences, Quaid-I-Azam University, Islamabad, Pakistan
e-mail: waseemkhattack@bs.qau.edu.pk

M. A. Avci
Department of Field Crops, Faculty of Agriculture, Selcuk University, Konya, Türkiye
e-mail: mavci@selcuk.edu.tr

S. Fahad (✉)
Department of Agronomy, Abdul Wali Khan University Mardan, Mardan, Khyber Pakhtunkhwa, Pakistan
e-mail: shah_fahad80@yahoo.com

S. Fahad et al. (eds.), *Cyanobacterial Blooms: Ecology, Evolution and Biogeochemical Impacts*, https://doi.org/10.1007/978-3-032-06042-6_17

cyanobacteria offer a sustainable, low-cost, and multifunctional solution for environmental cleanup and resource recovery. When aligned with eco-engineering principles and circular economy goals, their potential can be fully harnessed to support global environmental sustainability and climate resilience.

Keywords Cyanobacteria · Bioremediation · Environmental restoration · Genetic engineering · Wastewater treatment

17.1 Introduction

17.1.1 Overview of Cyanobacteria and Their Ecological Roles

Cyanobacteria, sometimes called blue-green algae, are one of the oldest and most important groups of microorganisms on Earth (Jassim et al., 2023). These photosynthetic prokaryotes have been present for more than 3.5 billion years, taking on a key role in shaping the biosphere of Earth. As one of the first oxygenic photosynthesizers, cyanobacteria were directly responsible for the Great Oxygenation Event, changing the composition of the atmosphere and facilitating the development of aerobic life. Cyanobacteria can be found in many environmental settings including freshwater lakes, rivers, oceans, deserts, and extreme habitats such as hot springs and polar regions (Cirés et al., 2017). They have tremendous metabolic flexibility and ecological adaptability. In aquatic systems, significant contributions to spatial and temporal primary production are made by cyanobacteria, which often comprise the largest portion of phytoplankton communities, particularly when resources are not limited. Cyanobacteria are the foundational group of food webs, and they are active participants in global biogeochemical cycles of carbon, nitrogen, phosphorus, and other essential nutrients. Many cyanobacterial taxa dominate in nitrogen-limited environments since they can fix atmospheric nitrogen in bioavailable form utilizing specialized cells called heterocysts (Mishra et al., 2021). Some cyanobacteria can also produce secondary metabolites, some of which are beneficial and some of which can potentially be harmful. These include bioactive compounds with pharmaceutical potential as well as cyanotoxins, potentially toxic to humans, other animals, and ecosystems in high concentrations, as occurs during harmful algal blooms (HABs). Cyanobacteria do more than produce these ecological products. They also form symbiotic relationships with fungi (forming lichens), plants (e.g., Azolla-Anabaena), and animals. Cyanobacteria embrace broad biological interactions. The structural variation among cyanobacteria, which includes unicellular forms, colonial forms, and filamentous forms, allows them to perform a range of ecological functions when interacting with others (biofilm formation, soil stabilization, nutrient retention, etc.). Cyanobacteria are acknowledged as dynamic beings with a range of potential biochemistries and ecological roles. They are being recognized not only as part of ecological problems, such closely associated with environmental issues like eutrophication, but also as delivery agents for environmental remediation (Akinnawo, 2023).

17.1.2 Bioremediation: Definition and Importance

Bioremediation is defined as a process that utilizes living organisms (usually microorganisms, fungi, and/or plants, or their enzymes) to detoxify, degrade, or remove hazardous materials, from the environment. Bioremediation is a cost-effective, sustainable, and environmentally friendly alternative to traditional physical and chemical remediation methods that can be costly, invasive, and potentially hazardous to ecological systems. Bioremediation can be utilized in a broad range of matrices, including soil, water, sediments, and the atmosphere; it also can address a variety of contaminants, including organic contaminants (e.g., petroleum hydrocarbons and pesticides) and inorganic contaminants (e.g., heavy metals and radionuclides) (Akinnawo, 2023). Bioremediation is based on the ability of organisms to degrade contaminants via the organism's natural ability to transform contaminants into less harmful or even nontoxic end products through biological processes. Biological processes can include biosorption, bioaccumulation, enzymatic degradation, redox processes, and metabolic transformations. Depending on the approach, bioremediation can be classified based on the strategy of either in situ (onsite) and ex situ (off-site) or intrinsic (natural attenuation) and engineered (stimulated) bioremediation (Butnariu & Butu, 2020).

Bioremediation has gained in importance over the last few centuries due to increased industrialization, intensified agricultural activities, expanding populations and urbanization, and the impacts of climate change, all resulting in growing environmental contamination. In particular, heavy metals (cadmium [Cd], lead [Pb], arsenic [As], and mercury [Hg]), excess nutrients (nitrogen and phosphorus), pesticides, oil spills, and pharmaceuticals are increasingly contaminating our ecosystems posing a risk to human and ecological health (Anas et al., 2024a). Remediation approaches employed with these contaminants are often untethered for support, typically struggling to remediate, especially for large-scale or complex contaminant mixtures. Inclusively, bioremediation has the potential to develop approaches that are site-specific, adaptive, and minimally invasive, as they can operate effectively under ambient environmental conditions (Grifoni et al., 2022). Cyanobacteria have emerged as powerful bioremediation agents based on unique physiological traits. They are naturally resistant to many pollutants, can grow rapidly in a vast array of conditions, and can produce extracellular polymeric substances (EPSs) which facilitate heavy metal sequestration and flocculation. As photosynthetic organisms, cyanobacteria can also grow without external carbon sources, essentially reducing input costs in a bioremediation scenario. In addition, because cyanobacteria are autotrophic, they can be used in terrestrial and aquatic remediation systems, creating significant opportunities for the ecological restoration of polluted systems (Kumar et al., 2020).

17.1.3 Environmental Challenges Addressed by Cyanobacteria

In the recent times, we are undergoing more anthropogenic destruction on the environment than ever before. Man-made compounds from industrial processes or chemicals introduced into the environment by agricultural practices or urban development had created a unique mixture of substances in ecosystems that undermine their stability and endanger a number of organisms within it, while exposing humans to harmful effects on health. Finding solutions to these issues is difficult, but cyanobacteria could serve as a natural and nontoxic alternative that is capable of removing many pollutants through bioremediation (Touliabah et al., 2022). Heavy metal pollution is one of the many problems that cyanobacteria can help resolve. Industrial activities, mining, and disposal of waste have polluted the soil and water with heavy metals like cadmium, arsenic, lead, mercury, and chromium (Anas et al., 2024b). Heavy metals cannot be broken down like organic chemicals; therefore, they can be persistent contaminants and bioaccumulate in the food web where they can subsequently cause chronic health problems (Fig. 17.1). Cyanobacteria can sequester heavy metals through processes such as biosorption and bioaccumulation where they can bind them to the outer cell wall or compartmentalize them in vacuoles. Cyanobacterial produced exopolymeric substances increase the capacity to bind metals and some strains of cyanobacteria evolved enzymatic systems which are capable of transforming metals to less toxic forms (Yadav et al., 2020).

Another important problem we can address is nutrient pollution. Excess nitrogen and phosphorus entering waterways occur from agricultural runoff, sewage

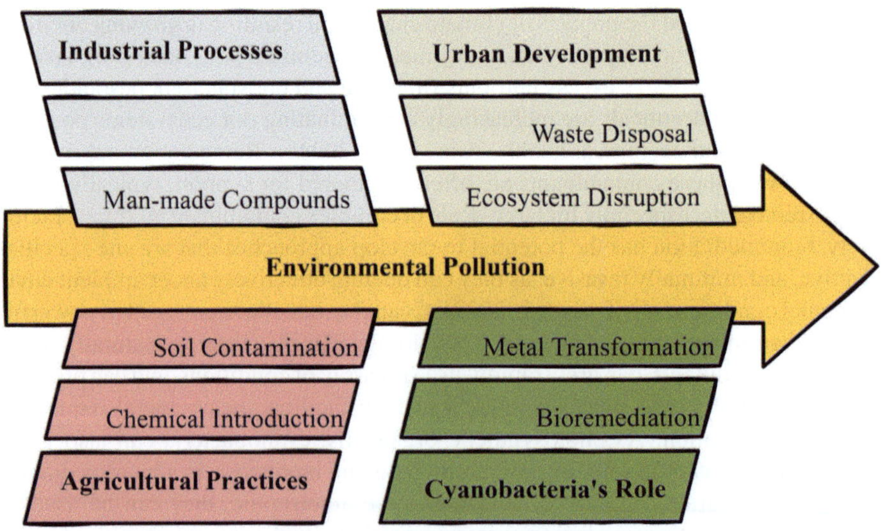

Fig. 17.1 Cyanobacteria as a solution to environmental pollution

discharge, and industrial effluents, resulting in eutrophication which can lead to algal blooms that cause a depletion of oxygen and aquatic dead zones (Wurtsbaugh et al., 2019). Interestingly, even though cyanobacteria tend to be bloom formers in these conditions, there are non-bloom former or engineered strains of cyanobacteria that can remediate excess nutrients through nutrient uptake and assimilation by nitrogen fixation. Constructed wetlands or algae ponds with cyanobacteria can also act as biological filters to intercept nutrient flow prior to sensitive water bodies. Degradation of pesticides is also on the rise, especially in the agricultural landscape. Many cyanobacteria have enzymatic pathways (e.g. hydrolases and oxygenases) that can degrade both organophosphate and organochlorine compounds. Their ability to biotransform toxic agrochemicals into less toxic intermediates can help facilitate bioremediation of contaminated farmlands and irrigation canals. Cyanobacteria can also jointly reduce organic pollutant loads, such as petroleum hydrocarbons, polycyclic aromatic hydrocarbons (PAHs), dyes, and pharmaceuticals. Some species can co-metabolize these organic pollutants in the presence of a primary carbon source, while other cyanobacteria can mineralize these pollutants under light-driven conditions. Furthermore, cyanobacteria could form biofilms that increase degradation efficiencies by pooling their biomass at pollutant interfaces (Wang et al., 2022).

In addition, cyanobacteria can help with radioactive waste treatment. Some strains of cyanobacteria can biosorb radionuclides (i.e., uranium and cesium) from contaminated waters; thus, they can decrease radionuclide mobility and accessibility to the living realm. Cyanobacteria could be useful for the treatment of nuclear wastewater (and potentially the rehabilitation of sites after nuclear accidents) (Thakur & Kumar, 2024). Cyanobacteria also positively contribute to carbon sequestration and promote soil productivity; they perform this by fixing atmospheric CO_2 and increasing organic matters associated with biomass accumulation, ultimately increasing environmental quality and assisting with solving climate change challenges. Their atmospheric nitrogen-fixing ability can accelerate soils' nutrient status and improve flexibility to sustainable agriculture, particularly as soil health improvement using cyanobacteria is particularly useful for the treatment of degraded soils and nutrient-deficient soils. Most importantly in the context of bioremediation, cyanobacteria are consistent with the principles of the circular bioeconomy. Biomass generated by the bioremediation system can potentially mean value-added opportunities, notably producing biofertilizers, biofuels, feed additives, and bioplastics, adding to profitability and avoiding waste. When integrative environmental cleanup with resource recovery is deployed, it provides economically viable and ecologically sustainable methods for managing polluted environments (Yadav et al., 2021a).

17.2 Mechanisms of Bioremediation by Cyanobacteria

Cyanobacteria have become effective and efficient bioremediation organisms due to their metabolic and physiological diversity, as well as ecological flexibility. Unlike many heterotrophic microbes, cyanobacteria are photoautotrophs that require only sunlight, carbon dioxide, and minimal nutrients for growth, making them advantageous and sustainable for large-scale bioremediation applications. This section describes the most important ways in which cyanobacteria assist in the cleanup of contaminated environments, such as nutrient removal, heavy metal sequestration, degradation of organic pollutants, and the potential benefits of bioremediation related to carbon sequestration and reducing climate change (Kumar et al., 2020).

17.2.1 Nutrient Removal (Nitrogen and Phosphorus Uptake)

High loading of nutrients—mainly nitrogen (N) and phosphorus (P)—from agricultural runoff, municipal effluent, and industrial effluents contributes to eutrophic conditions in aquatic systems. Nutrient loading leads to harmful algal blooms, hypoxic conditions (or low dissolved oxygen), and degradation of aquatic biodiversity. Cyanobacteria have a number of ways to assimilate and attenuate these nutrients from aquatic systems, which makes them critical elements for nutrient attenuation (Markou et al., 2014). Cyanobacteria can assimilate several sources of nitrogen such as nitrate (NO_3^-), nitrite (NO_2^-), ammonium (NH_4^+), and atmospheric nitrogen (N_2) using biological nitrogen fixation. When heterocyst-forming cyanobacteria like *Anabaena* and *Nostoc* enter the diazotrophic state, the nitrogenase enzyme converts nitrogen gas (N_2) to ammonia that can then readily be incorporated into amino acids and other cellular components. Because of their ability to fix nitrogen, cyanobacteria are particularly important and beneficial in nitrogen-deficient but contaminated systems (Saud et al., 2024). Equally important is phosphorus removal. Cyanobacteria take in phosphate (PO_4^{3-}) from their environment in order to make ATP, nucleic acids, and phospholipids (Bossa et al., 2024). Some cyanobacterial species exhibit luxury uptake of phosphate where they take in phosphate beyond current needs and then accumulate it in their cells in the form of polyphosphate. This luxury uptake has value in constructed wetlands and algal ponds where cyanobacteria are used, as biological filters, to remove phosphorus from effluents. Cyanobacteria also produce biofilms and microbial mats which trap and utilize nutrients in water, which increases endogenous retention and removal from the water column. Microbial mats may help to stabilize and retain nutrients by simply having an increased likelihood of detaining the nutrient to a surface area. Microbial mats also afford cyanobacteria protection against many environmental stressors and promote collaborative interactions with heterotrophic bacteria that may benefit nutrient cycling and stabilization further (Stal, 1995). If cyanobacterial systems were adopted more often, for example, in wastewater treatment facilities or in overly nutrient-laden water bodies, they would certainly reduce nutrient loads, reduce eutrophication, and contribute to ecosystem restoration (Zamparas & Zacharias, 2014).

17.2.2 Heavy Metal Sequestration and Detoxification

Cadmium (Cd), lead (Pb), mercury (Hg), chromium (Cr), arsenic (As), and nickel (Ni) are heavy metals that are persistent environmental contaminants of little biological importance yet are highly toxic. They are introduced into the environment through mining, electroplating, industrial effluents, and agricultural chemicals. In contrast to organic pollutants, heavy metals are nondegradable, and their environmental threats are derived from their bioaccumulation within the ecosystem. Cyanobacteria have a high capacity for binding and sequestering of heavy metals through several methods, including.

Biosorption: The cell walls of cyanobacteria contain negatively charged functional groups (e.g., carboxyl, hydroxyl, sulfate, and phosphate) that can bind positively charged metal ions. This is a passive process that can occur on both live and dead biomass and does not require any metabolic energy. Certain species of cyanobacteria such as *Spirulina, Anabaena,* and *Oscillatoria* have been shown to have good biosorption capabilities (Rakić et al., 2025).

Bioaccumulation: In contrast to biosorption, bioaccumulation involves the uptake of metals by the cyanobacteria and their subsequent internal storage. The metals bound by the cyanobacteria have the capacity to be bound to peptides (e.g., metallothioneins or phytochelatins) upon entering into the cell or being stored in vacuoles that may be present within the cyanobacteria. Once these metals are complexed, the metals are not harmful to the cyanobacteria (Cassier-Chauvat & Chauvat, 2014).

Extracellular Polymeric Substances (EPSs): Often cyanobacteria secrete EPS, thereby increasing metal binding and biofilm formation. EPS is a protective barrier as well as a storage of metal cations, giving the cyanobacteria containment by impeding diffusion, thereby also impeding toxicity (Nishanth et al., 2021).

Redox Transformation and Precipitation: Certain cyanobacterial species have redox reactions that convert metals to less harmful or less soluble species. For example, they can reduce hexavalent chromium (Cr^{6+}) to trivalent chromium (Cr^{3+}), which is less mobile and is less harmful. The metal-binding characteristics of cyanobacteria can also be useful in waste water treatment, mine drainage remediation, and biosorbent application, to name a few. Moreover, immobilized cyanobacterial mats and alginate-encapsulated beads are presently in development for continuous flow systems, which could allow for scalable solutions to heavy metal contamination (Tanmoy & LeFevre, 2024).

17.2.3 Degradation of Organic Pollutants (Pesticides, Hydrocarbons)

Environmental pollution caused by the accumulation of organic pollutants, such as pesticides, herbicides, dyes, hydrocarbons, and pharmaceutical compounds, has become one of the most serious issues. Organic pollutants can persist in the environment, can be recalcitrant to degradation, and can exhibit toxic effects on nontarget organisms. A primary resultant benefit of cyanobacteria on the material world is through bioremediation of organic pollutants via enzymatic degradation and co-metabolic transformation. Various cyanobacterial species have metabolic pathways for synthetic organic degradation, for example:

Degradation of pesticides: *Nostoc muscorum* and *Anabaena variabilis* can degrade organophosphates and carbamates when they produce enzymes called phosphatases and esterases that break down ester bonds yielding fewer toxic metabolites (Tanmoy & LeFevre, 2024).

Degradation of hydrocarbons: Cyanobacteria can metabolize aliphatic and aromatic hydrocarbons by co-metabolism, although not as efficient as some heterotroph, there are examples where cyanobacterial strains produce oxygenases and peroxidases that oxidize hydrocarbons to more water-soluble compounds, when there is a primary source of energy to metabolize (Yamini & Rajeswari, 2023).

Decolorization of textiles: Pollutants derived from dyeing, such as synthetic dyes, are major industry-related pollutants that have consequences for aquatic ecosystems. Species of cyanobacteria, such as *Oscillatoria* and *Lyngbya*, have potential to adsorb and enzymatically degrade synthetic dyes and thus show evidence of potential decolorization and detoxification. Bioremediation of organic pollutants occurs through further amplification with cyanobacterial consortia, due to the role of heterotrophic interactions with cyanobacteria to synergistically degrade pollutants. The cyanobacteria create oxygen and organic carbon, allowing the heterotrophs to degrade pollutants much more effectively, resulting in a stable and effective biodegradation system. These capabilities suggest the feasibility of applying cyanobacteria in bioreactors, constructed wetlands, and phytoremediation for removing a variety and complicated organic pollutants from contaminated sites (Bavithra et al., 2019).

17.2.4 Carbon Sequestration and Climate Mitigation

With atmospheric CO_2 quantities, continuing to rise owing to fossil fuel burning and land use changes, there is considerable need to develop biological systems capable of capturing and sustainably sequestering carbon (Farrelly et al., 2013). Cyanobacteria, being important actors in natural carbon cycling, are also candidates for active carbon sequestration. Cyanobacteria are photoautotrophic organisms which can convert CO_2

from the atmosphere into organic biomass using sunlight via oxygenic photosynthesis. This process of converting atmospheric CO_2 has the potential to reduce greenhouse gas sources while generating renewable biomass for bioenergy applications, biofertilizer, and bioplastics production; these aspects contribute to the circular bioeconomy. There are a few properties of cyanobacteria that are of significance when considering them in terms of carbon management including the following:

High photosynthetic efficiency: Cyanobacteria have faster growth rates and CO_2 fixation per unit area and time than terrestrial plants.
No requirement for land: Cyanobacteria can be cultivated in open ponds, photobioreactors, or lagoons that utilized wastewater; they do not have to compete with agricultural land for growth (Novoveská et al., 2023).
Adaptable growth conditions: In general, cyanobacteria can be grown in salt or brackish waters, on rooftops or in very dry areas.
Carbon capture and storage: Beyond growing biomass, cyanobacteria can also precipitate carbonates, especially under high-alkalinity conditions; in selected systems, the mineralization of carbon may present a longer-term carbon sequestration solution. Additionally, genetically modified cyanobacteria are being designed to take up carbon faster, produce more lipids or carbohydrates, and secrete this value-added product in the form of ethanol and bioplastics. This development combines carbon capture and sustainable production in a single process. At an ecological scale, cyanobacteria can also be reintroduced to restore fractured soils (e.g., sites polluted by industry) which (nitrogen-fixing strains) can rekindle soil fertility, vegetate a site, or store carbon in soil organic matter. In this way, not only can cyanobacteria act as the first line against pollutant removal, but they are also nature-based solutions for climate change adaptation as they sequester atmospheric carbon and restore ecological process (Anas et al., 2025).

17.3 Applications in Environmental Restoration

Cyanobacteria have a unique metabolic diversity coupled with a broad array of adaptations to extreme conditions and the ability to form symbiotic and self-sustaining systems that allow environmental restoration to use these microorganisms increasingly. Indeed importantly, these interstitial, photosynthetic prokaryotes can not only remove or neutralize variety of environmental pollutants but can also help recreate ecosystem structure and agency and hence ecological recovery. In this section, we explore a few uses of cyanobacteria toward environmental restoration including their roles in wastewater treatment, soil rehabilitation, recovery of eutrophic aquatic systems, and remediation of oil spills (Yadav et al., 2021b).

17.3.1 Wastewater Treatment and Effluent Purification

Cyanobacteria are promising organisms for treating municipal, agricultural, and industrial wastewater. Traditional wastewater treatment systems use chemical processes, energy-intensive systems, and complicated infrastructure. Cyanobacteria, on the other hand, provide environmentally beneficial and cost-effective technology for the removal of nutrients and contaminants from wastewater (Cuellar-Bermudez et al., 2017).

Nutrient Removal: Cyanobacteria remove excess nitrogen and phosphorus from wastewater by taking up nutrients into biomass. Nitrogen-fixing cyanobacteria (e.g., *Anabaena* and *Nostoc*) are capable of fixing atmospheric nitrogen into bioavailable forms through their nitrogenase enzymes, while some cyanobacterial species take up nitrate and ammonium. Similarly, phosphorus is also taken up, commonly as polyphosphate, and stored intracellularly. With the uptake of nitrogen and phosphorus, cyanobacteria ensure that eutrophying nutrients are less likely to be released into downstream environments (Wurtsbaugh et al., 2019).

Heavy Metal and Toxin Removal: Industrial effluents can contain toxic metals such as cadmium (Cd), lead (Pb), and chromium (Cr) (Kousar et al., 2025). There are multiple methods by which cyanobacteria can sequester these metals, and they can potentially bioaccumulate contaminants or biosorb them to the surface of cell walls. When bacterial biomass is continuously exposed to metals, they often remove metals from the solution around them, reducing their mobility and toxicity. Cyanobacteria produce a variety of extracellular polymeric substances (EPSs), which can also help in this process, since these EPSs can trap some contaminants and metals in the polysaccharide biofilm (Nishanth et al., 2021).

Organic Pollutant Degradation: Some cyanobacteria, through enzymatic or co-metabolic processes, are capable of degrading pesticides, dyes, and hydrocarbons. Their presence in wastewater treatment systems can help degrade recalcitrant organic pollutants that typical wastewater treatment systems do not remove.

Incorporation in Algal–Bacterial Systems: Cyanobacteria are often co-cultured with heterotrophic bacteria to create advanced wastewater treatment systems. Cyanobacteria provide oxygen through photosynthesis while the bacteria degrade organic matter. The relationship reduces costs associated with electric aeration and improves wastewater treatment.

Energy and Resource Recovery: The biomass harvested during treatment can be converted into biofertilizers, animal feed, and biofuels, thereby supporting a circular economy. In addition, some strains are capable of biosynthesizing bioplastics, or pigments, which can be considered high-value coproducts from the waste stream (Kaur et al., 2022).

17.3.2 Soil Rehabilitation and Desertification Control

Land degradation—and especially in arid and semiarid regions—is a serious global problem caused by deforestation, overgrazing, salinization, and climate change. Cyanobacteria can be an important factor for restoring degraded lands and for combating desertification through:

Biological Soil Crusts: In their natural desert and degraded land environments, cyanobacteria are part of biological soil crusts (biocrusts), which are communities of soil-dwelling organisms that include mosses, lichens, and other microbes, too. Cyanobacteria, e.g., *Microcoleus*, *Scytonema*, and *Nostoc*, have been shown to stabilize soils when they are present by secreting mucilaginous EPS substances which bind soil particles together, thus preventing erosion and increasing soil cohesiveness.

Nitrogen Fixation: Many cyanobacteria are diazotrophs; they fix atmospheric nitrogen and increase the proportion of bioavailable nitrogen contained in soils, thus increasing soil fertility in nutrient-depleted or sandy soils. Native flora benefitted from germination as a result of cyanobacterial soil applied treatments and the apparent recovery of nitrogen density in soil improved the subsequent growth of native vegetation and accelerated ecological succession (Dadzie et al., 2022).

Organic Matter and Carbon Input: Cyanobacteria aid in organic matter buildup in nutrient poor soils, by photosynthesizing carbohydrates and releasing carbon and nutrients in the form of biomass through lysis and death. As cyanobacteria grow and subsequently die, their biomass is a source of source of carbon and nutrients that increase microbial activity and the fertility of soil.

Water Retention and Microclimate Modification: Cyanobacterial mats have the capacity to retain moisture in soils and reduce surface temperatures, providing a more favorable microenvironment for seed germination and establishment. This quality of cyanobacteria makes them favorable in arid-zone reforestation and land reclamation or recovery attempts (Gufwan et al., 2025).

Biofertilizers/Green Manures: In agriculture, the inoculation of either soils or seeds with cyanobacteria (especially *Anabaena* and *Nostoc*) promotes soil health and productivity without the use of chemical fertilizers. This biofertilizer quality of cyanobacteria is especially important in areas of aquaculture (such as rice paddies) where the aquatic conditions favor cyanobacterial growth (Joshi et al., 2020). Cyanobacteria and their biofixing properties can be effective in bioreclamation of saline, alkaline, and chemically contaminated soils because they provide a sustainable and increased productivity and restore productivity to land potentially affected by industrial waste, mining, or saltwater intrusion.

17.3.3 Restoration of Eutrophic and Polluted Water Bodies

The eutrophication of lakes, rivers, and reservoirs caused by excess nutrient inputs has become a global issue, causing harmful algal blooms, hypoxia, and biodiversity loss. Our preconceptions are that cyanobacteria are the "bad guys" when discussing harmful algal blooms (HABs); however, selective and controlled deployments of non-bloom-forming or engineered cyanobacterial strains can be used to help restore balance in eutrophic water bodies.

Nutrient Uptake and Competitors: Some cyanobacteria are able to compete for nutrients with bloom-forming algae, while neither produces toxins nor scums. These cyanobacteria grow, contributing to nutrient uptake, particularly with phosphorus which is often the limiting nutrient in freshwater eutrophication (Hamilton et al., 2016).

Algal Turf Scrubbers and Phycoremediation: These systems utilize cyanobacterial biofilms that treat contaminated water by absorbing nutrients and metals. Water passes over a sloped surface that is colonized by cyanobacteria. As cyanobacteria and algae grow on surfaces, they filter pollutants out through the biofilm. The harvested biomass can be used as fertilizer or feedstock (Wang et al., 2022).

Reduction of Internal Loading: In certain lakes, nutrients may continue to leach from the sediments even after decreasing external inputs (Fig. 17.2). Cyanobacteria can decrease internal phosphorus loading, when used appropriately, by immobilizing phosphorus as biomass or by modifying microbial redox conditions that minimize phosphorus solubilization from sediments (Bormans et al., 2016).

Restoration of a Microbial Community: Cyanobacteria as beneficial taxa can also help re-establish local microbial community structure that leads to an appropriate balance of planktonic food web components, improving water clarity, oxygenation, and aquatic health (McKercher et al., 2024).

17.3.4 Cyanobacteria in Oil Spill Remediation

Oil spills are among the most devastating kinds of pollution for the environment and marine and coastal ecosystems. Conventional cleanup approaches (either via methods such as physical skimming, pumps, dispersants, or incineration) tend to be expensive, not always complete, and harmful to the ecology (Liu et al., 2024). Bioremediation provides a viable, environmentally friendly approach, particularly using microorganisms. Even though cyanobacteria are not oil degraders (usually hydrocarbonclastic bacteria perform that function), nevertheless, cyanobacteria can play critical auxiliary roles in bioremediation of oil spills, namely:

Oxygen for Hydrocarbon-Degrading Bacteria: Cyanobacteria perform photosynthesis and release oxygen that could support aerobic degradation by the oil-degrading

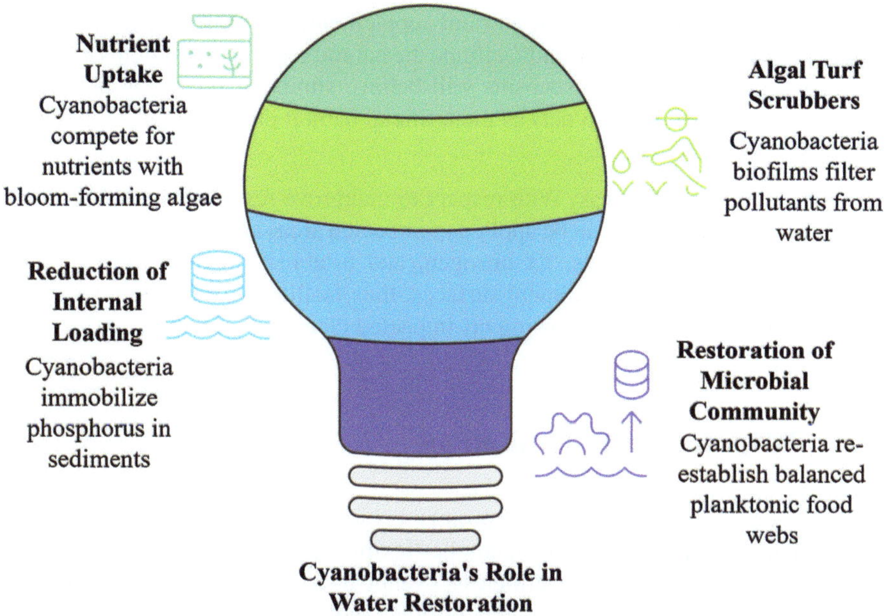

Fig. 17.2 Illustration of cyanobacteria's multifaceted role in water restoration, including nutrient uptake, reduction of internal phosphorus loading, pollutant filtration via algal turf scrubbers, and the restoration of microbial community balance to support healthy aquatic ecosystems

bacteria. This advantage of cyanobacterial-derived oxygen has an important role in providing an oxygen source in oxygen-depleted zones created by the presence of oil (Mazur-Marzec et al., 2024). Biodegradation Consortia: In oil-contaminated systems, cyanobacteria can establish consortia with hydrocarbon-degrading microbes. Their extracellular metabolites (such as sugars and amino acids) can stimulate the activity and diversity of bacteria in bioremediation efforts (Kulikova & Perminova, 2021).

EPS (Extracellular Polymeric Substances) and Biofilm Formation: Cyanobacteria produce exopolysaccharides which could emulsify hydrocarbons, increasing bioavailability of hydrocarbons (also increasing biofilm colonization of oil droplets by microorganisms), enhancing the speed and breadth of oil degradation (Borah et al., 2021).

Cyanobacteria with Direct Hydrocarbon Interaction: Some cyanobacterial genera possess low-level hydrocarbon degradation capabilities or are tolerant to elevated concentrations of hydrocarbons, making them resilient species in oil-impacted habitats. Some (*Oscillatoria*, *Synechococcus*) had achieved the degradation of alkanes and polycyclic aromatic hydrocarbons (PAHs) in laboratory studies (Ahmad, 2022).

Bioreactor and Mesocosm Applications: Naturally, cyanobacteria colonize on many surfaces in the ocean, and they can be included in floating bioreacted systems, or coastal mesocosms designed to treat oil spills. The reactors help maintain oxygen levels. A floating bioreactor with nutrient supplementation may help to reduce the load of pollutants/remediation and facilitate the natural recovery of systems impacted by oil and dispersed oil. Mesocosms will behave similarly, providing habitat for cyanobacteria to live while it acts as the artificial habitat phase reflects on pollutant levels (Sharma et al., 2021).

Restoration of Coastal Areas: With respect to mangrove restoration and salt marsh restoration, cyanobacteria can be applied to impacted areas after cleanup. Cyanobacteria can colonize sediments, fix nitrogen, and establish previous vegetation by colonizing and stabilizing sediment surfaces, thus facilitating long-term ecological resilience, in this situation following oil-impacted cleanup (Saud et al., 2024).

17.4 Advantages and Limitations

Cyanobacteria are gaining traction as promising biological agents for bioremediation in the environment due to their metabolic flexibilities, ecological robustness, and operational cost efficiency. Nevertheless, cyanobacteria possess many of the same potential benefits and limitations as any other biotechnological application. This section analyzes the characteristics, barriers, and performance of cyanobacteria for the purpose of bioremediation in relation to other biological agents, e.g., bacteria, fungi, and higher plants (Li et al., 2019).

17.4.1 Benefits of Using Cyanobacteria in Bioremediation

Cyanobacteria provide a number of ecological, operational, and economic advantages in environmental bioremediation, which is due to their biological characteristics, photosynthetic capabilities, and ability to thrive in a range of habitats (Singh et al., 2016).

(a) **Autotrophic Growth Reduces Input Costs**

Cyanobacteria are photoautotrophs, so they only need sunlight, carbon dioxide, and small amounts of inorganic nutrients to grow, as opposed to heterotrophic bacteria and fungi which need expensive organic substrates. For example, the significant advantages of this aspect of cyanobacteria are in the development of bioremediation systems in open pond systems or any natural water body, which has reduced costs by not requiring the use of organic substrates (Touliabah et al., 2022).

(b) **Multiple Contaminant Removal**

Cyanobacteria can also remove a wide range of contaminants, including nutrients (e.g., nitrogen and phosphorus), heavy metals (e.g., cadmium, arsenic, and lead), organic pollutants (e.g., pesticides, hydrocarbons, and synthetic dyes), and carbon dioxide (via carbon sequestration). Cyanobacterial will still use nitrogen, phosphorus, heavy metals, organic pollutants, and carbon to produce biomass and nutritious feed, but with the range of metabolic pathways present in cyanobacteria, they may be able to be delivered contaminant removal services that parallel an all-in-one biofilter capable of simultaneously addressing multiple contaminants (Vega et al., 2022).

(c) **Nitrogen Fixation and Soil Fertility Enhancement**

Nitrogen-fixing cyanobacteria like *Anabaena*, *Nostoc*, and *Calothrix* can add bioavailable nitrogen to soils, increasing value in low-nutrient or formerly disturbed soils. This makes cyanobacteria useful in farming, land reclamation, and desertification (Singh, 2014).

(d) **Resilience to Extreme Conditions**

Many cyanobacteria are extremophiles and can survive high salinity, pH, temperature, and heavy metal toxicity. Because of their resilient tendencies, cyanobacteria can thrive in contaminated, disturbed, or less-than-optimal ecosystems where other organisms fail to thrive (Chorus et al., 2021).

(e) **Valuable Biomass Production**

The biomass generated from cyanobacterial bioremediation systems can be processed into biofertilizers, animal feed, biofuels, pharmaceuticals, and nutraceuticals. For this reason, the involvement of cyanobacteria in these bioremediation and restoration activities provides a market and economic motivations (Gupta et al., 2012).

(f) **Ecosystem Engineer**

Cyanobacteria also assist in the development of biological soil crusts, the stabilization of soil sediments, and the inception of early successional vegetation in once disturbed ecosystems. Cyanobacteria do more than remove pollution; they are important in ecosystem restoration (Yadav et al., 2021b).

(g) **Sustainability and Climate Benefits**

Cyanobacteria capture atmospheric CO_2 through photosynthesis, mitigating greenhouse gas emissions. Their application thus supports broader goals of climate change adaptation and ecological sustainability (Scarano, 2017).

17.4.2 Challenges and Constraints (Toxicity Risks, Scalability)

While there are many advantages to using cyanobacteria as tools for bioremediation, there are also many challenges and limitations with their use. The most significant drawback of using cyanobacteria for bioremediation is the potential for the production of toxins. Some cyanobacterial species (e.g. *Microcystis*, *Anabaena*, and *Nodularia*) produce toxins such as microcystins (hepatotoxins), anatoxins (neurotoxins), and cylindrospermopsin (cytotoxin) (Ilieva et al., 2019). These toxins pose potential liability for aquatic organisms, livestock, and humans; particularly due to uncontrolled blooms releasing toxins into the environment. Therefore, it is particularly important to identify either a nontoxic strain of cyanobacteria or genetically engineered strain for bioremediation applications. Biomass for bioremediation purposes is beneficial for collection and resource recovery, but handling the contaminated biomass safely and disposing of the biomass are also a concern as the biomass likely may have accumulated heavy metals or persistent organic pollutants. Correct posttreatment or containment of the heavy metal or organic pollutants will also be important to managing secondary pollution. Cyanobacterial function is affected by environmental factors (light, temperature, pH, etc.), and their growth rates might be slower in temperate or shaded situations, which could limit how effective they are for bioremediation (Mumtaz et al., 2025).

Cyanobacterial bioremediation has been proven to work at lab or pilot scale, but when scaling to commercial or whole-ecosystem levels, there are extra technical problems: maintaining optimal light and nutrient conditions over large productions, controlling contamination or competition with other microbes, and preventing uncontrolled blooms (decreases in oxygen). Utilizing genetically modified cyanobacteria or implementing large algal cultures (over lakes, open waters) will encounter regulatory and social acceptance challenges, because of biosecurity, ecological balance concerns, and unknown side effects from using contamination. These issues will require a full risk assessment and uncertainty management strategy and utilizing a transparent process to discuss these issues with stakeholders (Settembre-Blundo et al., 2021). Cyanobacteria do not biodegrade complex organic pollutants as rapidly as some bacteria and fungi. The metabolic pathways some species have for degrading hydrocarbons, plastics, or pharmaceutical products are relatively slow and sometimes indirectly, which could limit application at some severely contaminated industrial sites (Truskewycz et al., 2019).

17.4.3 Comparative Analysis with Other Bioremediation Agents

To evaluate cyanobacteria as biodegrades for bioremediation purposes, it is essential to characterize the advantages and limitations of cyanobacteria relative to other commonly used biological agents, including bacteria, fungi, or plant (because a type of phytoremediation) (Table 17.1).

Since, while bacteria and fungi may sometimes have performance advantages over cyanobacteria (for example, for the degradation of complex organics or for functioning in the dark), cyanobacteria have a few assets that make them more suitable than the others for use in eutrophic aquatic systems, carbon sequestration, and resource recovery. In some cases, the use of cyanobacteria may even serve a role in a multi-trophic bioremediation framework, whereby the different bioremediation

Table 17.1 Comparative analysis of key bioremediation agents cyanobacteria, bacteria, fungi, and plants highlighting their energy sources, pollutant targets, environmental adaptability, scalability, and economic value to assess their suitability for diverse environmental remediation applications

Aspect	Cyanobacteria	Bacteria	Fungi	Plants (Phytoremediation)
Energy source	Light (photosynthesis)	Organic compounds (heterotrophic)	Organic compounds	Sunlight
Pollutant spectrum	Nutrients, heavy metals, some organics, CO_2	Wide variety, especially hydrocarbons	Effective for recalcitrant organics, dyes, and heavy metals	Metals, organics, nutrients
Growth requirements	Low-cost, low-input	Requires organic substrates	Requires substrates and specific growth conditions	Requires land, water, and time
Environmental tolerance	High tolerance to salinity, metals, and UV	Moderate to high	Moderate	Variable; often less tolerant
Toxin production	Some species produce harmful toxins	Rare	Rare	None
Scalability	Moderate; best in ponds or bioreactors	High; adaptable to bioreactors and soil	Moderate; better in solid-state fermentation	High, but slow process
Economic value of biomass	High (biofertilizers, biofuels, feed, pigments)	Moderate (enzymes, bio-products)	High for enzymes, but less market integration	Low to moderate
Application flexibility	Aquatic and soil systems	Mostly soil and water	Soil, sludge, water	Soil, water bodies

agents serve specific functional roles as part of a deployment of multiple organismal types (Abomohra & Ende, 2023).

17.5 Technological and Biotechnological Advances

Recent advancements in technology and biotechnology have broadened the potential of cyanobacteria as a bioremediation tool. These innovations are designed to maximize metabolic efficiency, enhance adaptation to the environmental conditions, and connect to larger-scale systems for pollution prevention. Advancements in gene editing, immobilization methods, and synergistic integration with algal and microbial methods or techniques are also showing promise for more targeted, adaptable, and sustainable options for employing cyanobacteria in the restoration of polluted habitats (Saud et al., 2024

17.5.1 Genetic Engineering for Enhanced Bioremediation

Genetically engineering is a strong optimization instrument and could be applied usefully to cyanobacteria for environmental bioremediation purposes. Genes associated with pollutant degradation, heavy metal binding, and stress (or tolerance) could be naturally factored into native cyanobacterial strains, such as synthetically designed model cyanobacteria through recombinant DNA technology in a synthetic biology framework (Chakdar et al., 2022). Additional genes, such as those coding for environmental degrading enzymes, such as peroxidases, monooxygenases, and hydrolases, can also be used to break down persistent organic pollutants, e.g., pesticides, polycyclic aromatic hydrocarbons (PAHs), and industrially based dyes. On the other hand, increased expression of naturally occurring metal-binding peptides, such as metallothionein or phytochelatins, would increase sequestration of toxic metals such as lead, arsenic, and cadmium. In addition, genetically modified cyanobacteria may be engineered to synthesize higher amounts of extracellular polymeric substance (EPS), leading to improved biosorption functionalities or for biofilm stability. Addressable gene editing tools, such as CRISPR-Cas9, allow a precise and reproducible way to modify genetic constructs of interest. This orthogonality invites the development of strains of use particularly for nontoxic, fast growing, and tolerant strains. These engineered strains can be designed not just for environmental cleanup but also for making desired and useful by-products, such as bioplastics, pigments, and/or instead produce, for example, biofuels, contributing to a circular bioeconomy (Engineering Biology Research Consortium, 2022).

17.5.2 Immobilization Techniques and Biofilm Applications

Immobilization and biofilm technologies are increasingly being developed to enhance the operational performance and environmental robustness of applications involving cyanobacteria. Immobilization is the mother of use of cyanobacterial cells in inert carriers, although retaining their metabolic activity. In this context, the encapsulation of cells can occur in calcium alginate beads, activated carbon, clay, or biopolymers. Immobilization provides improved biomass retention as well as increased contact time with environmental contaminants and allows the cyanobacteria to be reused in continuous type systems, such as bioreactors and/or wastewater treatment units. Immobilized cells are less susceptible to environmental variations and contamination compared to free cyanobacterial cells, which makes them highly appropriate for potential long-term use in natural or engineered conditions. Biofilm applications directly exploit the innate ability of cyanobacteria to attach to surfaces and develop communities, producing an external polysaccharide layer as they physiologically develop. Biofilms are very effective to capture pollutants and nutrients, with this being especially true of heavy metals and suspended particulates due to their inherent geometry and surface area to binding capacity. Biofilm-based systems (e.g., algal turf scrubber and trickling filters) improve nutrient uptake and pollutant degradation in flowing water conditions. Moreover, incorporating cyanobacteria with heterotrophic bacteria as mixed biofilms provides the self-renewing capabilities of oxygen produced by cyanobacteria through photosynthesis in a low-energy system capable of metabolizing complex organic pollutants. Immobilized and biofilm-associated cyanobacteria are also easier to harvest and manage post-treatment, reducing the risk of secondary contamination and simplifying downstream processing of biomass (Dreier et al., 2017).

17.5.3 Integration with Other Remediation Methods

Integrating cyanobacterial bioremediation with additional physical, chemical, and biological treatments is one of the most promising ways to improve the efficiency of these types of bioremediation approaches. This means using hybrid approaches to address a wide range of contaminants and enhance the robustness and performance of the whole system. One means of employing these hybrid, or multiphase systems, is to use cyanobacteria in conjunction with phytoremediation strategies, where the plants absorb and stabilize contaminants in the soil and cyanobacteria improve nutrient cycling and soil structure (especially in degraded soils and saline soils) (Mumtaz et al., 2025). In constructed wetlands, cyanobacteria can be used alongside the existing process to improve the removal of nitrogen and phosphorus, as well as metals, mainly due to cyanobacteria improving oxygenation and functioning as primary producers in wetlands. In AOPs, cyanobacteria may play a role as a pre-remediation or polishing remediation stage, which may help reduce the concentrations of pollutants and BOD before or after the chemical treatment stages. In the context of bioreactors, cyanobacteria could work in concert with heterotrophic bacteria or fungi to degrade complex organics or immobilize mixed pollutants, where cyanobacterial productivity may supplement or replace energy (or oxygen) and thus lessen the energy input to degraded organics. Additionally, the concept of integration may be adapted with nanotechnology by functionalizing with nanomaterials as a means to immobilize, or to work with, the existing catalytic activity of cyanobacterial cells, and developing high-efficiency bio-nano-hybrid systems for the removal of contaminants. These integrated approaches are not only more efficient but also adaptable to a variety of ecosystems and pollution profiles, making them suitable for large-scale environmental restoration initiatives (Grifoni et al., 2022).

17.6 Future Perspectives and Research Directions

Due to the increased global degradation of the environment resulting from anthropogenic industrialization and ecological disarray owing to climate change and land use change, the need for more efficient, scalable, and sustainable bioremediation technologies is at an all-time high. Cyanobacteria, through their unique physiological and ecological functions, present great potential in this area. However, to do this properly requires more research, advances in technology, and considerations of public policy. The emerging perspectives and future directions needed to scale up and institutionalize bioremediation with cyanobacteria (Venkatraman et al., 2024).

17.6.1 Innovations in Large-Scale Bioremediation

Moving from the laboratory level to the field is one of the largest challenges we face in cyanobacterial bioremediation. Future effort must be made to innovate the development of large-scale bioremediation systems that can function successfully in diverse environmental contexts. In our opinion, large-scale bioremediation system design will involve more than simply scaling up smaller designs; it will have to facilitate conditions for higher biomass productivity and oxidative metabolism that supports lengthy pollutant removal. To that end, the use of advanced photobioreactors, open raceway ponds, and modular constructed wetland systems, all with high biomass productivity and the capacity for extended pollutant removal, is imperative (Wei et al., 2024). The anticipated new designs, such as vertical photobioreactors designed for a degree of light penetration and subsequent attachment (to biomass), self-harvesting reactors (bioflocculation), and automated nutrient dosing systems, will make large-scale practical. Furthermore, it is important to develop these platforms with remote sensing technologies and automated monitoring systems to allow for ongoing monitoring of nutrient concentrations, biomass increase, contaminant degradation (Kaplan et al., 2024), and overall performance of the systems in real time. Remote sensing technologies, including drone surveillance and satellite imaging along with IoT-connected sensor systems, offer more effective monitoring of bioremediation activities over large areas, for example, polluted lakes, wastewater lagoons, and post-mining sites. However, there remains a need for further research to assess biomass harvesting, processing, and valorization as part of sustainable economic practice. For example, pairing bioremediation with bioenergy production (e.g., biogas, biodiesel, or making high-value compounds (e.g., pigments and biofertilizers) allows economically viable circular systems to develop. Constructing several field and demonstration projects in a variety of climate and socio-economic regions will be needed to validate and improve templates of these models, as well as context-specific implementation guidelines (Mitter et al., 2019).

17.6.2 Climate-Resilient Cyanobacterial Strains

With climate change at play in fast-tracking extreme weather with increased temperatures, droughts, salinization, etc., it is imperative to identify and develop climate-resilient cyanobacterial strains that can continue to function under these stressors. We need to examine our future paths of research and identify relevant next steps that include the isolation, screening, and engineering of strains that can sustain their higher bioremediation efficacy in fluctuating and degraded environments. We must also be mindful of the importance of trait development such as thermotolerance, salinity tolerance, desiccation, UV-ST, and other traits that are mostly relevant in the real-world bioremediation context. Climate change is causing rapid modifications to the biogeochemistry in many environments more broadly, which further complicates existing problems (Hutchins et al., 2019). Using advances in metagenomics, transcriptomics, and proteomics, we may be able to identify and characterize genes and metabolic pathways which provide resilience and pollutant-removal efficacy. We can then use these molecular resources and precedents and genetically engineer native or model cyanobacteria to create enhanced performers. We can also assemble synthetic microbial consortia, where cyanobacteria are between other stress-tolerant bacteria, archaea, or fungal microbes that create ecological stability and functional redundancy. Such synthetic consortia could complete complementary roles while additionally enhancing pollutant recovery and withstanding nontarget species invasion (Shanmuganathan et al., 2023). Field-based evolutionary adaptation experiments simulating climate change scenarios for cyanobacteria in the "real world" (e.g., elevated CO_2, under drought stress, and pollutant mixtures) can provide useful information about what will happen to their long-term viability in an ecosystem context. Creating strain libraries or genomic databases of cyanobacteria present in extreme habitats (e.g., deserts, hypersaline lakes, and hot springs) can also improve the inventory of options that can be selected for planned applications in extreme environments (Cirés et al., 2017).

17.6.3 Policy and Economic Considerations for Implementation

While cyanobacteria may have biological capabilities, the full integration of this technology as a natural system for environmental cleaning that is accepted by society will require largely governmental, regulatory, and development assistance. It is critical that governments, environmental regulators, and development agencies confirm that using cyanobacterial bioremediation as a technology is very much a green technology and is recognized and included in national and regional environmental action plans. It will include the establishment of unambiguous regulatory guidance for safe cultivation, release, and disposal of biomass derived from cyanobacteria relative to fabrication of any genetically modified strains or effort to use an open system.

There should also be additional policy support such as pilot project funding or subsidization for environmentally friendly technology and tax incentives for industries and municipalities adopting nature-based solutions to pollution. The integration of cyanobacteria systems into payment for ecosystem services (PES) could also provide native landowners and/or communities motivation to engage in cleanup efforts using algae-based systems (Jacobs, n.d.).

From an economic viewpoint, detailed cost–benefit analyses are necessary to document the long-term savings and additional value of the cyanobacterial remediation relative to chemical or mechanical techniques. The ability to co-produce biofuels, fertilizers, and other marketable biomass products from cyanobacterial biomass adds an economic dimension and is consistent with circular economy thinking. Public–private partnerships can accelerate innovation and put scalable solutions into the market by combining academic research with industrial capability as has been done in the past (Meissner, 2019). Capacity building, community awareness, and stakeholder engagement remain critical for practical implementation. Training, extension services, and citizen science can encourage local actors to function and maintain cyanobacterial operations, especially in developing areas with less access to the infrastructure and technical support. The cultural and ecological co-benefits of such systems (restoring landscapes, cleaner water, and improved health) can also help garner public support and acceptance (McNabb et al., 2024).

17.7 Conclusion

Cyanobacteria have proven to be very versatile and sustainable mitigators of pollution and environmental degradation for bioremediation, providing ecologically sound alternatives to address some of the most pressing global challenges of pollution, land degradation, and toxic water. Cyanobacteria can assimilate nutrients, sequester heavy metals, degrade organic pollutants, and fix atmospheric nitrogen; thus, they are easily viewed as natural detoxifiers for both terrestrial and aquatic systems. Additionally, as autotrophic organisms generally do not require as many inputs while being able to create biofilms and/or consortia with other microbes, cyanobacteria could therefore be adaptable to broad environmental conditions and contexts for remediation. The chapter has presented the fundamental means in which cyanobacteria provide bioremediation, through nutrient uptake, heavy metal sequestration, carbon fixation, and degradation of organic pollutants. The chapter has also presented real-world applications of cyanobacteria, such as wastewater treatment, rehabilitation of soils, restoration of eutrophic water bodies, and even oil spills. Challenges remain for implementing cyanobacteria for bioremediation including approaching issues of toxin production and other potential risks, scalability of systems, and unknowns of environmental conditions. As the travelling field of technological innovation amenable to cyanobacteria continues, while maintaining appropriate monitoring, an increase in efficacy and sustainability will develop. This includes innovations in genetic manipulation and immobilization of cyanobacterial systems, to developments in integrated cohabitation systems. Moving forward, the uptake of climate-resilient strains, effectively large-scale bioreactor designs, as well as just and supportive policies, will be essential in mainstreaming these nature-based technologies. Finally, cyanobacterial bioremediation is an auspicious route toward ecological restoration, developing a circular bioeconomy, and building climate-resilient infrastructures. It is through the conjunction of scientific innovation and sustainable policy, with people, that cyanobacteria can become a key resource for making a cleaner, more resilient planet.

References

Abomohra, A., & Ende, S. (2023). *Value-added products from algae: Phycochemical production and applications*. Springer Nature.

Ahmad, I. (2022). Microalgae-Bacteria consortia: A review on the degradation of polycyclic aromatic hydrocarbons (PAHs). *Arabian Journal for Science and Engineering, 47*(1), 19–43.

Akinnawo, S. O. (2023). Eutrophication: Causes, consequences, physical, chemical and biological techniques for mitigation strategies. *Environmental Challenges, 12*, Article 100733.

Anas, M., Khalid, A., Saleem, M. H., Ali Khan, K., Ahmed Khattak, W., & Fahad, S. (2025). Symbiotic synergy: Unveiling plant-microbe interactions in stress adaptation. *Journal of Crop Health, 77*(1), 1–21.

Anas, M., Khattak, W. A., Fahad, S., Alrawiq, N., Alrawiq, H. S., Abdelsalam, N. R., Jaremko, M., & Quraishi, U. M. (2024a). Mitigating heavy metal pollution in agriculture: A multi-omics and nanotechnology approach to safeguard global wheat production. *Journal of Hazardous Materials Advances*, 100584.

Anas, M., Khattak, W. A., Sarfraz, W., Khalid, N., Naeem, N., Farid, M., Saleem, M. H., Khan, K. A., Khalid, A., & Ahmed, T. (2024b). EDTA-enhanced phytoextraction: Leveraging Ricinus communis for the decontamination of Pb and Cd in agricultural soils. *Plant Growth Regulation*, 1–14.

Bavithra, G., Azevedo, J., Oliveira, F., Morais, J., Pinto, E., Ferreira, I. M., Vasconcelos, V., Campos, A., & Almeida, C. M. R. (2019). Assessment of constructed Wetlands' potential for the removal of cyanobacteria and microcystins (MC-LR). *Water, 12*(1), 10.

Borah, D., Gopalakrishnan, S., & Nooruddin, T. (2021). Carbohydrate biolubricants from algae and cyanobacteria. *Journal of Polymers and the Environment, 29*, 3444–3458.

Bormans, M., Maršálek, B., & Jančula, D. (2016). Controlling internal phosphorus loading in lakes by physical methods to reduce cyanobacterial blooms: A review. *Aquatic Ecology, 50*, 407–422.

Bossa, R., Di Colandrea, M., Salbitani, G., & Carfagna, S. (2024). Phosphorous utilization in microalgae: Physiological aspects and applied implications. *Plants, 13*(15), 2127.

Butnariu, M., & Butu, A. (2020). Viability of in situ and ex situ bioremediation approaches for degradation of noxious substances in stressed environs. In *Bioremediation and biotechnology, Vol 4: Techniques for noxious substances remediation* (pp. 167–193).

Cassier-Chauvat, C., & Chauvat, F. (2014). Responses to oxidative and heavy metal stresses in cyanobacteria: Recent advances. *International Journal of Molecular Sciences, 16*(1), 871–886.

Chakdar, H., Thapa, S., Srivastava, A., & Shukla, P. (2022). Genomic and proteomic insights into the heavy metal bioremediation by cyanobacteria. *Journal of Hazardous Materials, 424*, Article 127609.

Chorus, I., Fastner, J., & Welker, M. (2021). Cyanobacteria and cyanotoxins in a changing environment: Concepts, controversies, challenges. *Water, 13*(18), 2463.

Cirés, S., Casero, M. C., & Quesada, A. (2017). Toxicity at the edge of life: A review on cyanobacterial toxins from extreme environments. *Marine Drugs, 15*(7), 233.

Cuellar-Bermudez, S. P., Aleman-Nava, G. S., Chandra, R., Garcia-Perez, J. S., Contreras-Angulo, J. R., Markou, G., Muylaert, K., Rittmann, B. E., & Parra-Saldivar, R. (2017). Nutrients utilization and contaminants removal. A review of two approaches of algae and cyanobacteria in wastewater. *Algal Research, 24*, 438–449.

Dadzie, F. A., Moles, A. T., Erickson, T. E., Slavich, E., & Muñoz-Rojas, M. (2022). Native bacteria and cyanobacteria can influence seedling emergence and growth of native plants used in dryland restoration. *Journal of Applied Ecology, 59*(12), 2983–2992.

Dreier, C., Nielsen, A., Jørgensen, N. P., Kragstrup, T. W., & Meyer, R. L. (2017). *Host proteins determine MRSA biofilm structure and integrity.*

Engineering Biology Research Consortium. (2022). *Engineering biology for climate and sustainability: A research roadmap for a cleaner future.*

Farrelly, D. J., Everard, C. D., Fagan, C. C., & McDonnell, K. P. (2013). Carbon sequestration and the role of biological carbon mitigation: A review. *Renewable and Sustainable Energy Reviews, 21*, 712–727.

Grifoni, M., Franchi, E., Fusini, D., Vocciante, M., Barbafieri, M., Pedron, F., Rosellini, I., & Petruzzelli, G. (2022). Soil remediation: Towards a resilient and adaptive approach to deal with the ever-changing environmental challenges. *Environments, 9*(2), 18.

Gufwan, L. A., Peng, L., Gufwan, N. M., Lan, S., & Wu, L. (2025). Enhancing soil health through biocrusts: A microbial ecosystem approach for degradation control and restoration. *Microbial Ecology, 88*(1), 1–26.

Gupta, R. K., Choudhary, K. K., Kumar, M., Negi, A., & Rai, H. (2012). Bioremediation and cyanobacteria an overview. *BioNano Frontiers, 9*, 190–196.

Hamilton, D. P., Salmaso, N., & Paerl, H. W. (2016). Mitigating harmful cyanobacterial blooms: Strategies for control of nitrogen and phosphorus loads. *Aquatic Ecology, 50*, 351–366.

Hutchins, D. A., Jansson, J. K., Remais, J. V., Rich, V. I., Singh, B. K., & Trivedi, P. (2019). Climate change microbiology—Problems and perspectives. *Nature Reviews Microbiology, 17*(6), 391–396.

Ilieva, V., Kondeva-Burdina, M., Georgieva, T., & Pavlova, V. (2019). Toxicity of cyanobacteria. Organotropy of cyanotoxins and toxicodynamics of cyanotoxins by species. *Pharmacia, 66*(3), 91–97.

Jacobs, W. (n.d.). *Common factors behind success or failure of innovations.*

Jassim, Y. A., Awadh, E. F. A., & Al-Amery, S. M. H. (2023). A review of general properties of blue-green algae (Cyanobacteria). *Biomedicine and Chemical Sciences, 2*(2), 143–148.

Joshi, H., Shourie, A., & Singh, A. (2020). Cyanobacteria as a source of biofertilizers for sustainable agriculture. In *Advances in cyanobacterial biology* (pp. 385–396). Elsevier.

Kaplan, G., Yalcinkaya, F., Altıok, E., Pietrelli, A., Nastro, R. A., Lovecchio, N., Ieropoulos, I. A., & Tsipa, A. (2024). The role of remote sensing in the evolution of water pollution detection and monitoring: A comprehensive review. *Physics and Chemistry of the Earth, Parts A/B/C*, 103712.

Kaur, P., Singh, S., Ghoshal, G., Ramamurthy, P. C., Parihar, P., Singh, J., & Singh, A. (2022). Valorization of agri-food industry waste for the production of microbial pigments: An eco-friendly approach. In *Advances in agricultural and industrial microbiology: Volume 1: Microbial diversity and application in agroindustry* (pp. 137–167).

Kousar, S., Ahmad, D., Khattak, W. A., Anwar, M. M., Majeed, M., Shah, G. M., & Muhammad, M. (2025). Environmental resilience: Navigation of lead and beneficial elements in soil. In *Beneficial elements for remediation of heavy metals in polluted soil* (pp. 221–240). Elsevier.

Kulikova, N. A., & Perminova, I. V. (2021). Interactions between humic substances and microorganisms and their implications for nature-like bioremediation technologies. *Molecules, 26*(9), 2706.

Kumar, C., Chatterjee, A., Wenjing, W., Yadav, D., & Singh, P. K. (2020). Cyanobacteria: Potential and role for environmental remediation. In *Abatement of environmental pollutants* (pp. 193–202). Elsevier.

Li, H., Zhao, Q., & Huang, H. (2019). Current states and challenges of salt-affected soil remediation by cyanobacteria. *Science of the Total Environment, 669*, 258–272.

Liu, X., Sathishkumar, K., Zhang, H., Saxena, K. K., Zhang, F., Naraginiti, S., Rajendiran, R., Aruliah, R., & Guo, X. (2024). Frontiers in environmental cleanup: Recent advances in remediation of emerging pollutants from soil and water. *Journal of Hazardous Materials Advances*, 100461.

Markou, G., Vandamme, D., & Muylaert, K. (2014). Microalgal and cyanobacterial cultivation: The supply of nutrients. *Water Research, 65*, 186–202.

Mazur-Marzec, H., Andersson, A. F., Błaszczyk, A., Dąbek, P., Górecka, E., Grabski, M., Jankowska, K., Jurczak-Kurek, A., Kaczorowska, A. K., & Kaczorowski, T. (2024). Biodiversity of microorganisms in the Baltic Sea: The power of novel methods in the identification of marine microbes. *FEMS Microbiology Reviews, 48*(5), fuae024.

McKercher, L. J., Kimball, M. E., Scaroni, A. E., White, S. A., & Strosnider, W. H. (2024). Stormwater ponds serve as variable quality habitat for diverse taxa. *Wetlands Ecology and Management, 32*(1), 109–131.

McNabb, T., Charters, F. J., Challies, E., & Dionisio, R. (2024). Unlocking urban blue-green infrastructure: An interdisciplinary literature review analysing co-benefits and synergies between bio-physical and socio-cultural outcomes. *Blue-Green Systems, 6*(2), 217–231.

Meissner, D. (2019). Public-private partnership models for science, technology, and innovation cooperation. *Journal of the Knowledge Economy, 10*(4), 1341–1361.

Mishra, A., Rajput, S., Gupta, P. S., Goyal, V., Singh, S., Sharma, S., Shukla, S., Singh, A., Shukla, K., & Varma, A. (2021). Role of cyanobacteria in Rhizospheric nitrogen fixation. In *Soil nitrogen ecology* (pp. 497–519). Springer.

Mitter, H., Techen, A.-K., Sinabell, F., Helming, K., Kok, K., Priess, J. A., Schmid, E., Bodirsky, B. L., Holman, I., & Lehtonen, H. (2019). A protocol to develop shared socio-economic pathways for European agriculture. *Journal of Environmental Management, 252*, Article 109701.

Mumtaz, N., Ahmad, I. Z., Fatima, N., Ahmad, S. A., & Izhar, T. (2025). Bioremediation strategies for wastewater treatment utilizing cyanobacteria: Current insights and future directions. In *Biotechnologies for wastewater treatment and resource recovery* (pp. 241–254). Elsevier.

Nishanth, S., Bharti, A., Gupta, H., Gupta, K., Gulia, U., & Prasanna, R. (2021). Cyanobacterial extracellular polymeric substances (EPS): Biosynthesis and their potential applications. In *Microbial and natural macromolecules* (pp. 349–369). Elsevier.

Novoveská, L., Nielsen, S. L., Eroldoğan, O. T., Haznedaroglu, B. Z., Rinkevich, B., Fazi, S., Robbens, J., Vasquez, M., & Einarsson, H. (2023). Overview and challenges of large-scale cultivation of photosynthetic microalgae and cyanobacteria. *Marine Drugs, 21*(8), 445.

Rakić, I., Đurović, A., Kevrešan, Ž., Kovač, R., Kravić, S., Panić, S., Svirčev, Z., & Stojanović, Z. (2025). Exploring biosorption and bioaccumulation capacities of cyanobacteria Nostoc and Anabaena for remediation of heavy metals in wastewater. *International Journal of Environmental Science and Technology*, 1–18.

Saud, S., Nawaz, T., Hassan, S., Ur Rahman, T., Rasheed, M. N., Hussain, S., & Fahad, S. (2024). Nitrogen-fixing cyanobacteria and soil enrichment for a greener future. In *Environment, climate, plant and vegetation growth* (pp. 391–424). Springer.

Scarano, F. R. (2017). Ecosystem-based adaptation to climate change: Concept, scalability and a role for conservation science. *Perspectives in Ecology and Conservation, 15*(2), 65–73.

Settembre-Blundo, D., González-Sánchez, R., Medina-Salgado, S., & García-Muiña, F. E. (2021). Flexibility and resilience in corporate decision making: A new sustainability-based risk management system in uncertain times. *Global Journal of Flexible Systems Management, 22*(Suppl 2), 107–132.

Shanmuganathan, R., Kadri, M. S., Mathimani, T., Le, Q. H., & Pugazhendhi, A. (2023). Recent innovations and challenges in the eradication of emerging contaminants from aquatic systems. *Chemosphere, 332*, Article 138812.

Sharma, K. V., Sarvalingam, B. K., & Marigoudar, S. R. (2021). A review of mesocosm experiments on heavy metals in marine environment and related issues of emerging concerns. *Environmental Science and Pollution Research, 28*(2), 1304–1316.

Singh, J. S. (2014). Cyanobacteria: A vital bio-agent in eco-restoration of degraded lands and sustainable agriculture. *Climate Change and Environmental Sustainability, 2*(2), 133–137.

Singh, J. S., Kumar, A., Rai, A. N., & Singh, D. P. (2016). Cyanobacteria: A precious bio-resource in agriculture, ecosystem, and environmental sustainability. *Frontiers in Microbiology, 7*, 529.

Stal, L. J. (1995). Physiological ecology of cyanobacteria in microbial mats and other communities. *New Phytologist, 131*(1), 1–32.

Tanmoy, D. S., & LeFevre, G. H. (2024). Development of composite alginate bead media with encapsulated sorptive materials and microorganisms to bioaugment green stormwater infrastructure. *Environmental Science: Water Research & Technology, 10*(8), 1890–1907.

Thakur, A., & Kumar, A. (2024). Emerging paradigms into bioremediation approaches for nuclear contaminant removal: From challenge to solution. *Chemosphere, 352*, Article 141369.

Touliabah, H.E.-S., El-Sheekh, M. M., Ismail, M. M., & El-Kassas, H. (2022). A review of microalgae-and cyanobacteria-based biodegradation of organic pollutants. *Molecules, 27*(3), 1141.

Truskewycz, A., Gundry, T. D., Khudur, L. S., Kolobaric, A., Taha, M., Aburto-Medina, A., Ball, A. S., & Shahsavari, E. (2019). Petroleum hydrocarbon contamination in terrestrial ecosystems—Fate and microbial responses. *Molecules, 24*(18), 3400.

Vega, M. A., Scholes, R. C., Brady, A. R., Daly, R. A., Narrowe, A. B., Bosworth, L. B., Wrighton, K. C., Sedlak, D. L., & Sharp, J. O. (2022). Pharmaceutical biotransformation is influenced by photosynthesis and microbial nitrogen cycling in a benthic wetland biomat. *Environmental Science & Technology, 56*(20), 14462–14477.

Venkatraman, S., Rajendran, D. S., & Vaidyanathan, V. K. (2024). Development and scale-up of the bioreactor system in biorefinery: A significant step toward a green and bio-based economy. In *Biotechnological advances in biorefinery* (pp. 207–230). Springer.

Wang, C., Tan, Y., Zhu, L., Zhou, C., Yan, X., Xu, Q., Ruan, R., & Cheng, P. (2022). The intrinsic characteristics of microalgae biofilm and their potential applications in pollutants removal—A review. *Algal Research, 68*, Article 102849.

Wei, T., Zhao, Y., Zhou, M., Zhang, Z., Wei, Y., & Núñez, A. E. (2024). Initial concept and embodiment to develop modular constructed wetland: A unique and promising solution to sustainability transitions in water management. *Journal of Cleaner Production, 450*, Article 141912.

Wurtsbaugh, W. A., Paerl, H. W., & Dodds, W. K. (2019). Nutrients, eutrophication and harmful algal blooms along the freshwater to marine continuum. *Wiley Interdisciplinary Reviews: Water, 6*(5), Article e1373.

Yadav, A. P. S., Dwivedi, V., Kumar, S., Kushwaha, A., Goswami, L., & Reddy, B. S. (2020). Cyanobacterial extracellular polymeric substances for heavy metal removal: A mini review. *Journal of Composites Science, 5*(1), 1.

Yadav, G., Mishra, A., Ghosh, P., Sindhu, R., Vinayak, V., & Pugazhendhi, A. (2021a). Technical, economic and environmental feasibility of resource recovery technologies from wastewater. *Science of the Total Environment, 796*, Article 149022.

Yadav, P., Gupta, R. K., Singh, R. P., Yadav, P. K., Patel, A. K., & Pandey, K. D. (2021b). Role of cyanobacteria in green remediation. In *Sustainable environmental clean-up* (pp. 187–210). Elsevier.

Yamini, V., & Rajeswari, V. D. (2023). Metabolic capacity to alter polycyclic aromatic hydrocarbons and its microbe-mediated remediation. *Chemosphere, 329*, Article 138707.

Zamparas, M., & Zacharias, I. (2014). Restoration of eutrophic freshwater by managing internal nutrient loads. A review. *Science of the Total Environment, 496*, 551–562.

MIX
Papier aus verantwortungsvollen Quellen
Paper from responsible sources
FSC® C105338

If you have any concerns about our products,
you can contact us on
ProductSafety@springernature.com

In case Publisher is established outside the EU,
the EU authorized representative is:
Springer Nature Customer Service Center GmbH
Europaplatz 3, 69115 Heidelberg, Germany

Printed by Libri Plureos GmbH
in Hamburg, Germany